BULBS AROUND THE WORLD

Bulbous plants are found throughout the world. Marked on the endpaper map are the principle, but not the only, locations where various genera are to be found in the wild. Plants know no territorial boundaries, and genera listed under a country might well also be found in neighboring countries.

1. **Afghanistan**
   Allium
   Bellevalia
   Corydalis
   Crocus
   Eremurus
   Fritillaria
   Iris
   Notholirion
   Scilla
   Tulipa

2. **Argentina**
   Achimenes
   Begonia
   Eustephia
   Habranthus
   Herbertia
   Oxalis
   Tropaeolum

3. **Australia**
   Blandfordia
   Bulbine (eastern)
   Calostemma
   Crinum
   Eurycles
   Hypoxis (western)
   Moraea
   Wurmbea

4. **Bolivia**
   Begonia
   Camassia
   Hippeastrum
   Pyrolirion
   Rhodophiala
   Stenomesson
   Tropaeolum
   Urceolina

5. **Brazil**
   Alstroemeria
   Caladium
   Cypella
   Dioscorea
   Griffinia
   Habranthus
   Herbertia
   Hippeastrum
   Oxalis
   Rhodophiala
   Sinningia

6. **Bulgaria**
   Allium
   Gagea
   Iris
   Lilium
   Nectaroscordum
   Scilla

7. **Burma**
   Crinum
   Dioscorea
   Lilium
   Lycoris
   Nomocharis
   Notholirion

8. **Canada**
   Arisaema
   Brodiaea
   Camassia (B.C.)
   Erythronium
   Iris
   Lilium
   Stenanthium
   Trillium
   Zygadenus

9. **Canary Islands**
   Dracunculus
   Narcissus
   Pancratium

10. **Chile**
    Alstroemeria
    Brodiaea
    Calydorea
    Camassia
    Chlidanthus
    Conanthera
    Hippeastrum
    Ipheion
    Leucocoryne
    Oxalis
    Placea
    Tecophilaea
    Tropaeolum
    Zephyra

11. **China**
    Aconitum
    Allium
    Amana
    Arisaema
    Begonia
    Caloscordum
    Cardiocrinum
    Crinum
    Dioscorea
    Fritillaria
    Gagea
    Hemerocallis
    Iris
    Lilium
    Lycoris
    Nomocharis
    Nothoscordum
    Pinella
    Roscoea
    Scilla
    Trillium
    Tulipa

12. **Colombia**
    Caladium
    Caliphruria
    Dioscorea
    Eucharis
    Hippeastrum
    Phaedranassa
    Plagioliron
    Sphenostigma

13. **Corsica**
    Arum
    Brimeura
    Crocus
    Gladiolus
    Helicodiceros
    Hyacinthus
    Leucojum
    Pancratium

14. **Crete**
    Arum
    Bellevalia
    Biarum
    Chinodoxa
    Colchicum
    Crocus
    Cyclamen
    Gagea
    Tulipa

15. **Cypress**
    Crocus
    Cyclamen
    Iris
    Narcissus
    Tulipa

16. **Denmark**
    Anemone

17. **Ecuador**
    Begonia
    Caladium
    Callipsyche
    Dioscorea
    Leptochiton
    Phaedranassa
    Stenomesson

18. **Egypt**
    Colchicum

19. **Ethiopa**
    Acidanthera
    Allium
    Crinum
    Lapeirousia

20. **Europe**
    Genera that are widespread in Europe apart from genera listed under specific countries.
    Colchicum
    Dipcadi
    Eranthis (western)
    Erythronium
    Gagea
    Galanthus
    Gladiolus (southern)
    Hyacinthoides
    Iris
    Leucojum
    Lilium
    Lloydia
    Maianthemum
    Merendera
    Muscari
    Narcissus
    Ornithogalum
    Paradisea
    Pancratium
    Ranunculus
    Romulea
    Scilla
    Tulipa
    Umbilicus

21. **France**
    Anemone
    Arum
    Brimeura
    Crocus
    Dioscorea
    Fritillaria
    Hycinthus
    Iris
    Lilium
    Muscari
    Narcissus
    Scilla
    Tulipa

22. **Great Britain**
    Anemone
    Arum
    Leucojum
    Lloydia
    Narcissus
    Ornithogalum
    Scilla

23. **Greece**
    Androcymbium
    Anemone
    Biarum
    Colchicum
    Crocus
    Eranthis
    Fritillaria
    Gagea
    Leontice
    Lilium
    Muscari
    Muscarimia
    Narcissus
    Nectaroscordum
    Sternbergia
    Strangweia
    Tulipa

24. **Guatemala**
    Achimenes
    Begonia
    Calochortus
    Hymenocallis
    Milla
    Rigidella
    Tigridia
    Zephyranthes

25. **Guyana**
    Caladium
    Cipura
    Zephyranthes

26. **Himalaya Region**
    Arisaema
    Colocasia
    Fritillaria
    Iris
    Lilium
    Nomocharis
    Roscoea
    Sauromatum
    Trillium

27. **Hungary**
    Colchicum
    Lilium

28. **India**
    Arisaema
    Begonia
    Cardiocrinum
    Cryptocoryne
    Crinum
    Dioscorea
    Merendera

29. **Iraq**
    Hyacinthella
    Puschkinia
    Sternbergia
    Tulipa

30. **Iran**
    Bongardia
    Bellevalia
    Colchicum
    Corydalis
    Crocus
    Cyclamen
    Eremurus
    Fritillaria
    Gagea
    Iris
    Leontice
    Merendera
    Muscari
    Nectaroscordum
    Ornithogalum
    Pseudomuscari
    Puschkinia
    Sternbergia
    Tulipa

31. **Israel**
    Arum
    Bellevalia
    Crocus
    Gagea
    Hyacinthella
    Iris
    Lilium

BULBS

32. Italy
Aconitum
Allium
Arisarum
Biarum
Bellevalia
Colchicum
Crocus
Fritillaria
Iris
Lilium
Muscari
Narcissus
Nectaroscordum
Scilla
Sternbergia
Tulipa
Umbilicus

33. Ivory Coast
Crinum

34. Jamaica
Achimenes
Crinum
Dioscorea

35. Japan
Amana
Arisaema
Cardiocrinum
Eranthis
Erythronium
Fritillaria
Hemerocallis
Iris
ilium
coris
ellia
lium

sa
ne

38. Lebanon
Arum
Colchicum
Crocus
Cyclamen
Fritillaria
Iris
Lilium
Puschkinia

39. Libya
Cyclamen

40. Malaya
Arisaema
Colocasia
Cryptocoryne
Eurycles

41. Mediterranean Region
Arisarum
Arum
Biarum
Cyclamen
Dracunculus
Gladiolus
Gynandriris
Hermodactylus
Hyacinthus
Iris
Leontice
Leopoldia
Lilium
Lloydia
Merendera
Muscari
Narcissus
Ornithogalum
Pancratium
Pseudomuscari
Sternbergia
Tulipa
Urginea

42. Mexico
Achimenes
Bessera
Begonia

Bomarea
Bravoa
Callipsyche
Calochortus
Chlidanthus
Crinum
Commelina
Cooperia
Cypella
Dahlia
Hymenocallis
Manfreda
Milla
Nemastylis
Oxalis
Polianthes
Prochnyanthes
Pseudobravoa
Rigidella
Sphenostigma
Sprekelia
Tigridia
Zephyranthes

43. Mongolia
Caloscordum

44. Morocco
Narcissus
Ranunculus

45. Mozambique
Cyanastrum

46. Nigeria
Cyanastrum

47. New Zealand
Bulbinella
Pelargonium
Ranunculus

48. Panama
Achimenes

49. Paraguay
Achimenes
Cypella
Habranthus

50. Peru
Begonia
Bomarea
Caladium
Callipsyche
Camassia
Canna
Chlidanthus
Eucrosia
Eustephia
Hymenocallis
Mastigostyla
Pamianthe
Paramongaia
Pyrolirion
Stenomesson
Urceolina

51. Philippines
Eurycles

52. Sardinea
Brimeura
Gagea
Helicodiceros
Hyacinthus

53. Sicily
Bellevalia
Gagea
Iris
Muscari
Narcissus
Nectaroscordum
Sternbergia

54. Siberia
Caloscordum
Corydalis
Eranthis
Hemerocallis

55. South Africa
Agapanthus
Albuca
Amaryllis
Ammocharis
Anapalina

Androcymbium
Anoiganthus
Anomalesia
Anomatheca
Antholyza
Apodolirion
Babiana
Baeometra
Begonia
Boophane
Bowiea
Brunsvigia
Bulbine
Bulbinella
Ceropegia
Chasmanthe
Cipura
Clivia
Crinum
Crocosmia
Curtonus
Cyanella
Cybistetes
Cypella
Cyphia
Cyrtanthus
Dierama
Dioscorea
Dipcadi
Dipidax
Drimia
Eriospermum
Eucomis
Ferraria
Freesia
Galaxia
Galtonia
Geissorhiza
Gethyllis
Gloriosa
Gynandriris
Haemanthus
Hesperantha
Hexaglottis
Homeria
Homoglossum
Hypoxis
Ixia

Lachenalia
Lapeirousia
Ledebouria
Littonia
Massonia
Melasphaerula
Micranthus
Moraea
Neobakeria
Nerine
Ornithogalum
Oxalis
Pauridia
Pelargonium
Pillansia
Polyxena
Pseudogaltonia
Rhampicarpa
Rhodohypoxis
Romulea
Sandersonia
Scadoxus
Schizocarpus
Schizostylis
Scilla
Sparaxis
Spiloxene
Streptanthera
Synnotia
Syringodea
Tapeinanthus
Thereianthus
Tritonia
Tritoniopsis
Tulbaghia
Urginea
Vallota
Velthemia
Wachendorfia
Walleria
Watsonia
Whiteheadia
Wurmbea
Zantedeschia

56. **South America**
    Eleutherine
    Gelasine
    Nothoscordum
    Oxalis
    Solanum
    Trimezia
    Tropaeolum
    Zephyranthes

57. **Spain**
    Aconitum
    Allium
    Androcymbium
    Arisarum
    Arum
    Biarum
    Brimeura
    Bulbocodium
    Colchicum
    Crocus
    Dioscorea
    Gagea
    Hycinthoides
    Iris
    Lapiedra
    Leucojum
    Lilium
    Merendera
    Narcissus
    Pancratium
    Romulea
    Scilla
    Tulipa

58. **Switzerland**
    Anemone
    Anthericum
    Bulbocodium
    Gagea

59. **Syria**
    Arum
    Bongardia
    Bellevalia
    Colchicum
    Crocus
    Eminium
    Hyacinthella
    Iris
    Lilium
    Puschkinia
    Sternbergia
    Tulipa

60. **Tasmania**
    Blandfordia

61. **Taiwan**
    Lilium

62. **Tibet**
    Aconitum
    Cardiocrinum

Corydalis
Iris
Lilium
Nomocharis
Notholirion
Tulipa

63. **Turkestan**
    Allium
    Crocus
    Iris
    Ixiolirion
    Tulipa
    Ungernia

64. **Turkey**
    Allium
    Arum
    Bellevalia
    Biarum
    Chionodoxa
    Colchicum
    Corydalis
    Crocus
    Cyclamen
    Eminium
    Gagea
    Galanthus
    Gladiolus
    Hyacinthus
    Hyacinthella
    Iris
    Ixiolirion
    Leontice
    Lilium
    Merendera
    Muscari
    Muscarimia
    Nectaroscordum
    Ornithogalum
    Pseudomuscari
    Puschkinia
    Sternbergia
    Tulipa

United States

65. **Eastern section**
    Anemonella
    Arisaema
    Erythronium
    Iris
    Lilium
    Trillium

66. **Mid-Western**
    Camassia
    Lilium
    Odontostomum

67. **Southern States & S.E.**
    Canna
    Cooperia
    Eustylis
    Herbertia
    Hesperocallis
    Hymenocallis
    Lilium
    Manfreda
    Nemastylis
    Sphenostigma
    Zephyranthes

68. **Western & Southwestern**
    Bloomeria
    Brodiaea
    Calochortus
    Camassia
    Corydalis
    Dicentra
    Erythronium
    Fritillaria
    Iris
    Lilium
    Milla
    Nothoscordum
    Stenanthium
    Zygadenus

69. **Uruguay**
    Achimenes
    Calydorea
    Habranthus
    Herbertia

70. **U.S.S.R.**
    Anemone
    Bulbocodium
    Bellevalia
    Crocus
    Eremurus
    Fritillaria
    Gagea
    Galanthus
    Iris
    Lilium
    Lloydia
    Merendera
    Muscari

Scilla
Tulipa

71. **Venezuela**
    Hymenocallis

72. **West Indies**
    Canna
    Cipura
    Hymenocallis

73. **Yugoslavia**
    Bellevalia
    Crocus
    Gagea
    Hycinthoides
    Lilium
    Ornithogalum
    Sternbergia

74. **Zambia**
    Begonia
    Bowiea
    Cyanastrum
    Lapeirousia
    Moraea
    Pelargonium
    Sandersonia
    Scadoxus
    Walleria

75. **Zimbabwe**
    Anoiganthus
    Babiana
    Bowiea
    Gladiolus
    Hymenocallis
    Lapeirousia
    Ledebouria
    Moraea
    Ornithogalum
    Pelargonium
    Scadoxus
    Walleria

BULBS

Volume I, A–H

by
John E. Bryan

CHRISTOPHER HELM
London

© 1989 by Timber Press, Inc.
All rights reserved

ISBN 0-88192-101-7 (2 volumes)
Printed in Hong Kong

TIMBER PRESS, INC.
9999 S.W. Wilshire
Portland, Oregon 97225

Library of Congress Cataloging-in-Publication Data

Bryan, John E., 1931-
    Bulbs / by John E. Bryan.
        p.   cm.
    Bibliography: p.
    Includes index.
    Contents: v. 1. A-H.
    ISBN 0-88192-101-7
    1. Bulbs--Dictionaries.   I. Title.
SB425.B74   1989
635.9'44'0321--dc19   89-4684
                                              CIP

First published in Great Britain in 1989 by
Christopher Helm (Publishers) Ltd, Imperial House,
21-25 North Street, Bromley, Kent BR1 1SD

ISBN 07470-0231-2

CONTENTS

VOLUME I

Acknowledgments *vii*
Introduction *1*
Chapter 1: Overview *3*
Chapter 2: A Matter of History *5*
Chapter 3: Botany of Bulbs *11*
Chapter 4: Propagation *21*
Chapter 5: Cultivation *33*
Chapter 6: Bulbs in the Landscape *39*
Chapter 7: Growing Bulbs Out of Season *47*
Chapter 8: Pests and Diseases *55*
Chapter 9: Introduction to the Genera *61*
Alphabetical Listing of the Genera: *Achimenes* to *Hypoxis* *63*

VOLUME II

Alphabetical Listing Of the Genera: *Ipheion* to *Zygadenus* *219*
Appendix A: Families of Bulbous Plants *371*
Appendix B: Common Names *375*
Appendix C: Heights *381*
Appendix D: Colors *383*
Appendix E: Flowering Times *385*
Appendix F: Hardiness *387*
Appendix G: Growing in Containers *388*
Appendix H: Shade-Loving and Woodland Plants *389*
Appendix I: Cut Flowers *389*
Appendix J: Fragrance *390*
Glossary *391*
Bibliography *394*
List of Slide Contributors *397*
About the Botanical Drawings *398*
Index *399*

Dedication

To my daughers, Daphne and Jasmine,
for their love and support.

ACKNOWLEDGMENTS

It is now almost ten years since I started the project of writing a book on bulbous plants. Initially it was going to be a work of some 200 pages. The encouragement, advice and support of Richard Abel permitted the expansion of the original concept into this two-volume work.

Along the way—and at times I thought I would never finish the job—many people have been of great help. My dear friend Harriet Meyer Quarre has seen me through good times and bad, always with encouragement and advice, some of which I remember objecting to! Celia Wittenber has edited the manuscript, a tremendous job well done, and in other ways has been of immense help. Brian Mathew, Dr. Boussard, Chris Lovell, Daan Barnhoorn, Harry Hay, Professor Strid, David Ward, the staff of the Botanical Research Institute in Pretoria, Graham Duncan of the Kirstenbosch Botanic Garden, have not only allowed me to use their photographs, but in various letters have contributed valued information. To all of them I owe a sincere "Thank you."

In North Ameica. Dr. August de Hertogh, Dr. Robert Ornduff, John and Kitty Kohout, John Trager, Roy Klehm, Tom Abrego and Dr. Alfred Graf have all been most generous in sending photographs. I must also admit they have ben most patient regarding their return. I thank them for their help and their understanding.

The staff at various botanic gardens, herbaria and research institutes have been most generous with their time and assistance. I would like to particularly mention Jane Gates at the Helen Crocker Russell Library of Horticulture, and Dr. Bruce Bartholomew and Dr. Frank Almeda at the Botany Department of the California Academy of Sciences.

My personal friends in the bulb industry have also been of great help. Eddy McRae, a fellow student at the Royal Botanic Garden Edinburgh and with whom I worked at the Oregon Bulb Farms, gave me much valued information. Gordon Courtright and Jerry Davis answered many questions. To these fine horticulturists and others, I say "Thank you."

During the years, many personal friends, not connected with horticulture, have given me much encouragement. W.C.B., N.F., T.K., W.W.B., K.McL, F.L., C.L., J.H., M.N., W.L., I thank each of you!

Last, but by no means least, I owe a special "Thank you" to Jane Bliss, my secretary and Girl Friday. For several years she has put up with me—I was always asking her to check this, check that, make a list of this, see if you can find information regarding that. She performed all such tasks well, and remained cheerful.

Over a period of ten years, I have talked to many about this book. Many people have been generous with their time and with information. I thank them all. Without the many valued contributions people have made, the task of writing this book would have been much harder, and the journey from the start to the finish, much longer.

January 1989
Sausalito, California

INTRODUCTION

I cannot claim that my fascination with "bulbs," in the broadest definition of the word, came at the start of my career in horticulture. As an apprentice in a nursery located in Devon, England, I was introduced to the wonderful world of plants by having to water tomatoes in greenhouses for six solid weeks, each week being 48 hours long, for which I received the princely sum of 10 shillings, roughly $1.25, per week.

Having survived this period and having learned a great deal about the culture of tomatoes, I then was assigned the job of preparing bulbs for forcing. There were no special chambers with controlled temperatures and humidity at that time. Rather, we used the old method, still in use in many areas today, of plunging the bulbs into beds and covering them with soil and sand. The interest in the miracles that can be "forced" upon a bulb, in this case tulips, hyacinths, daffodils, and crocuses, was sparked then and has remained with me to this day.

While a student at the Royal Botanic Garden, Edinburgh, Scotland, and during postgraduate studies that followed at the Royal Horticultural Society's Gardens at Wisley, Surrey, England, at The Hague in The Netherlands, and in Paris, France, bulbs were never far from my mind, even though they did not demand all of my time. My interest was heightened when I first saw the magnificent colors of the bulb fields in Holland. This was reinforced when I came to the United States in 1961 to work at Jan de Graaff's, Oregon Bulb Farms near Portland. Here, the raising of many new lily hybrids and their introduction into commerce fascinated me.

Over the years it has been my good fortune to have known and worked with such great men in the lily world as Jan de Graaff, Earl Hornback, Harold Comber, and Edward McRae, with whom I was a student at the Royal Botanic Garden, Edinburgh, and who is presently the originator of many fine new hybrid lilies. My association with these fine horticulturalists and my past experience stood me in good stead when, while I was the Director of the Strybing Arboretum and Botanic Gardens located in Golden Gate Park in San Francisco, California, we undertook experimental plantings of many different types of bulbs, recording their growth patterns, time of flowering, height at flowering, etc.

Since that time I have visited many other parts of the world and have seen a great number of bulbs growing in the wild. To my mind this is the only way to understand the cultural needs of any plant and also to gain a greater perspective of the enormous numbers of varieties that are extant.

Despite all of this, you may well ask—Why another book about bulbs? Hasn't everything definitive concerning these beautiful plants been written already? It is my belief that there is a need for a book that will give the reader an insight into the culture of bulbs and the role they have played in the cultural affairs of mankind. I find their history fascinating, and I hope to convey and induce a comparable fascination in you.

This is not meant to be a series of monographs on each and every bulb. I believe that more bulbs are described here than has been done before, but I also know that I have by no means encompassed all of the known varieties and species. The selection of those described in detail is my choice alone, but the choices have not been easy to make. I have been guided by those plants that fanciers may be most likely to find in commerce, such as those available from such outstanding growers as Van Tubergen in Holland. I also have included those bulbs that people have a good chance of finding in cultivation and those that can be reasonably expected to be found in the collections of keen bulb enthusiasts. To complete the listings, I have entered those bulbs I deem worthy of introduction into the commercial world. No one person can hope to see all of the species and cultivars found in the far-flung corners of the Earth. Such would require several lifetimes. Where personal knowledge was lacking, I have relied upon the many experts of the various genera.

In many cases, the nomenclature is both confused and confusing. A great deal of work and thorough evaluation of the genera is needed. This is a mammoth task. As modern technology enables scientists to examine the various species in greater detail, as our understanding of plants increases, there are bound to be changes in the classification and the nomenclature of genera and individual species. Until such work is undertaken, this unhappy confusion will remain.

Where such confusion exists, I have chosen what I consider the logical approach. No doubt I will have disregarded many of the rules botanists would like to see imposed upon the nomenclature. However, I do not feel I have muddied the waters any more than they are presently. I have drawn upon my long experience with bulbs, with the much-enjoyed contact with gardeners during my work in television, with the publishing of a monthly newsletter, in lecturing, and, last but not least, my correspondence with many bulb lovers from all around the world.

I do not apologize for the style I use. Having written the "Green Thumb" column for five years in the *San Francisco Chronicle*, where space often was at a premium and there was so much to say, I tend to get to the gist quickly and not elaborate on points that the reader of intelligence can easily find out or already knows.

This book, then, is intended to give both the keen amateur and the professional an even greater appreciation of bulbs without having to purchase and peruse many different books. It is hoped that the good pointers and information contained in these two volumes will be enjoyed by and enrich the reader. Nothing I write here, however, can accomplish this as well as the thrill of watching bulbs grow and each year witnessing their dazzling displays of brilliance, often with so little effort on the part of the gardener.

If I have encouraged but one person to grow these wondrous plants who has never done so, if I have enhanced just a little the remarkable world of bulbs by an enlightened overview, or if I have added little pieces of historic information to your knowledge, then I will have achieved my purpose and the effort will have been more than worthwhile.

CHAPTER 1
OVERVIEW

Mention the word "bulb" to most people and a whole range of plants comes to mind—dahlias, gladiolus, irises, lilies, tulips, crocuses, and many others commonly known by this term. But are they "true" bulbs? And just what is a "true" bulb?

A true bulb is composed of leaves that are modified for storage and attached to a basal plate that can be considered a "squashed" stem. Crocuses are regarded as bulbs; in fact, they are not true bulbs but corms. Here, the base of the stem is used to store food, as is also the case with the gladiolus, which is also regarded as a bulb. In the genus *Iris* can be found true bulbs, as well as rhizomes, but a rhizome is actually a creeping stem swollen to enable foods to be stored. Many of the *Begonia* are considered "tuberous," the tuber being an underground branch modified for storage and capable of producing buds and roots. Not "true" bulbs but, to the world at large, "bulbs," "corms," "rhizomes," and "tubers" all fall under the umbrella term "bulb."

The diversity of color, flower form, size, habitat, and desirable growing conditions of bulbous plants rivals all other forms of vegetation. Certain *Crocus* species' flowers just manage to peep above the ground, while the towering shoots of *Dahlia imperialis* often will reach more than 16 feet.

Bulbs are found all over the world. *Nomocharis* is found above 12,000 feet in southwestern Szechwan Province, China; the English Bluebell in the wooded areas and glades of Britain; the *Watsonia* at the southern tip of Africa. *Camassia quamash* is at home in the wet meadows of the Pacific Northwest, yet subspecies are at home in dry hills.

The principal areas of the world, however, that are the common natural habitats of a high percentage of these plants are the Mediterranean; South Africa; the Middle East, especially Iran; Afghanistan; the southern U.S.S.R.; and the Pacific seaboard states of North and South America. To a lesser extent, not necessarily due to their unimportance but because fewer genera are indigenous, they are found in Japan, tropical and North Africa, Britain, and eastern North America.

Despite this great diversity, almost all bulbous plants have a common characteristic—a dormant period. This period in a bulb's life cycle is brought about by prevailing growing conditions, such as the heat and dryness of summer or the extreme cold (accompanied by snow) of winter. It is doubtful, however, that all activity actually stops, since many unexplained changes take place during this dormant, or resting, stage.

A crocus comes into flower as soon as the snows melt, the moisture being supplied by the melting snows. Such moisture allows for only a short growing season. The flowers are followed by the foliage, and, after it withers, the bulb enters the dormant state during the dry summer. In the fall, as moisture becomes available, the bulb produces roots and is thus able to flower again in the spring, when the snows melt and the temperature of the soil rises. Its growth cycle is thus complete.

A bluebell, enjoying the comparatively mild climate of Britain, will be brought to life by the warming soil of spring, accompanied by adequate moisture. The flowers and the foliage that sustain the strength of the bulb are produced well before the deciduous trees produce their canopy of leaves. In this way the bluebells can manufacture the necessary food before the sun is hidden by the leaves, and so you will never find them growing below evergreens. Underneath such a canopy, there is not enough light for them to complete their life cycle. Once the trees are in full leaf, the bluebells drift into a state of dormancy. Light, however, is not the only factor that triggers dormancy, although it is the principal one. The other factor is the gradual reduction in the amount of available moisture. The active roots of the trees remove the moisture from the top inches of soil, making it less readily available to the shallower-rooted bulbs. Thus, the bluebell has adapted to its environment.

Nerines, coming from a part of the world that enjoys winter rainfall, produce their leaves during that season. Often the rainfall will be sufficient to support the foliage for a number of months. During the drier summer months the bulb becomes dormant. In late summer the flowers come, then mature, and the seed ripens. With the advent of the winter rains, the seed germinates and grows into spring, becomes dormant when summer comes, and remains dormant until the rains come again. When the growing bulb, derived from the seedling, has reached the size at which it can produce a flower, it will do so at the end of the summer, well before the arrival of the winter rains. The foliage then follows, and the seedling, now an adult (bulb of flowering size), will continue with the pattern of its life cycle. What triggers the flower production? The accumulation of warmth in the soil causes chemical changes in the bulb, which in turn trigger the interior mechanism for flower growth.

Lilium, a genus unique to the Northern Hemisphere, exhibits a variety of flowering times. For example, *L. formosanum*, from a moist and subtropical habitat unusual for any *Lilium*, seems to flower in almost every month of the year. A dwarf variety of this species, collected at 8,000 feet (and more) above sea level, where the winters are cold, flowers in July–August and behaves like a typical bulb. The plants from the warmer climate (at lower elevations) are not long-lived, but seed production is prolific and they will flower in six months from seed. This almost continuous growing season and quick flowering apparently causes the bulbs to exhaust themselves. A large bulb is not produced; however, the species is able to perpetuate itself because of this rapid and good seed production.

Why, then, is a bulb, even if small, produced at all? We seem to be confronted here with a kind of "What came first, the chicken or the egg?" situation. Did the dwarf form come down from the higher elevaton, where the formation of a bulb is necessary for the species to survive, to the lower location, where seed production is more important, or did the reverse occur? Was it a natural transition, or was there some kind of massive land shift?

Questions asked; questions that remain unanswered, but certainly not for lack of trying on the part of the world's botanists, who are constantly experimenting, testing, arranging, rearranging—all with a view toward greater clarification and more exact classification.

I believe I can provide an answer to the question of the need for either a bulb or seeds. Due to the agreeable growing conditions throughout the year, the plant in the warmer climate does not need the production of a bulb in order to survive. It behaves more like a short-lived perennial in this climate so conducive to continual growth, with prolific seed production ensuring the continuation of the species. That a bulb is produced leads me to believe that at one time there was but one species, dependent upon bulb production for survival. Perhaps, in a matter of time—and who knows how many thousands of years that might be—this tropical form will lose its characteristics of a bulbous plant and become a short-lived perennial, or even perhaps a biennial.

Adapatation—that is the keyword here. That amazing ability of all living things to acclimate themselves to the constantly changing environment of this world.

The conditions bulbous plants enjoy in their natural habitats must be taken into consideration in order to grow them successfully in cultivation—we must duplicate in our gardens the environment they knew in the wild. Thusly will they become permanent residents of our gardens and not just short-term visitors.

Today's bulb industry is firmly established in those areas of the world where growers can duplicate by various cultural practices the climates the many genera have found to their liking and to which they have become adapted in the wild. Europe, The Netherlands, and, to some degree, England have climates that can support a vast bulb industry. The Pacific Northwest produces many lilies and daffodils, as well as bulbous plants native to that area. Japan produces a significant number of many kinds of bulbs, principally tulips, lilies, and the so-called minor bulbs. South Africa is an important center for anemones, ranunculus, freesias, hippeastrums, and chincherinchees.

With such diversity of form, with the apparent disappearance and reappearance of the majority of the bulbous plants each year, it is not surprising that they have entered into folklore. They were esteemed by the ancients. The properties of bulbs for curing the sick, prescribed by peoples of yesteryear, were numerous. Lilies and yarrows, boiled together in butter, produced an excellent salve for burns. Saffron in wine would relieve or cure the more unpleasant results of overindulgence in alcohol. Perhaps one of the more unusual was the idea that the sex of an unborn child could be determined by presenting the expectant mother with a lily and a rose—selecting the rose denoted a girl, the lily, a boy.

Mythology is replete with tales in which flowering bulbs play an important role. Hyacinthus, a beautiful Laconian youth, was killed by Apollo in a fit of jealous rage. The spilled blood of Hyacinthus was changed by Apollo into the flower we know as the hyacinth. Consumed by his unrequited love for Smilax, a young man named Crocus was so grieved that he died. The gods were so touched that they changed him into a flower bearing his name. Legends about the tulip, the narcissus, and many others are all familiar to us because of their places in mythology.

From early times bulbs have been part of the fabric of the human race. Their beauty has inspired mankind throughout the centuries—realized and appreciated in literature, paintings, and other art forms, and, most importantly, in our living art—our gardens.

CHAPTER 2
A MATTER OF HISTORY

How does one condense the mountain of information known about bulbous plants in the history of mankind? In mythology, folklore, literature, art? Almost impossible! Volumes and volumes have been written covering all of these subjects. How else would we have become familiar with the legendary characters in Greek mythology who have given their names to so many of our beautiful flowering bulbs—Crocus, Hyacinth, Narcissus? And there was Iris, believed to be the messenger of the gods, who led the souls of dead women to the Elysian Fields, where the flower bearing her name was planted on their graves. She also was the goddess of the rainbow, and her varicolored namesakes reflect this.

While Greek mythology is rife with these tales, the mythology of other countries brings us similar stories, an example being the story of the tulip. In Persian legend, Ferhad loved the young maiden Shirin, but, unfortunately, his love was spurned. Going into the desert to die, each tear he shed turned into a flower called "lale" in Persian, the flowering bulb we call the "tulip." This name actually is derived from the Turkish *tulbend*, "turban," so-called because of its resemblance to the traditional Turkish headdress. Eventually the word evolved to *tulipam*.

Mythology leads to folklore, which in turn leads to history, which is a compendium of all that is known of man's cultures, including literature and art. Very often it is difficult to distinguish between the real and the fanciful. But we do know that some of their legends are of some substance. For example the "old wive's tales" of the use of bulbous plants for food and medicinals are true, because some are still in use today.

Much of this information has been given us by the herbalists, that wonderful group of scientists, most of whom were physicians, who dedicated years and years to compiling texts on the medicinal qualities of plants. But they, too, were unable at times to distinguish between fact and fiction, so these fascinating tales became part of their learned treatises.

It was once thought that the liquid obtained by steeping the flower portions of the lily-of-the-valley in water would cure apoplexy, gout, rheumatism, sprains, poor memory, headaches, and on and on. We now use it as a slightly less potent substitute for digitalis in the treatment of heart disease and as a diuretic.

Mention the lily, especially *L. candidum,* and immediately several images come to mind—purity, chastity, innocence. In Christianity the lily is the symbol of the Virgin Mother, in the Semitic world, the symbol of Motherhood. It too is used primarily for medicinal purposes. The mucilaginous substance concocted when the bulb is cooked is employed as an external compress for tumors, ulcers, and inflammations; as an ointment, it is an emolient for softening corns and helping to heal scalds and burns. And let us not forget that the oil of the lily was used from antiquity to "cleareth ye face and maketh them without wrinkles." However, in a few countries, especially in the Far East, the bulb is eaten and enjoyed.

Dahlia tubers were used at one time to prepare a diabetic sugar. Although no longer popular as a sugar substitute, they are still useful in deriving chemical substances employed in medical testing of liver and kidney function.

It is not my desire to entertain you with any more of these fascinating details, but, perhaps, to whet your appetite to learn more and so send you to those books with full descriptions of bulbs in mythology and folklore.

So, on to history, both that of the written word and of art, for much of our knowledge is gleaned from the works of art of early man. Again, just a smattering, just enough to show you how the bulbous plants were and are important in our world.

Ancient relics reveal that as early as 2200 B.C. (and possibly much earlier) representations of lilies, tulips, irises, and other bulbs were found on walls, vases, and other artifacts. The Minoans of Crete depicted the lily (which is said to have been their sacred flower) in any number of ways; the tulip, seen on a black pottery jar outlined in white; the iris, painted on a number of walls, including their palaces; and the crocus, found on a pottery jug, in a fresco on a palace wall, as well as on the robes dedicated to one of their goddesses. Bulbs were memorialized on the thrones and tombs of early Egyptian pharaohs. Most of these relics were found in the Mediterranean region, for this was their home. At the risk of alienating some of my readers who may think levity has no place in a book of this kind, I must say that this is where their "roots" were.

In A.D. 50, Dioscorides, a Greek herbalist, presented his treatise *De Materia Medica*, in which he detailed the medicinal properties of about 600 plants and which was the leading text of pharmacology for more than a thousand years. An English translation of his work was made in 1655. Many herbals have been published since, each adding some new-found facts to the countless already known. And, as research continues, more books will come, with fresh ideas and data. The time when the inhabitants of the Mediterranean lands first lifted the various bulbs growing in the wild and planted them in their gardens marks their introduction into culture. While the first plants introduced usually are species, it is more than likely that the first major introduction into Western Europe, the tulip, was already hybridized. Gathered into the gardens where this genus is native, and with species from different areas being planted together for display, the opportunity for cross-hybridization was created. The seedlings that resulted from such natural cross-pollination undoubtedly produced many variations. That hybridization actually took place is supported by the fact that, by the 16th century, many different "sorts" were being grown. Note that the records define these "sorts" as distinct from "species." These would have been made, albeit by chance, long before the bulb was introduced into Western culture.

Today, more than 70 species of tulips are recognized, yet, only 50 years ago, the exact number had not been determined. This is yet another confirmation that hybrids were first introduced along with species. The bloodlines of certain varieties confuse breeders to this day.

In Islamic culture, the tulip was well regarded. Omar Khayyam mentioned it in his writing, prior to the 13th century. Species were collected by Mohammed Babur, the first Mogul ruler in India, in the early part of the 16th century. Pierre Belon, the French naturalist, traveling in Turkey in that same century, described the gardens as having many "red lilies"—in all probability a reference to tulips.

When Suleiman the Magnificent was besieging Vienna (1529), Emperor Ferdinand I of Austria wished to talk peace with him. As his emissary, he sent Augier Ghislain de Busbecq. This was indeed a fortunate choice for the world's gardeners. Busbecq admired the tulips he found growing in the gardens of Constantinople, bought a number of them, and shipped them back to the gardens of Ferdinand in Vienna.

The bulbs came under the care of Carolus Clusius. A short time later, he left Vienna to take an appointment as Professor of Botany at the University of Leiden in The Netherlands and his bulbs went with him. The Dutch not only admired the tulips, but, such was their love for them that they paid tremendous prices for just a single bulb. The "tulipmania" that engulfed Holland early in the 17th century was the high point of the passion to obtain these grand garden additions.

At this point, I would like to give a chronological list of when bulbs were first introduced in England. I chose this country as it has long been in the forefront of botanic science and horticultural enterprise, and so representative of when various plants were brought into general cultivation.

Year	Botanic name	Country of origin	Year	Botanic name	Country of origin
1562	*Tulipa*	Turkey	1759	*Chasmanthe aethiopica*	South Africa
1570	*Canna indica*	Central and South America	1759	*Hymenocallis speciosa*	West Indies
1570	*Iris xiphioides*	Pyrenees	1759	*Trillium erectum sessile*	Eastern North America
1576	*Narcissus hispanicus*	Spain	1767	*Albuca major*	South Africa
1587	*Moraea ciliata*	South Africa	1768	*Veltheimia viridifolia*	South Africa
1596	*Erythronium dens-canis*	Europe; Asia; Japan	1772	*Oxalis violacea*	U.S.A.
1596	*Gladiolus communis*	Mediterranean	1774	*Babiana plicata*	South Africa
1596	*Hyacinthus romanus*	Balkans; North Africa	1774	*Montbretia securigera*	South Africa
1596	*Leucojum vernum*	Central Europe	1774	*Tritonia lineata*	South Africa
1596	*Lilium martagon*	Central Europe	1774	*Vallota speciosa*	South Africa
1596	*Muscari comosum*	Europe	1778	*Achimenes coccinea*	Peru
1596	*Muscari moschatum*	Asia Minor	1787	*Hesperantha cinnamomea*	South Africa
1596	*Ornithogalum comosum*	Southern & Eastern Europe	1788	*Nerine curvifolia*	South Africa
1596	*Ranunculus asiaticus*	Orient	1788	*Wurmbea capensis*	South Africa
1596	*Scilla amoena*	Cyprus	1791	*Lapeyrousia corymbosa*	South Africa
1596	*Sternbergia lutea*	Europe	1793	*Homeria collina*	South Africa
1600	*Hermodactylus tuberosus*	Levant	1796	*Tigridia pavonia*	Mexico
1613	*Cyclamen europaeum*	Southern Europe	1798	*Dahlia variabilis*	Mexico
1615	*Pancratium illyricum*	Southern Europe	1800	*Eremurus spectabilis*	Siberia
1629	*Agapanthus africanus*	South Africa	1804	*Begonia evansiana*	China; Japan
1629	*Colchicum byzantinum*	Transylvania	1806	*Brodiaea coronaria*	Western North America
1629	*Paradisea liliastrum*	Southern Europe	1812	*Blandfordia grandiflora*	Australia
1629	*Polianthes tuberosa*	Mexico	1816	*Freesia refracta*	South Africa
1629	*Zephyranthes atamasco*	U.S.A.	1819	*Puschkinia scilloides*	Asia Minor
1658	*Sprekelia formosissima*	Mexico	1820	*Stenomesson croceum*	Peru
1664	*Arisaema triphyllum*	North America	1821	*Habranthus gracilifolius*	Uruguay
1677	*Hippeastrum reticulatum*	Brazil	1823	*Belamcanda chinensis*	China
1690	*Gloriosa superba*	Tropical Asia; Africa	1823	*Cypella herbertii*	Brazil
1702	*Eucomis regia*	South Africa	1823	*Merendera trigyna*	Caucasus; Iran
1731	*Haemanthus coccineus*	South Africa	1825	*Streptanthera cuprea*	South Africa
1731	*Zantedeschia aethiopica*	South Africa	1826	*Fritillaria ruthenica*	Caucasus
1732	*Crinum asiaticum*	Tropical Asia	1826	*Leucocoryne ixioides*	Chile
1739	*Allium victorialis*	Mediterranean Region	1827	*Herbertia pulchella*	Chile
1739	*Romulea bulbocodium*	Mediterranean Region	1837	*Camassia quamash*	Western North America
1750	*Watsonia meriana*	South Africa	1844	*Ixiolirion montanum*	Central Asia
1752	*Brunsvigia gigantea*	South Africa	1844	*Notholirion thomsonianum*	Afghanistan
1752	*Lachenalia orchioides*	South Africa	1851	*Eucharis candida*	Colombia
1754	*Alstroemeria pelegrina*	Chile	1852	*Cardiocrinum giganteum*	Himalayas
1758	*Lycoris radiata*	China	1852	*Sandersonia aurantiaca*	Natal
1758	*Sparaxis grandiflora*	South Africa	1853	*Littonia modesta*	South Africa
1758	*Zygadenus muscaetoxicum*	North America	1853	*Nomocharis nana*	Northern India; Tibet
			1854	*Caladium picturatum*	Peru
			1856	*Galanthus nivalis*	Europe

Year	Botanic name	Country of origin	Year	Botanic name	Country of origin
1860	*Arum hygrophilum*	Syria	1879	*Galtonia clavata*	South Africa
1864	*Schizostylis coccinea*	South Africa	1892	*Eranthis cilicica*	Greece
1869	*Bloomeria aurea*	U.S.A. (California)	1910	*Dracunculus vulgaris*	Mediterranean Region
1872	*Tecophilaea cyanocrocus*	Chile	1928	*Pamianthe peruviana*	Peru
1877	*Chionodoxa luciliae*	Asia Minor			

In 1493, Pope Alexander VI had given the lands of the Caribbean to Spain and Portugal, a gift confirmed in the Treaty of Tordesillas in 1506. Not unnaturally, England and France objected strenuously. Ships under the command of such great seamen as Sir Francis Drake and Sir Walter Raleigh sailed into the region known as the Spanish Main. It is possible that they returned with many plants, including the potato, the tomato, and tobacco, as well as one bulbous plant, the canna. However, who actually was responsible for the introduction of the latter is not known, although the recorded date of its introduction into culture is 1570. The choice of this particular bulb could have been due to its resemblance to the ginger plant, as spices commanded high prices at that time.

In 16th century Britain, growing unusual and different plants was in vogue. Many wealthy and influential people vied with one another not only in the creation of great gardens but also in planting them with new and unique species. The plants came from all over the Western world. Lord Zouche sent seeds from Crete, Italy, and Spain. Queen Elizabeth's envoy to France, Lord Hunsdon, sent home rare plants that he came across. John Gerard, a surgeon and head gardener to Lord Burleigh, Chief Secretary of State under Queen Elizabeth I, listed more than 1,000 plants in his 1596 publication, *Catalogus arborum fruticum, etc.*, all of which were purported to be growing in that garden. This accounts for the date of introduction as shown in my list, although they were introduced prior to this time. The date given, however, is the earliest on record for these plants. One year later, Gerard published his famous *Herball*.

The great expansion of trade that took place during the 16th and 17th centuries accounts for many bulbous plants being introduced during that period. One in particular, native to South Africa and introduced in 1587, could well have been the result of its particular properties—the *Moraea*. Thunberg writes in his *Kaffraria* that the roots of this plant were eaten by the Hottentots. They were roasted, boiled, or stewed with milk and appeared to him to be both palatable and nourishing, tasting much like potatoes.

South Africa, thanks to the voyages of such 15th- and 16th-century explorers as Bartholomeu Dias and Vasco da Gama, had fast become an important element in world trade. One of the lucrative "by-products" of the commercial business was the trade in human lives—slaves. It is highly possible that the *Moraea* was used to feed the slaves on the long sea journeys and so came into cultivation elsewhere purely by chance. The roots of this plant can be harvested and stored for considerable periods of time without spoiling. This would have made them an ideal food for sea voyages to distant lands.

Chance, however, played no part in the introduction of most plants into cultivation. After the Crusades, Britain became an integral part of Europe. During the reign of Henry VIII, seeds and plants accompanied state papers in the pouches of traveling diplomats. In 1551 William Turner, Doctor of Physics at Oxford, had been put in charge of the gardens at Sion House; the *Index Kewensis* credits him with being the first to record and introduce many plants.

The interest in plants by merchants, which resulted in the introduction of many species, including *Lilium martagon*, is recorded by John Gerard. Concerning one Nicholas Lete, a merchant who traded with the Levant, Gerard wrote that Lete was "greatly in love with rare floures." He also noted that Lete sent his servants into Syria and other countries for plants, to such an extent that he (Gerard) and the whole land was grateful to him.

William Harrison, Dean of Windsor, in his *Description of England* (1577), wrote that ". . . Strange herbs, plants, and annual fruits are daily brought unto us from the Indies, America, Ceylon, the Canary Islands and all parts of the world."

Such was the increase in the number of bulbous plants which had been introduced that John Parkinson, in his *Paradisi in Sole Paradisus Terrestris* (1629), mentioned daffodils, fritillarias, iacinthes, saffron flowers (crocuses), lilies, flowerdeluces, tulips, anemones, cyclamen, and muscari. He also was familiar with the blue agapanthus and *Ornithogalum aethiopica*. The names of those responsible for their introduction were not mentioned, except that "they were gathered by some Hollanders on the West side of the Cape of Good Hope."

That such plants came into cultivation is not surprising. The Cape Province of South Africa contains more species than the entire British Isles and is a veritable treasure trove of bulbous plants. The Dutch East India Company established a garden at the Cape in 1625 in order to supply its ships with fresh fruits and vegetables. The plants came under the scrutiny of a professional botanist, Paul Hartman, when he visited in 1672 and 1680. Hartman was Professor of Botany at Leiden, the home of Clusius, and while he himself did introduce some plants, his pupil, Hendrich Bernard Oldenland, did more. Oldenland, a Dane, became superintendent of the Cape Town garden and, together with his master gardener Jan Hartog, journeyed into the interior of this plantman's paradise and sent back to Europe seeds, bulbs, and plants.

Despite the richness of the known flora, the potential for use in European gardens was not fully exploited. For example, until 1747, there were few introductions from South Africa. In that year, Johan Andreas Auge, a German gardener who had studied in Holland, was appointed assistant gardener in Cape Town. He made many trips into the interior and put together collections of plants for sale, which were purchased by the crews from ships putting in at the Cape for supplies. The beauty of the flora became known, and the Cape became a favorite place for botanists.

One of these botanists, Carl Peter Thunberg, a pupil of the great Linnaeus, arrived in South Africa in 1772. Together with Auge, he undertook expeditions into the areas where the flora was prolific and varied, enriching the collection in the garden at Cape Town.

The arrival there of Scotsman Francis Masson, however, actually marks the beginning of the period of significant introduction of bulbous plants. Masson was sent to South Africa at the suggestion of Sir Joseph Banks, the most distinguished naturalist in Britain and the head of Kew Gardens. He was the first of many official collectors sent out by Banks, and he, too, teamed up with Thunberg, a most fortunate circumstance. The botanist and gardener made a good team. Among the bulbs they discovered were *Ixias, Lachenalias, Zantedeschia aethiopica* (the arum), *Ornithogalum thyrsoides* (the chincherinchees), and *Gladiolus*. We owe much to the efforts of these two men, as the introductions from South Africa in the 18th century were extremely significant.

Plants that had long been in cultivation in their native lands should have been prime candidates for introduction when those countries established communications with Europe, but this was not always the case. The dahlia is an example. The Aztecs had used the dahlia in many of their medicines. The plants, however, did not excite much enthusiasm in Europe until 1791, when Cavanilles published *Icones*. In 1789 the Marchioness of Bute was in Spain, where she saw the lovely flower for the first time. She sent one of the tubers to England, but it died. Then, in 1804, the dahlia was truly introduced and became popular, and all because of a bit of a scandal. It's rather fun to know that the history of bulbs includes something other than cut-and-dried facts.

Sir Godfrey Webster and his wife were living in Florence at the end of the 18th century. Lady Webster fell in love with a Lord Holland and eloped with him. Lord Webster finally divorced her, whereupon she married Holland, and they seem to have had a very successful marriage. In 1804 she saw the dahlia in Spain and sent some of them back to England. They thrived and she is given credit for having brought that lovely flower to Britain.

Sir Joseph Banks, whose work in the field of botany was outstanding and to whom gardeners all over the world owe a deep debt of gratitude, persuaded King George, in 1803, to sponsor William Kerr on a plant-hunting journey to China. Interest in China's flora was sparked by David Lance, who was in charge of the East India Company's factory in Canton. Kerr sailed in 1803 and by 1804 had sent back significant collections. Among the plants were *Begonia evansiana* and *Lilium japonicum*. Kerr was not credited with their introduction however, having fallen out of favor (possibly because he had become addicted to opium), and the credit went to Lance or the Company's directors.

Thanks to two missionaries in China, Pere Delavay and Armand David, the Western World came to know of that great genus *Nomocharis,* as well as many new species of genera already introduced. Their writings and discoveries made apparent the richness of the flora and the value of plants from this temperate region to European gardens. Many plant hunters subsequently explored the area, despite the great difficulties of travel in that turbulent part of the world. Among such men were Augustine Henry, after whom the *Lilium henryii* was named; E. H. Wilson, who introduced the *Lilium regale;* Robert Fortune; George Forrest; Francis Kingdon-Ward; and, in very recent times, Frank Ludlow and Major George Sheriff.

North America also contributed to the richness of bulbous plants in cultivation. David Douglas, who met such an untimely death in Hawaii, explored the West Coast, sending back many different species. Thomas Nuttal introduced *Camassia fraseri*. The lovely *Erythronium grandiflorum* we owe to the explorations of Lewis and Clark. William Bartram and John Lyon also introduced bulbous plants to Europe.

South America, the native habitat of a number of plants that we enjoy in our bulb gardens, also contributed, but not to the same extent as other parts of the world. Theodore Hartweg introduced *Achimenes longiflora* and Edward Frederich Poeppig introduced the amaryllis and alstroemerias, the first sight of which made him shout with joy. William Lobb and Harold Comber also were instrumental in bringing bulbs out of South America.

As so frequently happens, many of the first introductions did not survive. We owe a deep debt of gratitude to Comber, who reintroduced the alstroemerias—the grand hybrids we enjoy today were raised from his plants. These plants are finally receiving the attention they deserve from hybridizers and, with the many new colors being developed, are becoming increasingly important as cut flowers.

Flowers appeal to our esthetic values, to our appreciation of form and color. Throughout the ages, artists have endeavored to appeal to the human mind by capturing on canvas, in sculpture, and as engravings their grace, color, and charm, as well as to make a permanent record to which reference can be made at any time of the year. While early wooden engravings did not allow for great detail, the copper and steel plates that followed allowed for the finest of details to be captured. In many older books, the flowers were hand painted, inadvertent works of art, although they were produced mainly for purposes of identification. They were produced in limited numbers, primarily by members of religious orders, where most scholars were to be found at that time, and used primarily by them.

We have already spoken of the representation of lilies, irises, etc., in ancient times. During the Middle Ages, life was dominated, and in large part measured, by the tenets of the Christian religion. Woodblock illustrations in herbals depicted flowers for practical, everyday use in the identification of plants used mainly for medicinal purposes. When flowers were depicted in paintings or in other art forms, however, their principal role was symbolic, reflecting the current Christian interpretation of particular flowers. Thus, most paintings of the Annunciation included the Madonna Lily.

Duccio di Buoninsegna (1260–1339), in his painting of the Annunciation now in the National Gallery in London, placed a pot of lilies between Mary and the Angel, a symbolic usage in keeping with the age. Fra Filippo Lippi, when painting his Annunciation a hundred years later, portrayed a container of Madonna Lilies. He, however, showed them in a more naturalistic style now associated with the early Renaissance. The mode in which they were presented is evidence that, especially in northern Italy, plants were being grown, and grown well, in containers. The painting shows not just stems of flowers placed in a container but also an actual plant, complete with foliage. Such containers were placed most often on window sills and balconies. The custom of bringing plants and cut flowers into the home for decorative purposes did not really begin until the 19th century. Prior to that time, herbs and other aromatic plants were used indoors, not as decoration but for the fragrance they exuded when walked upon, which made dwellings fetid with the odors resulting from primitive sanitation more habitable.

Early in the 15th century, painters began to portray flowers in much greater detail. During the Renaissance the dominant influence of the Church began to decline, especially in the north of Europe. Depiction of still life, scenes of everyday, ordinary objects, including flowers, became an accepted and predominant motif of artists. Interestingly, acceptance of this form of art coincides with the introduction of the tulip into Holland. Thus, plants and flowers became common subjects for the painters of the Lowlands, although most often shown with other subjects.

Ludger Tom Ring, a Westphalian painter, did a still life in 1562 that was distinct from paintings done previous to that time—his subjects were irises and lilies.

Jan Brueghel (1568–1625), of the famous Flemish family of artists, painted magnificent portraitures of flowers. One cannot help but marvel at the diversity of flowers he assembled in a painting. Flowers that were in bloom at different times of the year are shown together. Summer-flowering lilies are mixed with tulips and daffodils, which bloom in the spring.

It must be remembered that artists would make sketches of the various flowers at the time they blossomed, and, when commissioned to produce a painting, draw upon the subjects they had sketched, assembling them to create a grand floral painting. Beauty, not adherence to factual representation of a particular season, was the rule that governed. For the same reason each flower was shown at its very best. No imperfect flowers, no wilted or faded flowers—even a fallen petal was shown in perfect shape. During this time it also was the custom, requested by art patrons, that the flowers be depicted against a dark, mostly black background. Not until the time of Jan van Huysum (1682–1749) were other backgrounds used.

French painter Philippe Rousseau (1816–1887) depicted crocuses and hyacinths on one of his canvases. They are shown not as cut flowers in a vase with each petal in perfect condition and in its right place, but more as we frequently find them—some a little past their prime, some simply thrust into a glass of water, some growing in soil in a small, clay pot. An interesting element of this same picture is a hyacinth growing in a hyacinth vase, with its roots in water; the shape of which is identical with the hyacinth vases used today. Against a brown background, with the flowers resting on a ledge, it looks for all the world like a corner of a work area where the flowers have been placed for a moment, not arranged for a painting—a very natural setting.

It should be noted that the subjects painted by Rousseau—crocuses and hyacinths—signified the end of the dominance of the tulip, lily, and iris in art. A greater variety of bulbs, from many different areas of the world, some quite exotic, were starting to be appreciated and grown for enjoyment both in the home and in the garden, which was reflected in the art of the period.

The tulip and daffodil reemerge in one work of French painter Paul Cezanne (1839–1906), but with a slight difference. The tulips are of a solid color and the daffodils have a red cup. They are not shown in a formal arrangement but thrust into a vase, as we do in our homes, where the flowers are loved for the qualities they hold and not just as stylized decorations.

The great European noble houses also appreciated the beauty of flowers. Not infrequently they selected them as part of their coat of arms, and their banners often depicted, in stylized form, the flower of a bulbous plant. The fleur-de-lis of France is perhaps one of the better-known examples.

"Consider the lilies of the field, how they grow, they toil not, neither do they spin, and yet I say unto you that Solomon in all his glory was not arrayed like one of these"—possibly one of the best-known passages in the Bible. Although many have tried to determine the precise flower to which Christ referred, there remains some doubt regarding the exact species. Early commentators thought the reference was to either *Lilium chalcedonicum* or *L. candidum*. Modern opinion leans toward the anemone, while others think it a general reference to wild flowers.

The desire to intertwine the symbolic with the practical can be seen in the intense discussion that has revolved around this point. Greater knowledge and understanding of the flora of the area where Christ taught has led some to pinpoint and identify the exact plants to which He referred. It is hoped that in such discussions dealing with the need to combine the practical with the symbolic the very beauty of the illustration used is not overlooked or does not go unappreciated.

Some of the world's greatest writers have used the flowers of various bulbs in telling their stories. The lily has been a favorite, both in a serious vein, as above, and in a more jocular fashion. In their operetta "Patience," Gilbert and Sullivan announce: "You will rank as an apostle in the high aesthetic band if you walk down Piccadilly with a poppy or a lily in your hand." And John Ruskin, in his *Stones of Venice*, wrote: "Remember the most beautiful things in this world are the most useless, peacocks and lilies for instance." Few gardeners would agree!

Although we generally associate the white-flowered lily with purity and sweetness (the yellow-flowered means falsehood, but no one seems to know why) and relate it to religion, it was only selected four times by Roman Catholic monks in assigning flowers to each day of the year and dedicating a flower to the Saints. The daffodil seems to have been the favorite, 12 species being mentioned—5 for March, 4 for April, and 1 each for February, May, and October. Of this flower, Shakespeare wrote, ". . . daffodils, that come before the swallow dares, and take the winds of March with beauty." John Masefield wrote, ". . . and April's in the west wind, and daffodils." Both were correct, as daffodils flower during those months in Britain.

It is interesting to note that the great authors were keen observers of nature. The beauty of daffodils moving in the wind also inspired William Wordsworth when he wrote possibly the best-known verse about flowers:

> I wandered lonely as a cloud
> That floats on high o'er vales and hills,
> When all at once I saw a crowd,
> A host, of golden daffodils;
> Beside the lake, beneath the trees,
> Fluttering and dancing in the breeze.

What better way to describe the ideal location to plant a mass of daffodils for naturalizing.

The second most popular flower dedicated to the Saints is the crocus, with 10. It seldom inspired the poets, however, despite its symbolic meaning—cheerfulness.

That writers should refer to the tulip in the 18th century is not surprising, considering the "tulipmania" that gripped Holland at the time. It is understandable that Samuel Johnson should write, "The business of a poet . . . is to examine, not the individual, but the species; . . . he does not number the streaks of the tulip, or describe the different shades in the verdure of the forest."

Lord Byron wrote, ". . . Heaven is free from clouds, but of all colours seems to be,—Melted to one vast Iris of the West,—where the day joins the past Eternity." Can any description better conjure the wide variety of colors found in the irises? Perhaps the meaning of "iris" in the language of flowers—"I have a message for you"—coupled with Byron's words, gives that meaning added significance.

In the calendar of the monks, February 14 was the day of the crocus, but in modern usage the rose has replaced it. Other bulbs mentioned are 4 days for the amaryllis, 3 for the tulip, 2 for the hyacinth and crown imperial, and 1 each for the anemone, snowdrop, cyclamen, star of Bethlehem, ixia, day lily, agapanthus, colchicum, arum, and ranunculus.

Thus have the bulbous plants become an integral part of the fabric of life.

CHAPTER 3
BOTANY OF BULBS

True bulb, corm, rhizome, tuber; what they are and how they function; their taxonomy and nomenclature.

True bulb

A true bulb consists of a stem and leaves altered or adapted for storage. The stem is compressed into a flattened plate, from which roots are produced and on which leaves, also modified for storage, are placed. These modified leaves are fleshy, because they are filled with the plant's food reserves. The actual placement of the leaves can differ. In the tulip, hyacinth, and daffodil, they are layered closely around each other, forming a tunicated bulb, i.e., the outermost leaves form a tunic around the bulb, with the outer leaves often dry and brown in color. In all other bulbs, the leaves are not wrapped but overlap each other, do not form a tunic, and are more succulent, e.g., the lily and fritillaria. Such bulbs are known as scaly bulbs.

Between the leaves, lateral buds rise from the basal plate and eventually form the daughter bulb(s) when the parent dies. The actual number of growing seasons required by such a daughter to achieve independence will vary. In the tulip (*Tulipa*), the parent dies after flowering; in the daffodil (*Narcissus*), the parent may live for more than two seasons, with daughter bulb(s) sharing the same basal plate.

From the basal plate, normally on the upper side in the center, the flower stem is found in a contracted, telescoped form. In some bulbs the embryonic flowers are formed the previous season, as in the hyacinth. In others, this stem will produce leaves and flowers on the same stalk during the season of growth. The lily is an example of the latter.

Not all bulbous plants arise from a true bulb, however. Included in this vast "family" are those produced from *corms*, *rhizomes*, and *tubers*.

Corm

A corm is a stem that is swollen and modified for storage. The normal shape is rounded, flattened on top, and slightly concave on the bottom. On the top will be found the young buds; roots will be produced from the basal area. Frequently a corm will produce a brown skin, not dissimilar in appearance to the tunic of a true bulb. Upon being cut, however, the corm will appear solid. The new corm will grow on top of the old one being formed at the base of the new stem, which is produced from the bud(s) on the top side of the corm. On the basal plate, young, small corms, known as cormels, will arise. The quantity of cormels produced often can be quite large. *Crocus*, *Gladiolus*, and *Watsonia* are corms.

Rhizome

A rhizome is an underground stem (sometimes breaking the surface of the soil) that is swollen from the apices (ends) of which shoots emerge. The roots are produced on the underside; side branches will be formed, which have leaves with roots on the undersides and foliage at their apexes. The most commonly recognized rhizome is the *Iris*, although some *Iris* species are true bulbs.

Tuber

A tuber also is a swollen underground stem, but not the base of a stem as in a corm. It is usually fleshy, rounded, and covered with scaly leaves, often minute and concentrated toward the top of the tuber. In the axils, eyes develop, which produce the stems. One example is the tuberous begonia. Another is the *Dahlia*, whose eyes or buds (often known as tuberous roots) are found at the base of the older stem(s).

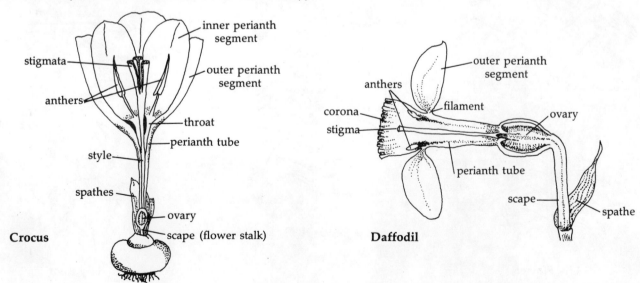

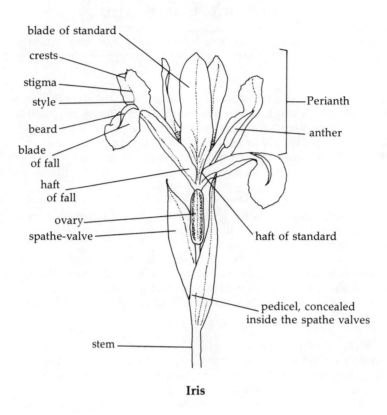

Iris

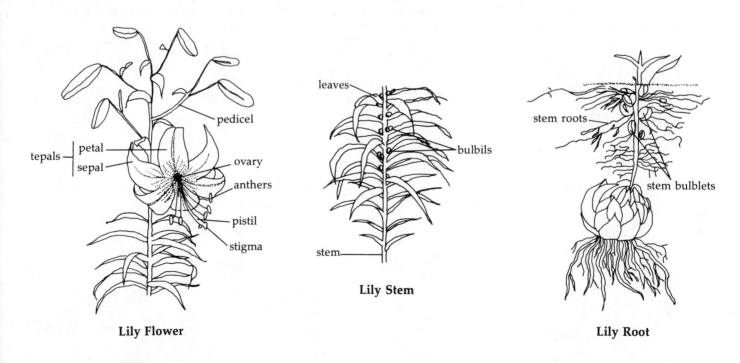

Lily Flower   **Lily Stem**   **Lily Root**

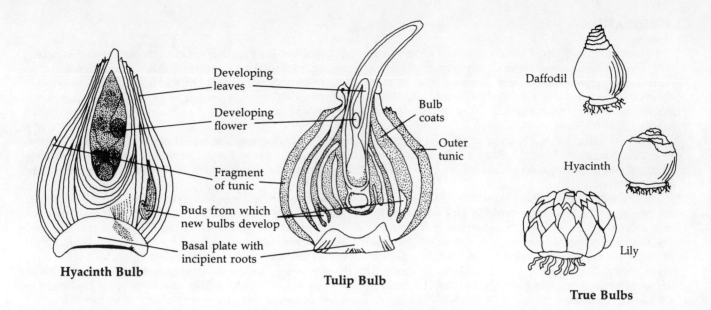
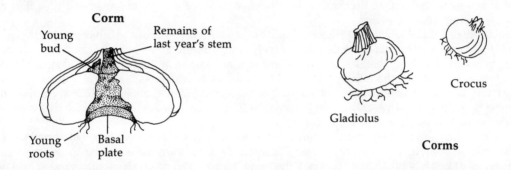
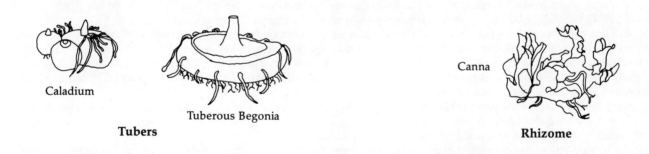
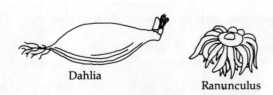

CLASSIFICATION
Taxonomy

In order to present the Plant (Vegetable)* Kingdom in a logical fashion, the examination of all its components is necessary. The grouping together of closely related plants has been accomplished by organizing them mainly according to reproductive characteristics and behavior. This is the science known as taxonomy. On the basis of this taxonomic observation, a natural system of classification was devised. Differences of opinion are found since the input has come from a number of eminent botanists. My own preference is for that system based on the work of German botanist Adolf Engler (1844–1930).

The basic structure is the same in all: Division, Class, Order, Family, Genus, Species. There are 13 divisions of the Vegetable Kingdom. Eleven include the Bacteria, Fungi, and Algae; one, Mosses and Liverworts; and one, Flowering Plants, the only one with which we will be concerned in this book and the one in which the greatest difference among experts occurs.

The Flowering Plants have been subdivided into Gymnospermae and Angiospermae. Conifers and cycads belong to the former, as the taxonomic criterion for Gymnospermae is that the seeds are not borne in an ovary. A further subdivision, called Classes, of the Angiospermae is made into the Monocotyledones (Monocots) and Dicotyledones (Dicots). The monocot embryos bear only one cotyledon (the first leaflike structure that appears after germination and sprouting). The true leaves, as a rule, are alternate, with simple form, parallel veins, showing no secondary thickening since their vascular strands contain no cambium, as in the dicots. The flowers usually are constructed of five alternating whorls of parts, each part being commonly trimerous. The dicot embryos bear two cotyledons. The leaves are net-veined, usually with a narrow base and a definite petiole. The flowers are pentamerous or tetramerous, with a distinct calyx and corolla.

These classes (and subclasses) are divided, in turn, into orders. Each order contains one or more families. Before proceeding, I think it is wise to understand the reasons for the importance of order and family and how they can be of great assistance in the identification of plants.

The order Liliales contains three families—Liliaceae, Iridaceae, and Amaryllidaceae. All have perianth segments of 3+3 (3 petals and 3 sepals), ovaries that are trilocular, carpels 3, and stamens 3+3 (Iridaceae 3+0). There are differences, however, between the three families. These differences are minor, however, when compared with other families. The need for breaking down the Liliales into the three families is as follows.

In Liliaceae, the ovary is superior, the stamens, petals, etc., being seated below the ovary. While the floral plan of the Amaryllidaceae is the same as the Liliaceae, the ovary is inferior, a difference of such significance that it merits being placed in a separate family, but not in a different order.

Those Liliales that have an inferior ovary and only three stamens are grouped in the *Iridaceae*. This, however, is not the only difference that can be seen. In Liliaceae and Amaryllidaceae, the petals are quite similar to one another, but in Iridaceae the outer series of petals is broad and recurved, the inner, narrow and erect. The actual number and pattern in which they are held on the flower, however, are not different from the Amaryllidaceae. Another refinement in the Iridaceae is that the style(s) is petaloid. The overall floral pattern is close to the other families of the order and thus it is placed in Liliales.

The most familiar plants in Liliaceae are the lily (genus *Lilium*) and the tulip (genus *Tulipa*); in Amaryllidaceae are found the amaryllis (genus *Amaryllis*) and the daffodil (genus *Narcissus*); and the Iridaceae contain the iris *(Iris)*, crocus *(Crocus)*, and gladiolus *(Gladiolus)*. See the Appendix for further discussion of these.

Recapping, then, we can see that there is justification for grouping in Orders. The juxtaposition of petals to sepals, the number of each part, the placement of all parts that make up the flowers—all unite the three families—Liliaceae, Amaryllidaceae, and Iridaceae—into the order Liliales. But, to be exact and, in the case being examined, due to the ovary being superior in one group of plants (the Liliaceae) and inferior in the others (Amaryllidaceae and Iridaceae), another breakdown in the classification can be undertaken and understood.

Other differences allow the Amaryllidaceae and Iridaceae to be separate. The former has 6 stamens, the latter, 3, with petaloid styles, hence the need for two families. Together, however, these three are more similar to each other than to other Families and thus are grouped in the same order, Liliales.

Before proceeding to the discussion of the genera (sing. genus), it will be useful to list the principal families into which bulbous plants have been placed. The list is not complete; it is everchanging in minor respects as botanists and taxonomists continue to study in greater depth the individual plants and the various genera. Shown with each family are representative plants. Including those cited above, the main families are: Araceae (arum); Begoniaceae (begonia); Compositae (dahlia); Genericeae (achimenes and sinningia [formerly gloxinia]); Primulaceae (cyclamen); and Ranunculaceae (anemone and ranunculus).

Having reached the family level in the classification of plants, it is self-evident that in order to identify exact plants within the family, greater differentiation is needed. The next step is to assign each plant to a genus.

---

*Having had my schooling in Europe and been awarded a scholarship from the Royal Horticultural Society, I tend to rely on that world-recognized authority and use the term "Vegetable Kingdom." Readers familiar with their publications and scholarship will appreciate the extremely high standards they maintain in all their undertakings. I have no wish to confuse you, so I must warn you that I use "Plant Kingdom" and "Vegetable Kingdom" synonymously.

Genus

This classification is used to include structurally or phylogenetically related plants. While the tulip and lily have the same arrangement of their flower parts, i.e., the same basic flower plan, there is a great difference between them. Some are quite obvious. The lily has a scaly bulb, the tulip has a truncated bulb; the lily has leaves on its stem, the tulip does not; in some lilies the petals are fused into a trumpet, with nectaries at their base, characteristics not found in the tulip. Although both are members of the same family, it is necessary to be more definitive about the classification in order to more fully distinguish between the two. So, we have the genus *Tulipa* and the genus *Lilium*. In the same way we can group together the erythroniums and the fritillarias, the alliums and the chinodoxas—all members of the same family (Liliaceae) but of sufficient difference, each from the other, to be classified each in a separate genus.

Having broadly classified the plants into genera, there still is a need to be able to identify the individual members of a particular genus. There are many tulips, there are many lilies—How to give each its own identity? The next step, then, is to give each a specific name—a species name.

Species

Using the floral diagram of the plant, we can determine that it is a lily of the genus *Lilium*. Taking two popular members of this genus, the tiger lily and the Easter lily, we can appreciate their similarities and their differences. The tiger lily is called *Lilium* (genus) *tigrinum* (now however changed) (species), the Easter lily, *Lilium longiflorum*. (See "Nomenclature," later in this chapter, for greater detail on the rules governing the specific names of plants.) The species name may be given on the basis of a certain characteristic, its native habitat, its discoverer—any number of ways to identify it as that species alone.

Examples of lilies with names that help describe them are *L. canadense,* which grows wild in much of Canada, and *L. japonicum,* a native of Japan. The Latin word "speciosus" means goodlooking, showy, splendid, and brilliant, and has been given to a lily that is all of these things, *L. speciosum*. Here, the ending of the species name has been changed to agree with the rules of grammar laid down by the governing body on nomenclature.

Subspecies

We have covered the main classifications, but provision has to be made for those plants that are almost identical but with some part of the plant that is consistently different. It is not necessarily a major part, such as the flower, but a minor part, such as the leaves, perhaps due to geographic location. While not meriting a distinct species name, they do merit another category, known as the subspecies.

There are not too many subspecies. One in the bulbous plant world that can be used to illustrate the classification is *Lilium canadense* subsp. *editorum*. Note that the word "Subspecies" is shortened to "subsp." or "ssp." and is not italicized but the actual name of the subspecies—*editorum*—is.

L. canadense grows mostly east of the Appalachian Mountain range. The subspecies is found in the Appalachians themselves and in drier situations, inhabiting rocky, woody slopes and open mountain meadows. The consistent, significant difference between this subspecies and the species itself is that the leaves of the subspecies are broader and not so tapered. This is enough of a difference to merit being given the rank of subspecies but not enough to merit full, specific rank.

Variety

It will be appreciated that, in the wild, variations occur among the seedlings of plants. Such variations may be but passing, seen on rare occasions but without consistency. When, however, consistent variations are found in permanent members of the Genus and Species, and, when cultivated, these differences can be propagated and not revert to the regular form of the Species, then such characteristics deserve recognition. For this departure from the norm the classification of Variety is given.

All too frequently, indeed, most persons who talk and write about plants err in the use of the word "variety." In botanic nomenclature, the correct use of the word "variety" is confined to a type of plant that occurs in the wild, without the interference of man. An example is a red form or color of flower appearing regularly in a species that is normally yellow. Such would merit the epithet *rubrum* and would be listed as var. *rubrum*. Again, this scientific or botanic name is italicized. If, however, the color is the result of man-made breeding, then it is correctly given the name "cultivar" (see below).

Cultivar (Hybrid)

In the commercial world, "variety" is given to a plant that is a man-made creation, often the result of careful breeding. We refer to 'Apeldoorn' as being a variety of the tulip, or 'Destiny' or 'Enchantment' as being varieties of the lily, but, correctly speaking, they are "cultivars."

The cultivar, or hybrid, is man-made, often following many years of experimental breeding, commonly with a definite end result or goal in mind. Many different species and other cultivars will have been cross-pollinated, the resulting seedlings being selected for the particular characteristic(s) desired, and all the thousands of remainders discarded! Cultivars, more familiarly referred to as hybrids, may be bred to realize the objective(s) of greater disease-resistance, more and better flower production, greater ability to withstand frost or wind, to last longer as cut flowers, or in some way to be superior to other cultivars listed in the catalogs of rival firms. The introduction into commerce of new cultivars often results in considerable financial reward for the breeder who has invested the time, knowledge, and energy necessary to come up with a successful end result.

Each year, new cultivars of tulips, lilies, dahlias, daffodils, etc., are offered to the public in the catalogs of commercial firms. Normally, only after testing in growing grounds and extensive comparisons with others already on the market are such new cultivars presented for sale. A name is given only after all of the many tests are passed. For example, *Lilium longiflorum* 'Croft' was introduced by Sydney Croft of Bandon, Oregon. He offered his plant, an Easter lily, to the commercial trade secure in the knowledge that he had a plant that was superior to any then on the market.

Clone

Should you purchase one of Croft's cultivars, or any other cultivar offered for sale, it is reasonable to expect that each one will be identical—a Clone.

Purchase a dozen tulips of the cultivar 'Clara Butt' and you will find that each flower produced by the bulbs is identical. How was this achieved? The propagation of the cultivar was undertaken without the use of cross-pollination; asexual methods were used. Sections or offsprings of the original bulb were used to increase the stock, each bulb, in effect, being a direct descendant of the original bulb. All such cultivars are known as "Clones." In the production of clones, the grower often will find that crossing certain parent plants results in consistently good seedlings, with but slight variations in color or form. Such crosses being worthy of distribution are offered to the public as "Strains."

Strains

Consistently produced seedlings, the result of crossing clonal parents, frequently are worthy of cultivation. The plants are not identical to each other, but their variation is within definite and defined limits. In lilies an example would be the 'Golden Splendor Strain'. When well established, the word "strain" is dropped and the bulbs offered as simply 'Golden Splendor'. The flowers of these bulbs are all of the same color, all of the same flower form or nearly so, and while the height varies it does so only within established limits. So it is possible to identify these plants as being 'Golden Splendor'. The flowering time could vary a little, and, should such strains be subject to growing under "forcing" conditions, the differences might well be more visible, but, under normal cultural conditions, the strain performs in a most consistent manner.

After a certain period of time, it might well be that the grower finds that by introducing new clonal parents the color is heightened, the vigor is improved, or some other feature of the plant is superior to the original strain introduced. In effect, these are "improved" plants, but, because the name, color form, and performance of the original strain are well known, and to avoid confusing the public, the newer bulbs might be offered under a name such as 'Golden Splendor Improved Strain'. Improvement can be ongoing and over the years the strain offered for sale will be far superior in every way to the original introduction, all, however, marketed under the same name—'Golden Splendor'.

It will be appreciated that hybridizers and growers use many species in the breeding of plants. Intermingled with the species used will be hybrids that have the desired characteristics the breeder wishes to obtain in the new creations. The picture becomes complex. The use of botanic or scientific names becomes cumbersome and impractical. The Botanical Classification no longer is a good way in which to classify the plants as there are too many cultivars, clones, and strains, all with mixed parentage and complicated bloodlines, involved.

To understand the form of a cultivar, strain, or other introduction and to more readily know the characteristics of the plants being discussed, another type of classification is necessary—Horticultural Classifications or Groupings. While such groupings can be determined by the parentage, they give greater emphasis to the flower form and shape. These classifications are adopted by societies such as the North American Lily Society or the National Dahlia Society, and, upon acceptance by a body like the Royal Horticultural Society of England, they become accepted as the classification used internationally by all who grow the plants that are the subject of the classification.

In some cases, dahlias are a good example, the grouping of the garden forms is straightforward and is used in catalogs, flower shows, etc. As will be seen from the following, with these classes being well defined, it is possible to place (categorize) all hybrids of dahlias on the market today. Such Horticultural Classifications also are in use for lilies, tulips, narcissus, and other genera where extensive breeding has made them necessary.

Show Single Dahlia. Class I*a*   *Single Dahlia.* Class I*b*   *Anemone-flowered Dahlia.* Class II   *Collarette Dahlia.* Class III

Paeony-flowered Dahlia. Class IV

Formal Decorative Dahlia. Class V

Informal Decorative Dahlia. Class VI

Show Dahlia. Class VII

Pompon Dahlia. Class VIII

Cactus Dahlia. Class IX

Semi-cactus Dahlia. Class X

Star Dahlia. Class XI

Nomenclature

Every known thing in the world has been given a name to distinguish it from anything else. And so it is in the Plant Kingdom, only more so, where a slight variation may result in a new genus or species. In the 18th century, Linnaeus, the famous Swedish botanist, devised the binomial system of plant nomenclature on which the modern system is based. This method consists of a two-part Latinized name (regardless of actual origin)—the first word, always capitalized, is the genus name; the second word, always lower-case, is the species name *both in italic type*. It then became possible for anyone in any part of the world to immediately recognize a given species. Common names will vary, from country to country, and even from state to state, as in the United States, but the scientific name will always remain the same.

In 1905 the modern rules were drawn up and agreed to at the Botanical Congress held in Vienna. From this Congress came the principal code governing the naming of plants—*The International Code of Botanical Nomenclature* (ICBN). There is a supplement that provides for names not in Latin form, that is, for cultivars, *The International Code of Nomenclature for Cultivated Plants*. These international Congresses are held in different countries on a regular basis to solve problems regarding nomenclature of plants.

The governing rules are simple in principle:
1. A plant can bear but one valid name.
2. The valid name is the earliest that conforms to the Code of Nomenclature.
3. The name must have been properly published, with a description of the plant.
4. The same name cannot have been used previously for any other plant.

Plants occurring in the wild are named by botanists after they have studied their structure and relationship with other plants. They can be designated species, subspecies, varieties, etc., as determined by the botanist examining the plant in question. As noted above, the botanic name is always Latinized, with the genus name first, always capitalized, and the species name second, always lower-case. Species names derived from a person's name, for example, *Wilsonii* (named for E. H. Wilson) or *Forrestii* (for George Forrest) can be capitalized. Originally provision also was made for names derived from a native name or from former generic names to be capitalized. Today it is generally accepted that all species names be printed in lower case. Thus, when reading older books, one may come across capitalized species names.

To render a name valid, it must be published with a full, authentic description of the plant, together with an illustration, in a dated publication. Like the name, the description must be in Latin. The date of first publication is the method for establishing the rank for priority of the name. As the very first name used is the only true one, a great number of name changes have resulted from better communications between botanists. Prior to the acceptance of the International Rules, other codes of nomenclature were in use. A name once bestowed upon a plant cannot be used for another plant in the same genus. There is an escape rule, however, which does allow for Historical Preference, as the confusion that would reign if older and older names were continually found for genera now accepted would far outweigh the advantages of being strictly accurate, in this regard, according to the rules.

The ideal is that any plant bear only one 'official' name. It is often left up to the International Horticultural Conferences as to which names are to be retained. An example of the need for such rulings can be found in the genus *Anemone*, of the family Ranunculaceae. Certain botanists place in this genus all plants that have regular flowers, petaloid sepals, no petals or nectaries, and many stamens and carpels. Others would separate out those species in this genus having distinct arrangements of bracts and similarly constructed carpels. Presently such species are all under *Anemone* but the botanists make a case for elevating the species *A. pulsatilla* to generic rank calling the genus *Pulsatilla* and placing into this genus those plants with similar arrangements of bracts and similar carpels. This would split plants presently in the genus *Anemone* away from that genus. Another group of botanists regard the plants presently known as *Anemone hepatica* as being worthy of generic rank, *Hepatica*, on the grounds that the leaves and flowers of that species are distinct from others in the genus *Anemone*.

Such splitting of genera is not unknown, but also is not always accepted by the International Congress. However, a list of genera sometimes divided by botanists has been drawn up by the International Congress, and this generally is adhered to in horticultural and botanic writings, often with reference being made to the other nomenclature.

Names of botanic varieties are always written after the name of the species as follows: *Narcissus triandrus* var. *albus*. Often, the listing in the literature will omit the word "variety" or the abbreviation "var." but, when three names are given, the third name is always that of a variety, unless it is a subspecies. The term "variety" or "var." can be dropped but "subspecies" or "subsp." cannot.

The name of a man-made cultivar must be placed in single quotes and cannot be Latinized and called a "fancy" name, e.g., *Lilium longiflorum* 'Croft', not *Lilium longiflorum croftii*.

A name given to one plant in a genus cannot be used for another in the same genus. Thus, if we refer to the lily 'Enchantment', we know to which hybrid we refer—an upright Asiatic—and the same name cannot be given to a trumpet lily. It could, however, be used for a new rose introduction, provided it had not been used previously for another rose. Cultivar names given to plants in a genus are kept by official registers throughout the world and referred to by those who wish to give a name to an introduction they are making. There is some difficulty in the selection of a name for plants as many that come quickly to mind have been used already.

Similar names cannot be used either. If 'Mary Lee' has been used, 'Maryly' would not be acceptable. The articles "A" or "The" also are to be avoided whenever possible. Thus, "Wedding" but not "The Wedding" or "Giant" but not "The

Giant," albeit this is not always adhered to.

Hybrids between genera are designated by listing the parents in alphabetic order, preceded by an ×, e.g., × *Chinoscilla* (*Chinodoxa* × *Scilla*).

It will be appreciated that the naming of plants by botanists and the introduction of new cultivars by growers and hybridizers, when named according to whim and not in keeping with the rules of nomenclature, can cause a great deal of confusion. While comparatively rare, instances of established names being given to plants resembling the original introduction do occur, so the buyer of bulbs should be aware that such practices are not unknown. The principal reason for such unethical introductions is that the grower can take advantage of the publicity and acceptance by the public of the original introduction, with a corresponding effect upon sales. While many people disregard the Latin names of plants, and many others are often overawed by scientific names, such names are essential. Without a specific nomenclature code, accepted around the world, it would be impossible to determine to which plant one is referring. With such specific naming of plants, accepted by botanists and horticulturists, there is no room for error or misinterpretation.

The placing of plants into families is logical. Knowing the floral formula of the families allows plants to be placed in that family. This automatically limits the genera in which the plant is to be found. By observing the differences between the genera in a family, the exact genus can be determined, then the difference between species comes into play so that exact identification can be made. Such processes may appear cumbersome, but the system has withstood the test of time and, without such widely accepted classification, the world of plants would be in a terrible mess.

CHAPTER 4
PROPAGATION

Seed, cuttings, division, scales, bulbils, bulblets, layering, cormels, meristem culture.

Bulbs lend themselves to a number of methods of propagation, many of them uncomplicated and fairly easy to do. Despite this, few home gardeners actually propagate their bulbs.

The most commonly practiced method used by home gardeners—lifting and dividing—stems from the cultural requirements of the bulbs. Irises, agapanthus, day lilies, and, to a certain degree, lilies and daffodils require lifting and dividing in order to continue flower production. While this is regarded as one method of propagation, it is essentially a necessary procedure in the life of a bulb.

Propagation is either asexual (vegetative) or sexual (by seed). Asexual methods ensure that the plants produced will be identical with the parent plant—in effect, cloning. Plants produced from seed may result in some variation, which is not always desirable. However, new cultivars (introductions) are almost always raised from seed. While many seedlings will be discarded, the possibility always exists that an improved plant will result, be it only one plant among thousands.

I say that cultivars are "almost" always raised from seed, because, on occasion, a bulb may "sport." The term applies to the phenomenon where a shoot (rarely the entire plant) will produce a flower of different form or color, which is desirable. To increase the stock of such a "sport," asexual methods must be used.

In breeding for new cultivars, the hybridizer will select plants that exhibit desired characteristics. These are mated by cross-pollination. Precautions must be taken so that no other pollen can fertilize the parent plants. This is done in a variety of ways. The simplest is to cover the fertilized flower with a paper bag; another is to top the fertilized stigma with an aluminum or other type of protective cap. In this way only known pollen, carrying the desired features sought, will fertilize the ovary. From such crosses, perhaps only one of the resulting seedlings will be found with the correct combination of characteristics the breeder is looking for.

The color and form of the flowers, the strength of the stem, and the appearance of the foliage are all critically examined. Should the new selection pass all of these tests, it also must exhibit resistance to disease and, most importantly, be easily propagated, otherwise it will be too expensive to raise in quantity for commercial purposes. Only after careful and stringent evaluation will a new cultivar be introduced into the marketplace.

It can be seen, then, that this type of propagation is a long and tedious process, generally resulting in only one success among the thousands of seedlings raised. Sometimes, however, the hybridizer will be fortunate and, from one such crossing, many new cultivars will be found.

SEED

If there is a disadvantage to raising bulbs from seed, it is the time it takes many of them to flower. Some, such as the begonia, dahlia, and ranunculus, can be in flower the same year as the seed is sown. Others, such as the tulip, daffodil, and most of the *Lilium*, take as many as 5–6 seasons before a flower is produced. This length of time discourages most gardeners. It should be pointed out, however, that if seed is sown each year there soon will be a succession of bulbs coming into flower each year. The exciting possibility always exists that a new and improved plant will pop up!

Soil Mixes

In recent years, "soilless" soil mixes have become popular. Prior to their advent, a standard mix for seed was: 2 parts sterilized top soil, 1 part peat moss, and 1 part sharp horticultural sand. (And I might add that this mix is still excellent.) The amount of sand is increased for plants requiring extra drainage; more peat moss is added if additional water retention is desired. The top soil is sifted through a fine sieve and small quantities sterilized by boiling ½ inch of water in a small pan, then adding the soil, allowing the mix to simmer for 15 minutes. For the home gardener who is raising small quantities of seed of many different types, this is still a fine soil mix and a good way to sterilize the soil.

There are many media available to the gardener today. Germinating mixes are sold prepackaged, sterilized, and ready for use. Those mixes using sphagnum peat moss mixed with perlite or vermiculite, or a combination of the two, are popular. I prefer perlite as it retains less moisture than vermiculite, a very important feature to be considered when mixed with the highly water-retentive peat moss.

In my opinion, seedlings that have germinated in a soil mix containing only a small percentage of soil have more robust roots and are more readily established when transplanted than those growing in a "soilless" mix.

Professional growers, when in need of large amounts of soil mixes, often develop their own special formulas. Peat moss mixed with perlite is popular, the exact percentage of each ingredient varying. Such media are light in weight (an advantage when shipping), retain moisture well, and allow the grower to fertilize with a broad range of products. Another advantage is that such mixes are sterile and thus no further sterilization is necessary, a money- and timesaving consideration.

Experimentation with various soil mixes is advised, until that particular one that performs well for the majority of crops has been determined. Several universities have conducted experiments with soilless mixes and their recommendations are well worth considering, with modifications as needed to adapt to a particular situation or crop.

While vermiculite is much used, soil mixes containing perlite will be found to warm easier. The disadvantage is that perlite is less retentive of moisture and thus greater care must be paid to watering. If watering is done with an automatic system, this becomes less of a problem. With such irrigation systems it is very important that the consistency of the soil mixes be uniform so that all of the plants in the media lose water in the same way.

To obtain well-mixed media, mechanical methods are an advantage. Throwing the ingredients into a large hopper, such as a cement mixer, is a quick and efficient way to ensure this. Such methods also help to eliminate poorly balanced mixes that can occur if unskilled labor is used.

Should ingredients contain weed seeds, fungi spores, or other harmful pathogens, sterilization is essential. A boiler capable of raising the necessary steam pressure or ovens that can bake the mix at the required temperature (180°F) will be needed. Temperatures above this will harm the good bacteria and organic matter contained in the soil. Time spent in ensuring that the soil mixes are sterile can make the difference between the success or failure of a crop.

A considerable price advantage can be obtained by ordering larger quantities. Although "bulk" deliveries are available in certain areas, and are necessary for the commercial grower, the smaller raisers of seed should consider the convenience in handling the materials in sacks and this should be weighed against any money saved by buying large amounts in bulk. Unless stored properly and protected from the elements, peat moss can become too wet and contaminated with weed seeds.

The principal points, then, to keep in mind are:

> Availability of materials used.
> Cost.
> Suitability for the majority of crops.
> Ease of storage.
> Weight.
> Moisture retention.
> Drainage.

Requirements for Germination

It is necessary to understand the requirements for germination of the seed before we go on to the actual sowing of the seed.

Nature provides the right conditions in the wild for species to be continued. Bounteous quantities of seed are produced, but only a small percentage germinates and even fewer of the resulting seedlings reach maturity. Nature can afford to squander some of its bounty; however, neither the home gardener nor the commercial grower can afford such luxury. It is essential that the highest percentage of seed sown results in plants suitable for the garden or for sale. Thus the elements necessary for germination must be provided to ensure success with the rather limited number of seeds available to the seedling raiser.

The three essential requirements are moisture, warmth, and air. Only when all three are met will the process commence. Light is a requirement only after the seedlings have germinated.

Moisture is needed to soak the outer seed coat so the root and shoot can break out of the hard coat. It also is needed to swell the cells inside the seed.

Warmth activates these cells, although the actual temperatures required will vary. Seeds from the Tropics need higher temperatures than those from more temperate climes. Providing the right temperature is essential—if too high, the seedling that develops might well become weak and etiolated; too low and there is no metabolic activity.

Air is necessary because the developing seedlings must breathe. Good circulation around them also helps keep fungus diseases under control.

Germination has occurred when the root and shoot emerge from the seed.

Sowing the Seed

The ideal method is to sow the seeds, allow them to germinate, and, without interruption, have them continue to grow into strong, healthy plants. Such conditions are provided by Nature in the spring, allowing the growth to continue with the lengthening days, then slowly reducing the growth rate by diminished light and lower temperatures as the year progresses.

The seeds of many bulbs, however, are not so accommodating. Many require a period of cold, as they receive in the wild. This causes changes inside the seed, breaking down, it would appear, certain hormones that inhibit germination. Exactly what changes take place is not known. It is a logical control, as many species of bulbs inhabit areas where the winters are severe. If the seeds germinated quickly, the seedlings would not be able to withstand the rigors of winter. The need to undergo a period of cold to break dormancy ensures germination in the spring, producing growth and the gaining of strength to survive the first winter after germination.

Nature, however, does sow the seed as soon as ripe, and, in the wild, usually waits for a cold period before germinating (tropical plants being one exception); therefore, some bulbs require such cold treatment before being sown by us. Among them are:

Allium	*Bloomeria*	*Chionodoxa*	*Erythronium*	*Hemerocallis*	*Leucojum*	*Nerine*
Anemone	*Calochortus*	*Crocus*	*Fritillaria*	*Hyacinth*	*Muscari*	*Scilla*
Arum	*Camassia*	*Cyclamen*	*Gladiolus*	*Iris*	*Narcissus*	*Tulipa*

If you are saving the seeds from your bulbous plants from temperate climes, sow them in pots when ripe and place them outside to enjoy the cold of winter. Come the following spring they will germinate. Certain seeds, however, will take more than one season to germinate if not treated in this manner. Among them are *Alstroemeria, Colchicum, Galanthus,* and *Sternbergia*. In warmer climates, where winter temperatures do not go below 40°F for any length of time, placing the seeds in the freezer compartment of a refrigerator will provide a substitute for natural conditions. It also is necessary to ensure that the seeds, once sown, enjoy steady, unfluctuating temperatures. Constant thawing and freezing can cause the soil to heave, thus dislodging the developing roots. In cooler climates, plunging the pots into the ground works well.

Protection against birds, mice, and other animals must be provided when placing the seeds outside. Placing wire mesh over the containers or setting the containers in a cold frame are two methods that serve the purpose nicely.

Most of the seed purchased from commercial sources becomes available in the early part of the year and is prepared ready for growing. Providing as long a growing season as possible is most beneficial. In the majority of cases this will require either a greenhouse or a heated frame. Few seeds will germinate below 50°F; most, however, prefer temperatures in the 60°–65°F range. If such conditions are not available, then sowing should be delayed. While the cost of a frame is not high and warming cables also may be purchased at reasonable prices, alternatives such as a window in the home or a warm basement with adequate light, using growing lamps if necessary, are satisfactory.

In commercial establishments, where considerable amounts of seed are sown, it is advisable to construct cold rooms where the precise temperatures needed by the various species to break dormancy can be maintained. Exact controls, proper handling of the seed, and the additional protection such structures provide against the ravages of mice and other pests give a good return on the investment needed for their construction.

A separate greenhouse or a separate section of one should be allocated for raising commercial quantities of seedlings for the production of bulbous plants. If large numbers are being sown, they can be planted in an open bed on a bench. The soil mix should be porous, uncompacted, light and fluffy, of a sandy texture, and free of all weed seeds. Excellent drainage and good cross ventilation must be assured. Bottom heat is an advantage. Some growers find it advantageous to install an automatic drip-watering system. While hand watering is preferred after germination, such a system can cut down on the labor needed for watering when the seedlings are growing well. Provision should be made for fertilizer to be fed through the system.

Every 10–15 days, it is wise to "fluff up" the soil mix between the rows. This increases the amount of air in the top layer of soil and cuts down the possibility of slime molds and other unwanted vegetative growth, which are contributing factors to fungal and disease problems. A further advantage of light, fluffy soil will be appreciated when the seedlings are being lifted for transplanting. The tender roots will come out easily, with a minimum of damage.

There is a definite advantage in additional light being available over the benches. This will help speed the growth of the seedlings should the days be cloudy, as is quite common during the early months of the year. Under controlled greenhouse conditions the grower, then, is able to duplicate with warmth and light the elements the seedlings would enjoy later in the year in their natural habitats. The longer the growing season provided, the larger the plants will grow and the more quickly they can be raised to commercial size. Those raised in optimum growing conditions right from the start will carry this growth advantage throughout their lives and make better adult plants.

Sowing in Containers

The containers, flats, or pots, be they of clay, plastic, or wood, must be clean. If not new, they should be sterilized before use. Fill the container with soil mix and firm gently. Ensuring that the surface is level is easily done by using a smooth wooden block or a piece of glass. The mix should be moist, holding the imprint of the hand when squeezed but breaking apart when touched; it should not be wet enough to wring water from it when squeezed. After gently firming, the level of the mix should be at least ¼" below the lip of the container.

Seed that is very fine, such as that of begonia (there are several million to the ounce), should not be covered. Larger seeds can be covered with their own depth of soil mix. In all cases, care must be taken that there is even distribution over the surface of the container. Do not sow too thickly; allow plenty of room for each seed—this will make transplanting much easier. The fine seeds should be watered in by placing the container in the water and allowing it to seep up through the bottom, until the surface is damp. Larger seeds can be watered by using a fine rose on the end of a watering can.

After sowing, the container should be covered with a sheet of glass or clear plastic. This must be kept level or drops of water will accumulate, falling into one location and causing that area to become too moist. The container then can be set in a place where ideally an even temperature of 60°–65°F at night can be maintained. Do not place where the sun will strike the container as this can cause the mix to dry too quickly. Check regularly, wiping away any excess moisture that has accumulated on the cover. As soon as any of the seeds have germinated, the cover must be removed. Maintain an even moisture content, however, giving additional water if necessary.

This cultural practice will enable dahlias, ranunculus, begonias, cannas, galtonias, agapanthus, anemones, gloriosas, and hippeastrums to produce sizable bulbs and, in some cases, to flower that summer.

If containers are used in the commercial production of seedlings, it is essential that standard-sized flats be used. These can be accommodated on benches in the greenhouse or in frames with a maximum utilization of available space.

In order to maintain consistency in sowing rates, it is advisable for one person to be responsible for all sowing. While the methods described above for home gardeners may seem tedious, it must be emphasized that the time taken to ensure that seed sown in containers gets off to a good start is time well spent.

In commerce, the construction of a simple frame inside the greenhouse should be considered instead of covering individual containers with glass or plastic. This will isolate the containers and provide a closed environment without the necessity of a cover.

Sowing in the Ground

While it is preferable to sow the seeds of bulbous plants in containers, certain species, such as lilies, irises, alstroemerias, tulips, and gladioli, can be sown outdoors in the ground. If this is your preference, work the soil well and, for best results, provide protection with a cold frame. Dahlias, canna, and summer hyacinths can be sown where they are to be grown. In cold areas this should be done after the danger of frost has passed; in areas where no frost is experienced, sow when the soil has warmed and nighttime temperatures are above 50°–55°F. In all cases, when sowing in the ground make sure the seed has received the necessary chilling required by that species. See individual notes on each bulb in Chapter or alphabetical listing.

The home gardener can sow the seeds of some species in those locations where they will be left to grow into adult plants. Commercial production does not allow for such propagation. While it is possible to sow lilies, for example, in the open ground and harvest mature, commercial-sized bulbs two or three seasons later, it is more economical to lift after one season, sort into sizes, and replant those from the open-ground sowing that are not fit to be sold. Dahlias, however, are marketed as small plants and offered for sale in convenient six-pacs or in 3- or 4-inch pots. Open-ground sowing is impractical for these plants. The trend with the majority of bulbs is toward clonal varieties. This renders seed sowing applicable only to those bulbous types that are sold as straight species or strains.

In the case of the professional grower, where monetary considerations are involved, it is especially important that correct horticultural practices be followed. The seed beds must be well prepared and of a fine tilth. They should be about 48 inches wide so that they can be tended from both sides. A good gap of 24 inches should be allowed between types. The seed should be covered lightly with soil after sowing, so lightly that some seeds can still be seen, indicating that those covered are only about ¼ inch below the soil. After germination, another top dressing can be given so that the seedlings are set firmly in the ground.

A ½-inch mulch of sawdust, spread evenly over the surface, can be applied when the seedlings are 1 or 2 inches tall. This will conserve moisture and keep down the weeds.

Labels identifying the specific seeds being grown should be placed at both ends of the bed, and records should be kept on the number of feet of each type sown. These records will allow for ascertaining production per foot of row. In this manner, more accurate seed sowing can be accomplished in following years, planting only that number of feet necessary to meet the production quota.

Speaking of recordkeeping, I feel this is a very important part of growing bulbs commercially. As shown above, much time, labor, and money can be saved recording seed numbers planted. The date the seed is received and the name of the supplier(s) should be listed. While all seeds may be first-class, invariably, strains from certain suppliers will perform better than others—earlier into flower, less mildew, quicker to germinate, etc. Some seed companies may offer different cultivars from others. Seed is not inexpensive and so it is advantageous for the grower to know that the supplier is of consequence when reordering.

GENERA DIFFICULT TO GROW

Two of the trickiest bulbs to raise from seed are lilies (genus *Lilium*) and begonias (genus *Begonia*)—lilies because of the diverse habits exhibited by the different species and hybrids; begonias because of the very fine seed. Due to their popularity, a more detailed discussion of their requirements seems in order.

Lilies

Four different habits are exhibited by lilies in germinating—hypogeal (below ground), immediate or delayed, and epigeal (above ground), immediate or delayed.

Hypogeal types do not produce a cotyledon that emerges above the ground. The food from the seed is transfered underground to a point between the true seed leaves that emerge above the soil and the root tip. This point becomes the miniature bulb, and the transference of the food is made before the seed leaves are produced. If the growth cycle is continuous, the name given is hypogeal immediate. If a cold, or incubation, period is necessary after the initial germination, it is known as hypogeal delayed.

Epigeal types send a cotyledon above the ground. The seed containing the endosperm, which nourishes the developing seedling, is either attached at the end of the cotyledon or discarded and left below ground. Just above the root tip a small node is apparent, which then becomes the bulb. When this has absorbed sufficient strength, true leaves are produced. Epigeal immediate types may be sown as soon as ripe, but this is not always practical. Epigeal delayed types require a definite cold period before germination will occur.

Fortunately, not all four types require different procedures. While the treatment required by the different types seems time-consuming, following the procedures outlined will bring seedlings to flowering size quickly and time actually will be saved in the long run.

It is advantageous to sow all types of lily seed in the spring, when growing conditions are best. This can be done either in a greenhouse or outdoors in specially prepared seedbeds.

The seeds of Asiatic and trumpet lilies are of the epigeal germination habit. The hybrids containing *L. sargentiae* blood, such as 'Pink Perfection', 'Golden Splendor', and 'Black Magic', are delayed types. While the seed of all of these types can be stored in the freezer to preserve viability, it is essential that the trumpet types containing the *L. sargentiae* blood receive at least 3 months of low temperatures before they are sown. In the spring, when the weather is warm and as soon as the ground can be worked, the seeds can be sown. They must be protected against late spring frosts and scorching sun during the summer and must be kept moist at all times.

The exotic hybrids, those between *L. auratum* and *L. speciosum* or those that have any blood of these species, belong to the hypogeal delayed group. *L. brownii* and those with *L. dauricum* blood are hypogeal immediate. After overwinter storage, the immediate types can be sown in the spring, at the same time as the epigeal types.

The hypogeal delayed types are stored in the freezer until May, when the seed is mixed with vermiculite and incubated. During that month the temperature should be maintained at 50°F; this is raised in June to 60°F and in July to 70°F. The latter temperature is maintained until 3 months before sowing, which is usually done the following spring. For this final 3-month period, the seedlings, which now will have developed a small bulb, must be kept at 34°F. They then can be sown in rows or broadcast, together with the vermiculite, in prepared seedbeds. With this method some flowers will be seen the second season and flowering-sized bulbs will be produced by the end of the third season. Like the other types, the seedlings must be protected against frost and too-strong sunlight.

After one or two seasons in the seedbeds, the plants are lifted and replanted so they can be cultivated in rows to bring them to flowering and commercial size. Any bulbs that have reached commercial size can be sold.

While, as I have said, the treatments described may seem time-consuming, they provide a high germination percentage and will lessen the time needed from seed to flowering bulb.

A gardener without experience in the raising of lilies from seed might well be confused by the different treatments needed to ensure good germination of the various types. Nothing is more gratifying and encouraging than a successful crop! For these first-timers, I suggest the lilies of the epigeal-type germination be tried first. The seed, after being harvested, is put into a sack and placed in the freezer. In the spring, after a minimum of 3 months in the cold, the seed can be sown.

A sterile container, at least 6 inches deep with a goodly number of drainage holes in the bottom, is lined with an inch of gravel. The seed-sowing mix, which should be fluffy, is added to a depth of 4 inches, which is then covered by a ¼-inch layer of milled sphagnum moss. The seeds are placed at 1-inch intervals on the surface of the moss and covered with another ¼ inch of the milled sphagnum.

The container is set in a pan of water to a depth that will let the water rise up through the container until the surface is moist. It is then lifted out of the water and the excess moisture allowed to drain. The container is then covered with plastic and placed in a warm location. In a little more than 2 weeks the seeds should begin to germinate. Remove the cover as soon as several seeds have germinated. The container should be moved to a spot where it can receive good, bright, but indirect, light.

After most of the seeds have germinated and the seedlings are an inch or so high, weak feedings of liquid fertilizer can be given. When all danger of frost has passed and the temperature at night is in the 40° range, the seedlings can be hardened off. That is done by placing them in a wind-free location outdoors and protecting them at night with a plastic covering. The plastic is lifted during the day and replaced again at night. After awhile, the sides of the covering are lifted during the night for a week and then the covering removed entirely, to be used only if an unexpected frost should occur.

In areas of little or no frost, the seed should be sown in March so that light and temperature are conducive to steady growth and the hardening-off period is not so critical. In no case should seedlings that have received protection of some sort be subjected to a great variance of conditions until hardening-off has taken place.

After the seedlings have reached 4–5 inches in height, they can be transferred to final growing locations or placed in individual containers, using a stronger soil mixture, but always with good drainage. If seedlings have been spaced too closely together, replanting or transplanting is essential. Care must be taken not to damage the fragile roots when separating the seedlings from one another. This is one of the reasons a fluffy soil mix is an advantage, making it easier to remove the young plants.

The home gardener might well be tempted to save seeds from other bulbous plants. With variations, the above method will be applicable to all seeds, the variable being the depth of coverage. Lily seeds are quite large, and hence are covered to a depth of ¼ inch. The finer the seed, the less the depth of coverage.

Certain bulbs do not produce as much growth as the lily in one season. For these, do not plant-out the first year; seedlings should be grown-on in individual containers until of size, which can be determined when they can be handled with ease, say about 2–3 inches high.

Begonias

As mentioned, begonia seed is very fine. Raising tubers from seed should not be undertaken if only a few plants are wanted. In such cases, purchasing the tubers will be less expensive.

From seed, the results can be most rewarding. Even if known to be within a given color range, the precise shades of the flowers are unknown, and it is entirely possible to obtain a spectacular display.

When ordering begonia seed, keep a couple of things in mind. The seeds remain viable for several years if refrigerated, but it is best to specify "new crop" when ordering. Also, because the seeds can be crushed in the mail, send along a mailing tube so the vendor can return them without harm being done to them, or perhaps the grower takes this precaution, knowing of the vulnerability of these fine seeds.

The seed pan for begonias can be shallow. Only an inch of growing mix is needed. In order to obtain tubers in one season, it is necessary to sow the seed no later than February; December is the most desirable month. The surface of the soil mix is made as even and flat as possible. The seed is sown by scattering it over the surface of the container. One method is to fold a file card or similar type of stiff paper, fill the crease with the seeds, carefully sprinkle over the container, and gently brush with a fine camel's hair brush to obtain even distribution. Great care must be taken in watering because the seed must not be covered with soil, which can happen if a coarse rose is used. The preferred method is to place the container in a pan of water and allow the water to seep to the surface. Drain and then cover with glass.

The ideal temperature for germination is 65°–70°F. Light is necessary for at least 14–16 hours per day. As natural sunlight is not available for that length of time, Gro-Lux lamps, placed some 10 inches above the containers, should be used. If the container is covered with glass, wipe off any condensation as necessary. After the initial watering, the container will need misting every three or four days.

The seeds will germinate in about 10–12 days. At this time the seedlings must be ventilated. Do so gradually so that in a week the glass can be removed entirely. The temperature can be lowered to 60°F at night as soon as germination is observed. When the seedlings are up and growing well, weak feedings of liquid fertilizer can be given with every third or fourth watering.

The seedlings will put out their first true leaves some 30 days after germination. They then can be transplanted to a richer soil, spaced some 2 inches apart. Additional light should be given throughout this period and discontinued only when there is approximately 14–16 hours of natural sunlight per day. Sown in late December, the seedlings will be large enough to transplant to individual pots in March and will reach good size for planting out in beds or in 8-inch pots by the time danger of frost is past, or in early May in areas where no frost is experienced.

Seedlings being grown in a mixture of sphagnum peat moss and perlite or vermiculite can receive weak feedings of fertilizer, but care must be taken not to overfertilize.

If being raised in containers, seedlings should be transferred to a growing-on soil mix as soon are they are large enough to handle. Many such mixes can be purchased. One that is satisfactory for most bulb seedlings is: 7 parts sterilized good top soil, 3 parts coarse peat moss, and 2 parts sharp sand. As always, use only sterile containers.

When transplanting, make sure the seedlings are not planted at a depth greater than enjoyed in the container. Space them out to give adequate room for growth in a 6–8 week period. After transplanting, coddle them a little for the first few days to allow time for the roots to take hold in their new home. Harden-off the plants gradually; keep moist but do not overwater.

Temperatures should be moderated so the plants develop sturdily and have enough light to avoid etiolation. In cold areas, place them outdoors only when the danger of frost is past; protect them if frost is expected. During the summer the young plants should enjoy the same conditions as mature plants.

The preceding conditions—soil, temperature, etc.—are applicable in all instances where begonias are raised from seed. The commercial grower need only adapt them to meet large-scale operations.

Transplanting the seedlings is a slow, tedious job requiring a great deal of patience, so it is best to space sowing times out over a period of time to ensure that the workers assigned to carrying out the job are not overwhelmed by the quantity of seedlings to be handled. This allows for a steady flow of seedlings. An easy method for handling the seedlings is the use of a wooden label with a notch cut in the end. The sooner the seedlings are transplanted, the better. If the grower is looking to get the best size in the shortest time, supplemental light will be needed.

After the initial transplanting and when the seedlings are growing vigorously, the developing tubers are lined out in nursery rows, spacing them carefully. They then can be harvested when of commercial size. Such growing-on should be done outside only where the weather is warm and winters are mild, or in a greenhouse.

ASEXUAL PROPAGATION

The only way a clone can be propagated to ensure that the resulting plant is the same as the parent is by asexual propagation. True species can be raised from seed since the bloodlines are fixed and they come "true" from seed. All cultivars of bulbous plants, if they are to be identical to the parent, must be propagated without the intervention of sex, i.e., without pollen from the plant's male organ and female ovaries. The reason is that the possible combination of chromosomes and genes, as contained in the male and female gametes, rarely if ever will match in such a way that a plant identical to the parent will be produced. The possibility of other combinations of genes being assembled is more common. Thus, unless a small part of the parent plant is detached and grown onto full size, there is no way an identical plant can be produced. The methods of asexual propagation are numerous and include division, cuttings, meristem culture, and bulblet and scale production. For the gardener, asexual propagation is the only way a favorite plant can be propagated to ensure an exact duplicate. For the commercial grower such methods are the only means for propagating clones.

STEM CUTTINGS (OTHER THAN CUTTING TUBERS, ETC.)

Two popular bulbs can be propagated by cuttings—dahlias and begonias. Dahlias lend themselves to this method easily; indeed, it is one of the ways in which commercial cultivars are produced in quantity. Begonias normally are raised from seed (see above), but cuttings provide an easy method of increasing the stock of a specific plant (species or cultivar).

Both amateur and professional should take particular care in the selection of "mother stock" plants. They should be free of virus, vigorous, and the flowers produced uniform and typical of the variety. In examining plants of any given species or cultivar, certain of them will be better than others. These are the ones to select and tag for separation at harvesting time. The inspection of the crop for selection should be undertaken at different times: at the beginning of the season when good growth has been made, again at the time when the plants are in full flower, and, lastly, toward the end of the season. The exact number of tubers needed for propagation is first determined in the spring. Then, after each field inspection, the number of plants is reduced to ensure that the exact number required and those of the finest have been selected. These selected bulbs are lifted ahead of the others and handled with extra care. They are labeled, tagged, and carefully stored to ensure that they are in the best condition possible when they are started back into growth.

Dahlias

While the method of producing cuttings is basically the same for both amateur and professional, the time production starts differ. If only a few cuttings of each plant are required, they should be ready in 6 weeks from the time they are first taken until of planting size. Those who wish just a few new plants should start the procedure in late March, since dahlias grow well only when the night temperature is in the 50°F range. This will give ample time to have rooted plants to set out in the garden by mid-May.

Commercial production should start much earlier. The use of a warm greenhouse or a series of heated frames will be necessary in regions that are not frost-free.

Roots selected during the previous growing season are lifted and stored until the middle of December. They then are washed and dipped in fungicide to rid them of any fungus spores.

In an area where good light is available, a bench or heated frame is prepared by creating a bed of peat moss some 6 inches deep. The tubers are nestled into the bed with the tops exposed to avoid etiolation. The air temperature should be 60°F at night, with the beds some 5° warmer.

In a week, shoots will be growing from the tubers and are gathered two or three times a week. They should be removed with a sharp knife, cutting close to the tuber but without removing any part of it. The cuttings should be 3 inches long. Each mother tuber, depending on the cultivar, should produce up to 20–30 cuttings. Some cultivars, however, will produce only 6–7, indicating a great difference in the various cultivars.

The cuttings are then prepared for placing into the rooting medium. I have found that dipping the ends of the cuttings in a root-promoting hormone (these products are sold commercially), after removing any lower leaves that might have been produced, is beneficial, as such hormones speed rooting time. No foliage should be covered by the rooting medium, and while various growers have their favorite media I find that the best is pure sharp, white sand or crushed pumice, into which the cuttings are inserted to half their length.

The rooting medium is placed on the bench to a depth of 8–10 inches and bottom heat given so that a temperature of 65°–75°F is maintained. The sand or pumice should not be compacted but rather fluffed up so the cuttings can be inserted without damage. After insertion water in and place a plastic tent over top. After a week, the flaps on the sides of the tent are lifted during the day to give ventilation. After another 3 days the flaps should be left open at night, and the tent removed completely after another 3–4 days have passed.

Nighttime air temperature should be kept at 65°F, with maximum light given at all times. Later in the season, in April/May, it will be necessary to provide shade during mid-day.

Some 2 weeks after being placed in the rooting medium, the cuttings should be well-rooted and can be removed for potting. Put each cutting into its individual pot containing a mixture of 7 parts top soil, 3 parts peat, and 2 parts sharp sand. The potted cuttings are placed in a frame, where a temperature of 55°F is maintained at night. They then are hardened-off, so that by mid-May protection is no longer needed, unless frost is expected. These cuttings are sizable plants ready for sale by that time. A tuber produced in 8 weeks from a cutting will have an approximate diameter of 1–1½ inches, with three or more tubers.

The mother plants, needed for the next season, are planted in the open ground, and by the end of the season will have produced large tubers. The renewal of the stock by the production of tubers from cuttings ensures the maximum production of flowers and lush foliage.

Begonias

There are two ways to take cuttings of tuberous begonias. While it is possible to divide the tubers (explained later), stem cuttings are a sure way to increase the stock. Cuttings will form a small tuber the first year and, by the end of the second season, will have reached commercial size.

The simplest method is to start the tuber into growth; then, when two shoots have reached a length of 3–4 inches, remove one. This must be done with a sharp knife, taking a sliver of the actual tuber at the same time. If by chance only one shoot is produced, no cutting can be taken. If the cuttings are taken early in the season, protection against frost must be given.

The other type of cutting is made when the plant is fully grown. Laterals usually will be produced from the first 3–5 nodes on the main stem. When 3–4 inches long, remove them from the stem, cutting close to the main stem, or cut right under the first node on the lateral. The number of cuttings possible will depend on the number of laterals produced by the tuber.

As always with soil mixes, growers will prefer one type over another. This is, indeed, an area where experimentation can be very useful. Costs and availability of the needed materials in a given area are factors to be considered. The following mixture is good: 2 parts sphagnum peat moss, 1 part vermiculite, and 1 part perlite.

The cuttings are inserted into the medium, but all leaves and bracts must be removed from that portion of the stem inserted into the mix. Care has to be taken during this procedure as the dormant bud in the axil of the leaf is easily damaged.

It is advantageous to have a misting system to help in the rooting. If this is not available, a plastic tent placed over the cuttings will reduce moisture loss and cut down on the need to water. The temperature of the mix should be around 65°–70°F from bottom heat. Air temperature can be about 5% lower.

The cuttings will root in about 30 days. When rooted, pot them up in a loose mix containing peat moss, perlite, and a small amount of soil. A slow-release fertilizer can be added because the cuttings soon will be looking for extra food. Be sure the rooted cutting is placed so that the second node is just at soil level.

While a small tuber will form the first year, the size can be increased by removing all flowers and keeping the plants warm and growing into the fall. Weak feedings of liquid fertilizer also will help development, and, by the end of the season, a tuber some 1 inch in diameter can be expected.

By the end of November dormancy should be encouraged by withholding water. The tuber then can be stored over winter and the process repeated again the following year.

DIVISION

Almost without exception, bulbs will need lifting and dividing after they have been in the same location for several seasons. If this is not done, overcrowding will reduce their vigor and they will not flower well. Such lifting and dividing is a cultural procedure; however, it is also a way in which stocks can be increased.

If propagation is the objective, division should be undertaken long before the bulbs become overcrowded. The bulbs should be lifted when the foliage has almost, but not quite, died down. Many horticulturists recommend waiting until the foliage has died completely; however, this can cause a problem as it is not so easy to find the bulbs if some of the foliage is not still attached.

After the bulbs are removed from the soil, they are sorted into sizes—small, medium, and large. If desired, the larger ones can be planted back at once to provide additional bulblets the following year. These should be spaced according to their mature size. Smaller bulbs are planted in nursery rows, but set as mature bulbs. They are cultivated and allowed to follow their natural life-cycle, but not allowed to flower until the bulbs have reached full size. The number of seasons required depends entirely on the size of the bulb when planted. During their growth season they should be fertilized, the first feeding being given when the foliage first appears and then at 3–4 week intervals. A balanced fertilizer, 10-10-10 or equivalent, is given the first two feedings. The third feeding should have less nitrogen, such as a 5-10-10 formula.

If space is not available for planting, the bulbs can be stored in a paper bag in a dry, airy location. A plastic bag may be used, but holes must be made in it so that air can circulate. The stored bulbs should be checked from time to time for mold or rotting. Before storing it is helpful to dust them with a fungicide. All damaged bulbs should be discarded, unless they are to be used for propagation by the "bulb cutting" method (see below).

Many lilies will produce quite large bulblets alongside the stem of the parent plant. These make excellent planting stock and should be treated as described above.

Rare bulbs that deserve special attention can be stored in boxes, planted in sand which can be slightly moist at first and then allowed to dry through the resting period. Spring-flowering bulbs benefit from a cold period during the winter. If soil temperatures are not below 35°–40°F for a period of at least 6, but preferably 10, weeks, it is advisable to induce this cold period by storing them in a refrigerator (not freezer) prior to planting.

BULB CUTTING

It is helpful to think of the actual structure of the bulb in understanding this method. A true bulb, made up of a basal plate which is a compressed stem flattened into a disc, is surrounded by leaves that are modified for storage. In the axils of the leaves attached to the stem of the plant, buds are to be found. It is not surprising, then, that tissue is found at the base of each leaf at the point where it is attached to the basal plate. Given the correct conditions, the tissue will produce buds capable of forming new shoots/plants.

Keeping this in mind, it is understandable then that true bulbs can be propagated by cutting the mature bulb into one-eighths, each portion carrying a piece of the basal plate. These sections then are subdivided by inserting a knife between the rings of leaves so that two or three are attached to each now smaller section of the basal plate. Midsummer is the best time for propagating the majority of bulbs by bulb cuttings.

These cuttings then are inserted into a mixture of peat moss and perlite or sand. They are placed vertically, with just the tips of the sections showing above the soil mix. Keep moist and maintain a temperature of about 65°F. New bulblets will form in about 6–8 weeks, the actual time varying according to the species being propagated. When the new bulblets are well formed, transplant them as you would seedlings.

This type of propagation has been practiced for years by growers of hyacinths. The bottom of the bulb is scooped out, exposing the base of the modified leaves. The response is a mass of bulblets that can be grown on to become salable stock in two or three seasons. The time saved by such a method is considerable and amounts to at least a year's growth ahead, when compared with raising plants from seed. Being an asexual method, each bulblet produced will be identical with the parent plant.

SCALES

The explanation of the production of true bulbs by bulb cutting makes understandable the production of lilies by the scaling method. The selected bulbs, chosen for their strength and vigor, are "scaled" by simply peeling off their outside scales. The complete bulb can be torn apart, much as you peel off artichoke leaves. If wanted, the number of scales taken off is restricted and the center of the bulb, with a fair number of scales left on, can be planted back to be grown on. Prior to scaling, commercial growers dip the bulbs in a fungicide bath with the water at 70°F.

The time of year this operation is carried out depends on the type of lily being propagated. Asiatic and Oriental hybrids are treated in the same way, except that the actual amount of time the scales spend in an incubator varies. Asiatics must have 6–8 weeks at 60°F, followed by 6 weeks of precooling. Orientals require about 12 weeks, 6 of incubation and 6 of precooling. Scaling should be timed so that the rooted scales can be planted outside in the spring when the ground is warming. Trumpets are incubated at 60°F for about 12 weeks and are planted without precooling. Should there be an interval between the time the scales are taken out of the incubator and planted, they should be stored at around 50°F.

All scales are placed in either boxes or plastic bags in a coarse grade of damp vermiculite. It is essential they be kept in darkness during the entire time they are being incubated and cold-stored. The vermiculite should be checked every two weeks to ensure that it remains moist.

Each scale will produce at least one (sometimes more) new bulblet at its base. In many lilies, especially those that have *L. tigrinum* blood, new bulblets also will form along the edge of the scale.

The scales are sown outdoors in beds, covered with 1 inch of soil. In the case of Orientals, some shading may be necessary during the first growing season. The beds can be left down for 2 years or lifted after one complete growing season. Commercial-sized bulbs can be obtained in two seasons, while the remainder are grown on in rows until of suitable size.

It will be appreciated that many new plants can be produced from one large bulb in a short time. Each plant will be identical to the parent bulb from which the scale was detached.

BULBILS (ABOVEGROUND)

Understanding that lilies can be propagated by scales and how and where they are formed, it is not surprising to find that certain lilies will produce small bulbils in the axils of the leaves on the aboveground stems. These bulbils can be harvested as soon as they are ripe, which is determined by the ease with which they are detached from their stems. They then are sown in flats with 4 inches of medium in the bottom and covered with ½ inch of soil or placed out in rows in shallow drills. When these have reached the size of small planting stock, some 1 inch in diameter, they can be planted out in field rows with several inches between each bulb. They are lifted and graded when they reach commercial size. Each season they must be examined and moved at the first signs of overcrowding. They are planted into their final positions when of sufficient size.

Some alliums also produce bulbils on the maturing flowerheads.

The use of stem bulbils of lilies in *commercial production* is rare; however, if maximum production is called for, they do provide a good means for quick increase. The bulbils are gathered by hand in late summer, as soon as they can be detached easily from the plants. Then they can be sown in shallow drills in well-prepared soil in the field or sown broadcast in beds. If early in the summer or in areas where there is a good growing climate into the fall, e.g., no frosts before November, sowing as they are gathered can be considered. If such conditions do not exist, the bulbils can be packed in vermiculite and sown outside in the spring as soon as the ground can be worked and the night temperature is above 45°F.

After a season of growth, the plants should be examined. If many are of good size, they can be lifted and replanted in regular production rows. If growth has been slow, they can be left in the same location for another season.

During the growing seasons, weed control is essential and the plants should receive a top dressing of a general fertilizer as soon as growth has commenced in the spring.

BULBLETS (UNDERGROUND)

Most lilies produce bulblets alongside the parent bulbs and on the portions of the stems underground. These can be separated and harvested in early fall when the top growth has begun to die back. In many cases sizable bulbs will be produced which may well flower the following season. If a weak stem with a solitary flower is produced, remove the flower and send the strength of the plant into producing a larger bulb.

The commercial grower can expect to harvest, from a crop of lilies, not only bulbs of commercial size, but also planting stock for the next year's crops. Many lilies will produce bulblets on the flowering stem on the portion underground. This is especially the case with Asiatic hybrids.

Generally two sizes of planting stock are needed, small and large. They are both planted back in the fall, harvesting and planting proceeding together. The small planting stock will produce some commercial-sized bulbs by the end of the next season but is normally relied upon for the large planting stock.

The large planting stock will produce a high proportion of commercial bulbs as well as a supply of smaller bulbs suitable for planting back for planting stock. The size of planting stock varies from grower to grower. Generally bulbs that are 1–2, 2–3, 3–4, and 4–5 inches (or metric equivalent) in circumference constitute large planting stock. There is always a temptation to sell the 4–5 inch size bulbs as they are commercially of a size that will produce a good flower the following season.

The following season's production requirements should be satisfied prior to 4–5 inch bulbs being removed for sale. It is such bulbs that when planted back will yield larger bulbs the following year, and will often bring in twice or more the dollar value of a 4–5 inch bulb.

During the growing season, the crops should be examined and the health and vigor of each planting recorded. It is from those with the highest rating that the planting stock should be selected.

These underground bulblets are easily separated from the stem. The general method of grading is for the bulbs to be placed on a conveyor belt which passes in front of workers standing on each side of the belt. The commercial bulbs are removed while the stems with bulblets attached pass down the line.

During the selection of planting stock, great care must be taken to ensure the correct labeling of the stock. All trays of one cultivar must be removed from the area so that mix-ups of cultivars do not occur. If at all possible such planting stock should be planted as soon as possible. If not feasible due to weather or other reason, the stock should be stored under refrigeration and kept covered to prevent loss of moisture from the bulbs.

Layering

Lilies that have a tendency to produce bulbils, including those with *L. tigrinum* and *L. speciosum* blood, and hybrids with *L.* × *hollandicum* and *L. dauricum* blood, can be layered. In this procedure, the stem is jerked from the bulb when the stem is just coming into flower. A trench 6 inches deep sloping to the ground level over a length of 18 inches is prepared. The lily stem is placed on the slope and covered with sandy soil. The top of the stem will be sticking out of the ground about 6 inches.

The early flowering cultivars will produce bulbils along the stem in about 8 weeks. As the season advances, the time needed decreases. Late-flowering types can be lifted in about 6 weeks. The bulbils are removed and treated as are seedlings; being lined in rows or grown on in beds.

CORMELS

Corms will produce a number of cormels, about the size of a pea or less, around the edges of the base. In some gladioli and other corms, a great number are produced. If increased cormel production is desired, plant the corm closer to the surface of the soil than is normal.

The cormels are harvested after the plant has finished flowering and the foliage has died or almost died back. The cormels are lined out in nursery rows, planted about 1½ inches deep. The actual time needed to produce a flowering corm from a cormel will vary with the species or cultivar being grown. In some cases a flower may be produced the following year; however, discretion must be used and the flower removed if the spike is not of good size. Delaying the flowering will produce a larger corm the following year.

MERISTEM AND TISSUE CULTURE

Horticulture has not been left behind with the advent of new technology. New methods of propagation are having a profound affect upon production.

The advantages of such modern propagation methods are found not only in greatly increased production in a short period of time, but, by selecting material prior to becoming infested with virus, clean, virus-free stock can be produced. Several techniques are used. Apical tissue, selected from the growing point of a shoot, prior to the cells being infested with virus, can be cultured. Alternatively, any part of a healthy plant can be propagated by tissue culture.

These small sections of the plant(s) being propagated are placed in vials on agar containing specific nutrients. The tissue then forms a callus—an undifferentiated (neither roots nor shoots) group of cells is formed. From one plant many "blocks" of tissue can be made. Grown under artificial conditions, they grow quickly. After reaching a suitable size (about the size of a marble) the vials, which up to now have been on a rotating disc, are removed. The growth is then divided into small sections and placed in separate vials. Given different nutrients in the agar upon which placed, the sections develop into small plants, both shoots and roots being initiated. These are grown on under laboratory conditions until large enough to handle.

Should production demand, the sections can be returned to the rotating discs and, in effect, an infinite number of new plants can be produced in a very short time. Tissue culture demands complex facilities and trained staff working in sterile conditions. It can be undertaken by amateurs in the kitchen if the gardener is prepared to observe strict demands for sterility and precise nutrient formula preparation.

Plants propagated by these methods are vigorous and free from disease, and, while the costs involved in building suitable facilities for this type of production are high, the production rate is such that more and more growers are turning to this method of propagation.

Plants produced by tissue culture are perforce clones. The grower, however, having produced a seedling of outstanding merit, no longer has to wait for several seasons before a new cultivar can be introduced into commerce; but by using such techniques he can rapidly increase his stock.

While employed extensively in the commercial production of lilies, techniques for producing other plants by the same methods are being developed. At the time of writing tulips have not been successfully propagated by these techniques, yet is is only a matter of time before the necessary formulas and techniques are developed. The effect such methods of production will have upon commercial horticulture can be appreciated. Clean, healthy stock and new varieties quickly brought into commercial production are but two benefits. However, no one has yet found a way to duplicate the ability of the male and female organs to produce new and better cultivars so the plant breeder will still be the person who produces new varieties. The time before introduction of them will, however, be shortened.

DIVISION AND CUTTING OF TUBERS AND ROOTS

Many tubers, among them begonias, cyclamens, dahlias, daylilies, eremurus, gloriosas, irises, polianthes, and zantedeschias, are propagated by dividing and cutting. As many of these require slightly different treatment, they will be dealt with in the sections devoted to each.

Begonia

Begonia tubers will produce more than one bud on the top of the tuber. When apparent, the tuber can be cut into sections, each with a bud. Large tubers often yield 4–5 sections. Avoid the temptation to cut into too-small pieces. The production of roots from the tissue that heals the cut surface is not large and thus care is needed. This operation is carried out in the spring.

Cyclamen

Cyclamens grown in the garden (not the florist type) will produce very large tubers that are treated in the same way as begonias. Small tubers also may be found when the clumps are lifted. This is done in late summer/early fall. The plants should be replanted as soon as possible.

Dahlia

Dahlias can be propagated by cutting apart the roots, but care must be taken to ensure that a section of the stem with an "eye" or bud is still attached to the root. Without this "eye" there is no possibility of the plant sending out a flowering shoot. The best time to do this is in the spring, setting out the now-divided roots at the normal planting time.

Daylily

Daylilies become very overcrowded, but this does not seem to seriously affect the flowering. The plants should be lifted in late summer or spring. Small sections of the clumps can be separated with a sharp knife. All parts planted back should have at least one "crown" or set of leaves containing a growing point.

Eremurus

Eremurus have unusual-looking roots, much like a starfish. They are very fleshy and break easily. Lifting and dividing the tubers is best done in late fall or early spring. Make each division by either untangling the roots or by cutting away with a knife those sections with a growing bud.

Gloriosa

Gloriosas, because of their requirement for warmer weather, are best divided in the spring. The sections with growing points attached can be cut with a knife.

Iris

Depending on the type, irises can be divided by just breaking apart the clumps of roots or by cutting the rhizomes into small sections, each with a growing point. The important thing to remember is to accomplish the work early in the summer, June/July being the best months. The entire plant is lifted out of the ground. The older roots will be dark in color and, depending on the length of time they have been in the ground, only the younger portions will have leaves and growing points. The older sections should be discarded. The young rhizomes, which are lighter in color, are separated by cutting them away from the other roots. For best results, the pieces should be at least 3 inches long. If the foliage is very long, it can be cut back so that it is about 6 inches in length. Place the new plants back in the ground as soon as possible and no later than the middle/end of August.

Polianthes

Polianthes are propagated after they have finished flowering. The tubers are removed from the soil; young portions with a growing point are removed or separated from the parent tuber. These are best stored in sand and planted out again the following spring.

Zantedeschia

Zantedeschia are lifted in late summer or early fall and the roots divided by cutting where necessary. The separated portions must contain a set of "eyes" or buds; these should be planted back as soon as possible. If this is not practical, they can be stored by planting in sand until the following spring.

Bear in mind that when propagating plants in any of the above manners no damaged or unhealthy plants should be used. All cut surfaces should be dusted with a fungicide to prevent rot from setting in. The greatest temptation is to propagate by cutting and dividing the tubers into such small portions that they will not perform well. While this will yield a greater number of plants, their ability to flower with the minimum of delay depends on the size of the portions being planted back. Small portions, if necessary and required, can be cultivated for another season to reach flowering size.

CHAPTER 5
CULTIVATION

Fertilizer requirements, after-growing care, storage.

Regardless of the plants grown, there is no substitute for thorough and proper cultivation. Bulbs, too, should be planted only after a good cultivation of the soil to a depth of about 12 inches, which is especially important if annuals have been grown in the same area previously. A general fertilizer (10-10-10) should be applied and raked into the surface. After such preparation, the soil should be allowed to settle for a day or two so that air pockets are eliminated. This is of particular importance with very sandy soil.

A good loam is the halfway point between clay and sand. A loamy soil will have the draining qualities of sand and the moisture-retention capacity of clay. Both types of soil are improved by the addition of organic matter. It binds the sandy soil, improving moisture retention and loosens clay, allowing air to enter and improves drainage. Well-rotted compost and well-rotted manure are both excellent organic additives. Avoid fresh manure and unrotted compost.

Many bulbs require deep planting. Indeed, the most common single mistake made is planting at an incorrect depth. The charts on p. 35 show the correct depths for a large number of bulbous plants. A rule of thumb for those not shown is to place the bulb at a depth of three times its height, e.g., a bulb 1 inch high should be planted with 3 inches of soil over it. This may seem a lot; however, it is what is needed when planting in good loam. The depth can be slightly less in heavy soil or slightly more in sandy soil. If in doubt, err on the deeper side.

PLANTING

There are many ways to plant bulbs. Bulb planters, which remove a core of soil, often have a marker on the side that enables a hole of correct depth to be made; or use a trowel. For small bulbs or large plantings, all the soil is removed from the area where they are to be set. This enables the gardener to space the bulbs evenly. The necessary bone meal (check directions on the bag) is sprinkled over the site and mixed with the soil before spacing and placing the bulbs. Then set the bulbs out, nestling them firmly but without excessive pressure, which can harm the basal plate or compact the soil. The earth is then replaced, covering the bulbs.

If a trowel or bulb planter is used, a small amount of bone meal should be mixed into the soil at the bottom of the hole. The bulb is then put in position and the hole filled with soil. The earth placed over the bulbs, by whatever method of planting used, should not be compacted. If annuals are to be planted over the tops of the bulbs, a planting board should be used to prevent undue compaction.

Water in to ensure that the soil is well placed around the bulbs. Applying a mulch after planting will preserve moisture and reduce the weed population. The taller the flower spikes, the deeper the mulch; i.e., 2 or 3 inches for taller bulbs, about an inch for small ones such as crocuses. Crocuses and other small bulbs should be covered with a finer grade of mulch than that used for daffodils or tulips. Mulch also serves as an insulator. It will minimize thawing and freezing of the soil, which causes heaving, most detrimental to bulbous plants. Heaving breaks the roots and forms air pockets under the bulbs. This is especially a problem for shallowly planted bulbs.

The bulbs require little attention until new seasonal growth appears above the soil. As soon as they break through, a general fertilizer should be given. This can be 10-10-10 or similar; however, excessive nitrogen should be avoided.

If you have a rainy climate, the fertilizer will be washed into the soil; if your spring weather is dry, the fertilizer will have to be watered in. If the soil is not friable, an additional feeding should be given some 3–4 weeks after the first feeding For spring-flowering bulbs, two feedings are sufficient.

Bulbs grown in nursery rows to increase their size will require two feedings. After planting and throughout their growth period, they must be kept free of weeds and moist, but not wet. Those being grown for size should not be allowed to flower, except if color is to be ascertained. If color recognition is a requirement, the flower buds should be removed as soon as possible, i.e., as soon as they are easily grasped, without damaging the plant.

In the flowering beds, spent flowers, together with the forming seeds pods, should be removed. During the spring, protect the plants from slugs and snails using a commercial bait.

After flowering, daffodil foliage becomes straggly and untidy. Bend the foliage over and slip a rubber band around it to keep it in place to improve the appearance of the garden. Daffodils being naturalized require the same basic treatment; however, the foliage is allowed to just ripen. As soon as it begins to turn brown, reduce watering. While it is preferable to let the foliage die completely before mowing the grass, it can be done safely without fear of damaging the bulbs.

Lifting & Storage

Daffodils, tulips, and hyacinths planted in the border should be lifted as soon as they become unsightly. There is no need to wait until the foliage has turned completely brown. Remove the bulbs from the ground with foliage still attached; it is not necessary to remove all of the soil. Allow the bulbs to ripen in a cool, airy place. When all of the foliage has died down, it will be easy to remove it and any remaining soil. The bulbs should then be examined and any unsound or bruised discarded. Next, they should be sorted according to size, smaller ones separated from those of flowering size and stored apart. As a precaution against fungus diseases, the bulbs should be dusted with a fungicide. All of the bulbs then are stored, at a temperature of around 63°F and where air can circulate freely. During the summer months, check periodically to make sure no rotting or spoilage is taking place.

Precooling

In colder climates, where soil temperatures are low, no precooling prior to planting is necessary. In warmer climates, bulbs that are to be set out in late October or November should be precooled prior to planting. This applies especially to tulips. Precooling should start 6–8 weeks before the soil temperatures are the same, or a little lower, than the precooling temperature of around 45°F. See also Chapter 7, page 47.

Planting Summer-flowering Bulbs

Summer-flowering bulbs should be planted some 2 weeks before the last frost is expected. In most of the United States this is the first week in May; at higher elevations, the middle to end of May. If the ground is cold the bulbs will not grow, so delay planting if there is any chance of a late frost or if the ground is not warming. In warm regions, planting can be done earlier, March/April being a good time. It must be remembered that summer-flowering bulbs are in need of good day length in order to grow well. Dahlias, caladiums, and begonias can be started into growth in the greenhouse or frame or even a well-lighted garage. These started bulbs should never be exposed to frost; however, starting them inside can speed the first flowering by several weeks. Care must be taken not to disturb the roots when planting or any advantage gained will be lost.

Some of the taller-flowering dahlias and gladioli in exposed locations may need support during the summer months. Such plants should be staked early in their growth cycle so the stems will remain straight. The stake should be placed behind the plant.

Gladioli can be planted at intervals during the season. Allow about 90 days to flowering and plant the last crop some 90 days before the first expected frost. In warmer areas, where little or no frost is expected, they can be grown throughout the year. About 120 days should be allowed from planting to flowering during the fall and winter months.

If large flowers are desired on dahlias, the number of buds allowed to develop per stem is restricted by removing most of the buds as soon as they are large enough to handle.

Those bulbs that are in continuous growth, such as dahlias and begonias, will benefit from feedings of liquid fertilizer each month. Feeding must be stopped some 6 weeks before the bulbs are to be lifted, which is just prior to the first frost. In warmer regions, where dahlias are left in the ground overwinter, a feeding of 0-10-10 given the first part of September will increase the hardiness of the bulbs and reduce the possibility of rotting in the ground.

Lifting Summer-flowering Bulbs

Bulbs to be lifted and stored overwinter should be allowed to ripen by withholding water toward the end of summer. The first frost often is a light one, enough to burn the foliage but not harm the stems or the bulbs themselves. The bulbs then should be protected and/or moved to a frost-free location. They should be allowed to dry. After the foliage has died it is removed and the bulbs stored.

Gladioli are stored in paper sacks with air circulation ensured by holes in the bags. Nets also can be used. Dahlias and begonias, as well as cannas and caladiums, should be stored in barely moist peat. In this way the roots will not lose any moisture and will be plump when planting time comes around again. Storing temperature for the summer-flowering bulbs must be above freezing but below that which will entice growth, around 40°–45°F.

As with the spring-flowering bulbs, any damaged or unsound ones should be discarded. Dusting with a fungicide also is advisable. At no time should dahlia roots or those of caladiums and cannas be allowed to dehydrate.

It also is wise to remember, when storing bulbs overwinter, to protect them from mice. Label the bulbs as to color prior to lifting to ensure good planting arrangements in the spring.

CULTIVATION OF LILIES

Culture during the growing season of lilies is not much different from the foregoing for summer bulbs. It is necessary to remove flower heads, however, once they have finished flowering. Should the flowers be cut for use in the home, remove as little of the stem as possible. The leaves manufacture food for the bulbs, so the more foliage left the better. While lilies do not like to sit in water, moisture must be made available during the growing and flowering season. Once they have finished flowering, watering can be reduced but not discontinued until the foliage has begun to ripen.

Lilies should remain undisturbed for years; however, there will come a time when, because of natural increase, the site will become overcrowded. When overcrowded, lift the bulbs early in the fall. The larger bulbs can be replanted at once; the smaller bulbs either planted or lined out in nursery rows for cultivation and eventual replanting. As the shoots of lilies are very brittle, it is a good idea to mark the location of the bulbs so that, when cultivating in the spring, the developing shoots are not damaged.

Other Bulbs

Trilliums, anemones, and other bulbs planted in the woodlands appreciate less moisture during late summer/early fall; duplicate, as much as possible, the rainfall they would receive in their natural habitats. These bulbs can be left undisturbed for many years and, indeed, should be moved only if cultivation or new plantings in their area will harm them.

Planting Charts

BASIC CULTURAL DIRECTIONS AND
PLANTING DEPTHS FOR SPRING FLOWERING BULBS

SOILS
Depths are indicated for average soils. In sandy soils plant a little deeper. In heavy or clay soils do not plant quite so deep. If in doubt err on the deeper side.

FERTILIZER
The addition of bone meal at planting time is beneficial. As soon as the bulbs show through the ground in spring feed with a general fertilizer.

HEIGHTS
These will vary according to variety. Lilies from 24 inches to several feet. Tulips from 8-10 inches to 30 inches. Ask your supplier for details of the varieties you purchase.

PLANTING TIME
Plant in September or October where there is frost during winter. In warmer climes plant in November or December.

FLOWERING TIMES
These will vary considerably according to variety. There are early, mid-season and late flowering tulips, daffodils and crocus. Lilies can be in flower from May to September.

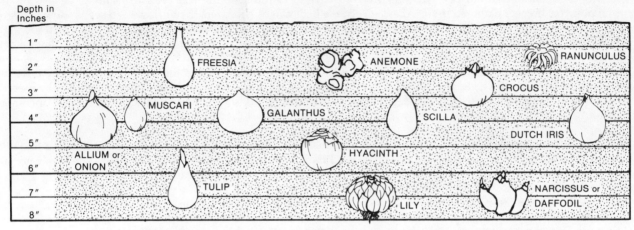

It is advisable to purchase bulbs while the selection is good; early planting can sometimes be a disadvantage. Bulbs are best planted when soil temperatures are low. This enables bulbs to produce adequate roots which results in a better flower production in the spring. If you purchase bulbs, you can safely store them by placing them in a paper bag (not plastic) and putting them in the vegetable section of your refrigerator. They should not, under any circumstance, be placed in the freezer section.

A long and splendid flowering border can be yours by selecting different types of bulbs. Tall growing as well as dwarf varieties are available. Bulbs are ideal container plants.

BASIC CULTURAL DIRECTIONS AND
PLANTING DEPTHS FOR SUMMER FLOWERING BULBS

SOILS
Soils should be well cultivated, moisture retentive not waterlogged. Maintain adequate moisture during summer. Depth shown for average soils, plant shallower in heavy soils, deeper in sandy soils.

FERTILIZERS
Apply fertilizer only after growth has emerged from the soil. Do not feed too much nitrogen or growth will be at expense of flowers.

HEIGHTS
Dahlias are available in taller and shorter growing varieties. Cannas are, for the most part tall growing. Begonias and Iris will vary depending on variety.

PLANTING TIME
Many summer flowering bulbs are harmed by frost. Plant 10-14 days before last hard frost. If planted prior to this, protect if frost is eminent.

FLOWERING TIME
Most summer flowering bulbs will start to flower some 6-8 weeks after planting. A succession of flowers can be obtained by staggering the planting dates of Gladiolus.

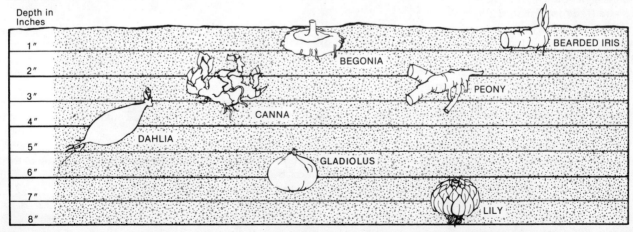

A long and splendid flowering border can be yours by selecting different types of bulbs. Tall growing as well as dwarf varieties are available. Bulbs are ideal container plants.

Bulbs that are naturalized or are permanent residents in the garden should be fertilized twice a year. An 8-8-8 formula is given in the fall and another in the spring. The time of application in the fall is important. It should be done when the night temperatures are dropping and are in the 50°F range. If the rains are not adequate for washing the fertilizer down to the level of the bulbs, water should be given.

The most common reason for bulbs not performing well in the garden, especially the spring-flowering ones, is too much moisture during the late summer, when most prefer a drier "resting" period.

CULTIVATION ON A COMMERCIAL SCALE

The advice I have offered the home gardener regarding the cultivation of bulbs also applies to the commercial grower, there being no substitute for good cultivation of the soil prior to planting.

While weeds are usually easy to manage in the home garden, weeding can become quite expensive in the field-scale operations of a commercial nursery. Growers are well advised to fumigate the soil prior to planting, destroying any harmful pests or diseases harbored there. Alternatively, the beds can be prepared, the buried weed seeds allowed to germinate, and then destroyed with a flame gun prior to sowing. In either case destruction of weeds before actual sowing is a time-saving operation, as they must be removed by hand if allowed to grow among germinating bulb seedlings—definitely not a time- and labor-saving chore.

Keep in mind that the use of machinery is far more economical than hand labor for most operations. The distances between beds, width of the beds, and width of the paths should be designed to accommodate the machinery used.

Subsequent cultural procedures differ, depending on the crop raised, so it is best to look at the various operations separately.

Seed Beds

After the soil has been fumigated as directed on the label of the fumigant used, the beds are worked to a fine tilth, using a rotovator. The depth need not be great. The seed then is sown and covered with a fine tilth. Obviously no pre-emergent herbicide should be used on the beds themselves. Headrows and paths can be treated, if desired, or a mulch can be laid in the rows to inhibit germination.

The developing seedlings must be kept moist, but not wet. Irrigate using overhead sprinklers that distribute a fine droplet spray that will not disturb the soil. Spraying against boytritis must be undertaken as the seedlings grow. The control of fungal diseases should be an ongoing program as they are devastating in a seed bed. Aphids, too, must be controlled as they can infest the developing crop with virus in addition to weakening the plants. It is of the utmost importance that seedlings be protected throughout their development.

After the seedlings have become well established and the first true leaves appear, fertilize with a balanced fertilizer, 10-10-10 or similar. It is best applied while the foliage is dry and then watered into the ground afterward. Fertilization must be ongoing throughout the growing season, done at regular intervals. The number of applications will vary, depending on the type used.

In exceptionally warm regions seed beds may require protection from the sun. Shade cloth mounted on suitable supports and placed above the beds is recommended. The structures should be placed at a sufficient height to allow for weeding, and they should be wide enough to shade the entire bed, especially during the height of the day; e.g., in the Northern Hemisphere they should extend over the southern edge of the beds. Lath frames also can be used, but handling and construction are more expensive.

Weeds and other vegetative growth can harbor many pests and diseases, and should be kept under control by spraying as necessary. Some growers find it advantageous to sow a grain crop surrounding the blocks of seed beds as a buffer against these infestations.

Rows and Beds for Small Bulbs

Rotation of the bulb crops is absolutely necessary. If several different types (e.g., bulbs and perennials) are being grown, a rotation system is not difficult to implement. If only bulbs are being grown, however, allow the ground to lie fallow after two or three crops. The fallow period can be used to grow a grain crop or, if carefully planned, advantage can be taken of the season to sow a crop to be used as green manure, plowing it into the ground to improve soil quality.

Planting usually is carried out by mechanical means. Good tilth of the soil will make for straight rows and, in consequence, ease later control of pests, diseases, and weeds. The spacing between rows must allow for good cultivation, so the exact width of the rows will be determined by the machinery used. When using mechanical cultivation, care must be exercised not to damage the vegetation or roots.

A number of soilborne diseases attack bulbs. *Fusarium* is one; another is "Ink Spot" (*Mystrosporium adustum*), which likes irises in particular. It is essential, therefore, that the soil be correctly fumigated prior to planting. While some of these may not be seen during the growing season, they can manifest themselves after harvesting and during storage. Bulb spoilage at later stages is often attributed to incorrect cultural conditions in the field.

Apply a general fertilizer some 3–4 days prior to planting, working it into the soil to obtain the desired tilth. A pre-emergent herbicide can be applied, except on the very smallest of the planting stock. Prior to using such controls on the smallest stock, experiment to ascertain the susceptibility of the crop to the material being used. The value of such trials cannot be overemphasized as the climate, soil texture, and quality, as well as structure, can be determined only by such means. Keeping any crop free of weeds is essential, and, while hand weeding is expensive, it is preferable to the partial or total loss of a crop due to the incorrect use of chemical controls. While chemical weed controls can be used on the headrows and waste areas surrounding the production fields, great care must be taken when applying them and no spray allowed to drift onto the crop.

Pretty much the same regimen should be followed as recommended in the "Seed Bed" section above: moisture given during the growing season, preferably by overhead irrigation; pest, disease, and weed control; generally three feedings of fertilizer, determined by the length of the growing season. (Crops that mature and whose foliage dies by midsummer will require only two feedings at the most; so it is a complete waste to apply such fertilizers if the crop is unable to benefit.)

During the growing season, as the plants come into flower, the rows or beds must be checked at least once a week to rogue out plants not true to type. Care must be taken to remove all of the plant, including bulbils and stem bulbs. If this is not done, these can find their way into planting stock saved from the row, causing a severe problem.

After this procedure has been performed, the flowers and buds must be removed. There is debate among growers as to whether the removal of the flowers and buds actually increases the size of the bulbs produced. Experiments with the particular group being grown can help to determine this in each case. Another point to be considered is whether the fallen petals and spent flowers will cause rot or fungal problems. This should be carefully considered prior to making a decision about debudding/deflowering. In any case it is a good idea to allow a certain percentage (about 2%) of the crop to flower fully so that the authenticity of the species or cultivars can be checked. Catalog descriptions can be made and prospective buyers can see a true selection of flowers. A number also can be cut for exhibition and display.

Harvesting

The date of harvesting will be determined by the crop itself. Most crops are harvested by mechanical bulb diggers that pass under the bulbs, lifting the entire row with soil. The bulbs then either are dropped back onto the ground or onto a conveyor belt, from which they are placed into containers for transport to the grading and packing sheds. These operations must be handled with care to avoid physical damage.

The condition of the soil at harvesting time is important. Bulbs are more likely to be bruised if the soil is very dry rather than moist. Irrigation performed several days prior to harvesting makes this operation easier and less hazardous to the bulbs.

Top growth, too, can be a problem, so it is best to cut and remove it prior to the actual harvest operation. Remember to correctly identify and label the bulb containers. Because this is a very critical operation, only a very responsible person should be put in charge.

Invariably some bulbs will be left in the field after harvesting. Unless they can be positively identified, they are best discarded. The mixing of stock of one variety with another can prove to be extremely expensive in the roguing and sorting required in following years. If there is still uncertainty or a reluctance to discard bulbs that are still viable, they can be presented as a mixed variety and offered at reduced prices.

CHAPTER 6
BULBS IN THE LANDSCAPE

Naturalizing bulbs, use in the garden, woodlands, borders, and containers.

Among the many plants we grow in gardens and woodland areas, few attract the eye or add focal points as effectively as bulbous plants. To achieve such points of interest the correct bulbs must be selected. The type chosen is perforce compatible with the other plants, but can vary greatly in height and time of flowering. Crocuses planted in a lawn can add much interest early in the year when often there is little other color in the garden. Daffodils, planted close to a rose bed, will give color before they face the competition of the roses. Bluebells are superb in woodland settings where they will enjoy shade during the summer. While effective when planted among shrubs, their foliage when the flowers have finished can be messy, not a problem in woodland settings, but unsightly in more formal borders. A shrub border composed of spring- or early summer-flowering shrubs can have the season of interest and color prolonged by the addition of summer-flowering lilies that can tower over the shrubs. Well-placed containers of flowering bulbs add color to decks and terraces. All things considered, bulbs are among the most versatile of plants. By choosing wisely the gardener can complement his garden, prolonging those periods of interest and adding different and often exotic forms and colors, not to mention fragrance.

The finest example of a "bulb garden" is undoubtedly the Keukenhof Park in the Netherlands. Situated in the heart of the bulb-growing district between Haarlem and Leiden, this park of some 70 acres is at its best from late March through May. Over six million flower bulbs are on display. This magnificent display by the Dutch Bulb Growers is rightly world renowned. Each spring the growers vie with each other in presenting, as part of their sales promotion, established favorites and the latest cultivars of spring-flowering bulbs. All who have an interest in bulbs should visit these gardens.

Such plantings on a grand scale are not possible in a home garden. Nor do many Park Departments have sufficient funds today to make them. Bulbs can be used to good effect in combination plantings with spring-flowering annuals, but, with modest expenditure, spectacular naturalized plantings can be formed. Correctly placed, they can be eye-catching and not appear foreign to the landscape. The picture of a gentle slope under deciduous trees with a wide range of daffodils, varying in color, flowering times, and heights, is one that always will be enjoyed. With correct maintenance, such plantings will increase naturally each year. They are indeed a sound investment.

NATURALIZING BULBS IN GRASS

Not all bulbs are suitable for naturalizing in grass. The more effective are the species and cultivars of *Narcissus* among taller-growing grasses, and *Crocus,* especially planted in a lawn or shorter grasses, the former being the most commonly used. As mentioned deciduous trees on a gentle slope or undulating ground add much to the beauty of such plantings. Preferably the trees will be high branching so the view is not obstructed. Not only do the branches form interesting patterns of sunlight and shade, they also serve as the framework that confines the eye to the area. Evergreen trees or trees with branches reaching to the ground provide an effective backdrop to plantings of bulbs.

Location

Several factors must be kept in mind when considering the location for naturalizing bulbs. Crocuses require viewing from a closer range than daffodils. Small-flowered daffodils make a showing when seen from a little farther away, and the large-flowered cultivars from even a longer distance. Bluebells, being less imposing as individual plants, look best in drifts under small trees or planted among or in front of large shrubs.

A naturalized scene of bulbs is more effective on a slope or on undulating ground than on level ground. Even if only slight, a rise will afford better drainage and present a better staging than flat land; the eye is then able to appreciate all of the planting. The same applies to undulating ground, where the plantings should be on the slopes and crests, not in the valley. The direction in which the slope faces is of lesser importance; it should not be exposed to much wind, which can flatten the flowers, especially if the wind is accompanied by rain.

Soil

The type of soil, while important, is not critical, but extremes are not suitable. It is difficult to establish bulbs in heavy clay without ameliorating the soil; the same applies to very sandy soil. Above all other conditions of the soil, free drainage is essential, and if necessary the installation of a drainage system should be considered.

On flat areas, the importation of good quality top soil is often a solution to create at least mounds if not an entire slope. Such moundings must be in scale with the surroundings. The bulbs should spill over the mounds and flow with the contours created. It is a distinct advantage when creating such mounds to know the number and type of bulbs that will be planted—then the mounds can be sculptured accordingly.

The quality of the soil should also determine the selection of cultivars to be planted. The poorer the soil, the more vigorous the growth habit.

Culture

To maintain the vigor of bulbs being naturalized in grass, the foliage should be allowed to mature. Unfortunately this means that among grasses the site of the planting can become unsightly by the end of spring and early summer. The combination of long grass and the ripening foliage of the bulbs means the area will look unkempt, another reason for careful site selection. Close to the home or in an area where there is much foot traffic, the gardener will be torn between making the area tidy and mowing the grass, or allowing the bulbs' foliage to ripen. Unless such untidiness can be tolerated, naturalized plantings should be located away from the home. It should be remembered, when selecting an area for naturalizing bulbs, that time is an essential element in the culture of bulbs: time to allow foliage to ripen and also time to allow the bulbs to increase naturally. Allowing them to stay for years undisturbed will finally result in a grand display.

Attention to the culture of the bulbs each year, while minimal, is on no account to be neglected. The basic operations are as follows. Allow for a last cutting of the grass just as the first bulbs appear through. As soon as a number of bulbs has appeared, give them a dressing of 12-12-12 or similar "balanced" fertilizer. The spring rains should wash the fertilizer into the soil; if no rain is expected, water it into the soil. In heavy soils such feeding is not so essential as on sandy soils but is still recommended. As soon as the flowers have passed their best, they are better removed. Here again it is not essential, but it is preferred.

Slugs and snails like a habitat of grass; the tender buds of the spring flowers provide a tasty meal. Slug and snail control should be practiced.

PLANTING
Crocus, in lawns

Crocuses are planted some 3–4 inches deep. Cut the turf so that it can be rolled back, exposing the soil, much as you would turn back a carpet. The exposed area of soil should then be forked over to a depth of some 6 inches and a dressing of bone meal should be added, forked lightly into the prepared soil. The bulbs are then set into position, following a random pattern, no straight lines and in clumps of 15–25, spacing the bulbs some 3–5 inches apart. Here again the spacing should not be regular, but varying; some, but not too many, being placed only an inch or 2 apart. If the soil is not moist, water the bulbs in before turning back the grass to cover them. If moist, turn the grass back, firm it well, and water.

Mow the grass before rolling it back to plant. No harm will be done to the area by light foot traffic, but, as soon as the first shoots are noticed, the area should be protected by staking it off. Such protection can be removed as soon as the bulbs have started to flower.

Daffodils, in grassy areas

Planting daffodils with a trowel is not a difficult job, especially if the ground has been dug over beforehand. Some authorities recommend lifting out the plugs of grass and replacing them after the bulbs are planted. This is not necessary in the majority of cases as rough grasses will fill back in quite quickly. Having a clear area above them will allow the bulbs to grow with less competition, especially the first year.

In sandy soil the daffodils should have at least 6–8 inches of soil over their tops; in heavier soils 5–6 inches will suffice. If in doubt err on the deeper side. Incorporate bone meal into the bottom of the soil before setting in the bulbs. This operation is made easier by working over the soil at the bottom of the holes; this can be done while incorporating the bone meal, which should not be in direct contact with the bulbs. This softer area of soil allows the bulbs to be nestled into the soil without damaging the basal plate.

The soil should be moist at planting time. Water if the ground is dry, and then watered in after planting. As the depth is considerable, rain cannot always be relied upon to accomplish this operation.

While bone meal is slow acting, certain trace minerals are quickly available, and these are the essential elements contained in bone meal. Application of a general fertilizer in the spring should follow the first emergence of shoots from the ground.

Additional bulbs can be added each fall as the years go. Follow the same procedures for each planting.

Bluebells

The easiest way to accomplish plantings of bluebells under small trees, the ideal location for the plants, is to use a fork to dig over the ground. Once so prepared, the use of a trowel becomes less of a chore. Setting the bulbs 2–3 inches deep and *en masse* is preferred as it looks more natural. Spreading the area with a good mulch of leaf mold will cover the traces of disturbance to the soil and also provide good conditioning to the soil. While bone meal is not essential when planting bluebells, it is recommended, and another feeding of a balanced fertilizer as soon as the shoots appear in the spring is preferred, especially for the first season or two after planting. After that these bulbs are capable of fending for themselves.

NATURALIZING DAFFODILS
Desired forms of planting

This will vary according to the individual taste of the gardener. It is strongly recommended that in areas where daffodils are to be planted a spade be used to turn over the proposed planting areas. These areas should be grouped so that each colony of bulbs forms a "cloud" upon the area of grass. As much attention should be given to areas left unplanted as those areas to be planted. Viewed from a distance, the planted areas should not obscure one another;

Flower shows are great places to see the many different kinds of bulbs. Here, *Crocus* and *Cyclamen* display varied colors and forms.
AUTHOR

The man who can be called the "Father of the Bulb Industry" Clusius is seen here with *Fritillaria* and the tulip species named in his honor, *Tulipa clusiana*. AUTHOR

Plate 1

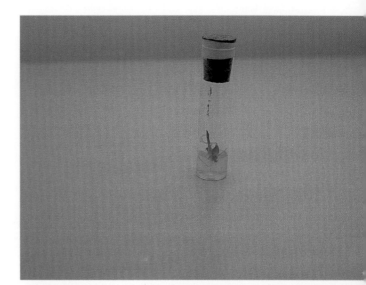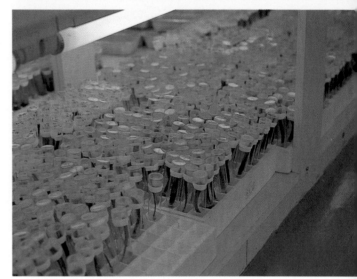

Tissue culture. Modern methods of propagation have replaced older methods in raising virus-free plants, and have speeded the multiplication process as well. Plant tissue is confined to the test tube, and, under controlled conditions, growth is encouraged. Light and temperature are controlled, as is the food given the plant. AUTHOR

Field production Ranunculus. To satisfy the need for thousands of plants each year, fields of bulbs are grown, the majority of which need special culture and irrigation. Here the production of *Ranunculus* on a large scale, with irrigation pipes in place, can be seen to advantage. AUTHOR

Plate 2

Begonia cuttings. Rooting of *Begonia* cuttings is not at all difficult. In the special rooting soil mix the roots are produced quickly, but care in handling is very important. AUTHOR

True bulb. The lily bulb is a true bulb. Here is a good example of how many of these plants produce stem roots, i.e., roots above the bulb, showing the need for deep planting. AUTHOR

Selection of hybrids. After many years of trial and error, the day comes when it becomes possible to select from among many seedlings those that deserve to be separated, given names, and introduced into commercial production. Here the author and Jan de Graaff are shown selecting new cultivars of lilies. AUTHOR

Lily hybridizing. Countless flowers are pollinated and then the stigmas covered with foil caps to prevent unknown pollen from interfering with the work of the hybridizer. AUTHOR

Plate 3

Iris fields. Large areas of land are used in the production of bulbous plants. Here fields of iris cultivars are being grown. They will be watched carefully and every measure taken to ensure that all are true to name. AUTHOR

Freesia fields. Before too long these bulbs will be enjoyed by thousands. AUTHOR

Ranunculus. As can be seen, these plants are true in color. Removal of "rogues" is done each day, making sure those plants lifted for sale are indeed true. AUTHOR

Oriental Hybrids. When seed is sown, the flowers produced can vary. From these beds, selections will be made and those worthy of introduction will be named and production of a new cultivar begins. AUTHOR

Plate 4

Tulip Fields. The entire area around Lisse in Holland is a patchwork of color in spring. The picture will change as the flowers are removed, strengthening the bulbs. AUTHOR

Planting *Lilium longiflorum*. Easter lilies are grown on a large scale in California and Oregon. The bulbs fall into the drills, and while some effort is made to place them upright, even those that are on their sides grow well. AUTHOR

Harvesting *Hippeastrum* cultivars. Harvesting bulbs is a labor intensive operation. While machines can do this work, the quality of the bulbs is superior with hand harvesting, a worthwhile operation, when individual bulbs fetch a high price. AUTHOR

Production fields of *Freesia*. Bulbs are produced in enormous quantities. It is interesting to note that the plants on the edges of the beds, having received more light, have finished flowering, and are making greater growth than those in the centers of the beds. By harvest time, all will be of equal size. AUTHOR

Plate 5

Commercial production. Large acreages of lilies are now grown in the Netherlands. Here the famous cultivar, 'Enchantement', is being grown in what used to be the bottom of the Zuider Zee. AUTHOR

Many strains of lilies, such as 'Golden Splendor' are raised from seed. In the foreground are seed beds; these will be lifted in the fall and planted out in rows. In two or three seasons from time of sowing, good commercial, flowering size bulbs are obtained. AUTHOR

Large scale production does not mean that care is not taken to keep weeds under control and there is no substitute for good cultivation, in the field or in the garden. Here a crop of lilies in Oregon prove by their vigorous growth that they are healthy and well cared for. AUTHOR

After the stock is checked, to make sure it is true to name, the flowers are removed. The bulb then makes its maximum growth, and not waste its energies on producing seed, which is not, (in this case) needed. AUTHOR

Plate 6

Sawdust is an effective mulch; it keeps weeds down and conserves moisture. Here it is being used on beds of young lily bulbs, in Oregon. The important point is to get such mulches in place early; in the season. AUTHOR

A great number of Narcissus are produced in California; this is a photo of daffodils near the coast in northern California. AUTHOR

Propagation by scales, is one of the quickest ways to increase the stock of lilies. When correctly spaced on the soil, they are covered; important, or they will dry out. AUTHOR

The small bulblets, taken from the lilies when they are harvested, are often planted back in beds. Lifted in the fall, commercial sized bulbs will be marketed; the remainder planted back for next years crop. AUTHOR

Plate 7

Lycoris aurea. These lovely flowers deserve a spot in any garden; they look best when planted in bold groups. AUTHOR

Tuberous Begonias. Each year, at the Chelsea Flower Show in London, the display of Tuberous Begonias is sensational, commanding attention from the many thousand visitors, who spend many hours at this annual flower show. AUTHOR

Ixia **production.** A small section of the commercial fields of these great plants, grown in beds. Such mass plantings are a joy to behold. AUTHOR

Crocosmia. Bulbs deserve consideration for a place in mixed borders. WARD

Combination of colors. With the wide selection of forms and colors, almost any combination is possible. Here, the flowers of the *Muscari* hide the stems of the *Narcissus,* and compliment each other. When considering such, attention must be paid to height and time of flowering. AUTHOR

Colchicum **'The Giant'.** Coming around a corner and finding a little pocket of color is always a pleasant surprise. WARD

Canna **hybrids.** Large, bold plantings of a single color are effective. They provide quite an attraction in parks. AUTHOR

Plate 9

Tulips and English Daisies. Tulips combine well with many winter and spring flowering annuals. At Kew Gardens the red tulips, the white English daisies, and the blue of the Forget-me-not, make a patriotic spring display. AUTHOR

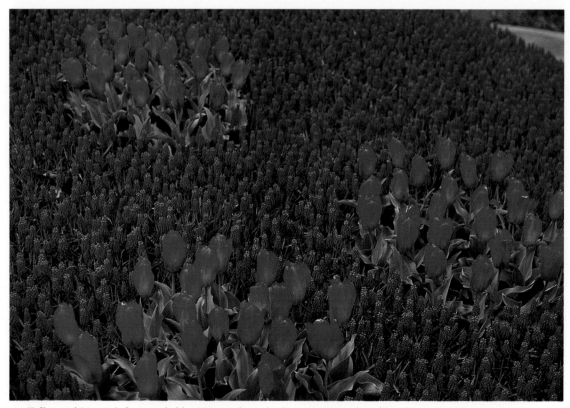

Tulips and Muscari. Surrounded by *Muscari*, the red tulips are outstanding, like islands in a blue sea. AUTHOR

Cardiocrinum giganteum. When well established, these plants are always eye-stoppers in a woodland setting. WARD

Erythronium **'White Beauty'.** These plants, with the addition of a few leaves, make an attractive exhibit. They show how effective small groupings of plants can be when set off by rocks and grass. AUTHOR

Cyclamen neapolitanum. The bright colors of these plants add much to woodland settings, making the green appear even more lush. WARD

***Crocus* in lawns.** Unfortunately such vistas as this in The Hague are not common, yet, growing *Crocus* in grass is not difficult, and can be very effective. AUTHOR

***Muscari* and Daffodils.** Each year, this river of color is a favorite sight at the Keukenhof Gardens. AUTHOR

Trillium grandiflorum. Bold groupings alongside a path, wandering through woodland, is an ideal way to show the beauty of these plants. AUTHOR

***Dahlia* Test Garden.** These dahlias, in Golden Gate Park, San Francisco, are much admired by visitors. AUTHOR

Many gardeners do not replant, in display beds, tulips previously used for that purpose, prefering to plant in bold groupings, where the individual size of the flowers is less important. Such plantings make a grand display. AUTHOR

Narcissus poeticus. It takes a lot of effort to achieve the lovely effect that nature achieves so easily. Here an alpine meadow filled with *N. poeticus* shows what should be, and can be, achieved with mass planting. WARD

Mixing bulbs together. It is of the utmost importance that bulbs that flower at the same time, be used when such designs are planted. If the tulips had flowered at a different time than the *Muscari*, this would have been a disaster area, and not a pleasant sight to enjoy. AUTHOR

Plate 13

Planting amongst mature trees. Scale is always important. If the beds of bulbs had been smaller, the effect would not be as attractive. Here at Keukenhof, scale has been taken into account; the result—a pleasant picture. AUTHOR

Arisaema wallichianum. and *A. thunbergia* var. *urushima* are species that make attractive plantings in the woodland garden. WARD

Plate 14

Bulbs grow well in containers; here *Crocus* are being grown in pots designed for them. 'February Gold' Daffodils are at home in many types of containers, but wooden ones always seem more correct. AUTHOR

Tulipa multiflora. 'Georgette'. The advantage of planting cultivars of this species, is that you obtain a greater number of flowers, even though individually, they might be a little smaller. A greater impact from a limited surface area, is an important point. AUTHOR

Narcissus need deeper containers than tulips. Selecting the right size of container is important, not only for the culture of the bulbs, but so that, when in flower, the proportions of flowers with the container are attractive. AUTHOR

Plate 15

Lilies in an arrangement. Lilies are superb cut flowers, long lasting, fragrant and aristocratic. Herman Wall, one of the great flower photographers, with whom I had the pleasure of working for several years, took this photo in the early 1960s. But lilies are timeless. HERMAN WALL

Keukenhof. This renowned garden in Holland is visited by thousands each spring. Display gardens like this enable gardeners to see, in natural settings, the great beauty and fantastic colors obtainable from bulbs. AUTHOR

Hemerocallis **in the landscape.** While one seldom sees borders filled with just a single type of plant, the Day Lilies hardly need companion plants; they provide color throughout the summer. KLEM NURSERY

Plate 16

distinct separations are allowed, with here and there a few bulbs set just a little apart from the larger plantings. This duplicates the cloud effect in the sky.

Another way to achieve the desired effect is to mark out the principal area of planting and then just let the bulbs fall from the hand in a random pattern, planting them where they fall. At all times straight lines should be avoided as this will not create the "natural" look desired.

Cultivars of daffodils to be considered

The general mixture of daffodils offered by nurseries and sold often as Mixture of types, cultivars for Naturalizing, is most suitable. This will provide the desired mixture of different heights, colors, and times of flowering. The mix can be augmented by the purchase of bulbs of known colors and heights and time of flowering, and this should most certainly be considered after the first season of flowering, when inadequacies in the mix are more easily determined. The newest and most expensive cultivars will give no better an effect than tried-and-true cultivars that are often much lower in price. Good stock at a fair price is a good rule of thumb when purchasing daffodil bulbs for naturalizing. Vigorous species should not be overlooked. *N. bulbocodium* is an ideal species for such treatment. However, it is not a tall-growing plant and thus should be planted in front of other, more vigorous cultivars.

By the addition of cultivars over the years, a rich planting can be achieved that will give color for a considerable period of time.

PROFESSIONAL AND COMMERCIAL LANDSCAPING

General considerations

The professional landscaper has to think along the same lines as the homeowner, so the foregoing remarks are valid. But other considerations should be kept in mind as well.

Nothing looks as good as naturalized bulbs in the spring—the effect can be a knockout. But the care of the bulbs when they emerge from the ground and after they have flowered is critical to achieve the same effect in succeeding years. The amount of maintenance needed each year must be ascertained and the plantings undertaken only when proper maintenance is available. As the care is sometimes not understood, it should be outlined in detail and given to the person undertaking the work.

The bulbs themselves are easy to obtain, but they should be ordered well ahead to avoid delays. Storage of large quantities of bulbs on the site is not always ideal, as there should be little delay between the arrival of the bulbs and their planting.

Site selection

Planting areas can be defined, but again it is advantageous to look at the site and the surroundings carefully. The removal of certain vegetation to allow drifts to disappear into or under the vegetation can heighten the drift effect. In some cases the addition of shrubs with complementary flowering times—such as the flowering quince, *Chaenomeles japonica*, and various cultivars, in red or pink—can produce a superb background or focal point when located among such naturalized plantings.

Culture

A sharp lookout at the time the bulbs are emerging from the soil is necessary and will require one or more inspections of the site. The grass should be cut as late as possible. If the mower is set at 1 inch or so in height, the emerging bulb growth will be unharmed. As the bulbs grow more quickly than the grass, the end result of the display of flowers carried well above the grass will look at their best. As soon as the planted area is well defined, the remainder of the grassy areas should be mown so that the bulbs are seen clearly. This mowing must be done with care so that the bulbs are not damaged.

Size of plantings

While the homeowner will be able to add to the collection of bulbs each fall, it is in the first year that the work of a professional landscaper will be judged. The plantings must be bold and consideration given to closer planting than the homeowner would be likely to undertake. It is also wise to consider plantings of just one cultivar. A mass planting of 250 'King Alfred' daffodils will most certainly be noticed as they come into flower. However, if 50 each of five different cultivars were planted, the effect would not be dramatic. While perhaps fine for the home garden, it would be lost in the scale of a commercial development. If more than one cultivar is desired, those cultivars selected should flower at exactly the same time. This is but one reason, albeit the principal one, among others that need to be considered.

In commercial plantings the benefits of planting larger quantities of one cultivar, or several identical in development, do not end with effect alone. Having the bulbs flower and mature at one time, having them emerge from the ground at one time, means the necessary culture can be carried out at one time. They are thus more amenable to commercial-type care—only one application time for fertilizer, etc. After they have finished flowering all can be removed at once or the grass cut at one time, rather than waiting for the different flowering times and maturity dates.

Containers

When considering large plantings in commercial developments, containers with the same cultivars should be considered. These containers can be placed to give a perspective to the plantings and the continuance of the theme being established. This allows the theme to be carried into other parts of the property and ties the plantings into the overall scheme of the property.

Consideration can, in some instances, be given to growing the bulbs in containers. When the bulbs are of interest (and the flower buds well formed with a few flowers in bloom), plunge the containers into the ground in a suitable medium such as bark chips, peat moss, or the like. The effect is instantaneous and the installation and removal a comparatively simple operation. Should turf need to be removed, it can be lifted and stored, laid out where care can be given in another location. Replacing the turf after the bulbs are removed should not present a problem, because in springtime the turf will have good climatic conditions.

While plunging containers into the ground would probably not be cost effective in the summer, the cooler conditions of spring are more favorable and the display of color would last for a considerable time. If desired, the containers could be replaced with others planted with late spring or early summer annuals.

NATURALIZING ON SPARSELY COVERED GROUND

This subject has not received much attention. Generally such areas are poor in plant foods, or else there would be more vegetation. The addition of organic matter when planting and forking in more organic matter each fall will make naturalizing bulbs possible. If there exist certain native species of bulbs, additional bulbs of that species should be planted. Attention should be given to cultivars and varieties of these native bulbs as well as closely related genera. Many sites, perhaps barely covered with shrubs planted in their own pockets of soil to hold a slope, can be given added interest by a few more bulbs.

Commercial aspects of bulbs on sparsely covered ground

The professional landscaper and home gardener is often confronted with having to plant on sparsely covered ground, be it level or sloping. Bulbs can provide added interest. All too frequently, shrubs and groundcovers that are used to cover such areas do not look attractive until they have had a season or two to become established, and good growth made. Spring- and summer-flowering bulbs will add interest within a short space of time; for example, bulbs planted in the fall will flower the following spring.

If irrigation is provided to the area, then attention must be given to the choice of bulb planted. Many bulbs require a resting period after their foliage ripens, and during that time dry conditions are needed or rotting of the bulbs will occur. Such bulbs could not then be relied upon to give more than the first season of color. Daffodils are good for such locations. While they do prefer to be drier toward the end of summer, many of the stronger-growing cultivars, such as 'King Alfred', 'Dutch Master', etc., do not readily succumb to rots, and prove quite reliable in giving two seasons of color. Even these strong-growing cultivars might not give a good performance the third season, however.

Dahlias need summer moisture and are a good choice for such areas, but adequate fertilizer should be given if strong plants are needed. A slow-release fertilizer is a good tool for commercial landscapers faced with such conditions. The range of heights and colors in dahlia cultivars is a wide one, but bold plantings of one color will provide a more stunning impact than a mixture of colors. While the color should be the same, a mix of dwarf and taller-growing plants provides greater interest than plants of the same height.

While slow-release fertilizers are useful, the professional is well advised to give attention to an ongoing feeding program. In poor soils, organic fertilizers will be more beneficial to the soil than inorganic. The drainage also must be attended to as no plants will survive if they are growing in pockets of waterlogged soil. This is of greater importance on clay soils than on sandy soils.

Bulbs in the woodland

Trilliums, daffodils, lilies, and cardiocrinums are superb for planting in woodland areas. The bluebells are at home in such surroundings but must be planted in large quantities to be effective. *Cyclamen neapolitanum* does well at the base of a large tree, if planted on the shady side, but not too close to the tree or to a path. They should be seen and admired but not walked upon. If these plants like their home, they can spread with surprising rapidity. I have seen areas of many square yards covered in just a few years, without any attention to the plants at any time.

The species or cultivar of lilies for such woodland plantings should be selected carefully. The upright, flowering Asiatic types do not look at home in this kind of setting; they are too stiff and formal. While large-flowered trumpet types can be used, they do not look, in my opinion, as much at home as the pendant types such as *Lilium martagon, L. pardalinum, L. hansonii, L. kelloggii, L. wardii,* and such hybrids as the 'Bellingham Hybrids' and those from other American species having the same pendant flower form.

When planting these lilies, make sure they will have room to spread; plant a sufficient number so they will be a feature. The ideal location is an area where the lower part of the stems are shielded from the sun while the flower heads are in some sun and can be seen against a background of foliage.

The tall-flowering *Cardiocrinum* species (formerly *Lilium giganteum*) likes to have good moisture during the summer. While not ideal for the heart of a woodland area, these plants are ideal for a glade in the woodland. They take several years to flower but the display of lovely white trumpet flowers is worth waiting for. It is because of the size of these plants (they can reach well over 72 inches in height) that they are better suited for an open glade. In addition, these plants will form large colonies, so ample room should be provided to ensure plants can remain undisturbed for years.

Snowdrops and snowflakes also have their place in the woodland garden, but should be near a path so they can be admired. Being a little more formal in appearance, they are ideal for the fringes of the woods. When planning their planting locations consider having groupings of them alongside the paths leading to the woodland area. This will soften the abruptness of just having them in the woods.

The Winter aconite is another plant that does well in woodlands, but do allow room for this to wander. This also applies to the bluebells. While these plants will give color the first spring after planting, the result of letting them remain undisturbed for years is a massive and most natural-looking planting.

Autumn- and spring-flowering crocuses as well as the colchicums deserve consideration, but only around the perimeter of woodland gardens as these need sunlight to perform well.

Several anemones grow wild in the woodlands of Europe but they do not like too much heat in summer unless moisture is available. Given these conditions they are lovely. *Anemone nemorosa* and *A. quinquefolia,* native to the United States, are excellent for the woodland as is the somewhat rarer *A. appenina,* which is better suited to warmer temperatures.

Trilliums are great woodland plants. *Trillium sessile,* with its dark crimson petals, will spread well in favored locations; but *T. grandiflorum* perhaps is the best, provided the soil remains damp throughout the year. The pure white flowers, taking on hues of pink/purple as they age, are a delight. *T. cernuum,* native to the eastern United States, is a pleasant plant that prefers quite cold winters. If you are growing these plants, get at least 4 inches of soil over them when setting them out in the fall. The soil in which they are planted should contain a high percentage of organic material.

Fritillarias also look at home when grown in woodlands. There are several named clones of *Fritillaria meleagris* offered for sale. They do require summer moisture and will not do well without it. These are at their best with smaller-leaved shrubs; the scale is then correct.

It will be seen from the above that bulbs that look at home have, for the most part, pendant flowers. The grape hyacinths (*Muscari* species) do not look out of place on the edge of woodland areas, but they should receive good sun; these are not effective unless planted *en masse.* They grow well, multiply easily, and are worth considering along the borders surrounding the woodland area.

With all woodland plantings of bulbs, it must be remembered that competition offered by vigorous shrubs and groundcovers can be severe. Occasional additional feedings will be appreciated by bulbs that face such competition. The gardener should also keep in mind the growth habit of other plants in or near the locations where bulbs are planted—additional pruning of these plants may be needed to protect the bulbs.

Commercial considerations for the woodland garden

While the home gardener will be able to plant any or all of the bulbs mentioned above in woodland areas, the professional gardener and landscape architect should start by growing those bulbs that are of easier culture in commercial landscapes. Daffodils, lilies, and bluebells are almost certain to give good effects. Trilliums, being a little less showy in flower and slightly more fragile and expensive to obtain, should be added later when variation and increased interest is needed. *Cardiocrinum* will not flower for a number of years so also must be considered as a later addition, except where funds are available for continued maintenance.

Where there is much foot traffic, care must be taken in placing bulbs alongside the paths. It is wise to consider such plantings only when the bulbs can be protected in some way. They do not take too well to being walked upon. An effective way of giving protection is to line the paths with tree trunks. These should be no smaller than 6–8 inches in diameter. They should not be much over 14–18 inches in diameter or they will appear too massive. A good effect is achieved by leaving a small space between logs that are used, say a distance of some 18 inches. In this area the smaller bulbs can be planted and this type of planting breaks the line of the logs and a more pleasing effect is achieved. The logs should be of such a size that they can not be easily moved.

Cyclamens, colchicums, and autumn-flowering crocuses are well suited for bordering paths. The planting should not be in a regular pattern but varying in depth back from the path. Anemones can stand a little foot traffic, but for all intents and purposes all bulbs planted in public woodlands should be protected from foot traffic. Planting bulbs in front of or close to shrubs also will afford them protection.

In larger open areas, a grouping of logs, placed as if they had fallen onto one another, can offer a natural and pleasant looking setting for bulbs and provide protection as well. Ferns make effective companion plants in this setting.

Muscari can be planted, but if strictly woodland plants are to be grown and an authentic woodland planting is desired, they should not be taken into account as they are not woodland plants. They can be effectively used at a little distance from the entry to such woodland areas, leading the eye to a planting of bluebells, which are native woodland denizens.

Size of plantings

Large bold plantings are recommended in woodlands, especially if the trees are of good size. Nature would not have allowed the trees to have reached such a size before some colonies of woodland bulbs had become established, hence the need to keep in mind the scale of the woods. However, smaller colonies are not to be ruled out, but they should appear as if they are offshoots of the larger. It is a good idea to watch the pattern of shade formed by the trees and plant in the sunlit areas. On sunny days the colors of the plantings will be enhanced. The sunlight will not fall in regular patterns, making the plantings look natural.

In woods the surface of the ground is covered with fallen leaves and a number of smaller branches. These should be left in place and only removed if they impede the development of the bulbs. If they must be removed, scatter them in areas where they will do no harm; there is no need to remove them from the site. Scattering leaves over a planted area is suggested as this will hide the bulbs from potential disturbance by the public.

While the first impulse is to scatter bulbs throughout the entire area, it is more attractive to concentrate them in specific areas. This is again duplicating nature, since often once a plant finds an ideal location it will increase rapidly. By all means tempt people to enter the woods by placing some color where it can be seen from the outer limits, but hold back the main plantings so they are indeed a surprise. There is another reason for such planting. If hidden the plantings will be soon discovered by those who love plants, but the casual walker will not be so interested and another small measure of protection to the plants is achieved.

BULBS IN THE GARDEN

No garden is too small to have at least a few bulbs. Tulips, gladioli, dahlias, lilies, irises, alstroemerias, and nerines can be planted and look "at home." The smaller the area, the smaller the size of the plant; scale always is of importance.

While bulbs are frequently recommended for the perennial border, not all bulbs are suitable. The struggle for space and root competition may be too much, with the possible exception of dahlias, gladioli, and rhizomatous irises and certain types of lilies. If no other space is available then other bulbs can be grown in the perennial garden, but most deserve a place where they can be appreciated for their individual beauty, not mixed in and having to compete with the perennials. If lilies, such as the upright flowering Asiatics, trumpets, or Oriental hybrids, are grown, then sufficient quantities must be planted so they make a good showing in their own right. They should also be staged against complementary foliage, preferably not competing with perennials of equal height and strong color. They should be planted to take center stage when in flower.

The tuberous begonias are so lovely that a shady (but not deep shade) location should be provided. The base of a tree in a lawn is a good place, as well as containers and hanging baskets.

In my opinion, tulips, daffodils, large-flowered ornamental onions, bulbous irises, gladioli, freesias, anemones, and ranunculus always look their best as components of an annual border. They give a look of permanence even though their companions are annuals. Pink tulips above a bed of forget-me-nots; daffodils with pansies; large-flowered onions towering above taller-growing annuals; irises and gladioli, used not only for their color but also for their stiffer foliage, form grand contrasts with looser-flowering annuals. Not that the latter do not have a beauty of their own, but just as a fine piece of jewelry looks better on a lovely woman, so do these bulbs need such a platform to give of their best and undoubted value.

The hyacinth, the most fragrant of the spring-flowering bulbs, can be used in many ways. Place individual clumps near a window or close to an entrance, or in containers on the deck. Be sure to bring them into the home if the weather is inclement, and keep in mind the adaptability of bulbs for use indoors, even if only for a special dinner party or when guests are expected. The bulbs can be returned outside afterwards to continue their growth. Being indoors for too long a period will shorten their length of being at their best. Color and fragrance are as much a part of the charm of bulbs as any other plant. But it does not stop there. In some bulbs the beauty of the faded flowers should be remembered, for example, the flowering heads of certain ornamental onions.

Bulbs for cut flowers

Bulbs grown for cut flowers are best grown in a location where, if they are cut, obvious gaps in their display are avoided. Planting them very close together so that the bulbs almost touch will allow for a selective cutting to take place with enough flowers left to remain attractive. An effective way to plant in this fashion is to open up a trench to the correct depth and place the bulbs into position.

Daffodils, irises, tulips, gladioli, nerines, amaryllis, lilies, freesias, dahlias, anemones, and ranunculus are all excellent cut flowers. Alstroemerias are becoming more popular as new and colorful hybrids reach the market. Ornithogalums, while not commonly grown by the home gardener, will provide long-lasting flowers for the home and are worth making space for in the garden. Watsonias, while not a popular bulb for cut-flower production, deserve to be more widely grown by home gardeners, especially if they have an area where these lovely plants can become established. Calla lilies, while mainly associated with weddings and funerals, actually present a lovely display in the home, especially when mixed with other flowers, and they last a long time.

When cutting flowers for the home, the less foliage removed from the plant the better; as previously noted, the leaves manufacture the plant's food, and, if you want the bulbs to perform satisfactorily again, do not remove any more foliage than is absolutely necessary. The use of preservative substances in the water of cut flowers is now common. Bulbs do appreciate and last longer if arranged in copper containers. If not available, then dropping a few copper coins into the water does help them last longer.

Soilless culture

Modern lifestyles favor smaller townhouses and apartments. Certain bulbs, types of *Narcissus, Hyacinthus* cultivars, and *Crocus,* can be grown indoors without soil. Special vases are available, or you can try a dish filled with pebbles and then planted with bulbs with more pebbles added to keep the bulbs standing upright. After planting the bulbs water should be added so that it almost but not quite touches the base of the bulbs. They should then be placed in temperatures in the 45°–50°F range until there is a good production of roots. Only when these roots have been produced and some top growth noticed should the bulbs be brought into the home and given normal room temperatures. During the rooting period, it is best to keep the bulbs in the dark. This method of growing bulbs without soil is explained in greater detail in Chapter 7, page 52.

Containers, which come in a wide assortment of shapes, depths, and colors, can be used effectively to grow bulbs. The danger lies in the container not being of sufficient size to hold enough soil to maintain fairly constant temperatures. If the container is too small and then warmed by the sun, the soil temperature will reach a point which harms the bulbs. If there is no room for a large container, then do keep the smaller ones shaded to avoid extremes of temperature.

Drainage is essential; use a porous soil mix with a good organic content. The container should be of sufficient depth to give the bulbs the desired thickness of soil cover, as if they were in the open ground. The culture in containers and soil mixes is described in Chapter. I wish to emphasize here that having only a very small amount of room—be it veranda, deck, or balcony—should not preclude a homeowner from the pleasure of growing bulbs, even if the selection is limited.

COMMERCIAL ASPECTS OF PLANNING AND PLANTING WITH BULBS

The professional gardener and landscape architect will find the preceding remarks applicable to many aspects of commercial planting and planning. However, there are other points that should be borne in mind by the professional.

Annuals are not so expensive as bulbs. When planning large beds of bulbs, be they for spring or summer color, the most effective use will be made if the bulbs are planted *en masse* and edged with annuals which complement the colors. Or, for very large beds, consider islands of bulbs surrounded by annuals. Care should be taken in the selection of cultivars so that the beds are in flower at the same time and at the same height, or alternating lower- and taller-growing cultivars. Mixed colors are best avoided when a public display is the objective. While such can be appreciated by an individual in his own garden, the passing foot traffic experienced in larger areas demands a colorful impact at one time.

Very little effect will be obtained by having just one or two types of bulbs in flower at one time unless *en masse*. While a good effect can be obtained by having some earlier-flowering bulbs in flower, after a climax of color, small plantings look insignificant.

Very effective plantings can be obtained with such plants as *Bellis, Primula, Viola, Myosotis*. These give good value for the money, but the addition of tall-stemmed tulips growing through them gives a display sure to catch the eye of the public. Such plantings tend to be more formal in appearance, and the tulips or other bulbs used should, in these cases, be planted in a formal style, equally spaced and in a distinct pattern.

Summer-flowering bulbs can be effective in beds. Surrounded by lower-growing types the larger cultivars and showy flowers of *Dahlia* are superb. Gladioli, when used for summer color, should be regarded as extra color to 'perk up' a bed. They should not be relied upon to provide color by themselves as they flower for a short period of time. Hence, the *Gladiolus* should be used to provide additional, not basic, interest. Their swordlike foliage provides contrast when, for example, they emerge above a planting of taller-growing marigolds. Such plantings give pleasure all summer long, when the form of the foliage is almost as effective as the flowers themselves.

A massed display of gladioli is effective in itself if timed for a certain date. As a guide for such plantings allow 90 days for them to come into flower for early and late plantings and 70–80 days for midsummer. Such a planting can effectively hide a bare wall where the area for planting is only a foot or so in width. In such locations it is difficult to obtain annuals of sufficient height without crowding the walkways. Under such conditions the colors and cultivars should blend well with their background, and as timing is all-important the use of a single cultivar is to be preferred.

In the perennial border, bulbs planted in bold clumps add variation during the summer months. When planning, be sure to consider other plants in the border to achieve the maximum effect of contrasting foliage textures and colors. As the effect of such plantings is dramatic, so is the bleak look after they have passed their best. The professional should consider growing other bulbs in containers to be plunged into position when the first are removed. Remember that, while plants coming into flower is an exciting time, their going out of flower is not. Better to err on the late side than to be too early with such plantings.

Vandalism can be a problem, especially with hanging baskets that can be reached easily by younger people. Place them up high where an effort is required to reach them and use the cheapest possible containers to avoid too much of an attractive nuisance. Firmly attach the containers to their support.

CHAPTER 7
GROWING BULBS OUT OF SEASON

Preparation of bulbs for forcing, forcing bulbs for the home and by the professional grower.

INTRODUCTION

Many types of bulbs are used by greenhouse growers to provide "out-of-season" color for retail florists. A number of firms offer in their various catalogs bulbs the homeowner can have in flower at Christmas. In order to obtain such out-of-season flowering, the bulbs are said to have been "forced." Before bulbs can be brought into flower, certain treatments must be given to the bulbs to advance their development. This development inside the bulbs, normally accomplished by Nature, brings the bulbs into flower at their usual flowering time. To have bulbs in flower out-of-season the grower and/or producer either speeds the development for earlier flowering or retards the development for later flowering.

PREPARATION OF THE BULBS FOR OUT-OF-SEASON FLOWERING

The forcing of bulbs for the market has been long practiced and is most profitable. Rising fuel costs have given much impetus to searching for treatments prior to planting to ensure good performance from bulbs. Providing warmth alone does not always mean success. Some bulbs have not performed in a satisfactory manner; others just have not produced flowers at all.

It has been noted over the years that bulbs from areas such as Cornwall, and in particular the Isles of Scilly, come into flower as much as two weeks before similar bulbs, even the same cultivars but that were produced in cooler climates. Such earlier flowering was an obvious advantage. The question was—Why do these bulbs flower earlier? Research was undertaken to answer this question. Among other things it was discovered that the development of the embryo flower continued during the dormant or rest period of the bulbs, especially among true bulbs.

It has long been a common practice to give *Narcissus* bulbs a warm-water treatment to control eelworm. The critical temperature (111°F for three hours) was found to damage the bulbs, resulting in poor flower production if temperatures were not exactly maintained. By prewarming the bulbs and treating only after the flowers were formed, losses were reduced. Thus it was established that warmth and warm-water treatment, at a particular time in the formation of the flowers, had a decided effect on flowering performance.

Genera perform differently when forced. There are differences between genera as to their specific requirements of treatment in order to take advantage of their natural processes within the bulbs. There are also differences, even though to a lesser degree, between cultivars within a genus. The dates on which the bulbs are required to flower also will alter the treatment. An example is perhaps in order.

Before cooling, tulips should be warmed for 2 weeks at temperatures between 68° and 72°F. For very early flowering, warming at 94°F for one week prior to the 2-week treatment is beneficial. For *Narcissus* the treatment is 4 days at 94°F, followed by 2 weeks at 63°F before cooling. Such treatments are known as "curing." By such warmings followed by cooling, bulbs are "prepared" for early flowering by "forcing." Even if not forced in a greenhouse or other structure, "prepared" bulbs will flower earlier, even if grown outdoors.

Bulbs that, during their growing period, have received good growing conditions, including both moisture and temperature, are of course much easier to treat as described above. By the same token, it is best if all the bulbs so being treated have reached the same stage of development. For this reason bulbs from Holland, parts of England, and the Pacific Northwest are preferred for forcing, as in these regions large quantities of bulbs are produced under almost identical conditions. In Japan, by contrast, climatic conditions can vary considerably from one area to another. This does not have any detrimental effect on the bulbs' performance in the garden, but under the stress of forcing irregular flowering is more likely to occur. If it were possible to establish that all the bulbs from Japan (or any other region except those mentioned) had received identical growing conditions and were all at the same development stage, then such would be suitable for treatment and subsequent forcing.

The standard process for preparing bulbs for forcing is:
1. Curing by warming for short or long periods.
2. Cooling for varied periods, depending on genus, at 48°–50°F for 4–9 months.
3. Note that cooling below 45°F can have adverse effects on the bulbs.

It can be seen that warming, when followed by cooling, *induces* flowering. Warmth in itself, if given for long periods, will *delay* flowering. Thus the producers of flowering bulbs can help the grower achieve earlier-than-normal flowering times as well as delayed flowering times. In this way crops for specific dates can be planned and the appropriate treatment given. This is a tremendous advantage, especially in light of the costs of fuel to heat greenhouses.

The basic principles for treatment are:
1. The flowers must be formed within the bulb prior to cooling.
2. Temperatures must be maintained within specific and strict limits. (These specific and strict limits will vary slightly for cultivars of any genus.)

The purposes of warming before cold treatment are varied and include:
1. Making sure the flower is fully formed before cooling.
2. Retarding flower production for flower sales after the normal season.

With these principles in mind, we can now look at specific requirements for bulb (in the wide sense) crops. Growers should obtain exact and up-to-date information about the cultivars to be grown, and adapt schedules as needed. The treatment of the bulbs when they are brought into the greenhouse is mentioned on page 51.

Hyacinths

Growers cure these bulbs in a number of ways. One is by soil warming, i.e., heating the soil by deploying electric cables or warm water pipes in the soil. Flower bud development is induced by temperatures in the range of 68°–78°F. Certain cultivars, among them 'L'Innocence', need temperatures toward the higher end. Curing these bulbs for Christmas flowering, which is about the earliest these bulbs can be obtained in flower without undue expense, requires lifting them in the middle of June, when the foliage is still green. The bulbs are allowed to dry and are then cleaned and graded as needed. Then, for a period of 7 weeks, the bulbs are exposed to exact temperatures for set periods of time: 14 days at 86°F, then 3 weeks at 78°F, then 2 weeks at 73°F. After this the bulbs are stored at 63°F until the last week of September. Hyacinths for later forcing, i.e., well after Christmas, are lifted in mid-July and then stored at 78°F for 2 months, then at 63°F, as for Christmas-flowering bulbs.

For garden hyacinths the curing time is mid-July to mid-August, after which they are stored at 55°F.

Narcissus

In most cases, the flower production of *Narcissus* is accomplished by the natural heat of the soil, but must be in the range of 68°–72°F for at least 2 weeks so that precooling can start in the second or third week of August. Precool at 48°F for 8 weeks. Those cultivars with colored cups require less time. As timing is critical, growers are advised to obtain exact information regarding each cultivar if they plan to precool the bulbs themselves.

Narcissus being grown for very early flowering are subjected to somewhat different treatment, as warmer temperatures are required to encourage flower formation within the bulb. Bulbs for early forcing are lifted in the second week of July and, after drying and cleaning, are given 4 days at 94°F, followed by 14 days at 63°F. These treatments are followed by cooling the bulbs at 48°F until the first part of October, when they are planted.

Tulips

Stamens and carpels must be well formed before any cold treatment is given. This is ascertained by cutting open the bulbs and examining the development of the embryo flowers. To make sure this stage of development is reached it is frequently necessary to prewarm the bulbs. This is accomplished by giving the bulbs temperatures in the 68°–72°F range for 2 weeks, timing it so that the bulbs can receive 8 weeks at 48°F before being planted.

Tulips for very early forcing require another type of treatment. The bulbs receive 7 days at 94°F before 14 days at the 68°–72°F range.

There are a number of variations on the warming and cooling treatments: one is a short period at 93°F for 5 days in mid-July, followed by 14 days at 78°F, then 21 days at 62°F.

Remember that the treatment best suited to each cultivar varies. However, as a general rule, those within the same grouping, e.g., Darwin tulips, require the same treatments. A grower wishing to force tulips into flower is advised to talk with the supplier and determine which cultivars are available that have received the necessary treatment(s).

Muscari

These bulbs need about 6 weeks at 48°F from the first part of August to September.

Iris

These need a warm period from the time they are dug until they are placed in cold temperatures. This is usually done by giving the bulbs some 21 days at 83°F, followed by some 35 days at 50°F, after which they are ready for planting.

In Holland the Wedgewood iris often is given 14 days at 93°F, 3 days at 104°F, 2 weeks at 63°F, and then 6 weeks at 48°F. Bulbs of the same cultivars from warmer climates will not require the same treatment. Ten days at 95°F, followed by the cooling treatment, is sufficient to cure these bulbs. Dutch iris have been much used by florists out of season. These bulbs can be retarded by keeping them in a temperature of 86°F. This temperature is maintained from the beginning of September until some 6 weeks before planting; for the last 6 weeks they are kept at 63°F. The schedule for such late flowering is: plant in the last 10 days of June for September flowering; plant in the first week of July for October flowering; plant at the end of July for November flowering. For the 6 weeks immediately before planting the bulbs receive a temperature of 63°F; from the time of harvest, a temperature of 86°F.

Summary

By manipulating the temperature bulbs receive, the grower can program the crops for a long period of flowering. Research to perfect temperature control programs continues, so the above can only familiarize growers with the basic principles involved. It must be emphasized that the requirements are very exact: variations of only a few degrees either up or down frequently can lead to poor crop performance.

BULBS FROM SOUTHERN HEMISPHERE

The potential of producing out-of-season flowers from bulbs grown in the Southern Hemisphere, where there are a number of bulb-producing firms, has not, in my opinion, been fully explored. It should be. However, the question of the public's acceptance of tulips and *Narcissus* and other spring-flowering bulbs in summer or fall is another question entirely. The needed treatments and the costs involved would be offset by the cost of the freight. As the harvesting of the bulbs is some 6 months before/ahead of the time in the Northern Hemisphere, the prospect is interesting.

FORCING BULBS

The two most important conditions in forcing bulbs are temperature and time. For the correct development of the flower buds inside true bulbs and the proper treatment of corms, tubers, etc., certain and definite stages of development inside the bulbs have to be reached.

The complex treatments given by growers have been outlined in principle earlier in this book. These are essential to bring bulbs to the state of development which will permit them to be subjected to the cultural conditions for forcing and, most importantly, to perform well when so subjected.

The principal crops forced are the *Narcissus*, tulips, hyacinths, irises, and, to a lesser degree, the lilies. While there are similarities in treatment, they will be discussed individually in some detail.

HANDLING AND STORAGE

There are handling and storage conditions common to the majority of bulbs being forced. They must always be handled with care. They are living plants, even if dormant. Too much weight can squash them, causing serious damage and providing easy entry for fungus diseases which cause rotting. Remove all packages from the containers, boxes, etc., in which they arrive and arrange them so air can circulate between them. Examine the bulbs to make sure none is damaged: any that are should be discarded. Then store them at 63°F if at all possible; however, between 55°–65°F is acceptable. Storage of lilies is best at some 38°F; below 32° will cause freezing of the bulbs, and temperatures above 45°F will encourage sprouting in the boxes. Lilies also differ from other bulbs in that, being made of fleshy scales, they can dry out and shrivel if exposed to drying conditions.

All bulbs received from growers will have the floral buds already developed and can be treated through controlled temperature as soon as the particular program being followed commences. The exception to this is any bulbs received in August. Depending on type (see Preparation of Bulbs) these may need temperatures at 63°F to form floral buds, and the treatment may take several weeks.

If bulbs of the types mentioned above are to be stored for long periods of time, say a month or more, they are best placed in trays with a wire-mesh bottom to ensure good air circulation and should be held at 63°F with as little variation as possible. Temperatures above 63°F can seriously harm them; lilies are, as mentioned above, treated differently.

Holding Back Flowering

The term "forcing" is given to describe the culture of bringing the bulbs into flower ahead of their normal flowering time. Another term describes plants being brought into flower *after* their normal flowering time: "held back." While not used extensively with spring-flowering bulbs—tulips, daffodils, etc., as these seem to have little sales appeal out of season (but this is slowly changing)—it is a procedure used quite extensively for gladiolus and lilies, and to a lesser degree for anemones, ranunculus, and irises.

For the majority of these, holding back is a simple operation. The bulbs or corms are stored in well-ventilated storerooms at temperatures of some 50°F, generally on mesh-bottomed trays. Lilies are kept at lower temperatures. Periodic checks for sprouting of lilies held in storage for such late plantings must be made. Trumpet lilies are not easy to hold back, but just want to grow, and as soon as shoots are seen emerging from the bulbs they must be planted. Asiatic hybrids can be held back for a considerable length of time, but it is best to plant no later than May or June.

The lily bulbs will deteriorate by shrinking a little but will still produce a good number of flowers, although with a lower bud count than usual. A period of rooting in moderate temperatures is necessary; allow temperatures above 55°F only after root formation has taken place. Where there is adequate storage space, bulbs should be planted and allowed to root well in storage before being brought into the growing areas. With careful attention to planting dates, Asiatic hybrids can be available for marketing throughout the year; however, the need for longer day length in the fall of the year and the costs of such lighting have to be balanced against the prices obtainable in the marketplace. For such prolonged storage, larger-sized bulbs, 5–6 or 6–7 inches in circumference are preferred.

Do not store bulbs in areas where fruits, especially apples, are kept. The ethylene gas given off by the fruit is harmful. If humidity can be controlled it should be in the 60–70% range.

Soil mixes

Bulbs can be grown in many different kinds of soil mixes. The bulb itself contains the nutrients it needs to grow and flower. Moisture, air, and support must be supplied. A good mix is equal parts of soil, sand, and peat moss. The soil should be weed-free and not have been used for bulbs before. If in doubt sterilize the soil to ensure that no disease pathogens are present.

The soil mix must drain well but retain moisture at the same time. A wide range of pH is suitable, as the bulbs will grow between 5.5 and 7.5. Highly acid or alkaline mixes cannot be used. Do not use too fine a mix because the roots need air at all stages of growth.

Containers

Pots, both clay and plastic, can be used. Wooden flats are preferred if the bulbs are being used for cut-flower production. In all cases adequate drainage is essential. While the size of the existing hole in a clay pot is normally alright, some plastic pots may need to have additional holes cut into them. The depth should not be less than 3–5 inches, but deep pots are not necessary. The exception is the stem-rooted lilies, which require pots at least 8–10 inches in depth. *Narcissus* produce more roots than the other bulbs generally forced, so they should be given containers of at least 6 inches in depth. In all instances used containers should be cleaned thoroughly; if new clay pots are used, they should be soaked prior to use and allowed to dry before filling with soil.

Planting in containers

The bulbs should be planted in a loose mixture, with the top of the bulb level with the rim of the pot. Fill the containers with soil to the necessary level for the bulbs to be at the desired height when placed on the soil. Bulbs should not touch each other but be separated by about half their width. Tulips should be planted with the flat side of the bulb facing the outside of the pot. From this side the largest leaf is produced, and better air circulation can be obtained between the stems. Bulbs to be used for cut flowers can be planted closer together, allowing ¼ inch between each bulb. If a particular cultivar has dense foliage, however, more space is needed.

Care must be taken never to damage the basal plate of a bulb. Two problems can arise if they are pushed too hard into the soil: the first is that physical damage can result; the other is that the soil underneath the bulb can become compacted and a good root growth pattern will be discouraged because the roots then have to force their way through. The preferred method for planting is to loosen the soil mix, place the bulb in position, cover with soil, and water in. After such watering the bulbs should be buried to at least ¾ of its depth.

Rooting bulbs

Various bulbs have different requirements while rooting prior to being exposed to warmer temperatures. This is not so critical for the home gardener as for the commercial grower. A week's delay in flowering may be annoying to the home gardener, but such delays can be disastrous for the commercial grower. For this reason many of the latter construct special growing chambers, where the necessary temperatures can be maintained. In many instances the areas are separate, with the range being at least 32°–55°F. For the home gardener, a second-hand, home-style refrigerator will serve the purpose. Two other conditions should also be established—air and darkness. The bulbs being grown in growing chambers will need to be watered from time to time. Keep in mind when constructing the chamber that it should be disinfected and easy to clean.

The effectiveness of outdoor rooting sites depends on climate variations. In most parts of the country, outdoor areas are adequate, the exception being warmer regions which do not reach the low temperatures required in early fall. The schedule of forcing must take this into account. The selected site must drain well and, where possible, not have been used to grow bulbs before.

The desired temperatures for outdoor rooting are the same as those for indoor controlled areas. Maintaining such temperatures is difficult and in some cases will require additional protection from cold, while others may require being protected from the sun. In a number of places, temperatures may drop, inhibiting the production of roots despite the added protection. Plants such as hyacinths and tulips require warmer temperatures prior to forcing—about 48°F—and darkness to ensure good root production in such areas.

If the following temperatures can be maintained, then many bulbs can be forced satisfactorily using an outside rooting area: 50°–55°F the first 2–3 weeks after planting, 48°–50°F until the middle of November, 35°–41°F from then to the time the bulbs are brought indoors—but never allow the temperature to drop below 32°F, protecting with a tarp or straw or plastic if necessary.

The outside bed where the bulbs are to set should have a 3–5 inch layer of sharp sand or gravel on the bottom to ensure good drainage. The pots are then placed in position, and any air spaces between the pots filled in with soil or sand. A layer of sand is then put over the pots to a depth of 2–3 inches. Soil to a depth of 3–5 inches is placed over the sand. It is a good idea to put salt hay or a tarpaulin over the top of the area if much rain or severe cold weather is expected.

The reason for the sand on top of the pots is a practical one. When the pots are to be brought indoors, the sand separates easily from the top of the pot and the bulbs are not disturbed. If just covered with soil which becomes wet, this operation can be difficult without doing damage.

Use a soil thermometer to monitor the soil temperature and check the bulbs for moisture. Water if they become dry.

FLOWERING FOR SPECIFIC DATES

The date at which the bulbs are required to flower determines: the timing of their being planted, the length of time they are rooting, and the time they are brought into the warmer temperatures to flower. All bulbs must be well rooted before higher forcing temperatures are given. For practical reasons it is not possible to give exact starting and flowering dates. Even if the times of starting the bulbs and desired flowering were broken down into separate weeks, a grower must exercise skill and experience in bringing the bulbs to flower at the desired time. For this reason, normal flowering times are divided into three periods: early flowering, December 15th–February 1st; mid-season flowering, February 2nd–March 15th; and late flowering, March 15th–May 15th.

The keen and experienced home gardener will be able to time, with considerable accuracy, the flowering of bulbs being grown out of season. But there is a difference in having the bulbs in flower over a period of time when compared with the pinpoint accuracy the commercial grower must achieve for a continuous flow of flowers for the market. Having the bulbs in the home to enjoy is a far cry from having to meet market deadlines. Accurate records kept over a period of years will help greatly. Reading the following schedules which a commercial grower must follow will help the home grower as well as demonstrate what an exact science horticulture has become.

GUIDELINES FOR COMMERCIAL FORCING OF BULBS

The schedules given below are meant to be a guide only to a few representative types of bulbs. Experience and availability for growing space will determine the exact dates as well as the prospective flowering dates to be targeted.

Narcissus

About August 20th. — Bulbs are precooled for 6–7 weeks at 48°F.

1st week October. — Bulbs are planted and placed in the growing room or are plunged outside: 48°F given for 5 weeks, then temperature can be lowered to 41°F if well rooted.

Last week November. — First pots are brought into the greenhouse. Temperature 60°F at night, daytime only a little higher; pots brought in at weekly intervals.

Mid-season flowering.

1st weeks October. — Bulbs are planted and given a temperature of 48°F for 4 weeks, longer if roots are not well developed. Temperature is then lowered to 41°F.

January 1st. — Bulbs are brought into the greenhouse starting 1st week in January. If the bulbs still in the root-growing temperature of 41°F are starting to make top growth, lower the temperature by 4–5°F.

Late flowering.

Last week November. — Bulbs are planted, placed in growing room for 3–4 weeks, then 4–5 weeks at 41°F, then held at 35°F.

Last week February. — Bulbs are brought into the greenhouse at weekly intervals or as needed: about 24–26 days are needed to flower.

Hyacinths

Last 10 days September. — "Prepared" bulbs are planted. While not all cultivars are amenable to such treatment, and others may require less, most cultivars require treatments, even if for flowering in the latter part of the "early season." After planting the bulbs are placed in a temperature of 48°F. Those that are kept for the end of the period (mid-Jan. on) can be dropped to 41°F if root production is good and some growth of the shoots seen.

December 1st. — Bulbs are brought into warmer temperatures for flowering. It is very important that during the first 4 days the pots are inside, the temperature be kept between 50°–55°F. During this time the pots are not watered. After this period temperature can be raised to around 70°F and water given. As soon as color is seen on the flowers, the temperature is dropped to around 65°F.

Mid-season flowering.

1st week October. — Bulbs are planted, then given 3 weeks at 48°F, then temperature dropped to 41°F.

1st week January. — Bulbs are brought into the greenhouse, the temperature kept at 50°–55°F and no water as mentioned for early season, then given same temperatures as early-season bulbs.

Late flowering.

2nd week November to the 1st week of December. — Bulbs are planted; those for the latter part of the season are potted one month later; bulbs given 21–24 days at 48°F then the temperature dropped to 35°F.

Last week February. — Bulbs are brought into the greenhouse, same initial treatment as other bulbs, as mentioned above. The temperatures in the greenhouse, however, need not be so high for forcing, 65°F at night being sufficient.

If the shoots of the hyacinths are growing longer than 3–4 inches, temperatures should be dropped at once to 41°F. Forcing requires constant attention to the progress of the bulbs both in the growing area and in the greenhouse.

Tulips

Early flowering.

Last week August. — Bulbs given precooling for 6–7 weeks at 41°F; may not be necessary for bulbs planned to flower toward the end of early season, i.e., after mid-January. It is recommended for late-flowering cultivars and Darwin Hybrids.

1st week October. — Bulbs are planted; precooled bulbs and those cultivars not needing precooling can be planted 2 weeks earlier; after-planting temperature at 48°F; those for late December flowering (including Christmas) kept at this temperature. Those bulbs for mid-January flowering, drop the temperature to 41°F after 3–4 weeks.

Last week November.	Bulbs needed for the 1st part of the period brought into the greenhouse, those for 2nd half start right after Christmas, always at weekly intervals. |//

Mid-season flowering.
1st week October.	Darwin Hybrids and late-flowering cultivars given 4 weeks precooling prior to this date. Precooled bulbs and those not needing it planted and kept at 48°F for a month; when roots appear through bottom of the pots temperature is lowered to 41°F. Plants for pots need no precooling and are planted 15 days later.
2nd week January.	Bulbs brought into the greenhouse; temperature in the 65°F range; if higher or lower temperatures are planned, the time bulbs brought inside should be adjusted accordingly.

Late Flowering.
2nd week November.	Bulbs are planted for cut-flower production. Those for pot culture planted 2 weeks later. Both at 48°F for 3 weeks, then 41°F for 5 weeks, then 35°F.
Last week February.	Bulbs brought into the greenhouse as needed; allow 3 weeks to flowering time.

Note that in many areas outdoor temperatures may be too low for adequate root growth. Should root development be insufficient, the bulbs should be brought into warmer temperatures, about 48°–52°F and kept in the dark to promote root growth. It cannot be emphasized too strongly that a good root system is essential before the bulbs are brought into the greenhouse.

Summary

Experience will help you determine the various differences in the behaviors of bulbs. Thus the cultivars used and the temperatures at which the greenhouses are operated will govern the dates of the procedures, which may often change significantly. The above information is a guide to be changed as needed for individual growers and their production schedules. The amount of sunlight as well as temperature will have an effect on flowering, including height of the stems.

HOME GARDENERS

The procedures the nursery owner must follow, due to the need to get the crop to harvest on time, are much more complicated than those the average home gardener (should) need follow. But one rule of thumb all must follow is that bulbs must be of good quality—sound, with no soft or rotted portions and the basal plate in good condition.

A total of 13–18 weeks is regarded as the optimum time for precooling bulbs. This includes the time the bulbs spend potted up in a growing temperature, that is, the best temperature for root production. Bulbs being grown for cut flowers will produce longer stems if the period leans toward the 18 weeks, shorter stems on those with just 13 weeks.

Pot the bulbs as described on page 23, then place them in a cool, dark location. The temperature should be below 50°F, preferably in the 45°–48°F range. They must be kept moist during this period.

In cooler climates, placing the pots in a trench, on gravel or sand and covered as described previously, is beneficial. In milder climates, an old refrigerator can be used. Keep the bulbs cool, moist, and in the dark until the roots are well formed. At no time should they be subjected to temperatures below 35°F.

If it is necessary to protect the bulbs from mice or other rodents, chicken wire can be used. When well rooted, the pots can be brought inside the home. Given normal inside temperatures and good light, the plants can be expected to flower in about 3–4 weeks. Remember, if you wish to slow down the development of a container of bulbs, placing them in lower temperatures and lower light conditions will help. Conversely, the process can be speeded up by increasing the amount of light the plants receive and giving a more favorable temperature.

Once a bulb has been forced in the ways described, it cannot be counted on to perform well again. For this reason the bulb should be discarded. If you have a large area where, for example, bulbs are being naturalized, plant them there. They will recover from forcing in a couple of seasons, providing the correct cultural conditions for such bulbs are given. See Culture, page 33.

The home gardener should consider *Crocus, Muscari,* and *Iris reticulata* good forcing prospects, in addition to the bulbs regularly forced by the commercial grower. The choice of cultivars is important. Some are more reliable than others and more tolerant of the "general" treatment given by the home gardener.

The following schedule for flowering times will be of assistance to the home gardener:

To flower in January.	Plant the first part of October.
To flower in February	Plant in the middle of October.
To flower later.	Plant in the first part of November.

Forcing without soil and without cooling

I do feel that bulbs should be given cool conditions when they are being started into growth. This, however, is not essential for all bulbs. Soil also is not necessary for certain bulbs; they can be grown in gravel or small stones. As has been mentioned the bulbs have within them the nutrients they need as well as the flowers, albeit in embryo.

If bulbs are to be grown without soil, then they must be grown in containers that will hold water. A layer of rocks or gravel an inch deep is placed in the container. The bulbs are set into position, spaced about an inch apart. The container is then filled with more gravel/rocks so that the shoulders and necks of the bulbs are above the medium. Water is then added until it is just about, but not quite, to the bottom of the bulbs. Throughout the growth of the plants, the water is kept at this level. Topping up will have to be done from time to time.

For best rooting, a temperature of around 50°F should be maintained. Lower temperatures down to 45°F would be even better. Light should be kept low, darkness being ideal. In 3–4 weeks the bulbs will have produced a great number of roots.

The best bulbs for almost certain success are the 'Paper-white Narcissus' (*Narcissus tazetta*) and its yellow cultivar 'Soleil d'Or', together with the 'Chinese Sacred Lily' (*N. tazetta orientalis*). All can be grown without elaborate cooling.

Hyacinths, and they should be those that have been "prepared," also can be grown in specially formed glasses. The bulb is held by the glass and water is added so that it is just below the level of the bulb. The container is then put in a cool, dark location until a substantial number of roots are produced. This will take anywhere from 4–10 weeks. As soon as the roots are abundant and the shoot is some 3–4 inches high, increase the light and temperature. This method is not only attractive but of great interest as well. The water does need to be topped off from time to time and the addition of a piece of charcoal into the water will keep it sweet. Bulbs once flowered have to be discarded, as they completely exhaust themselves and their food supply when grown in water alone.

There are many ways in which bulbs can be enjoyed. I would encourage the home gardener to try growing these wonderful and colorful plants and to get them into flower out of season. Even if the first attempt is unsuccessful, it will provide a good conversation piece. Should it flower—and it should if quality stock was purchased and the procedures outlined above were followed—much pleasure and delight will be yours.

CHAPTER 8
PESTS AND DISEASES

Without a doubt, bulbs should be relatively free of pests and diseases when purchased, and after planting such a condition will be realized if good hygiene is practiced. Bulbs grown poorly, with little or no attention to correct and timely cultural practices, will not be able to maintain their vigor. Those subjected to such carelessness are much more susceptible to attack from pests and diseases.

Even with good culture, several pests need to be expected almost as a matter of course—the aphids and those perennials, the slugs and snails. Few gardens will be free of these no matter how many precautions are taken and how high the standards of cleanliness. The greenhouse horticulturalist knows that whiteflies, mealy bugs, and cutworms can be a problem, although they can be bothersome to bulbous plants grown outdoors as well.

Aphids, while they can cause some distortion in a crop, are dangerous mainly because of their ability to spread virus diseases. Once a bulb has contracted a virus, it is worthy only for discard. This is especially true of lilies. Slugs and snails can wipe out seedlings and even mature bulbs if the emerging leaves are eaten, robbing the bulb of much food. Since the foliage is the food manufacturing plant for future seasons, the bulbs so attacked will be weak at best and thus in turn more vulnerable to other pests.

The gardener should not be discouraged by the lengthy list of pests and diseases to which bulbous plants can fall prey. Seldom will all of the possible enemies be present in one garden. In addition, as I have stressed, correct cultural practices will greatly reduce the number and their spoiling effects.

The range of products available to control pests and diseases is large. Each year new products are introduced, and each year certain products are no longer on the market. Whenever possible, spraying should be the last resort in any control program. Elimination by washing with water and the use of soaps and biologic controls should always take precedence over the use of chemicals. Should a pest or disease threaten to wipe out a planting, the alternatives are stark: spray, or lose the plants being attacked. The gardener must make a choice. The choice today, with the increased awareness of the possible damage to both humans and the environment by the use of certain chemicals, rests with the individual.

At all times, when chemical controls are being used, the label must be read carefully. The product must have been cleared by the proper authorities for use on the plants being sprayed, and strict attention given to the dilutions and strengths recommended. **Always consult the label of any product for specific application information.**

If there is any doubt in your mind regarding the appropriateness of a product for use on any given plant, information should be obtained from a fully qualified person, such as your county agricultural commissioner, extension agent, licensed pest-control operator, or a member of the staff of your local nursery.

PESTS
Ants

As a general rule, ants do not cause much damage to bulbous plants. They generally are found on plants where there already are colonies of aphids or other insects which exude honeydew, on which the ants feed. Should ants be seen on plants, the grower is advised to look for another pest.

Ant nests can be bothersome but unlikely to be troublesome. Eradication of the nest is the most effective control. Once the nest is located, boiling water poured directly on it will take care of the problem. Should the problem be severe, however, seek professional help.

Aphids

One of the most common pests is the aphid. Their rate of increase is appalling. It is not uncommon for 20 generations or more to be produced in one year. Certain generations are produced parthenogenetically, that is, without fertilization of the female by the male. There are both winged and wingless forms in the life cycle of each of the many species of these insects, and they vary in color. Winged females are fertilized by males in the fall. They lay eggs, which remain dormant overwinter, emerging as nymphs in the spring. These mature, giving birth to living young parthenogenetically. The winged forms can spread rapidly. Toward the end of summer, males are produced that fertilize the females, which in turn lay the eggs that overwinter.

The aphids, being sucking insects, weaken the plants by disrupting the necessary flow of food inside the plant. This can cause distortion, stunting, and loss of vigor. Leaves often will yellow when the attack is severe. The honeydew secreted by these insects is much appreciated by ants and also forms a good growing medium for certain sooty molds.

The first defense should be a jet of clean water to dislodge the pests from the foliage. Ladybugs will help control minor infestations. Should a severe attack occur, Malathion, Diazinon, Sevin, or other insecticides must be used. Always follow the instructions shown on the label and follow up with treatment to control the eggs that will have hatched after the first spraying. Several applications may be needed if the infestation is really bad. Another precaution is to check for any aphids that may be present on bulbs you are purchasing. Often a colony will be found under the tunic of a bulb and clustered around the growing point. These are not the bulbs you want to buy. If you do not discover the insects until after you have bought the bulbs, either return them for fresh ones, or dip them in or dust them with a suitable insecticide prior to planting.

Green Peach Aphid *Myzus persicae*

Known in Great Britain as greenfly, this pest, unlike many aphids that can vary in color, is always green. They are larger than the majority of the common aphids, frequently being over ¼ inch in size. Iris, tulips, and lilies are among the bulbs most frequently attacked, causing the same problems and damage as the other aphids. Control methods are the same as for the other aphids.

European Corn Borer (*Ostinia nubilalis*)

If, at the end of the season, a great number of corn stalks is left on the surface of the ground, they might well provide the necessary refuge for the European corn borer to overwinter. The caterpillar emerges in late summer/early spring and will attack plants that are similar to corn. This applies to those with a definite stalk. For this reason the most commonly attacked are dahlias and gladiolus, two of the popular Summer-flowering bulbs. The larva is light pink in color and about ¾ inch long. The egg-laying female moth, yellow-brown in color, is mostly nocturnal and, therefore, hard to control. Cleanliness certainly is one of the best controls, but, should the caterpillar be found, the affected plants should be removed at once and discarded.

Stalk Borer (*Papaipema nebris*)

This pest is not so selective as the corn borer and can be found on a great number of plants; but, as with the corn borer, the summer-flowering bulbs are the most susceptible. Dahlias, irises, and, to a certain extent, the later-flowering lilies are the most likely to be attacked. This pest is found mainly east of the Rocky Mountains in the U.S.A. The caterpillar is one inch or somewhat longer in length and pale yellow with a purplish band, which colors are quite marked when the caterpillar is young, becoming fainter as it ages. Spray with Sevin or a similar product to eradicate.

Japanese Beetle (*Popillia japonica*)

While not a problem in the western part of the United States, there are few regions east of the Mississippi free of the troublesome Japanese beetle. The grub, some ½ inch long, overwinters in the soil. In the spring it moves toward the surface and feeds on the roots of grasses, devouring them. During early summer the adult beetle emerges and continues to feed until late summer or early fall, completely eating away buds, flowers, and, in the case of the tender stalks of dahlias and cannas, the stems as well. The beetle is visible to the naked eye, being up to ½ inch long and bronze in color. Control of a limited attack simply requires removing them by hand. Traps also are a valid control, but if there is a severe attack, Malathion can be an effective insecticide.

Wireworms (*Melanotus* species)

These pests, common throughout North America, are the larvae of a beetle. They spend their lives in the soil and sometimes in decaying or rotted wood. The larvae eat into the roots of dahlias, gladiolis, and other bulbous plants, devouring the bulb. The result is the collapse of the plant. As they are seldom seen, the best method of eradication is to turn the soil frequently (but not too often) so they are exposed to their natural enemies; this also discourages the laying of eggs.

Nematodes

There are many different species of nematodes. Seldom are they visible to the naked eye, but they are abundant in most soils. Not all are harmful, however, some being responsible for the breakdown of organic matter in the soil. They are one of the main pests responsible for root decay, which in turn can cause deformed foliage, even causing the leaves to split and bulbs to have much browned tissue. If you have such a problem remove the bulb together with a sample of the soil that surrounds it and have the bulb and soil examined by the local authority. If a severe infestation exists the only recourse is soil fumigation, which should be done by experts and should be standard procedure when bulbs are bieng raised commercially.

Narcissus Fly

As its name implies, this fly attacks the *Narcissus*. However, it can also be found at times on the amaryllis, galanthus, and leucojum. The adult lays eggs in the neck of the bulb as the foliage dies down. The larvae eat the bulb underground, destroying the center. If any foliage is produced the following spring, it will be at best very weak and yellow in color.

The practical method of control is to keep the surface of the soil disturbed as the foliage dies, covering the hole leading to the neck of the bulbs. Any bulbs attacked should be destroyed. The soil can be dusted with Diazinon or other products recommended and listed as controls.

Whitefly

Whiteflies are a common pest. They are found more often in greenhouses than outdoors in northern latitudes, but can be a problem on various bulbous plants in the warmer latitudes, especially during the summer months. They are seldom noticed unless in large numbers. The wings are covered with a powdery substance which gives them both their color and their common name.

There are several genera, *Trialeurodes* being the most common. The adults are quite small, varying in size from 1–3 mm. Adult females lay light-yellow eggs on the undersides of leaves, being attached to the leaf surface by a stalk. During incubation the color may change. The time needed to hatch varies according to the temperature, from 1 week in very warm weather to 3 weeks in cooler conditions. After they hatch, the nymphs wander over the lower surface of the leaves until they select a position to which they remain fixed. At this stage they start to feed by piercing the leaf, obtaining sustenance from the sap of the plant. After 3–4 weeks the nymph pupates and eventually the adult emerges. During the late spring and summer months, whiteflies can be found in any of the various life-cycle stages. Overwintering is in either the nymphal or pupal stage. Innumerable generations are produced in the course of one year.

The plants are weakened as the sap is removed by this pest. Yellowing of the foliage and drying of the leaves are the end results if an attack is left unchecked. Whiteflies excrete honeydew so fungi such as sooty mold can grow, which in turn limits the functions of the leaves.

When plants are infested, a chemical product should be used to control, according to the directions on the label.

Mealybug

These insects are most commonly found on bulbous plants that have persistent foliage, such as the cliveas. The name was given because of the waxy, white, mealy secretions that cover their bodies. Ants frequently are attracted to these secretions and sometimes aid in the distribution of the pest by transfering them from plant to plant.

The most common place to find this pest is on young foliage, especially in a confined area, such as inside a sheath at the base of the leaves. Generally, clusters of the insects are found together, and many generations are produced in one year. They feed on the plant by drawing sap, which weakens the leaves and causes distortion of the shoots in severe cases.

Denatured alcohol applied to the pests by means of a cotton swab will clear away small infestations, but, in severe cases, an insecticide should be used.

Plants grown in greenhouses are more likely to be attacked than those grown outdoors.

Cutworms

The larvae of various nocturnal moths make up a group of pests known collectively as cutworms. One, the variegated cutworm (the larva of *Peridroma saucia*), is found worldwide. It attacks many different plants, favoring beans, cabbage, corn, and tomatoes, but it will infest various bulbous crops also. This pest cuts the stems at ground level, but it also will climb the stems to feed on the buds and leaves. It spends the winter as a naked, brown pupa in the soil; adult moths emerge in the early spring; eggs are laid and, in about a week, hatch into plump, smooth caterpillars. These feed for many weeks, become over one inch in length, and then burrow into the soil to pupate. Three or more generations can be expected in one year.

Hand picking and destroying the caterpillars is the most effective control, but chemical controls also can be used.

Thrips

There are a number of species of thrips, generally named for the plant on which each feeds. Two that are especially troublesome are the gladiolus thrips (*Taeniothrips simplex*) and the onion thrips (*Thrips tabaci*). The latter also will be found attacking dahlias and gloxinias, as well as other garden flowers.

These pests are equipped with rasping mandibles that shave off the outer layers of the leaves so they can feed on the sap-containing tissues. This causes strips of yellow-to-brown foliage, and, on gladiolus leaves, a silvery appearance. Flowers, if produced, are distorted.

The onion thrips cause white blotches to appear on the leaves. The tips then turn brown and become distorted. Eventually, the whole plant will topple. Any bulb produced will be misshapen.

As with most pests, clean cultural procedures, including the eradication of weeds, are the best ways to keep attacks from thrips to a minimum. If weeds are left to grow, the insects will feed happily on them, returning to attack their favorite crops as soon as they are planted.

Because they are a common problem, especially on gladiolus corms, the corms should be dusted before storing and again prior to planting. Bulbs should not be planted in locations where known infestations of thrips exist or have existed unless effective control is made. During the growing season, Malathion is one insecticide that will give control.

Mites

These ubiquitous pests are found on humans, animals, and plants, in the soil—in everything—but only a few attack bulbs. The spider mites, often called "red spiders," actually can be green, yellow, and brown, as well as red. They attack a great number of bulbs. In warmer climates the early spring-flowering bulbs are not frequently infested. In colder climates, where they flower at a later time and where the temperatures rise more quickly in the spring, the range of bulbs affected can be wide. In each case the damage is the same. The mites suck the nutrients from the leaves, often destroying the tissue so that the leaves have a dry, yellowish appearance. On the underside of the leaves, webs can be seen. A good way to determine if this pest is present is to hold a piece of white paper or card under a leaf, tap it, and then examine the paper. If a number of very small spots start to move, you can be pretty sure mites are present. Hosing the foliage may reduce the population, but spraying with an insecticide such as Malathion is a more positive control.

Although the spider mite will attack cyclamen, the more likely insect to be found on this plant is the cyclamen mite. Indoors this is a particular pest as the plants will not grow, or, if attacked when of larger size, the flowers will become distorted and the plant will display symptoms much like those of a spider mite infestation.

Both of these pests flourish during the dry summer months. Syringing of the foliage with cold water deters them, but, if badly attacked, the plants are best discarded.

Slugs and Snails

These are familiar to all gardeners. Clean cultivation assists greatly in reducing their number, but, as a precaution, suitable baits should be used as soon as the bulbs emerge from the ground in the spring. Care also should be exercised when the bulbs are in flower as snails like to climb the stems and devour the flowers. Picking these pests off the plants by hand when they emerge in the evening probably is the most effective control. A dusting of diatomaceous earth may be needed in more serious attacks.

Animals

A number of animals consider bulbous plants a tasty meal. Mice, moles, gophers, and other burrowing animals munch on the bulbs; rabbits, deer, and squirrels will devour the top growth. While there are a number of products on the market designed to deter such pests, the best control is a physical barrier, such as wire mesh placed around the bulbs in the ground or a suitable type of fencing aboveground to protect the aerial portion of the plants. But, as we all know, these creatures are not easily discouraged.

DISEASES

The majority of fungus diseases flourish when there is poor air circulation and high humidity. Dirty containers, unwashed storage areas, and weeds act as hosts for the diseases, contributing greatly to the problem. Stored bulbs should be dusted with fungicide. Any plants suspected of having been infected are best discarded rather than taking a chance on growing them, with the possibility that they in turn may pass on the disease to other plants.

As with insecticides, care should be exercised when purchasing any product to be used as a fungicide, making sure it has been cleared for use on the plant being treated. The directions on the labels must be read and followed carefully regarding strength of application, times and frequency of spraying, etc.

Botrytis (*Botrytis* species)

During the cool, damp summer weather of temperate climates and the cooler winters of subtropical climates, small yellow or orange-brown spots may be noticed on the leaves of plants. Bulbs are no exception; both the actual bulbs and the foliage are vulnerable. In a short space of time the spots become larger, until the entire leaf or surface of the bulb is covered and a gray mold becomes evident. Because of the damage to the living plant material, the plant succumbs. Should the bulb survive for another season, the foliage produced will be weaker and again succumb to attack. Bulbs showing any signs of these spots should never be planted.

Their liking for shade and moisture make the tuberous begonias the most frequently attacked, as are lilies, especially if planted among shrubs, which reduce air circulation. These, however, are not the only bulbous plants that fall victim to botrytis; dahlias and tulips, although to a lesser extent, also are subject to attack.

During the growing season, especially if the temperature is cool and moisture high, spraying with Benlate, Ferbam, or other fungicides cleared for use on the particular plant being treated, will control the fungus. Several sprayings generally are required.

Mildew

Both downy mildew and powdery mildew attack certain bulbs; however, the latter is less common than the former. Downy is like cotton in appearance and is most prevalent when the temperature is low and humidity high. Manzate is an effective control.

Powdery mildew (*Erysiphe* species) is very common, especially on the West Coast of the United States. It attacks a very wide range of plant material. The spores are spread by the wind and can be especially bothersome in shade areas. The fungus is first seen on the undersides of older leaves as small, white spots, followed by a weblike appearance that spreads to cover the entire underside of the leaf. When the attack is severe, the leaf yellows, then browns, and thus is useless to plant. The plant is weakened and performance reduced.

Benlate will control this disease but more than one application generally is necessary. Follow directions on the label regarding rates and frequency of application.

Onion Smut (*Urocystis cepulae*)

While more prevalent in the northern latitudes and infrequently found in warmer climates, this disease can attack ornamental onions, as well as the vegetable forms. Dark blisters appear on the young shoots and the plant is destroyed. If this disease is common in your area, avoid planting ornamental onions and be sure to discard any affected plants. Use a fungicide to control, following directions.

Verticillium Wilt (*Verticillium albo-atrum*)

This disease can live a long time in the soil, which is its habitat. It attacks a wide range of plant material, ranging from maple and elm trees to dahlias. Like many soil-borne organisms, it enters the plants through the roots and causes the vascular system to malfunction. The result is that the plant slowly dwindles away and dies. Verticillium also can be the cause of plants wilting under stress, such as warm spring days. Although the plant will recover in the cool of the evening, the wilting gradually increases and the entire plant wilts and dies. Fumigation of the soil and good rotation are means of keeping this problem in bounds.

Damping-off Disease (*Pythium debaryanum*)

This is another soil-borne disease. It will attack the young seedling just at soil level, causing the stem to become pinched in appearance, often turning black as the tissue dies. The disease is encouraged by lack of air circulation and excessive dampness, especially when coupled with low light, so that the seedlings do not grow quickly.

The best control is to use sterilized soil, thin seedlings as soon as they are large enough to handle, and keep seedlings growing well, with just the right amount of water, but not too moist. Seeds can be dusted prior to sowing with Arasan;

young plants can be sprayed with Ferbam, Zineb, or similar products.

Any plants that are attacked should be removed and discarded, along with any infected soil, and all containers meticulously cleaned.

Virus

All plants are susceptible to attack by a virus. In tulips the symptoms are the "breaking" or mottling of the colors in the flowers; in lilies the plants are weak and flowers often aborted. In all plants attacked, the foliage is mottled and, when held up to the light, irregular streaking is seen. There is little that can be done, so the plants are best lifted and discarded.

Virus diseases are spread mainly by aphids and other sucking insects that travel from plant to plant. Growers now have the means to produce virus-free plants by using the meristem or tissue culture methods of propagation (see p. 30–31).

SUMMARY

With the great number of pests and diseases that can attack the bulbous plants in the garden and greenhouse, a person may easily become discouraged. There is no need for such despair. The problems found in growing bulbs in general are much more likely to be caused by incorrect cultural conditions than by the onslaught of any pest or disease.

If pests or diseases attack your plants, check the actual damage that is being wrought. Aphids can be washed off with water if only a few leaves are infected. Their removal by hand might well suffice to bring the bulbous plants to the flowering stage without irreparable harm having been done. Good hygiene in the garden, the removal of debris and weeds, the correct cleaning of containers, both of wood and of other materials, can go a long way in prevention of major outbreaks of devastating infestations by pests and diseases.

Chemical controls often are the only remedy for plants that have been severely attacked. They should be used with discretion unless, in the case of commercial growers, financial disaster threatens. Accurate determination of the problem is necessary. Only then can correct and effective measures be taken. In all cases, as I have pointed out repeatedly, the directions on the product labels must be followed scrupulously in order to avoid another type of damage being done to the plants (and possibly to humans). And the other caution I will stress again is that the products have been cleared for use in the control of pests and diseases on the plants that have been attacked.

COMMERCIAL PRODUCTION

The foregoing list of pests and diseases can attack commercial crops just as well as those grown by the home gardener. In the latter case the loss of a few bulbs to pests and disease will not cause a calamity; distress, perhaps, but nowhere near that a commercial grower can experience, especially financially.

Many pests and diseases can affect the production of high-class commercial bulbs. Soil-borne pests such as nematodes (which can devastate a crop) and diseases should be controlled by using fumigants. A thorough soil analysis should be undertaken prior to planting and every precaution taken to ensure the soil is free from all possible pests and soil-borne diseases. Rotation and good cultural practices will keep problems to a minimum. But both in the raising of crops and in the forcing of bulbs into flower a regular spraying program must be followed.

Attention must be given to the health of the bulb at all states of its production. Diseased bulbs should never be planted. During the growing period any severely attacked bulbs should be removed and destroyed. Roguing should be carried out to remove any plants attacked by virus. Bulbs that have been heavily bruised and physically damaged in harvesting also should be discarded.

Fusarium is a special problem with tulips. Its common name, sour disease, describes one of the symptoms of attack—the bulbs smell sour. A white mold will be found growing on the skin of the bulb with soft tissue underneath. Such bulbs must be discarded; in fact, all soft bulbs—that is, soft to the touch—should be discarded. Healthy bulbs are firm. Bulbs that are much lighter in weight than the others should be discarded, as invariably they have been consumed by a disease.

Penicillium mold, which is blue-gray in color, is not harmful but should be controlled by dipping the bulbs in a suitable fungicide before planting. Such a mold is indicative of poor aeration in storage and, when seen, steps must be taken to correct the problem before other, more serious problems occur.

Fusarium basal rot should be looked for at all stages of the bulbs' culture. The basal plates of the bulbs are attacked and they become discolored and soft. Such bulbs must be discarded and care taken not to select planting stock from such.

The commercial grower must maintain a constant surveillance program of the stock and take remedial action as soon as the first signs of problems are noticed.

PHYSIOLOGIC DISORDERS

These disorders are not attributed to pests or diseases but to the plants being exposed, during the growth cycle, to environmental conditions that are harmful and place stress on them. Temperatures that are too low or too high, and poor ventilation in storage, or exposure to ethylene are contributing factors. Preventive measures and attention to detail at all phases of growth will alleviate these problems.

Certain problems become more apparent when the bulbs are subjected to stress such as occurs in the forcing of bulbs. The most common are the following:

"Tulip blindness." While it has been found particularly in the cultivars 'Yokohama,' 'Apricot Beauty,' and 'Demeter,' it also can occur in other cultivars and should be watched for by forcers. The plants simply fail to produce any flower parts. The cause is not yet known but the problem seems related to the larger sizes of bulbs used for forcing, 14 cm and up. It is advisable to use 12–13 cm size, not larger.

"Stem topple" is the name given to the collapse of the stem below the flower. It can occur before or shortly after flowering. It is brought about by the use of excessively cold treatment and temperatures above 68°F in the greenhouse. The cause is the lack of calcium being translocated up the stem to the fast-developing flowering stem above the last leaves. It can be ameliorated by using Calcium nitrate at the rate of 2 lb. per 100 gallons in the irrigation water.

'Flower blasting' is the failure of the bulb to produce a good flower despite all of the flower parts being present; in other words, it is aborted. Such a problem can be brought on by different events in the life of the bulb. It may be initiated as early as August or September or as late as when the flower begins to color. Causes are several: high temperatures during transportation of the bulbs, lack of aeration, and/or the presence of ethylene gas, which might well have been produced by bulbs infested by fusarium. Incorrect water balance is another cause and thus it is essential that at all stages of growth transportation and storage of bulbs be carefully watched.

Hyacinths are susceptible to one disorder known as "spitting." This is the separation of the flower stalk from the basal plates while being forced. It occurs when the bulbs have been frozen, as can occur if they are rooted in open ground and control of the temperatures received is not watched carefully. Protect such bulbs from freezing and, should they freeze, thaw them very slowly.

Daffodils are subject to a form of flower abortion when they are forced at too high a temperature. The problem is less likely to occur when the temperature remains below 65°F, but the exact reason has not yet been determined. Abortion of flowers is the most common with bulbs being forced out-of-season and is the general result of incorrect rooting conditions. Most commonly, insufficient time is allowed for rooting or the temperature during this phase of culture is too high or too low. It is imperative that temperatures be watched carefully and kept within the limits mentioned in Chapter 4, page 23.

CHAPTER 9
INTRODUCTION TO THE GENERA

This section is an alphabetic listing of all the genera of bulbous plants, worldwide, that I have been able to ascertain. I have attempted to maintain a certain order in the presentation, which may vary from time to time and case to case due to the amount of information available. Certain plants produce tuberous roots which carry the plants over the periods of stress they endure in the wild. Knowing where to draw the line in such cases is difficult. *Dicentra,* as an example, produces fleshy tubers on the roots. These tubers will not produce shoots if entirely separated from the plant. Attached to a portion of the rootstock above the tubers, they will produce shoots and a new plant is formed. Without the tuber, the plant would have greater difficulty in making and continuing growth.

Another example is the *Dahlia.* Here the tuber without a portion of the stem which has on it an eye will just rot away. The eye on the stem, without the backup support of the tuber for the supply of nutrients prior to the establishment of a new root system, also would be most likely to perish.

Should or should not these types of plants be included in a listing of bulbous plants, in the broad sense of the word? This is a good question, and my answer is found in the listing. Those genera where some or all of the species depend to a greater or lesser degree on the tubers produced on the roots are included.

Some genera, such as the *Solanum* and *Dioscorea,* to which the potatoes and the yams belong, are of great economic importance. While such are listed, the amount of information included is limited. I feel that in such cases books on crops of economic importance and those on food crops should be consulted for greater depth of information on their culture, and indeed for their descriptions. In many cases the flowers are not significant and thus are not likely to be grown in any collection of bulbous plants.

Another omission from this listing is the genera in the Orchidaceae. Here, again, separating out those species with truly bulbous rootstocks is difficult. Terrestial and ephiphytic plants are involved. Where should the line be drawn? This family is worthy of separate treatment and again few of the genera and species involved would be at once classified by the majority of gardeners as bulbous plants.

Two other genera deserve mention. They are *Clivea* and *Hemerocallis.* One might ask, if *Agapanthus* is included why not these as well? It is my opinion that these two genera are a little more removed from truly bulbous plants. This again is my opinion, but in support of this view the literature describing the *Agapanthus* rootstock is frequently such as "with thick rhizomes" whereas for *Hemerocallis* it often reads "the roots are more or less tuberous" and for *Clivea,* "has fleshy roots with certain species with bulb-like parts." Thus "with thick rhizomes" I feel on safer ground for their inclusion compared with the other descriptions. Both are included however as in the broad sense they can be classified as bulbous plants.

Rhizomes do present a problem. Here again I have used my judgment. If a plant has a rhizomateous rootstock (as distinct from a rhizomateous habit of rooting) it is listed. I feel the distinction is clear and just.

The listing has been set up in the following manner:

Each entry is headed by the genus and the family to which it belongs, followed by the derivation of the name, native habitat, brief history or other interesting information, and noteworthy characteristics which distinguish the genus from others which are closely allied. Common names are given; in some cases these apply to the entire genus and in others to particular species within the genus.

This general introduction is followed by a section on "Culture," which covers the requirements for successfully growing the plants. If any special requirements are needed by a particular species in the genus, these are given under the entry of that species.

"Propagation" is next, followed by "Pests and Diseases" if the plants are subject to any unusual problems. Then follows a discussion of "Uses" which refers to *mode d'emploi* of the genus in the garden, home, or greenhouse.

The final entry consists of several parts. The first and primary portion lists those species I consider the most important. A description of each species is given. The species I have listed were selected for several reasons:

1. They are well established in cultivation.
2. They are available from commercial sources, or are fairly easy to obtain from specialist firms—not always in bulb form but sometimes as seed.
3. They should be considered by every gardener as they have distinct landscape merit.

There then follows a table of those species that are quite frequently encountered in the literature. Included are those that deserve consideration, although of lesser importance than the primary species, and those species of recent introduction, some of which may indeed become popular in the not-too-distant future. Height, time of flowering, and country of origin are given, as is the date of introduction, if known. Sometimes the species may not yet have been "introduced" as known still in the native habitat—only occasionally cultivated in country of origin.

In spite of some gaps, the information given will allow the reader to understand the character of the plant. The gaps are due to lack of available information, or of information readily available to me. When a synonym is listed, a reference to the more valid name is given.

The reader is advised to turn to the genus in question. The species will be listed in the general or primary list or in the last listing. If a synonym is listed, it will refer to the more valid name.

Flowering times are listed and here, too, there is a difficulty. Many from the Southern Hemisphere will vary in time of flowering in the Northern Hemisphere. I have given the expected dates. If grown in the colder parts of the country the plants would no doubt flower a little later than indicated; if grown in comparatively warm climates, where there is little or no frost in winter, then an earlier date can be expected. In fact this also applies to bulbs that are at home in the Northern Hemisphere. One could not expect a bulb that is grown in an area where the ground is covered with snow in the winter to flower at the same time as a bulb being grown in Southern California. Such is particularly true of bulbs flowering in the spring but then differences diminish as the seasons and months advance. Plants from the Tropics are often in flower for long periods of time. Indeed, they are sometimes always in flower. Thus, determining the time of flowering in colder climes is difficult to establish. Where possible, information that will be of value is given. If I am not sure and information is not available, then I have not stated a flowering time.

The dates of introduction are given where known. In some cases they have not been introduced into cultivation—but are well known in the wild and deserve introduction.

Many genera are in a state of flux, the botanists themselves not agreeing on the nomenclature. In such cases I decided to use the name under which the genus will most commonly be found in the literature, with cross references from names used by other authorities. Some plants have been moved back and forth between genera many times. In some cases it may be many years before the dispute is finally settled.

In recent years there have been many introductions from such areas as the eastern Mediterranean countries, notably Turkey and nearby countries. The number of species introduced in such genera as *Allium* and *Colchicum* is large. While many of these have been given specific names, I doubt if all of them will remain valid. No doubt certain species will be found to be but geographic variants of species previously introduced.

The number of books describing the flora of a particular area also gives rise to a plethora of species. While no doubt some of the species described are valid and will remain so, there seems to be a tendency to look for very minor differences in species. According to the rules of nomenclature such differences justify specific rank being given, but one questions the benefit to gardeners and horticulturists. Indeed, all concerned with plants might be better served by the grouping together of many very similar species rather than the opposite.

The supplementary listings of species, which close my entries of the genera, are sometimes ponderous and lengthy. In certain instances they could have been made longer. However, I feel that, at this time, such would not add greatly to the reader's knowledge. While extensive, such listings as are made make no claim to be complete.

The list of cultivars given includes those that are of proven value and those that I feel will become "standard" cultivars in the future. Here again I have made a judgment call, based on my experience and conversations with experts in the field.

The information that follows will, I hope, add to the enjoyment and knowledge of all who love bulbous plants.

Among the bulbous plants are some that belong to genera that also include species that are not bulbous. An example is *Aconitum*. Those species with distinct tuberous rootstocks I have listed; however, those species that are not tuberous are not included nor is reference made to them.

ACHIMENES — GESNERIACEAE

The derivation of the name is unknown. It could possibly be from *achaemenis* ("a magic plant"), used by Pliny, or from the Greek *cheimaino* ("to suffer from cold" or "to be wintry"), a reference to the warmth required by this genus during the growing season. The genus is comprised of some 30 species, all herbaceous perennials, but not all in cultivation, native to Central and South America. The tubers/rhizomes, usually light brown, look like caterpillars and are 2–3 in. long. The leaves are opposite or in whorls of three, mostly toothed, growing on the slender stems, which may reach 24 in. in length. The very showy flowers, sometimes solitary, sometimes in clusters, are borne in the leaf axils. Flowers have a narrow corolla, which terminates in a blossom that is 1–2½ in. in diameter, with 5 lobes and 4 anthers. It has several common names, including Magic Flower, Nut Orchid, and Widow's Tears.

CULTURE Must have warm temperatures throughout growing season. If planted outdoors where night temperatures fall below 60°F, growth will be arrested, so plant in a cool greenhouse or indoors in regions with cool climates. Must be grown in filtered sunlight or light shade, never exposed to direct sun. Grows well under artificial light, needing 14–16 hours of light. Should never be allowed to dry out during growing season.

Plant tubers/rhizomes in a flat or similar container, February through April, depending on available warmth and light. Sandy soil mix should contain one part good topsoil or prepared planting mix, one part sharp sand, and one part sphagnum peat moss. Cover with 1 in. soil mix and water in. Daytime temperature 65°–75°F, nighttime no lower than 60°F. Keep moist; do not overwater. Plant no more than 10 good-sized tubers in a 10 in. pot. Do not overcrowd. Good rule-of-thumb: one tuber per one in. of container.

When shoots are about 2 in. high, transplant to containers or outdoors in areas of constant warmth. Strengthen soil mix but keep free-draining—one part well-rotted leaf mold, two parts peat moss, and one part good topsoil or packaged soil mix blended with two parts sharp sand. Nip out tips to encourage bushiness. At the same time start regular feeding schedule of liquid fertilizer, preferably using half-strength dilutions but doubling number of applications; maintain throughout growing period.

Number of flowers produced will diminish toward end of summer. Reduce moisture, so by end of summer or early fall growing medium is dry. Lift tubers, shake off soil, and store at temperatures never below 50°F in dry peat moss or perlite. Hold until growth cycle recommences in spring.

PROPAGATION Propagation is relatively simple. Simply break previous year's tubers into pieces, about ½ in. long, and plant. Can be broken while plant is dormant. Also can be raised from seed but will not reach good flowering size until at least second season.

PESTS & DISEASES Light syringings with water beneficial and will help keep most common pests, thrips, and spider mites, under control. Aphids also like these plants, so take proper control measures if they should appear.

USES Excellent house plants; well worth extra care needed to get them off to a good start. Great for hanging baskets, providing delicate colors and lush growth.

SPECIES & CULTIVARS Most *Achimenes* grown are hybrids. There are several species, however, that gardeners might wish to grow.

A. erecta (syn. *A. coccinea*; *A. rosea*). Native to Jamaica; introduced 1778. Long, thin, reddish rhizomes. 18 in. in height. Small, brilliant crimson flowers; flowering Aug.–Dec. Foliage dark green.

A. heterophylla (syn. *A. giesbrechtii*; *A. ignescens*). Native to Guatemala; introduced 1842. Very large rhizomes. Plant quite hairy. 12 in. in height. Dark reddish-orange flowers with yellow throat; flowering in August. Leaves are coarsely toothed-ovate, medium green.

A. longiflora. Native of Panama to Mexico; introduced 1841. Pear-shaped, scaly, white rhizomes. 12 in. in height. Flowers vary in color from almost pink to lavender and from pale to dark blue; throat usually white, sometimes with spot of yellow or violet. Some forms are pure white (*alba*), crimson (*grandiflora*), and reddish-violet (*rosea*). Type species has large blue flowers, carried in profusion. Flowering in July. Foliage coarsely toothed, 3–4 leaves being carried in whorls.

A. tubiflora (syn. *A. dolichoderia*).. Native to Argentina, Paraguay, and Uruguay; introduced 1845. Large, potato-like tubers. 24 in. in height. White, waxy flowers with light fragrance. Summer flowering. Leaves opposite, pointed and oblong in shape.

COMMENTS The most readily available and widely grown cultivars are:
- 'Cascade Cockade.' Deep pink, white eye.
- 'Cascade Evening Glow.' Deep glowing salmon.
- 'Cascade Fairy Pink.' Baby pink, white eye.
- 'Cascade Fashionable Pink.' Pink.
- 'Cascade Great Rosy Red.' Rose.
- 'Cascade Violet Night.' Purple-blue, white eye.
- 'Elke Michelssen.' Deep orange-red.
- 'English Waltz.' Salmon.
- 'India.' Moorish blue.
- 'Paul Arnold.' Blue-violet.
- 'Quick Step.' Blue.
- 'Viola Michelssen.' Rosy red.

ACHIMENES

A. andrieuxii Native to Chiapas, Mexico. Introduced 1959. Violet with white throats, 3–4 in. in height.

A. antirrhina (syn. *A. atrosanguinea, A. foliosa*). Native to Mexico. Introduced 1845. Reddish-orange, red dots on throat, tube yellowish with brown or purple dotted lines. 15 in. in height. July–Aug. flowering.

A. atrosanguinea syn. of *A. antirrhina*.

A. bella Native to Mexico. Introduced 1961. Violet with 3 yellow lines, corolla bell-shaped. 3 in. in height.

A. candida Native to Mexico-Panama. Introduced 1848. Creamy white, red at base, flecked reddish brown with yellow & lines of red. 6–18 in. in height. Autumn flowering.

A. cettoana Native to Mexico. Introduced 1959. Violet or purple-blue. 12 in. in height. July flowering.

A. coccinea see *A. erecta*.

A. dolichoderia see *A. tubiflora*.

A. dulcis Native of Mexico. Introduced 1961. Milk white with yellow spot, lilac dots on throat. To 24 in. in height. Aug. flowering.

A. ehrenbergii (syn. *A. lanata*). Native to Mexico. Introduced 1896. Lavender, lined & spotted orange in white throat. Compact habit. Midsummer to Autumn flowering.

A. fimbriata Native to Mexico. Introduced 1959. White with patterns of violet. 4 in. in height. Sept. flowering.
A. flava Native to Mexico. Introduced 1961. Yellow, tube with maroon dots inside. 18 in. in height. July flowering.
A. foliosa syn. of *A. antirrhina*.
A. giesbrechtii see *A. heterophylla*.
A. glabrata (syn. *A. gloxiniflora*). Native to Mexico. Introduced 1844. White or lavender flushed, yellow in throat, sometimes dotted purple. To 18 in. in height. Midsummer flowering.
A. gloxiniflora syn. of *A. glabrata*.
A. grandiflora Native to Mexico, Central America. Introduced 1842. Red-violet with white throat. To 24 in. in height.
A. hirsuta syn. of *A. skinneri*.
A. ignescens see *A. heterophylla*.
A. lanata syn. of *A. ehrenbergii*.
A. mexicana (syn. *A. scheeri*). Native to Mexico. Introduced 1850. Purplish or bluish, white inside. 12–14 in. in height. July flowering.
A. misera Native to Guatemala. Introduced 1848. White. 10 in. in height.
A. patens Native to Mexico. Introduced 1845. Deep reddish purple, flushed yellow or dotted purple inside pale throat. 12 in. in height. Aug. flowering.
A. pedunculata Native from Mexico to Honduras. Introduced 1840. Orange & yellow blotched with crimson, rhizomes whitish. 36 in. in height. Late Summer flowering.
A. rosea see *A. erecta*.
A. scheeri syn. *A. mexicana*.
A. skinneri (syn. *A. hirsuta*). Native to Guatemala. Introduced 1848. Red or rose, throat bright yellow, dotted. To 30 in. in height. July–Sept. flowering.
A. warszewicziana Native from Mexico to El Salvador. Introduced circa 1848. White with purple spots on tube. 10 in. in height. Summer flowering.

ACIDANTHERA — IRIDACEAE

From the Latin *akis* ("point") and *anthera* ("anther"), referring to the sharp, pointed anthers. *Acidanthera* is native to the southern and tropical areas of Africa. As with many genera, confusion exists as to whether this is a true genus, or whether it is part of *Gladiolus*, which it so strongly resembles. The main differences seem to be that *Acidanthera* have straight perianth tubes, as opposed to the bent ones of *Gladiolus*, and have a lovely fragrance, which *Gladiolus* does not. At one time, 20 species were assigned to this genus. W. Marais (*Kew Bulletin* 28, 2:311, 1973), recognized 9 species, all from tropical areas and not usually found in cultivation.

A. bicolor was the first species listed and introduced as a *Gladiolus*, 1896. Subsequently, a stronger variety was found, which was designated *A. b. murielae*. Catalogs began listing them as separate species, while, in actuality, the corms offered for sale were *A. b. murielae*. Popular usage as well as reluctance on the part of growers to list them otherwise without a firm basis for doing so may be the main reasons why at least these two species are still placed in *Acidanthera* rather than *Gladiolus*. Botanists, however, have placed several species, including *A. bicolor, A. b.* var. *murielae*, and *A. tubergenii* 'Zwanenburg', in the species of *Gladiolus callianthus*. (See also *Gladiolus*.)

These summer-flowering plants have strap-like leaves. The flowers are 2–3 in. in diameter, with equal perianth segments. The petals at the base form a perianth tube that can be 4–5 in. long. The petals are generally white, with a colored blotch at the bottom.

CULTURE Cultivated in same manner as *Gladiolus*. Corms best planted in spring after all danger of frost is past, 4–6 in. deep and 6–10 in. apart. Outdoor plantings grow best in well-drained garden soil, in full sun. Do not overwater during summer. No additional feedings necessary provided soil is in good health. Foliage dies down in early fall. Corms can be lifted and stored overwinter or left in the ground if temperature does not drop below 45°F. Should be lifted and divided every 3–4 years. Planting close to house provides added protection. Flowers late summer and, unless protected, blossoms will be harmed by early frost.

If planting in container, make sure the container is large enough for 4 in. of soil above and below the corm; use well-drained soil, in full sun.

PROPAGATION Each corm produces a goodly number of cormels around parent. Collect when lifting parent and grow on in nursery rows until flowering size, which takes at least 2 good growing seasons.

PESTS & DISEASES No special problems.

USES Beautiful flowers, marvelous vertical accents in floral arrangements, providing light, delicate fragrance. Attractive and aromatic house plants. Can be grown outdoors in containers but moved indoors if temperatures fall below 45°F at any time. Excellent garden plants, especially in warmer climates.

SPECIES
A. bicolor var. *murielae*. Native of Ethiopia; introduced 1896. Globose corm. Grows to height of 36 in. Leaves strap-shaped. Flowers 3 in. or more in diameter with distinct purple spot at base of petals, carried in two ranks, usually 6–8 per stem. Flower tube often 3–4 in. long. Flowering Aug.–Sept.

ACIDANTHERA
A. aequinoctialis Native to Sierra Leone. Introduced 1893. White with a purple spot at base (corm flattish, globose, large). 36–48 in. in height. November flowering.
A. bicolor see *Gladiolus callianthus*.
A. candida Native to tropical E. Africa. White, (corm Gladiolus-like). 18 in. in height. June–Aug. flowering.
A. capensis Native to S.W. Africa. White, (corm about ¾ in wide). 12 in. in height. July flowering.
A. crispa Native to South Africa. Rose, (corm narrow, ovate). 12 in. in height. June flowering.
A. c. var. *grandiflora* Flower usually 4 in., large.
A. c. var. *parviflora* Smaller flower.
A. c. var. *pectinata* Leaves not crinkled.
A. laxiflora see *Gladiolus ukambanensis*.
A. pallida Native of South Africa. Rose (corm globose). 12 in. in height.
A. tubergenii 'Zwanenburg' see *Gladiolus callianthus* 'Zwanenburg'.
A. viridis Native of South Africa. Green (corm flattish). 12 in. in height. July flowering.

ACONITUM — RANUNCULACEAE

The name was used by Theophrastus for a poisonous plant. There are some 60 species, not all of them with tuberous roots. All of the aconites contain a poisonous narcotic alkaloid, either aconitine or pseudaconitine. Care must be taken that they are not within reach of animals, particularly cattle, or mistaken at any time for an edible root.

The species are hardy, and among them can be found plants most attractive for late summer/early fall color. For the most part they prefer light shade. Foliage is usually palmately lobed. Flowers are held in a terminal raceme and get their common name of monkshood from the upper sepal, which is concave and helmet-shaped. Of the other four sepals, 2 are much broader than the others. There are 5 petals; of these the 2 uppermost have long claws and hooded tips, the lower 3 are small or sometimes missing. There are numerous stamens. All species are from the Northern Hemisphere and widely distributed.

CULTURE The plants are of easy culture. They require moisture throughout the year and perform best in light shade. Garden soil is adequate but they do appreciate organic matter. Should be planted some 2–3 in. deep; the distance apart will vary according to the height to which the plants grow. Some light feeding in spring can be given but feeding should not be heavy. Do not let the plants become dry.

PROPAGATION Lifting and dividing the plants in the fall or spring is the best way to increase the stock. Seed should be sown in spring and flowering-sized plants will be obtained by the end of the second season. Seed should be barely covered and given temperatures in the 45°F range at night. Outdoor sowing is fine, but sowing in a frame to provide a certain amount of protection is preferred, especially to prevent young seedlings from being too wet in areas where rain is frequent and heavy.

PESTS & DISEASES No special problems.

USES Good plants for the shady parts of the garden where they will get bright light but not excessive sunlight. They grow well in containers but these must be of good size as the plants do not like to be overcrowded.

SPECIES

A. anthora. Native to the Alps, Pyrenees, and Carpathian mountains; introduced 1596. It is a variable species with regard to color. Its tuberous roots produce an upright, leafy stem reaching some 24 in. in height. The foliage is 5–7 lobed and each of the lobes is in turn cut into linear segments. The flowers are yellowish or blue, the upper sepal is arched, the spur has a hook, and the flowers are carried on "S"-shaped pedicels. The following are some of the variants that have been noted: var. *atrovirens* from the Pyrenees has green leaves with orange-yellow flowers; var. *coeruleum* from Galicia has blue-violet flowers; var. *nemoroseum* from the Ukraine is not as tall growing, and while the leaves are not as cut they are hairier then the type. All flowering in July.

A. coriophyllum. Native to the Yunnan Province of China. Tubers are up to 4 in. in length. The stem is covered with hairs and can reach 36 in. in height. The leaves are roundish and some 4–6 in. wide, and are also quite thick and tough. The flowers are greenish yellow, over an inch in length, with the helmet being 1 in. long. Flowers carried in a loose raceme on stems that often branch at the base. Fall-flowering.

A. forrestii. Native to the Yunnan Province of China; probably introduced in 1915. Tubers are 2–4 in. in length. Reaches 24–60 in. in height with a thick stem clothed in more or less hairless, tripartite leaves and producing a dense raceme of deep purple-blue flowers on short pedicels. The axis of the flower head is covered with spreading yellow hairs, which also are to be seen on the flowers, but to a lesser degree. The helmet is almost an inch in length and the plant flowers in mid- to late summer. While not widely cultivated it deserves to be more commonly grown. This is one of the species grouped in the *A. napellus* group.

A. hookeri. A dwarf species native to Tibet. The small tubers produce masses of very fine roots and plants reach about 8 in. in height. Not many leaves are produced, and these are rather fleshy, without hairs or nearly so, roundish in shape, and divided into some 3–5 lobes. The flowers are blue or deep purple, and, while many plants produce only one flower, well-established plants will have a raceme of several flowers. Summer-flowering.

COMMENTS *A. napellus* is a collective name given to several species all having the following characteristics: tuberous roots, leaves palmately cut and often into fine segments, these being long, and narrowed to the entire part of the leaf. The helmet of the flowers is curved outward, hemispherical or arched. The color is purple, reddish-purple or white. *A. forrestii* belongs in this grouping. The other grouping of tuberous species is *A. cammarum*. The difference between *A. napellus* and *A. cammarum* is that in *A. cammarum* the leaves have short lobes and the undivided area of the leaves is larger than in *A. napellus*. Other differences are that the stems of *A. napellus* group are not branched but usually are in *A. cammarum*, and if branched the lateral branches carry racemes smaller than the terminal raceme, while in *A. cammarum* they are almost of equal size. The species in each group are included in the following table.

ACONITUM

A. cammarum group.
A. gracile Native to Italy, Hungary & eastern Europe. Pale blue or violet. Summer flowering.
A. judenbergense Native to E. Alps, Bulgaria. Purplish, reddish, variegated or white. Tall. Summer flowering.
A. rostratum Native to Switzerland. Purplish, reddish, variegated or white. 24 in. in height. June flowering.
A. variegatum Native of E. Alps. Introduced 1597. Violet to white. To 72 in. in height. Summer flowering.
A. paniculatum Native to S. Europe. Violet. Tall. Summer flowering.

A. napellus group.
A. anglicum Native to England. Long in cultivation. Blue-lilac. To 36 in. in height. May–June flowering.
A. angustifolium Native to S.E. Alps. Violet-blue. To 24 in. in height. Summer flowering.
A. brachypodum Native to W. China. Introduced 1906. Bright blue. To 24 in. in height. September flowering.
A. chinense Native to China (*A. faurei* in Japan). Deep bright blue. 60–72 in. in height. Summer flowering.
A. delphinense Native to S.E. France. Violet. Tall. Summer flowering.
A. faure: see *A. chinense*
A. ferox Native to Himalayas, N. India. Pale dirty blue. 36–72 in. in height. Summer flowering.
A. firmum Native to Yugoslavia. Purplish, reddish purple, or white. 48–72 in. in height. June–Aug. flowering.
A. formosum Native to Yugoslavia. Violet. 12–24 in. in height. Summer flowering.
A. funiculare Native to Bhutan (Himalayas). Introduced 1916. Lilac. 18 in. in height. Summer flowering.
A. naviculare Native to Sikkim. Reddish blue. 4 in. in height. Summer flowering.

A. neomontanum Native to European Alps. Violet. 36 in. in height. June–Aug. flowering.
A. pyramidale Native to France. Violet. 36+ in. in height. Summer flowering.
A. rotundifolium Native to W. Central Asia. Greenish white, violet veins. Summer flowering.
A. souliei Native to W. Yunnan, China. Pale yellow. 10–24 in. in height. Summer flowering.
A. × *stoerckianum* (*A. neomontanum* × *A. variegatum*) Native to Switzerland, Austria. Violet. 36 in. in height. Summer flowering.
A. venatorium Native to Upper Burma, W. China. Deep violet. 36 in. in height. Autumn flowering.

Minor Listing:

A. gmelinii Native to Siberia. Introduced 1817. Cream. 24 in. in height. July flowering.
A. heterophyllum Native to the Himalayas. Introduced 1874. Pale yellow & deep blue in front. 24–48 in. in height. Aug. flowering.
A. japonicum Native to Japan. Introduced 1790. Blue or violet, tinged red. 36–48 in. in height. July–Sept. flowering.
A. kashmiricum Native to Kashmir. Introduced 1930. Deep blue. 4–12 in. in height. Aug. flowering.
A. noveboracense very similar to *A. uncinatum*. Native to Catskill Mts. of New York to Wisconsin & Iowa. Violet-blue flowers.
A. orientale Native to Caucasus to Persia. White, suffused yellow. 48–72 in. in height. July–Aug. flowering.
A. reclinatum similar to *uncinatum* but with white flowers. Native to Southern Appalachia.
A. uncinatum Native to N. America. Introduced 1768. Lilac. 4–8 in. in height. July flowering.
A. violaceum Native to the Himalayas. Blue to white or variegated. 4–12 in. in height. Autumn flowering.

AGAPANTHUS — AMARYLLIDACEAE

From the Greek *agape* ("love"), *anthos* ("flower"). Despite its common name, lily of the Nile, this genus is native to South Africa. It has also been called drooping agapanthus and bell agapanthus. It has long been admired by the people of its native land and is cultivated extensively in their gardens. The Bantus used it in magic and medicine. A Xhosa bride would wear pieces of the root to ensure fertility and easy childbirth. The roots were powdered to provide medicine for a number of disorders.

The genus with 10 species is one of the oldest in cultivation in Western gardens. It was recorded in the catalog of the Leyden Botanic Gardens of 1687, and in 1692 it flowered at Hampton Court. It has been studied by many South African botanists, and a new classification was published in the *Journal of South African Botany* (Vol. IV), based mainly on the work of Frances Leighton. Despite all the work that has been done, botanists generally believe that *Agapanthus* is not a true bulbous plant, even though it frequently is regarded as such due to its fleshy, thick rootstock. Gardeners and nurseries place it in the bulbous plant group, regardless of what the botanists say.

Agapanthus is divided into two main groups—evergreen and deciduous. The deciduous species generally are hardier than the evergreens. Even if the foliage is killed by frost, the creeping, fleshy rootstock will survive to 26°F. The leaves of all species are strap-like, linear, and arching, varying in length. The heights of the different species also vary, but most are quite tall. The flowers, carried on strong stalks in a many-flowered umbel, are most often blue, although there are some white species, with 6 stamens. Both types bear flowers with either long or short perianth tubes. *Agapanthus* is summer-flowering and the blossoms are long-lasting.

CULTURE When established, *Agapanthus* will withstand drought, but for best production of flowers and foliage it should have abundant summer moisture, as in its native habitat and, preferably, full sun. Will grow in quite deep shade but flower production is reduced. Not fussy in regard to soil, but, for best results, use rich, well-drained soil. Does well alongside a stream; however, rootstock should be above water level with fleshy roots receiving abundant water.

Plant about 18–24 in. apart in beds, a little closer together in pots. Will grow well in containers, even if overcrowded, provided adequate moisture and fertilizer are given. Roots should be set at a depth so they are buried with green portion of stem appearing just above soil level. Three feedings of granular fertilizer per year is adequate: early spring, early summer, and late summer. Ideal formulas: 1st feeding, 14-10-10; mid-feeding, 10-10-10; final feeding, a little higher in phosphates and potash, in 5-10-10 range. Liquid feedings should be applied with every other watering at half recommended dilution.

PROPAGATION Easily propagated. When plants become overcrowded, lift clumps and divide fleshy roots with sharp knife or spade. Clean, remove dead foliage, and replant roots. Best done in spring.

PESTS & DISEASES Seldom attacked. Overcrowding can produce some stem rot. Slugs and snails find foliage attractive. Both problems can be avoided by good garden cultivation.

USES Excellent border plant; can be grown in background in herbaceous border, making ideal foil for red and yellow flowers. Ideal container plant; needs protection in colder areas where winter temperatures drop below 26°F; move containers indoors to spot where they will receive good light. Cut flowers last well in water. Dried flower heads used in floral arrangements.

SPECIES & CULTIVARS

A. africanus (syn. *A. umbellatus*). Native to southwestern Cape Province; introduced 1629. *A. umbellatus* not valid name as plants to which it was applied were described some 30 years earlier. Many plants sold today as *A. africanus* or *A. umbellatus* actually *A. praecox* ssp. *orientalis* according to reclassification by Leighton. *A. africanus* is an evergreen; flowering stems 24 in. in height, foliage strap-like, 18–36" in length, deep blue flowers, flowering Sept.–Oct. White form, *A. walshii*, is very similar to *A. africanus*, but a distinct species. *A. a.* var. *atro-caeruleus*, has dark violet flowers.

A. campanulatus (Bell Agapanthus). Introduced 1934. Found at higher elevations of eastern Cape Province and highlands farther north in Natal and Lesotho. Deciduous. Height varies from 18–24 in. Pale to deep blue flowers with spreading perianth segments. Flowering July–Sept. Leaves erect, 18–24 in. long, 1–1½ in. wide, strap-shaped.

A. caulescens. Native of Natal and Transvaal, where it grows in highlands; introduced 1901. Evergreen. Reaches more than 48 in. in height. Flowers bright to deep blue. Leaves strap-shaped, 20–36 in. long produced from a distinct stem.

A. inapertus (Drooping Agapanthus; syn. *A. weillighii*). Native to Transvaal; introduced 1893. Deciduous. Probably tallest growing, reaching up to 6 ft. Flowers of exceptional deep blue color hang downward. Leaves are 18–24 in. long, 1½ in. wide from a distinct stem some 2–3 in. long.

A. praecox (ssp. *orientalis* Common Agapanthus). Often sold as *A. orientalis*, *A. africanus*, and *A. umbellatus*, which are incorrect names. Native to southern and eastern parts of South Africa; introduced circa 1630. Popular garden plant. Evergreen, up to 5 ft in height. Leaves 3 ft long, 3 in. wide, may be yellow-striped, variegated, etc. Flowers pale to deep blue, sometimes double, with as many as 100 flowers in head, petals 2–3½ in. long. 'Rancho Dwarf,' 24 in. high with pure white flowers, and 'Peter Pan,' also a dwarf with exceptional blue flowers, are excellent selections of *A. p.* ssp. *orientalis* introduced in recent years.

AGAPANTHUS

A. coddii Native to N.W. Transvaal, S. Africa. Mid to pale blue. 42 in. in height. Aug. flowering.
A. comptonii Native to E. Cape Province. Sky blue to deep blue (evergreen). 24–48 in. in height. Sept.–Oct. flowering.
A. dyeri Native to Transvaal, Mozambique. Bright blue. 18–36 in. in height. July–Sept. flowering.
A. insignis Hybrid of garden origin. Pale blue or lavender. Summer flowering.
A. nutans Native of Natal, Zululand. Pale blue. 36 in. in height. Aug. flowering.
A. umbellatus see *A. africanus.*
A. walshii see *A. africanus.*
A. weillighii see *A. inapertus.*

ALBUCA — LILIACEAE

Albucus was the name originally given by the Greeks to a white species of *Asphodelus*, which subsequently was raised to generic rank. Common names include Soldier-in-the-box, Sentry Box, and Sentry-in-the-box. Very little has been written about this genus of about 30 species, of which only 11 are worth consideration, and there seems to be considerable confusion about it. Van der Spuy lists 3 main species in her book *Wild Flowers of South Africa for the Garden*, while Batten and Bokelmann, in their book *Wild Flowers of the Eastern Cape* illustrate 4 species, of which 2 are named. Found primarily in the Cape Province and Natal in South Africa. Thunberg writes in *Kaffraria* that the succulent stalk of *A. major* is mucilaginous and is chewed by the natives, as well as by travelers, to quench thirst.

Albuca species inhabit sandy soil, like full sun, and flower in late spring or early summer. The small, scaly, white bulbs are not hardy but can be grown in warmer climates where there is only occasional frost. Good plants for the cool greenhouse. Full grown, bulbs are quite large, so, if they are being grown in a container, be sure it is a good-sized one.

Plants grow to an average height of 3 ft, varying by species. The lanceolate leaves are quite long, up to 20 in., and are basal, the flower spike rising above the foliage. For the most part, the flowers are yellowish-white, with either a green or deep red stripe. Some have a pleasant fragrance. Each stalk will carry many flowers, the outside petals of which are stiff, while the inner petals often are held closely together. Though not joined, they form a tube covering the anthers and stigma.

CULTURE Bulbs planted late fall/early spring, 1– in. deep, 8–10 in. apart in sandy, well-drained soil, in full sun. In containers use a porous soil mix of organic top soil, peat moss, and sufficient quantities of sharp sand, to ensure very good drainage in late summer. Temperatures no lower than 45°F at night. Weak feedings of liquid fertilizer after growth has commenced; soil kept moist. Reduce water after flowering so bulbs go dormant by end of summer. When foliage has died down, keep bulbs dry until growth starts again.

PROPAGATION Separate offsets when bulbs are dormant. Plants produce a lot of viable seed though not generally available from catalogs, obtainable from South African nurseries, and a listing of such can be obtained by writing to Kirstenbosch Botanic Garden, Cape Town.

PESTS & DISEASES No special problems.

USES Not commonly grown, though deserve greater recognition. Easily obscured by other foliage, so place in garden where the low-growing flowers can be appreciated. Do well in containers; set bulbs only 1–2 in. deep; 8–10 in. apart.

SPECIES

A. canadensis (syn. *A. major* and *A. minor*). Native to the southwestern Cape area of South Africa, despite name; introduced 1768. Linnaeus first described it and knew of its origin but gave name because Jacques Cornutus (1605–61) had listed it in a book of Canadian plants. Reaches more than 2 ft. in height. Flowers pale yellow with greenish overlay and green stripe on outside of each petal. Flowering March–May. Leaves narrow, almost cylindrical, 20–24 in. long, and sharply pointed, light green in color.

A. circinata. Found on sandy coastal dunes of Natal. Fragrant flowers are shades of green and pale mustard. Height 12–18 in.; blossoms at Christmastime in native habitat; June/July in Northern Hemisphere. Leaves 12–20 in. long, cylindrical, sparse. Reported to be easy to cultivate.

A. humilis. Native to Lesotho in the Drakensburg Mountains; introduced early 20th century. Flowers white with green stripes; inner petals have no stripes but tipped with yellow. Lowest growing species, only 6–8 in. high, so is excellent for rock garden in warm climates or as a small pot plant when cultivated in greenhouse. Leaves cylindrical, 10–12 in. long, only 1 or 2 produced.

A. longifolia. Native of Eastern Cape Province; introduced circa 1897. Fleshy bulb 3½ in. long. Narrow leaves more than 2 ft long. Flower stalk about 2 ft high. Many flowered raceme. Upright flowers, white banded with green. Spring-flowering.

A. nelsonii. Native to Natal; introduced 1880. Reaches more than 5 ft in height. Leaves 2 in. wide. Upright flowers, white striped with dull red. Probably prettiest of species, described as "most massive and striking." Blossoms in summer.

ALBUCA

A. acuminata Native to South Africa. Introduced circa 1897. Dull green with whitish margin. 24–36 in. in height. Spring flowering.
A. altissima Native to Cape Province, South Africa. Introduced circa 1897. White & green. Bulb 1½–2 in. across, flattish round. 18–24 in. in height. Flowering Oct. in wild, April–May N. hemisphere.
A. aurea Native to South Africa. Introduced 1818. Pale yellow, greenish stripe. 24 in. in height. June flowering.
A. cooperi Native to Cape Province, South Africa. Introduced circa 1897 Pale yellow-green. 24 in. in height. Flowering Oct.–Dec. in wild, April–July N. hemisphere.
A. crinifolia Native to South Africa. White with reddish brown stripe. To 30 in. in height. Spring flowering.
A. fragrans Native to South Africa. Yellow, tinged green. 20–30 in. in height. April–May flowering.
A. major see *A. canadensis.*
A. minor see *A. canadensis.*
A. setosa (syn. *A. baurii, A. pochychlamys*) Native to Transvaal, South Africa. Introduced circa 1897. White with greenish stripe. 12–16 in. in height. Flowering July–Sept. in wild, Feb.–April N. hemisphere.
A. spiralis Native to Cape Province, South Africa. Introduced 1948. Pale yellow-green. 16+ in. in height. Flowering July–Sept. in wild, Feb.–April N. hemisphere.

ALLIUM — AMARYLLIDACEAE

Allium is the ancient Latin name for "garlic." There are many important food crops in the genus—chives (*A. schoenoprasum*); garlic (*A. sativum*); leek (*A. porrum*); onion (*A. cepa*); spring onion (*A. fistulosum*); and shallot (*A. ascalonicum*). There are more than 500 species in the genus; we will list only those grown in the garden for decorative purposes and those offered for sale commercially. It is only in the 20th century that most ornamentals were introduced. Without exception the species cultivated and found in catalogs are from the Northern Hemisphere; those from the Southern Hemisphere are regarded as unworthy of cultivation.

Although the onions are placed in the Liliaceae because the flowers have superior ovaries, there are some botanists who feel that, because the flowers are umbels, they should be in Amaryllidaceae, while still others feel they deserve their own family, Alliaceae.

The bulbs vary in size, generally in proportion to overall height when full grown; however, all are fleshy and similar to the onion of the kitchen. Some, such as *A. giganteum*, are smooth and shiny but, when cut, display the typical "onion rings."

It must be noted that unless care is exercised some species can become so well established that they turn out to be pests and are difficult to eradicate.

The plants, for the most part, have the distinctive onion smell, both in the foliage and bulb. Leaves are cylindrical and hollow. Flowers are individually small or, at best, medium-sized, but are in an attractive umbel, which can reach considerable size, as in the case of *A. giganteum*. The flowers have 6 tepals, being free or slightly joined at the base. The umbel is enclosed in a sheath, often skin-like with membranous leaves, especially when young.

CULTURE Should be given rich, well-drained soil, although invasive species will thrive almost anywhere. Plant where they can wander at will. Do not set too deep, preferably at or above soil level, so sun can ripen bulbs. Bulbs native to the Himalayas require more moisture during summer than others; those from eastern Asia and Europe require adequate moisture, but, if in doubt, err on dry side.

Only one species is a good pot plant—*A. neapolitanum*. Pot in fall in sandy soil mix. Treat as you would daffodils, i.e., cool temperature, about 45°F at night until well rooted, then increase by 10–20 degrees.

PROPAGATION Easiest way is by offsets produced by larger bulbs. Some species produce bulbils in inflorescence which can be sown in containers, covering lightly with sandy soil mix. Most reach sufficient size for planting out in a few months but might need another season or two before flowering.

Raising species from seed is not difficult. It is best sown in the spring. If only small quantities are to be raised, then sowing in a container is to be preferred. A sandy soil mix with humus for moisture retention should be used and the seeds sown thinly over the surface, then barely covered with the soil mix. The seeds should be watered and then placed in an area of good light with temperatures in the 55°F range at night. After the seeds have germinated, keep moist but with good ventilation.

When the seedlings are large enough to handle, they can be placed in small containers or transplanted to other containers spacing the seedlings some 2–3 in. apart. During the summer they can be placed outdoors. When the foliage dies down the small bulbs can be allowed to dry and then retained in the container for planting out in the following spring, or in milder climates planted in the fall. Sizes suitable for planting out should be obtained in two seasons, frequently after one growing season.

If large quantities are to be raised, then preparing a bed outdoors in a sunny location is preferred. Such areas should be free of weed seeds, because when the seeds germinate it is not easy to distinguish them from grass seedlings. The soil is best augmented with some sand worked into the top inch or two of soil and the seed barely covered. Keep moist and allow to become drier only when the foliage has died down. If of suitable size the bulbs can be planted at the end of the growing season. If they need to be grown on, line them out for another year's growth in a well-drained, sunny area for another year of culture.

PESTS & DISEASES Onion fly, *Delia antiqua*, is a serious pest of the onion cultivated for food, *Allium cepa*. It also may be found on ornamental species. The attacks occur in spring and early summer, the maggots of the flies moving from bulb to bulb, reducing them to a decayed, liquid mess. The first sign of attack is the foliage turning yellow, then becoming whiter before falling and lying on the ground. The flies lay their eggs in the neck of the bulb or on young leaves. Within a few days the young maggots find their way into the bulbs and start their destruction. In three weeks the maggots are full size and pupate in a chestnut-brown puparia which remains just a few inches deep in the soil. Three broods, or more in warmer climates, are produced the last of the season overwintering in the soil. Infested bulbs should be destroyed, and in any area where the pest is observed should be sterilized; the use of an insecticide is recommended if such attacks on vegetable onions have been observed. The preventive sprays should be used as the growth emerges in the spring.

Certain thrips, causing distortion to the foliage, flecking, etc., can be a problem and as soon as any mottled foliage is seen the cause should be investigated and an insecticide used. The problem is more prevalent in warmer temperatures. This same thrip, *Thrips tabaci*, while occasionally found outdoors is more of a problem if the plants are being grown in a greenhouse.

Onion or Downy Mildew, *Peronospora destructor*, is more of a problem on vegetable onions but can attack ornamental onions as well. The leaves show damage at the tips, which then shrivel, making the plants unsightly. The fungus will overwinter in the bulbs and in the soil. If such attacks are noticed, the bulbs should be lifted at the end of the growing season, dusted with a fungicide, and the soil sterilized. The spraying of the foliage during the growing season may be necessary, using a fungicide.

White Rot, *Sclerotium cepivorum*, another problem of vegetable onions, may also attack ornamentals. The foliage of the plants yellow in the summer months, the roots of the plants rot, and the bulbs are covered with white, fluffy mycelium. Attacked plants should be treated with a fungicide applied onto the foliage during the growing season and the bulbs lifted and dusted before being replanted.

Smut, *Urocystis cepulae*, causes dark, blister-like streaks on the foliage and bulbs. This is more of a problem on young plants and use of a fungicide as a preventive is suggested if the presence of this problem on food crops of onions is known. The soil also should be sterilized.

While such pests and diseases are not so common on ornamental onions as on vegetable crops of onions, the grower is well advised to keep an eye open for such problems and as soon as noticed take the appropriate remedial action.

USES Ornamental *Allium* are interesting garden plants. Large-flowered species are attention-getting accents when set among annuals. Some species are excellent for covering even poor ground, but care must be taken to keep them under control. Flower heads are unusual and attractive in floral arrangements. Dried flowers also are excellent in arrangements, but stalks should be allowed to ripen fully before cutting.

SPECIES & CULTIVARS

A. aflatunense. Lovely species from central China; introduced by Van Tubergen from northern Iran, circa 1900. 30–36 in. in height. Dense, spherical umbel of purple-lilac flowers. 'Purple Sensation,' a selection from species, has deep violet-purple flowers. Excellent cut flower. Flowering late spring/early summer. Leaves 4 in. wide, 20–24 in. in height.

A. **atropurpureum.** Native to Bulgaria, Romania, and western Turkey; introduced 1800. Reaches 3 ft in height but strong stems do not need staking, even in exposed positions. Small, purple flowers produced early summer. Leaves narrow, lanceolate, 18 in. in length.

A. **caeruleum** (syn. *A. azureum*).. Native to bleak areas of Siberia; introduced 1830. Deep blue flowers which reach 2 ft in height. Leaves linear 10–18 in. long. Midsummer-flowering.

A. **cernuum** (syn. *A. recurvatum*).. Native to North America, mainly the Allegheny Mountains; introduced circa 1800, 8–18 in. in height. Pendant flowers. Light pink blossoms early in summer. Leaves several, linear and flat 1/8 to 3/16 in. wide, 6–10 in. long.

A. **christophii** (syn. *A. albopilosum*). Native to Turkestan; introduced 1901. Reaches 24 in. in height. Produces some of largest individual heads of any *Allium*, often 10–12 in. in diameter. Good cut flower. Color purple-metallic blue. 3–7 leaves, strap-shaped, 1 in. wide, 20 in. long. Early summer-flowering.

A. **cowanii** (form of *A. neapolitanum*). Native to southern Europe—Asia Minor; introduced 1828. 24 in. in height. Flower heads whitest of species. Flowering toward end of spring.

A. **flavum** (syn. *A. webbii*).. Native to southern Europe; introduced 1753. Only 12 in. in height. Bell-shaped flowers of glistening yellow. Midsummer-flowering. Good rock garden plant.

A. **giganteum.** Native to central Asia; introduced 1883. One of the most spectacular *Alliums*. Very large flower heads can reach melon-size and stand 48 in. above ground. Foliage 2 in. wide and often 30 in. or more long. Excellent summer bedding plant, especially when placed so lilac-blue flowers contrast with summer-flowering annuals. Flower heads often saved after flowering and used in dry arrangements.

A. **karataviense.** Native to central Asia, especially Turkestan; introduced 1878. Has most attractive foliage of genus, 4–6 in. wide, glaucous blue, and spreading over ground. Flowers white with hint of rose, carried on strong stems that are 6 in. in height. Flowering May–June.

A. **macleanii** (syn. *A. elatum*). Native of Bokhara; introduced 1882. Reaches 36–40 in. in height. Foliage broad, shiny, deep green. Lilac-purple flowers above foliage. Similar to *A. giganteum*. Late spring-flowering. Rare, but offered in some catalogs.

A. **moly** (Golden Garlic). Native of eastern Spain and southwestern France; introduced in 17th century, long in cultivation. Bright yellow flowers reach height of 12 in. above metallic blue-green leaves, which are flat and 2 in. wide. Flowers July. Excellent for both naturalizing and cutting. Once planted will spread rapidly so should be given room to expand.

A. **narcissiflorum** (syn. *A. pedemontanum*). Native to the Alps of Italy. Flowers bright rose, 6 in. in height. Early summer-flowering. 4–6 leaves, basal, linear, 1/8 in. wide.

A. **neapolitanum.** Native to northern Italy; introduced 1788. Frequently found growing in grassy areas. While also widespread in the Mediterranean region, its name, which means the Onion of Naples, most likely also indicates its primary habitat. It is regarded as one of the finest of the species due to its lovely white flowers and the absence of the strong onion aroma. Blooms in the spring, sending up strong flowering spikes that will reach up to 24 in. in height. The flowers are pure white and carried in umbels which are rather loose, with the individual flowers being up to 1 in. in diameter; as many as 30 flowers per umbel. While opening widely the flowers retain a cup shape, the tepals not quite reaching the horizontal position. The stamens are shorter than the tepals and are held in the center of each tepal making them quite prominent, contrasting well with the white flowers. The foliage is strap-shaped, keeled, 3/4 in. in width and 12–18 in. in length. Not the hardiest of the species, it will need protection in areas where temperatures drop to the low teens. Makes an excellent pot plant and can be forced, the bulbs treated like Narcissus *A. n. grandiflorum* is somewhat larger than the type.

A. **oreophilum.** (*A. ostrowskianum*). Native to Turkestan; introduced 1831. Dwarf onion, only 4–6 in. in height. Carmine-pink flowers. 'Zwanenburg,' introduced by Van Tubergen, has deeper colored flowers. Midsummer-flowering.

A. **pulchellum** (Sometimes listed as a subspecies of *A. carinatum*.). Native to the Mediterranean region. Reaches 24 in. in height. Very narrow foliage. Flowers reddish-carmine. 'Tubergen' differs from species in that it is only 12–18 in. in height and flowers are violet-pink. Late summer-flowering.

A. **rosenbachianum.** Native to northern Afghanistan; introduced in 1894. White stamens contrast with dark violet flowers on 24-in. spikes. *A. r. album* silvery-white. Early summer-flowering. Foliage linear, lanceolate, to 2 in. wide, 20 in. long.

A. **schubertii.** Native to Palestine; introduced 1843. 12–24 in. in height. Leaves distinctive in rose-pink flowering variety—wavy and 1 in. wide. Summer-flowering.

A. **siculum** (syn. *A. dioscoridis*; now generally listed in the genus *Nectaroscordum*). Native to southern France and Sicily; introduced circa 1700. Needs full sun and little moisture during summer. 30 in. in height. Unusual species with light greenish-white or maroon flowers with green reverse to petals. Var. *dioscoridis* (honey garlic) has white flowers tinged with green and red. *A. bulgaricum* is similar to *A. siculum*.

A. **sphaerocephalon.** Very widely distributed, from Great Britain, throughout Europe, to Iran; introduced 1594. Purple-crimson flowers carried on 36-in.-high stems. Appreciates summer moisture. Foliage cylindrical, hollow, 24 in. long. Flowers June–Aug.

A. **stipitatum.**. Native of Turkestan, Afghanistan, and central Asia. Vigorous species that produces large heads of lilac-purple flowers on stems that may reach 48 in. in height. Flowering late spring. Excellent cut flower species. Strap-like basal leaves 2 in. wide.

A. **subhirsutum** (syn. *A. ciliatum*).. Native to southern Europe and Ethiopia; introduced 1753. White, sweet-scented flowers on 8–12 in. stems. Flowering late spring or early summer. Often used as cut flower, being one of few *Allium* forced into flower. Linear leaves, to 1/8 in. wide, flat, 8–10 in. long.

A. **triquetrum.** European species; introduced 1789. Has become a rampant weed in some parts of the United States, especially where mechanical cultivation takes place. Due to rapid propagation, should be grown only where spreading can be restricted. Has attractive white flowers with green striped petals. Reaches height of 18 in. Stem unique among *Allium* in that it is three-sided. Flowering late spring. Leaves basal, linear, dark green, 10–15 in. long to 1½ in. wide.

A. **ursinum.** Native to Europe and Asia; long in cultivation. Reaches height of 12 in. White flowers in spring. Strong garlic odor. Naturalizes well, even in quite dense shade. Likes summer moisture.

ALLIUM

A. acuminatum Native to Western parts of Northern California & British Columbia, Idaho, Colorado, Arizona. Introduced in 1840. Rose-purple. 6–10 in. in height. May–June flowering.
A. acutiflorum Native to Europe. Purple. 5–20 in. in height. June–Aug. flowering.
A. acutangulum syn. of *A. angulosum*.
A. akaka (syn. *A. latifolium*) Native to Iran. Introduced 1830. Rose. 2–3 in. in height. Spring flowering.
A. albidum (syn. *A. denudatum*, *A. flavescens*) Native to Siberia & Southern U.S.S.R. White. 10–12 in. in height. Summer flowering.
A. albopilosum see *A. christophii*.
A. algirdense Native to N. Iraq & S.E. Turkey. White. 4–12 in. in height. Aug.–Sept. flowering.
A. alexeianum Native to Turkestan. Introduced 1889. Whitish. 2–4 in. in height. Summer flowering.
A. altissimum Native to Russian Central Asia, N.E. Iran & N.W. Afghanistan. Pinkish-purple. 36–60 in. in height. Summer flowering.
A. amabile Native to N.W. Yunnan. Introduced 1922. Magenta-crimson. 4–8 in. in height. Autumn flowering.
A. ambiguum syn. of *A. obtusum*.
A. amblyophyllum Native to Turkestan. Introduced 1885. Lilac. 12 in. in height. Summer flowering.
A. amethystinum (syn. *A. descendens*). Native to Central Europe. Purple. 10–50 in. in height. May–June flowering.
A. ammophilum (syn. *A. flavescens*). Native to Serbia, Bulgaria. Golden yellow. 8–12 in. in height. May–June flowering.
A. ampeloprasum Native to S. Europe, W. Asia. Introduced 1753. Purple to whitish. 36–72 in. in height. May–June flowering.
A. ampeloprasum var. *porrum* (syn. *A. porrum*). Segments with green or reddish midrib.
A. amplectens Native to California. White to pinkish. 8–20 in. in height. March–June flowering.
A. anceps Native to Nevada to Southern Oregon. Pale rose with purplish veins. 3–4 in. in height. April–June flowering.
A. anceps var. *lemmonii* Native to Northern California, Western Nevada, S.W. Idaho. Whitish to pale rose. 4–8 in. in height. May–June flowering.
A. angulosum (syn. *A. acutangulum*) Native to Europe to Siberia. Lilac-purple or white. 8–18 in. in height. June–Aug. flowering.
A. ascalonicum Native to Syria. Introduced 1546. Pink, lilac. 9 in. in height. Summer flowering.
A. atrorubens Native to Western U.S.A. Dark-red, purple. 3–6 in. in height. May–June flowering.
A. a. var. *inyonis*. Pale with dark midveins.
A. atroviolaceum Native to Europe to Afghanistan. Introduced 1846. Purple. 40 in. in height. May–June flowering.
A. attenuifolium Native to Western South America. White to pink. 6–15 in. in height. Spring flowering.
A. austinae syn. of *A. campanulatum*.
A. azureum see *A. caeruleum*.
A. backhousianum Native to the Himalayas. Introduced 1885. White. 36–48 in. in height.
A. balansae Native to Turkey. Introduced 1882. Lilac-pink. 2–3 in. in height. August flowering.
A. beesianum Native to Yunnan, Western China, Szechwan. Blue. 16 in. in height. Late Spring flowering.
A. bidwelliae syn. *A. campanulatum*.
A. bisceptrum Native to N.W. U.S.A. Rose-purple. 4–12 in. in height. May–June flowering.
A. bolanderi Native from Southern Oregon to Northern California. Rose-purple to white. 4–10 in. in height. May–June flowering.
A. breviscapum Native to Iran. Rose, leaves narrow, hairy beneath. 2–3 in. in height. Spring flowering.
A. brevistylum Native to Montana, Colorado, Utah. Dark rose. 12–24 in. in height. Spring flowering.
A. breweri Native to California. Introduced 1882. Deep rose. 1–3 in. in height. July flowering.
A. bulgaricum (Now generally listed in the genus *Nectaroscordum*). Native to E. Bulgaria, W. Turkey, Cyprus. Introduced 1873. White, tinged green & red. 30 in. in height. Summer flowering.
A. bullardii syn. of *A. campanulatum*.
A. bulleyanum Native of W. China. Deep maroon. 24 in. in height. Summer flowering.
A. burlewii Native to California. Pinkish purple. 2–3 in. in height. May–July flowering.
A. cabulicum Native to Kabul. Introduced 1892. Whitish. 3–4 in. in height.
A. callidycton Native to Iraq, Iran & Turkey, Caucasia. Introduced 1843. Pink. 6–12 in. in height. Aug. flowering.
A. callimischon Native to Greece & Crete. Papery white. 6–12 in. in height. Sept.–Nov. flowering.
A. c. ssp. *callimischon* Native to the Peloponnese. Unspotted.
A. c. ssp. *haemostictum* Native to Crete. Introduced 1978. Red spotted petals.
A. calocephalum Native to Iraq. Introduced circa 1975. Creamy white. 12–30 in. in height.
A. campanulatum (syn. *A. austinae, A. bidwelliae, A. bullardii*). Native to California. Introduced 1880. Bright rose. 2–3 in. in height. July flowering.
A. capitellatum Native to Turkey & Iran. Introduced 1846. Greenish-white or brownish pink. 4–12 in. in height.
A. cappadocicum Native to Turkey. Introduced 1882. Pinkish-white. 8–12 in. in height. July flowering.
A. cardiostemon (syn. *A. trilophostemon*) Native to Turkey, Iran, Caucasus. Introduced 1840. Blackish maroon. 6–16 in. in height. June–July flowering.
A. carinatum Native to W. & S. Europe. Introduced 1753. Purple. 12–24 in. in height. June–Aug. flowering.
A. caspium Native from Iran to U.S.S.R. White or pinkish. 8–10 in. in height.
A. chamaemoly Native to Mediterranean Region. White. 2 in. in height. Nov.–Feb. flowering.
A. chlorurum (syn. *A. tauricolum*) Native to Turkey. Introduced 1882. Violet, veined. 2–3 in. in height. August flowering.
A. chrysantherum Native to Turkey, Iran, Iraq, Syria. Introduced 1882. Greenish yellow or yellow. Ovary black. 12–40 in. in height. April–June flowering.
A. ciliatum see *A. subhirsutum*.
A. circinatum Native to Crete. Whitish striped pink. 2½–2¾ in. in height. March–April flowering.
A. colchicifolium Native to Iraq, Iran, Turkey. Introduced 1859. Whitish or green, segments veined. 2–8 in. in height. May flowering.
A. commututum Native to S.E. Europe. Introduced 1854. Pinkish. 20–40 in. in height. May–June flowering.
A. concinnum syn. of *A. obtusum*.
A. controversum syn. of *A. pyrenaicum*.
A. cratericola Native to California. Pale to dark purple. 1–3 in. in height. March–May flowering.
A. crenulatum Native to Vancouver, B.C., Washington & Oregon. Rose-purple. 3 in. in height. Summer flowering.
A. crispum Native to Central California. Reddish purple. 4–12 in. in height. March–May flowering.
A. cupanii Native to S.E. Europe, Turkey, Balkans. Introduced 1810. Pink or nearly white. 5–10 in. in height. May–Oct. flowering.
A. cyaneum Native to Kansu, China. Introduced 1890. Blue, bell-shaped. 6–12 in. in height. Summer flowering.
A. cyathoporum var. *farreri* Native to Kansu, China. Introduced 1914. Red-purple. 12–15 in. in height. Summer flowering.

A. davisiae Native to California. Pale with red midveins. 8–16 in. in height. April–May flowering.
A. decipiens Native to Turkey, S. Russia. Introduced 1830. Pale pinkish-purple or whitish. 8–20 in. in height. June flowering.
A. denudatum syn. of *A. albidum*.
A. descendens syn. of *A. amethystinum*.
A. dichlamydeum (Variety of *A. serratum*). Native to N. Coastal Ranges of California. Deep rose-purple. 4–12 in. in height. May–July flowering.
A. dilatum (similar to *A. guttatum*). Native to S.W. Crete. White with green line on each tepal. 6–20 in. in height. May–June flowering.
A. dioscoridis see *A. siculum*.
A. douglasii Native to U.S.A. Rose-pink. 8–10 in. in height. Summer flowering.
A. dregeanum Native to Namaqualand, Karoo. White to pink. 28–32 in. in height. Summer flowering.
A. elatum see *A. macleanii*.
A. ellisii Native to Iran, Asia Minor. Introduced 1903. Bright rose. 12 in. in height. July flowering.
A. erdelii syn. of *A. orientale*.
A. ericetorum syn. of *A. ochroleucum*.
A. eriophyllum syn. of *A. longisepalum*.
A. falcifolium Native to Western U.S.A. Introduced 1882. Pale rose. 2–3 in. in height. Summer flowering.
A. fallax syn. of *A. montanum* var. *petraeum*.
A. farreri syn. of *A. cyathoporum* var. *farreri*.
A. fetisowii Native to Turkestan. Introduced 1879. Rosy–lilac. 24 in. in height. Summer flowering.
A. fimbriatum Native to California. Introduced 1928. Rose to purple. 2–3 in. in height. Summer flowering.
A. flavescens syn. of *A. albidum* & *A. ammophilum*.
A. flavidum Native to Central Asia. Bright yellow. 12 in. in height. Summer flowering.
A. flavum var. *minus* Native to S. Europe, Turkey. Bright yellow. 3–4 in. in height. Spring flowering.
A. fragrans syn. of *Nothoscordum inodorum*.
A. frigidum Native to Greece, Turkey. Introduced 1853. Pink or white. 6–10 in. in height. July–Sept. flowering.
A. geyeri (syn. *A. pikeanum*). Native to Western U.S.A. Pink. 3–12 in. in height. Summer flowering.
A. glaucum syn. of *A. montanum*.
A. globosum syn. of *A. saxatile*.
A. glumaceum Native to Central Turkey. Introduced 1882. Purplish rose. 4 in. in height. Summer flowering.
A. grandisceptrum syn. of *A. unifolium*.
A. guttatum (syn. *A. margaritaceum*). Native to Mediterranean, S.E. Europe, Turkey. Introduced 1809. Whitish with purple blotch. 6–20 in. in height. May–Aug. flowering.
A. haemanthoides Very much like *A. akaka*. Native to Turkey & Iran. Introduced 1875.
A. haematochiton Native to California. White to rose. 4–12 in. in height. March–May flowering.
A. hickmanii Native to California, White to pink. 4–6 in. in height. April flowering.
A. hirtifolium Native to W. & S.W. Iran. Introduced 1840. Purple & white. To 48 in. in height. Spring flowering.
A. hoeltzeri Native to Turkestan. Introduced 1884. White, anthers red. 5–7 in. in height. Spring flowering.
A. hoffmanii Native to Northern California. Purplish-pink. 1–2 in. in height. June–July flowering.
A. howellii Native to Central California. Pale lilac-violet to deep rose. 4–12 in. in height. April–May flowering.
A. hyalinum Native to Central California. White or pinkish. 4–12 in. in height. March–June flowering.
A. inderiense similar to *A. narcissiflorum*. Native to S.E. Russia. Bright rose. Early summer flowering.
A. insubricum Related to *A. narrcissiflorum*. Native to Lombardy Alps, N. Italy. Rose, anthers orange. 6–12 in. in height. Summer flowering.
A. jajlae syn. of *A. scorodoprasum* ssp. *jajlae*.
A. jesdianum Native to Iraq & W. Iran. Purplish violet. 20–30 in. in height. May–July flowering.
A. jubatum Native to Bulgaria, Turkey. Introduced 1918. Outer segments purple, inner paler pinkish. 12–30 in. in height. May–July flowering.
A. kansuense Native to Kansu, China. Introduced 1889. Violet. 4–12 in. in height. Late summer flowering.
A. kharputense Native to Turkey & W. Iran. Introduced 1982. White. 10 in. in height. May flowering.
A. kunthianum Native to Turkey, Iran, Caucasus. Introduced 1935. Pale pink with purple midveins. 6–10 in. in height. Early summer flowering.
A. latifolium syn. of *A. akaka*.
A. lehmannianum Native to Aral Desert. Rose-pink with purple midribs. 2–3 in. in height. Summer flowering.
A. lemmonii Native to N. California. Whitish to pale rose. 4–8 in. in height. May–June flowering.
A. libani Native to Lebanon, Syria. Pinkish. 6 in. in height. Late spring flowering.
A. longicuspis Native to Turkey, Iran, U.S.S.R. Introduced 1875. White or pink. 16–44 in. in height. Early summer flowering.
A. longisepalum (syn. *A. eriophyllum*). Native to E. Turkey, N. Syria, Iraq, S. & W. Iran. Introduced 1846. Pinkish. 14 in. in height. May–June flowering.
A. loratum Native to Kashmir. Creamy white. 3–6 in. in height.
A. macnabianum Native to N. America. Deep magenta. 12 in. in height. Summer flowering.
A. macranthum (syn. *A. oviflorum*). Native to Sikkim, N. India. Introduced 1883. Mauve-purple. 6 in. in height. Summer flowering.
A. macrum Native to Oregon, Washington. White, purplish veins. 2–3 in. in height. Summer flowering.
A. magicum syn. of *A. multibulbosum*.
A. mairei (syn. *A. yunnanense*). Native to Yunnan, W. China. Pale rose, pink dots. 4–8 in. in height. Autumn flowering.
A. margaritaceum syn. of *A. guttatum*.
A. melanantherum (syn. *A. thracicum*) Native to Bulgaria, N. Greece, E. Yugoslavia, Turkey. Introduced 1883. Pink with reddish midvein on segments. 6–16 in. in height. Summer flowering.
A. membranaceum Native to N. California. Whitish to pink. 6–14 in. in height. May–July flowering.
A. meteoricum Native to Albania, N. & Central Greece, S. Yugoslavia. Pink or white, anthers orange. 6–10 in. in height. June–Sept. flowering.
A. mirum Native to Afghanistan, Iran. Introduced 1964. Brownish purple or whitish. 6–10 in. in height. May–June flowering.
A. montanum (syn. *A. glaucum*, European form of *A. senescens*) Native to Europe, N. Asia. Introduced 1823. Lilac-purple. 8–20 in. in height. Summer flowering.
A. m. var. *petraeum* (syn. *A. fallax*). Dwarfer than type, 6 in. in height.
A. monticola Native to S. California. Pink. 2–6 in. in height. May–July flowering.
A. moschatum Native to S. Europe. Introduced 1753. Pink or white. 6–10 in. in height. June–Sept. flowering.
A. multibulbosum (syn. *A. magicum*) Related to *A. nigrum*. Native to Mediterranean. White with reddish veins. 12–24 in. in height. May–June flowering.

A. murrayanum Close to *A. acuminatum* Native to North America. Rose-purple. 8–14 in. in height. May–June flowering.
A. mutabile Native to North America. Introduced 1824. White, turning rose. 12–24 in. in height. July flowering.
A. myrianthum Native to Iraq, Iran, Turkey. Introduced 1844. White. 10–30 in. in height. May–June flowering.
A. neriniflorum see *Caloscordum neriniflorum*.
A. nevadense Native to Western U.S.A. White or pale rose. 3–6 in. in height. April–June flowering.
A. nigrum Native to Mediterranean region. Introduced 1762. White, green midribs to pinkish-purple. 20–40 in. in height. April–June flowering.
A. noeanum Native to W. Iran, Iraq, Syria, S.E. Turkey. Introduced 1875. Pink. 6–12 in. in height. May–June flowering.
A. nutans Native to Siberia & Central Asia. Pink. 18–36 in. in height. June–July flowering.
A. obliquum Native from Romania to Siberia & N.W. China. Yellow. 18–36 in. in height. April–July flowering.
A. obtusum (syn. *A. ambiguum, A. concinnum*). Native to California. Greenish white, purple-rose midveins. 2–3 in. in height. May–July flowering.
A. ochroleucum (syn. *A. ericetorum*) Native to Italy. Yellow to whitish. 6–12 in. in height. Summer flowering.
A. odorum syn. of *A. ramosum* & *A. tuberosum*.
A. oleraceum Native to England, Central Europe to Siberia. Whitish tinged with green, pink or brown. 10–32 in. in height. July–Sept. flowering.
A. orientale (syn. *A. erdelii*). Native to Syria, Turkey, E. Aegean, Mediterranean region. White or reddish segments with green midvein. 12–18 in. in height. April–June flowering.
A. ostrowskianum see *A. oreophilum*.
A. oviflorum syn. of *A. macranthum*.
A. oxyphilum Virginia form of *A. cernuum*. Flower pure white or pale pink.
A. pallens Native to Turkey, S. Europe, Mediterranean. Introduced 1762. White on pink, anthers yellow. 4–16 in. in height. May–July flowering.
A. palmeri Native to New Mexico. Rose-pink. 8 in. in height. Summer flowering.
A. paniculatum Native to Turkey, Iraq, Iran, Mediterranean region of S. Europe. Introduced 1759. Purplish rose. 8–20 in. in height. June–Oct. flowering.
A. paradoxum Native to Caucasus, Northern Iran. Introduced 1823. White. 6–12 in. in height. April–May flowering.
A. parciflorum (syn. *A. pauciflorum*). Native to Corsica, Sardinia. Introduced 1888. Purple. 4–10 in. in height. May–Oct. flowering.
A. parishii Native to S. California & W. Arizona. Pale pink. 4–8 in. in height. April–May flowering.
A. parnassicum Native to S. Greece. Purplish rose. 4–10 in. in height. June–Oct. flowering.
A. parryi Variety of *A. fimbriatum* Native to California. White to rose-purple. 3–6 in. in height. June–July flowering.
A. parvum Native to Oregon, Idaho, Utah, Nevada. Rose-purple or pink with purple midrib. 1–2 in. in height. April–July flowering.
A. pauciflorum syn. of *A. parciflorum*.
A. pedemontanum see *A. narcissiflorum*.
A. pendulinum (syn. *A. triquetrum pendulinum*) Native to Corsica, Italy, Sardinia, Sicily. White with green stripe. 18 in. in height. March–June flowering.
A. peninsulare Native to California. Deep rose-purple. 8–16 in. in height. March–June flowering.
A. pervestitum Native to South Western Russia to Southern Ukraine. Whitish or yellowish (sometimes tinged pink). 12–32 in. in height. June–July flowering.
A. phanerantherum Native to Turkey, Syria, Iraq & Iran. Introduced 1882. White with green vein, sometimes purplish. 10–34 in. in height. Summer flowering.
A. phthioticum Native to S. Central Greece. White. 6–22 in. in height. April–June flowering.
A. pikeanum syn. of *A. geyeri*.
A. pilosum Native to S. Greece—Cyclades. Lilac. 4–6 in. in height. June–Oct. flowering.
A. platycaule Native to California, Oregon. Deep rose. 1–5 in. in height. May–Aug. flowering.
A. polyastrum Native to Western China. Magenta-rose. 24 in. in height. Summer flowering.
A. porrum syn. of *A. ampeloprasum* var. *porrum*.
A. praecox Native to California. Rose-purple. 8–20 in. in height. March–April flowering.
A. proponticum Native to Turkey. Introduced 1977. Pink to reddish purple. 36+ in. in height. June–Aug. flowering.
A. protensum Native to Afghanistan & Russian Central Asia. Brown. Less than 24 in. in height. May–June flowering.
A. pruinatum Native to Portugal. Pink to reddish purple, sometimes white. 8–16 in. in height. June–Aug. flowering.
A. przewalskianum Native to Kansu, China. Introduced 1889. Rosy-lilac. 6–10 in. in height. Summer flowering.
A. pseudoflavum Caucasian form of *A. flavum*.
A. pulchellum var. *valdensium* Native to Maritime Alps, France. Bright rose. Smaller than type. Late summer flowering.
A. purdomii syn. of *A. cyaneum*.
A. pyrenaicum (syn. *A. controversum*) Native to Pyrenees Mts. of E. Spain. Whitish. 18–36 in. in height. June–Aug. flowering.
A. pyrrhorhizum a close ally of *A. mairei*.
A. ramosum (syn. *A. odorum*) Native to Siberia. White. 10–20 in. in height. June–July flowering.
A. recurvatum see *A. cernuum*.
A. regelianum Native to W. & Central Russia. Pink to reddish purple. 12–36 in. in height. June–Aug. flowering.
A. regelii Native to N.E. Iran, Afghanistan & Russian Central Asia. Pink or dark purple. 36–40 in. in height. May–June flowering.
A. reticulatum syn. of *A. textile*.
A. roseum Native to European Mediterranean region. Introduced 1752. Bright pink. 6–22 in. in height. April–June flowering.
A. rotundum Native to Turkey, Europe, Syria, Iraq, Iran, Caucasus. Introduced 1762. Bicolored: outer segments deep purple; inner pinkish purple. 12–32 in. in height. May–July flowering.
A. rouyi Native to S.W. Europe. Yellowish. 5–10 in. in height. May–Oct. flowering.
A. rubellum Native to S.E. Russia. Pink or white, anthers yellow. 6–10 in. in height. June–Sept. flowering.
A. rubens (similar to *A. geyeri*) Native from Urals to E. Siberia. Rosy violet. 3–12 in. in height. July flowering.
A. rubrovittatum Native to Crete & Karpathos. Reddish purple with pale margins. 12–32 in. in height. May–July flowering.
A. rupestre Native to Crimea, Caucasus, Turkey. Introduced 1812. White or pink with purple anthers. 4–16 in. in height. May–July flowering.
A. sabulosum Native to S.E. Russia, Central Asia. Greenish or whitish. 6–20 in. in height. May–Aug. flowering.
A. sanbornii Native to N. California. White or pink. 10–12 in. in height. June–Sept. flowering.
A. sardoum Native to Sardinia. White. 12–36 in. in height. June–Aug. flowering.
A. sativum (Garlic)

A. saxatile (syn. *A. globosum*) Native to S.E. Europe. Introduced 1798. Deep pink or whitish. 4–14 in. in height. July–Sept. flowering.
A. scaberrimum Native to S.E. France. White. 36–72 in. in height. May–July flowering.
A. scabriscapum Native to Iran, Iraq, Russia, Turkey. Introduced 1854. Yellow with green vein on segments. 5–20 in. in height. July–Sept. flowering.
A. schmitzii (Chive Group) Native to Portugal. Lilac to pale purple. 4–18 in. in height. May–Aug. flowering.
A. schoenoprasum (Chives) Widespread in Northern Hemisphere. Introduced 1753. Lilac or pale purple. 4–18 in. in height. May–Aug. flowering.
A. schubertii Native to E. Mediterranean. Introduced 1843. Pink. 12–24 in. in height. April–May flowering.
A. scorodoprasum Native to Europe, Asia Minor. Introduced 1753. Lilac to purplish. 16–36 in. in height. May–June flowering.
A. s. ssp. *jajlae* Native to Crimea & Caucasus. Reddish-purple. 24 in. in height. June–July flowering.
A. scorzonerifolium Native to S.W. Spain, Portugal. Bright yellow. 9 in. in height. June–July flowering.
A. semenowi Native to Russia, Kashmir. Introduced 1884. Yellow. 4–12 in. in height. Summer flowering.
A. senescens Native to Europe. Pale pink to purple. 4–30 in. in height. July–Sept. flowering.
A. serratum Native to N. California. Pink to purplish. 12 in. in height. March–May flowering.
A. sibiricum robust variety of *A. schoenoprasum*.
A. sikkimense Native to W. China, N. India. Blue-purple. 4–12 in. in height. Summer flowering.
A. sindjarense Native to Turkey, Syria, Iran, Iraq. Introduced 1875. Pink to purplish, dark stripe on segments. 4–10 in. in height. Summer flowering.
A. sipyleum Native to Crete, Greece, Turkey. Introduced 1844. Pinkish, darker stripe on segments. 4–10 in. in height. June–Oct. flowering.
A. siskiyouense Native to S. Oregon, N. California. Rose. 1–3 in. in height. June–July flowering.
A. stamineum Native to Balkans, Turkey, Iran, Iraq. Introduced 1859. Pale pinkish-purple, darker midveins on segments. 4–12 in. in height. May–Aug. flowering.
A. staticiforme Native to S. Greece, Crete, Turkey, Aegean Ils. Introduced 1809. Whitish or pinkish with red veins on segments. 4–12 in. in height. May–June flowering.
A. stellatum Native to Eastern Central U.S.A. Purplish to rose-pink. 12 in. in height. Autumn flowering.
A. stocksianum Native to S. Afghanistan & Beluchistan. Rosy-violet. 2 in. in height. July flowering.
A. stracheyi Native to Kamach. Red-purple. 12 in. in height. Summer flowering.
A. suaveolens Native to Europe: France to Yugoslavia. Pink to whitish. 8–18 in. in height. June–Aug. flowering.
A. subvillosum Native to Balearic Isles, Greece, Spain, Portugal, Sicily, N. Africa. White. 4–12 in. in height. March–May flowering.
A. suworowi (Similar to *A. rosenbachianum*) Native to Central Asia. Dark mauve-purple. 24 in. in height. Summer flowering.
A. talijevii Native to S.W. Russia. White. 12–36 in. in height. June–Aug. flowering.
A. tanguticum Native to China. Lilac. 8–10 in. in height. Summer flowering.
A. tardans Native to Crete. Pink. 12 in. in height. June–Aug. flowering.
A. tauricolum syn. of *A. chlorurum*.
A. textile (syn. *A. reticulatum*) Native to Western N. America. White to pink. 4–8 in. in height. May–June flowering.
A. thracicum syn. of *A. melanantherum*.
A. tibeticum Native to Tibet & Kansu. Deep blue or purple. 6 in. in height. Summer flowering.
A. tolmiei Native to Pacific northwest. Pale rose. 2–5 in. in height. April–June flowering.
A. trachycoleum Native to Europe, N. Africa to Turkey. Introduced 1753. Pink to reddish-purple, sometimes white. 32 in. in height. June–Aug. flowering.
A. tribracteatum Native to California & Oregon. Pale rose. 2–5 in. in height. April–July flowering.
A. tricoccum American counterpart of *A. ursinum,* which see.
A. trifoliatum Native to Mediterranean: France to Sicily, Crete & Israel. White or pinkish. 6–18 in. in height. April & May flowering.
A. trilophostemon syn. of *A. cardiostemon*.
A. tubergenii Native to Asia Minor. Introduced 1900. Grayish white. 24 in. in height. Summer flowering.
A. tuberosum (syn. *A. odorum*) Native to India to Japan. White. 10–20 in. in height. July–Oct. flowering.
A. unifolium (syn. *A. grandisceptrum*) Native to California. Introduced 1873. Rose. 8–24 in. in height. April–June flowering.
A. validum Native from N. California to British Columbia. Introduced 1881. Rose or white. 12–30 in. in height. July–Sept. flowering.
A. victorialis Native to E. Europe: Germany to the Himalayas. Introduced 1739. White or yellowish. 12–24 in. in height. May–Aug. flowering.
A. vineale Native to Europe & N. America. Introduced 1753. Pinkish, reddish or greenish. 12–36 in. in height. June–Aug. flowering.
A. viride Native to N. Iraq, Iran, Caucasus. Introduced 1928. Green with white margin. 24–60 in. in height. June–Aug. flowering.
A. waldsteinii Native to N.E. Italy, Crimea, N. Romania, Russia, N. Yugoslavia. Dark purple. 12–32 in. in height. May–July flowering.
A. wallichii Native to Himalayas. Rose-purple. 6–30 in. in height. Aug.–Sept. flowering.
A. watsonii Native to Oregon, California. Rose-purple. 3 in. in height. Aug.–Sept. flowering.
A. webbii see *A. flavum*.
A. yosemitense Native to Sierra Nevada Mts. Pale rose. 8–12 in. in height. June–July flowering.
A. yunnanense syn. of *A. mairei*.
A. zebdanense Native to E. Mediterranean: Syria to Israel. Introduced 1859. White. 8–24 in. in height. May–June flowering.

ALOPHIA — IRIDACEAE

While some authorities regard this as a separate genus, most now include these plants in the genus *Herbertia*, which see. It seems unlikely that the genus can be as widespread as often listed, that is, from Texas to Argentina and Chile. The species *Alophia drummondiana* is also referred to as *A. drummondii*, syn. with *Cypella drummondii* and *Herbertia drummondii*. It would seem that placing them all in the genus *Herbertia* would simplify the question. Thus, in my opinion, a closer look at this genus, as well as *Herbertia* and *Cypella*, is merited.

ALSTROEMERIA — ALSTROEMERIACEAE

Linnaeus named this genus in honor of his friend Baron Clas Alstroemer (1736–94). There are some 50 species of *Alstroemeria*. Commonly known as the Peruvian lily, even though few species actually are native to Peru, being more commonly found in Chile and Brazil. While long known in cultivation, these South American plants have been hybridized extensively only in the last 20 years. It was the reintroduction of *A. ligtu* by Harold Comber some years ago that initiated the modern development of these remarkable plants. A new and different type of marketing has developed with this increased interest. The breeders lease the plants to other growers, thus enabling them to keep a tight control over their cultivars.

The white, fleshy roots (rhizomes) are brittle, a characteristic that must be remembered because, until well established, care must be taken when transplanting. One species, *A. haemantha*, furnishes an edible farine when the roots are dried and crushed. These midsummer-flowering plants reach a height of 2 ft or more. The thin, linear, lanceolate leaves, 2–4 in. in length, are carried on the stem of the flower spike. The flowers are corymbs, although a few species, not of great horicultural value, are solitary. They are mostly zygomorphic, nontubular, and brightly colored. They are surprisingly hardy but do need protection, such as a thick mulch, in areas where the temperatures drop to near 0°F. For this reason they should be planted at a good depth.

CULTURE Handle roots with care until plants are well established. Quite hardy, especially when protected by winter mulch. Surprisingly easy to grow in wide range of soils. Adequate supply of moisture necessary during growing season. Good drainage and full sun also essential if they are to flourish. Fleshy roots can withstand drought but performance will be poor. Plant about 1 ft apart and some 8 in. deep outdoors. Can be left undisturbed for years. When overcrowding occurs lift and separate and replant tuberous roots with crowns. Very fine bed at University of California's Berkeley Botanic Garden, which is rotovated, chopping up the roots. Plants seem to thrive with such treatment but not recommended until beds have been established for at least 7 years. Some shade necessary in warmest climates to preserve pastel colors of newest hybrids.

Container planting requires pots at least 8–10 in. deep with good drainage. Soil mix should be well-draining type. Stems quite long so will require staking.

PROPAGATION Can be raised from seed as soon as possible after ripening to obtain quick and good germination; older seed may take several seasons. Established plants can be lifted in early fall, as soon as foliage begins to die back. Roots with crowns can be separated and planted back. Can be treated similarly in spring if early fall lifting and dividing is not possible.

PESTS & DISEASES Young shoots favorite food of slugs and snails. Mites will attack foliage, especially in greenhouses; can be kept in check by frequent syringings with water.

USES Ideal plants for a slope, where, once established, will increase, especially if adequate summer moisture and good drainage available. Excellent cut flowers and container plants. Care should be taken in purchasing new, named hybrids as some that are good greenhouse cut flower varieties may not be suitable for gardens.

SPECIES & CULTIVARS

A. aurantiaca. Native to Chile; introduced 1831. 36 in. in height. Flowers numerous, often as many as 40–50 in one head, as much as 1½ in. in diameter. Outer petals blunt, tinged red with green tip; inner petals pointed, orange streaked with red. Foliage on the flowering stem lower 3–4 in. long, lanceolate, gray-green beneath. Many cultivars offered by nurseries in wide color range. *A. a.* var. *lutea* is yellow; 'Orange King' is orange. *A. aurea* is a variety of *A. aurantiaca*—golden form.

A. ligtu. Native to Chile; introduced 1838. Not so tall as *A. aurantiaca* and produces fewer flowers per stem. Color range large. Outer segments of flowers vary from pale lilac to reddish to white; inner petals usually yellow with purple, foliage thin, narrow, linear, 3 in. long. Var. *angustifolia* has narrower leaves with pink flowers; var. *pulchra* has larger, more purplish flowers. *A. ligtu* is known as St. Martin's flower.

A. pelegrina. Native to Chile; introduced 1754. Not so tall-growing as other species; only 12–14 in. in height. Flowers longer than other species; extremely varied in color, from rose to lilac to yellow. The pure white form, var. *alba*, was introduced 1879, and is commonly known as Lily of the Incas. Undoubtedly will take back seat with advent of new hybrids. Of value mainly to hybridizers. Leaves lanceolate, up to 2 in. long on flower stem.

A. pulchra (syn. *A. tricolor*). A native of Chile that reaches to 18 in. in height and flowering in summer. Several flower heads are produced, each with 2 or 3 flowers. The foliage is linear, lanceolate, dark green. The outer perianth segments are white, broadly obovate; the upper inner segments are narrow and longer, dotted with yellow on the white and tipped with red or reddish-purple, this color also can be found sometimes at the very tips of the outer perianth segments but is often faint. The yellow on these narrower perianth segments is sometimes more prominent toward the apex and can give the impression of a separate band of color. The plant was first named *A. tricolor*, because of this rather elusive character of having three bands of color.

ALSTROEMERIA

A. bicolor var. of *A. ligtu*.
A. brasiliensis Native to Central Brazil. Reddish yellow, spotted brown. 36–48 in. in height. Summer flowering.
A. campaniflora Native to Brazil. Introduced 1932. Green, dark spots. 36 in. in height. Summer flowering.
A. caryophyllea Native to Brazil. Introduced 1776. Scarlet or scarlet striped. 8–12 in. in height. Summer flowering.
A. chilensis Native to Chile. Introduced 1842. Blood-red or pink, lined yellow. 24–36 in. in height. Summer flowering.
A. haemantha (syn. *A. simsii*) Native to Chile. Introduced 1822. Red & yellow spotted lilac. 30–36 in. in height. June–July flowering.
A. hookeri (dwarf form of *A. ligtu*) Native to the Andes. Introduced 1836. Pink. 6 in. in height. Summer & early Autumn flowering.
A. pallida a form of, or nearly related to *A. ligtu*.
A. psittacina syn. of *A. pulchella*.
A. pulchella (syn. *A. psittacina*) Native to N. Brazil. Introduced 1829. Reddish. 36 in. in height. June–Sept. flowering.
A. pygmaea Native to Straits of Magellan, South America & the Andes. Whitish. Dwarf. Summer flowering.
A. revoluta related to *A. ligtu* Native to Chile. Purple. Summer flowering.

A. rosea form of, or related to *A. ligtu*.
A. simsii syn. of *A. haemantha*.
A. spathulata Native to Chile. Reddish with yellow blotches. 12 in. in height. Summer flowering.
A. tricolor syn. of *A. pulchra*.
A. versicolor Native to Chile. Yellow with purple spots. 12 in. in height. Late Summer flowering.
A. violacea Native to Chile. Bright lilac. 12–24 in. in height. Early Summer flowering.

AMANA — LILIACEAE

The genus name is derived from the Japanese for the species *Amana edulis*. There are three species in the genus, two of which are native to Japan, and the other one is native to China. *Amana* accommodates plants very similar to *Tulipa* yet with specific differences: an elongated style (as long as the ovary), perianth segments are long and narrow, and most noticeably linear, leaflike bracts are located on the stem about an inch below the flower. If there are but two bracts, as in *A. latifolia*, they are arranged opposite; if more than two they are equally spaced around the stem.

The rootstock is a small bulb with quite an elongated neck and a dark brown tunic that is felted within. All species are dwarf, seldom over a few inches in height, and flower in late winter or very early spring. The flowers are like tulips in overall appearance except that, when open, the pointed tepals are somewhat boat-shaped and narrow and their bases barely overlap each other. None of the species is of great horticultural value.

CULTURE The bulbs should be planted in late summer in a well-drained, sandy soil, in a sunny location. Set bulbs just over an inch in depth and some 3–4 in. apart. They are not fully hardy and require some winter protection against frosts. In areas where winter temperatures fall below 26°F they should be given the protection of a cold frame. Plants should be kept moist during the growing season with water being withheld when the foliage starts to die back in late spring, and bulbs kept on the dry side during the summer months.

PROPAGATION When plants are established they can be lifted in early summer, after the foliage has died back, and bulbils separated from the parent bulbs. This should be done only after the bulbs have established themselves and have been growing for at least 2–3 seasons. Seed can be sown in summer or early fall, in a sandy soil mix, barely covering the seed. The temperature should be some 55°F minimum at night but warmer temperatures would not be detrimental. Seedlings should be transplanted into larger containers as soon as of suitable size and grown on. Care must be taken not to overwater during the resting period the following summer.

PESTS & DISEASES No special problems.

USES For the collector only.

SPECIES & CULTIVARS

A. edulis (syn. *Tulipa edulis*). Native to Japan and China; introduced 1903. In the wild this plant is found growing in grassy meadows, often along river banks. Flowers are gray-greenish-white and some 1½ in. in diameter when open. Tepals are some ¼ in. wide at the base, less than 1 in. in length and pointed; exterior is a dull purple, confined to the central portion and leaving a white margin that is very narrow. Flowers are carried singly on a flowering stem that reaches 3–5 in. in height but one bulb may produce more than one flowering stem. Two or three green bracts are produced on the stem and shield the flower bud but during flowering the bracts are some 1–1½ in. below. Foliage is green, erect at first then trailing, sheathing the flowering stem at the base; generally only 2 leaves produced, 6–8 in. in length, ¼ in. in width. Flowering February–March, or April.

A. graminifolia (syn. *Tulipa erythronioides*). Native to China; introduced 1919. Very similar to *A. edulis*, but flowering earlier in January; color on the exterior of the tepals also is more pronounced, and the leaves are the same length but wider, ½ in. Other characteristics similar.

A. latifolia. Regarded by certain authorities as *A. edulis* var. *latifolia* is native to Japan. The difference between this species and *A. edulis* is that the foliage is much shorter, 3–5 in. in length, and wider, up to ½ in. in width. In its native habitat it is found growing in sunny locations in woodlands. Tepals pointed, some ¾ in. in length and ¼ in. wide at the base. Flower color is white with purple exteriors; the purple is arranged in closely spaced thin lines. Flowering February–March and dormant during the summer months.

AMARYLLIS — AMARYLLIDACEAE

Amaryllis is a Greek feminine proper name. The genus *Amaryllis* is monotypic now, with but one species—*A. belladonna,*—perhaps one of the most beautiful flowers extant. *Amaryllis longifolia* and *A. falcata* are synonyms of *Cybistetes*, separated because in *Cybistetes*; the flowers are zygomorphic. This is not to be confused with the very popular Christmas-flowering *Hippeastrum*, which, along with a number of other plants, such the nerines, lycoris, sternbergias, and crinums, was once included in this genus. There is still some doubt about the correctness of the inclusion of *A. belladonna*, and, indeed, when the genus is examined, the *Amaryllis* might become *Callicore rosea*. The reasoning behind this is that *Hippeastrum equestre* was first described as *Amaryllis belladonna*. However, that is for the botanists to decide, and the plant we are concerned with here is well known as the belladonna lily.

A. belladonna comes from South Africa, where it grows wild around the southwestern Cape. The bulbs are brown, large, and rounded. It has deciduous foliage, produced after the flowers. The leaves are strap-like and remain green throughout the winter in milder climates. They then die down and the bulb becomes dormant. In late summer the flower stalk emerges and the cycle is repeated. The flowers are large and sweet-scented. The perianth tube is short-flaring to the 6-petaled flowers, which vary from white to deep pink.

There have been a number of intergeneric hybrids using *A. belladonna* as one of the parents. Among the other genera used are *Nerine*, (*Amarine* of catalogs), *Crinum*, and *Brunsvigia*. Because of the potential of these lovely flowers, more cultivars and intergeneric hybrids can be expected to be introduced in the future. Such hybridization could be a very interesting hobby for the home gardener.

CULTURE Bulbs planted in May/early June, with neck just at soil level. In colder climates, where winter temperatures fall below freezing and remain there for a number of weeks, plant much deeper, with as much as 5–6 in. of soil over them. Can survive and grow quite well even where winter temperatures reach 10°F. Should be given extra protection in such locations, such as lightweight winter mulch at least 3 in. deep. Soil must be well-drained in all locations. Essential that bulbs remain absolutely dry during summer after leaves die down. Water again as soon as flower spike appears. Plant in full sun, except hot, arid places, where they appreciate some shade. Ideal soil is sandy, with humus. As bulbs will remain in place for many years, dig out area and prepare their special soil mix—equal parts of good top soil, sharp sand, and peat moss or well-rotted compost. Do not compact anymore than necessary to set firmly in ground.

Little feeding required. Dressing of 10-10-10 or similar formula applied as flowers begin to fade and foliage emerges is beneficial. Bulbs found in abandoned plantings in northern California have survived and thrived due to exact weather formula needed—moisture in fall and spring, with hot, dry summer. Follow this pattern when growing bulbs in other climate zones. If summer moisture is abundant, shelter bulbs with sheet of glass or plastic but allow free access of air.

When grown in pots, use very porous soil mix, do not water until flower spikes appear and then only very little. Give additional water when leaves appear and continue until foliage begins to die down. Then allow pots to dry out.

PROPAGATION Parent bulbs allowed to grow undisturbed. Lift in spring only if offsets needed for new plantings, which is preferred method of propagation. Seed can be taken from pod even if not completely ripe. Can be sown in flats. Germinates readily. Place near parent bulbs so can have same required cycle of sun and water. When seedlings are large enough, after first full season after germination, lift and plant in desired locations.

PESTS & DISEASES No special problems.

USES Ideal for planting among low-growing shrubs provided dry conditions prevail during summer. Excellent cut flowers, with potential for commercial production. Good container plants but uninteresting after foliage dies down and until flower spikes develop in late summer.

SPECIES & CULTIVARS

A. belladonna. Native to South Africa; introduced 1712. Strap-like leaves dark green to 1 in. wide to 18 in. long produced in late fall/early spring. Strong flower stalks reach a height up to 30 in. Fragrant, trumpet-shaped flowers, as many as 10 per stalk, produced Aug./Sept. Colors variable, generally pink but var. *blanda* has large, white blossoms and *A. b. spectabilis* has rose-colored petals that are white on the inside.

Some cultivars offered by various nurseries:
'Cape Town'. Deep rose-red. 'Hator'. White flowers.
'Kewensis'. Pink with yellow throat.
'Rubra'. More red than pink.
'Windhoek'. Lovely rose-pink with white center.

In the literature reference will be made to:
A. b. var. *maxima.* A strong growing form.
A. b. var. *pallida.* Rose-pale pink flowers.
A. b. var. *purpurea.* Purplish flowers.
A. b. var. *rosea.* Rose colored flowers.
A. falcata: see *Cybistetes longifolia.*
A. hallii: see *Lycoris squamigera.*
A. humilis: see *Nerine humilis.*
A. longifolia: see *Cybistetes longifolia.*
A. sarniensis: see *Nerine sarniensis.*

Many species of *Amaryllis* are listed in literature. The following gives their correct names, see also the entry for *Hippeastrum*.
A. advena: see *Hippeastrum advenum.*
A. aulica: see *Hippeastrum aulicum.*
A. bagnoldii: see *Hippeastrum bagnoldi.*
A. barlowii: see *Hippeastrum roseum.*
A. bifida: see *Hippeastrum bifidum.*
A. elegans: see *Hippeastrum solandriflorum.*
A. leopoldii: see *Hippeastrum leopoldii.*
A. miniata: see *Hippeastrum miniatum.*
A. psittacina: see *Hippeastrum psittacinum.*
A. reginae: see *Hippeastrum reginae.*
A. reticulata: see *Hippeastrum reticulatum.*
A. rutila: see *Hippeastrum rutilum.*
A. solandriflora: see *Hippeastrum solandriflorum.*
A. striata: see *Hippeastrum rutilum.*
A. vittata: see *Hippeastrum vittatum.*

AMMOCHARIS— AMARYLLIDACEAE

The name given is derived from Greek *ammos* ("sand") and *charis* ("beauty"), this being a description of its habitat and truly beautiful flowers. At one time a number of species were assigned to this genus but many have been transferred to *Crinum*. *Ammocharis falcata* is a synonym of *Cybistetes longifolia*, which see. The principal species remaining in the genus *Ammocharis* is *A. coranica*, introduced circa 1800, a native of much of southern Africa with the exception of the southwestern Cape. For this reason it might be regarded as a monotypic genus. The flowers vary in color, from pink to deep rose with as many as 20 flowers carried in a large umbel. The flowers are tubular at the base and then the perianth segments separate to give an open flower with the tepals each curling back. From the mouth of the flower the stamens emerge and are quite prominent with the stigma and style being carried past them by as much as an inch. It is interesting to note that, unlike many other plants, the pedicels do not lengthen or change position when the flowers drop and the seed is produced. The height of the flower stalk is some 10–12 in. and the stalk leans when the flower head is in full bloom.

The foliage is strap-shaped and can reach a length of some 20 in. but often is much shorter in the wild due to the grazing of cattle. The leaves remain quite close to the ground as plants with only some 6 in. of leaf are commonly seen. The flowering time is November to January in the native habitat, early to midsummer in the Northern Hemisphere. The bulbs are rounded and some 4–6 in. in diameter. They dislike being disturbed so frequently the bulb will not flower for several seasons after being moved or planted.

Natives use a thick paste prepared from the cooked bulb to repair cracks in their cooking vessels.

CULTURE The bulbs will not stand temperatures lower than 35°F but can be protected with heavy mulch. In areas with temperatures below 32°F for any period of time the bulbs should be considered subjects for a cool greenhouse. Free drainage must be given as these bulbs are at home in very sandy soils. The bulbs require bright sunlight and moisture from the time the foliage emerges in the spring until flowering time. After flowering little or no water is required.

Bulbs should be planted in late winter or early spring and planted so the necks of the bulbs are at ground level. Space 8–10 in. apart and do not disturb for several years. Some liquid fertilizer can be given in early spring when growth is first noticed. If being grown in containers a sandy, friable soil mix should be used. Plant 3 bulbs to a container some 24 in. in diameter and with a depth of some 12–16 in. Again they should be left undisturbed.

PROPAGATION The bulbs are slow to produce offsets. Seed is the best means of propagation, as it germinates quite readily. Seed should be sown as soon as it is ripe and the seed capsules are opening well. Barely cover the seeds with a sandy soil mix and keep moisture level quite high. Temperature of 55°F at night is preferred. Transplant to individual containers as soon as large enough to handle. While growing keep moist, allowing the plants to become dry as foliage starts to die back. Seed can be sown in location where plants are to grow, yet it is better to improve this soil by adding sand and some peat moss to the area. Seed should be sown thinly in such areas, spacing seed several inches apart.

It will take several seasons for the bulbs to flower from seed. Those raised without transplanting will flower earlier than those starting life in containers.

PESTS & DISEASES No special problems if allowed to become dry after growing season has finished and plants have finished flowering.

USES Value lies only in these being rare in cultivation, but *A. coranica* has merit in warm climates as a bulb for the border or as a container plant.

ANAPALINA — IRIDACEAE

The name is derived from the Greek *anapalin* ("reversed order"), referring to the bracteoles being longer than the bracts. A small genus native to South Africa and producing lovely summer flowers that are in the red to pink shades.

The rootstock is a globose corm covered with fiber. Leaves are linear and pointed, clasping the lower part of the stem of the flowering spike. The flowers are tubular, often with the perianth segments separating only at the end of the perianth tube. Sometimes the petals flare with the uppermost petal hooded and the lower having distinct markings. The flower spike is often one-sided. After the flowers pass, the seed capsule expands to considerable size and is often colored and most attractive. The plants vary in height from 18–24 in. and are summer-flowering.

CULTURE The plants need sun and well-drained soil The species are not fully hardy but *A. nervosa* is the hardiest. They can be planted in the spring, lifted in the fall, and stored overwinter in a frost-free location, and planted again in the spring. They should be planted 3–4 in. deep and spaced some 10–12 in. apart. They make good container plants but must have adequate moisture during the spring and early summer months. On established plantings, some fertilizer should be given as soon as growth commences. In warm climates where there is little or no frost the corms can be left in the ground overwinter.

PROPAGATION A great number of little cormels are produced. These can be removed from the parent corms and grown on; they will reach flowering size in two seasons of growth. Seed can be sown in the spring in moderate heat, 55°F at night in a well-drained soil mix, the seed being only just covered. The seed should be kept moist but not wet. As soon as germination has taken place good light is needed. The seedlings can be grown on in the container where sown but should be transplanted to allow good air circulation if the seedlings become crowded. They can be set outside during the summer months and kept moist. Transplanted in the second spring the corms should be large enough to plant the following year.

PESTS & DISEASES No special problems.

USES Good plants for the sunny, well-drained border. They can be used in containers and make good cut flowers.

SPECIES

A. coranica. See description above.

A. heterostyla (syn. *Crinum heterostylum* and *C. parvum* of gardens). Native to East Africa; introduced 1937. Dwarf, rarely over 6 in. in height; leaves 12–15 in. in length, up to an inch in width. Flowers white with pink stripe on reverse of tepals, tubular at base, which is up to 2 in. in length, then tepals divide to give flowers some ½ in. in diameter. 3–10 flowers in the umbel; flowering in February.

A. tinneana (syn. *Crinum bainesii, C. lastii, C. rhodanthum, C. thruppii, C. tinneanum*). Native to eastern and southwestern Africa; introduced 1899. Strap-shaped leaves 12 in. in length, 1 in. in width, sometimes larger; flowers pink to carmine with distinct white line on inner surface of tepals. Narrow perianth tube; each of the up to 30 flowers in the umbel is some 3–4 in. in length; tepals curl at the tips. Plant varies from 1½–9 in. in height.

COMMENTS These bulbs are rare in cultivation. The validity of the species *A. tinneana* and *A. heterostyla* is doubtful. *A. coranica* is a fine plant and worthy of cultivation in warmer, dry climates.

SPECIES

A. caffra (syn. *Chasmanthe caffra, Petamenes caffra*). Native to the eastern Cape Province; introduced 1928. Found growing on peaty hillsides and grasslands, however, most commonly near the coast. The rootstock is a globose corm covered with fiber. There are 3 leaves, sometimes as many as 6, and they clasp the lower part of the stem of the flowering spike. The leaves are 15 in. in length and narrow, being only ⅓ in. in width; 3 prominent nerves are seen on the leaves. The flowers are arranged on either side of the flower spike. The flowers are red, 3 in. or more in length. The lower ⅓ is a tube, the 3 lower perianth segments then separate and are reflexed, 2 of the upper continue to extend remaining on the same plane as the tube and then reflex, while the uppermost petal extends beyond the other and is concave, not reflexed; the stamens lie close against this upper petal. The stigma extends beyond this upper segment and splits into three. There are often as many as 25 flowers per spike. After the flowers fade the seed pod expands and is pink and balloon-shaped. This is a most attractive species, it flowers in spring and summer but may be found at almost anytime of the year.

A. longituba. Native to the Indian Ocean side of South Africa. Often found growing more inland than the previous species. The foliage is like the previous species but the leaves are produced after the flowers. The flowers are pink, as many as 25 per stalk, and the corolla tube is long, often 2 in. or more. The flowers are carried on three sides of the stalk. After the corolla tube the segments separate, the 3 lower flaring downward and in the center of each there is a dark carmine stripe on a band of golden yellow; the tips of the segments are carmine. This color is also found in the upper segments at their tips; the top segment is concave, hiding the stamens and stigma. It flowers in March in the wild, late summer in the Northern Hemisphere.

A. nervosa (syn. *A. revoluta*). Native to the Indian Ocean side of South Africa. Found growing in grasslands, most commonly on slopes. The most hardy *Anapalina* species. The corm has many fabrous scales. The leaves are few, linear, and stiff, with prominent ridges, up to 18–20 in. in height and tapered along their length. The flowers are arranged in a spiral on the stem to a height of 20 in. or more. The color is salmon-pink with the lower part of the corolla tube being orange. The perianth segments recurve a little at the tips but are all of more or less the same length. The stamens are longer than the perianth segments, often by as much as ½ in. Flowering Oct.–Feb. in the wild, late spring through summer in the Northern Hemisphere.

A. revoluta see *A. nervosa*

A. triticea. Native to the Cape Peninsula. Found growing in dry, stony ground on the upper slopes of the mountains. In the wild the flower spikes are some 6 in. in length with 25 or more flowers; in cultivation the spikes will be larger. The flowers are bright red, tubular for 1 in. or more, then the 6 segments go their separate ways, the lower, broader petal recurving, the next 4 not reflexing as much and the uppermost petal being concave, with the stamens protruding from the flowers. The leaves are linear; 2–4 are produced at the base of the plant, and the lower part of the blade is rolled to give an impression of being stalked. The height of the plant is some 18–20 in. with the leaves being a little lower than the flowering spike. It flowers in January, midsummer in the Northern Hemisphere.

COMMENTS These plants deserve to be in cultivation and must rank among the species most likely to succeed when they are introduced and distributed. They are easy to grow, and, while the color range is limited, hybridizing might well unlock other colors. They are good in so many ways that it is a shame more are not cultivated in our gardens.

ANAPALINA

A. burchellii similar to *A. triticea*. Native to S. Cape Province. 2–3 in. in height.
A. pulchra similar to *A. nervosa* with leaves broad from base, it lacks an elongated upper tepal.
A. revoluta see *A. nervosa*.

ANDROCYMBIUM — LILIACEAE

The name is derived from the Greek *anèr* ("man") and *kumbion* ("cup"). There are some 9 or 10 species, found mostly in and around the Mediterranean on the Iberian Peninsula, North Africa, and the Mediterranean climate zone of South Africa in Cape Province.

All *Androcymbium* are produced from a small corm, generally tunicated, with a dark brown or black tunic, and somewhat elongated, not unlike a tulip bulb.

These plants are so low-growing, seldom more than 6 in. high, often in grass, that they can easily escape notice. Frequently the bracts hide the flowers completely, so that one has to bend down and open them in order to get a good look at the flowers themselves.

The most noticeable characteristic is that the leaves lie in a rosette on the ground. The bracts, which are more upright, are variously colored, mainly white or greenish-white; some species have purple venation in the bracts. Inside the bracts are found the flowers, nestling in a cup formed by the bracts. This accounts for one of its common names—Little-men-in-a-boat.

The flowers for the most part are greenish-white, and, when open, have a diameter of about an inch. The 6 tepals, which do not join to form a tube, have one distinguishing point—they have quite noticeable swollen yellow glands at the base. The 6 stamens are attached to the lower part of the tepals. Each flower carries 3 styles, which are free from one another, and the flowers, whose number will vary, are carried in an umbel, which is almost sessile. The flowers are found at the bottom of the bract.

CULTURE Almost all of the species are found in sandy soils, most frequently near coastlines. Being from Mediterranean climate regions, where the sun is hot and there is very little shade, they require moisture during the winter months but little or none during late spring and early summer. These conditions must be duplicated in the garden if the plants are to survive—well-drained soil, with more sand than humus; full sun, with a good, dry resting period.

The corms, best planted in the fall, should be placed just below the surface of the soil, spaced 10 in. apart. The plants require very little care and should be given water only if there are no rains. Temperatures should be in the upper 30°F range as the plants will not withstand temperatures below 35°F. Humidity should be kept low, and it is thus difficult to grow them in a greenhouse unless they are associated with other dry-atmosphere-loving plants. Light intensity should be as high as possible. Any fertilizer given should be applied as soon as growth commences, but hardly any is required.

As one can see from the cultural conditions needed, these plants are best suited for warmer regions. If grown indoors in a greenhouse, the difficulty is in keeping down the humidity.

PROPAGATION In the wild there is a small natural increase by the corms producing cormels around the parent. These can be separated and grown on either in a seedbed, provided the temperatures are in the 55–66°F range, or sown in a very sandy mix containing light, sharp sand, in a container and barely covered by the mix. Such propagation is done as soon as the foliage has died down in early to midsummer.

The plants also produce seed, which can be sown as soon as ripe in well-drained, sandy soil, whose moisture level is low but not bone-dry. The seed will ripen very quickly in hot, dry weather. As the seed capsules start to turn brown, examine them daily to see if they are beginning to open. Care must be taken when harvesting because the capsules split open from the bottom.

When large enough to handle, transplant the seedlings into other containers or individual pots. As soon as the foliage starts to die down, reduce the amount of water to almost nothing and allow to go dormant with warmth. The following spring, if the plants have made good growth, they can be set out into the garden. If plants are to be grown for another season in containers, use those that are at least 6 in. deep so that there is minimum disturbance when the plants are finally put into the ground.

PESTS & DISEASES There should be few problems from pests, except that slugs and snails may attack in the early spring as growth commences. However, because this genus prefers hot, dry conditions, these pests should not cause trouble if the correct procedures are maintained.

USES The rock garden; warm, sunny borders where there is no winter frost; or areas difficult to water—in such locations these plants can provide a great deal of interest. If grown in containers, they should be at least 10 in. deep and wide enough for the leaves to spread over the surface of the sandy soil mix required.

SPECIES

A. capense. Native to South Africa; while the date of introduction is not recorded it is most likely to be circa 1820. Found growing in the eastern parts of the Little Karoo and in the Outeniquas and Tsitsikammas hills, which are just inland from the Indian Ocean in southeastern South Africa; in the coastal areas it is found growing among short grass. The long leaves, up to 10 in. in length, are broad at the base, a little over an inch, narrowing to very pointed tips. Normally there are 3–4 such leaves with wavy edges which hug the ground. The pale-green bracts are produced in the center and cover the flowers. The bracts are up to 2–3 in. in height and as wide, or wider, than the leaves. The flowers are an insignificant cream color, 6–8 held tightly together. In the wild flowering in Sept.–Oct.; March–April in the Northern Hemisphere.

A. ciliolatum (syn. *A. fenestratum*). Native to South Africa. Found growing in Namaqualand on the western Atlantic seaboard in sandy soil. Although the habitat might be quite moist in spring, it enjoys a dry season late in summer. The leaves, usually 2, are lanceolate and broad with minute, fringed margins; they reach up to 6 in. in length and lie on or very close to the ground. The bracts, which reach a height of 5–6 in. are pale green, sometimes almost white. There are usually 3, sometimes 4 bracts, as broad at the base as the leaves, about 3 in., that open to expose the cluster of many white flowers. Flowering time is early spring or Sept.–Oct. in its native habitat. See also *A. melanthoides*.

A. eucomoides. Native to the Cape Peninsula of South Africa; most likely first grown in cultivation circa 1800. It is found growing in poorer, sandy soil. The actual size of the plant varies and, in better soils, will almost double in size. The light-green leaves tend to be more upright than in some of the other species and have a wavy appearance. Most commonly 2 or 3 leaves are produced, 5–6 in. in length, broader at the base, tapering to a sharp point, and slightly keeled. The bracts are shorter and stubbier than the leaves and form a cup for the pale greenish-white flowers. Flowering July–Oct. in its native habitat; the latter part of winter/very early spring (Dec.–Mar.) in the Northern Hemisphere. Rare.

A. europaeum. Native to the Cabo de Gata area of Spain; date of introduction not recorded. The corm is more elongated than other species and has a black tunic. Like its South African relatives, it is found growing in sandy soil or in sparse grass, but unlike them, it is lower-growing, rarely reaching more than 2 in. in height. The glossy green leaves reach 4–5 in. in length, and are relatively numerous, with as many as 10 being produced. They are about ½ in. wide at the base, tapering along their length to a sharp point. The bracts are somewhat lighter in color and hold in their centers the flowers, which can vary from white to pink-striped mauve, sometimes with speckles of darker color. The flowers of this species are larger than most of the others, being up to an inch in diameter, and the number varies from 1–5 or more. Flowering time is Dec.–Feb. Very rare in cultivation.

A. fenestratum see *a. ciliolatum*

A. gramineum (syn. *A. roseum*). This species is very similar to *A. europaeum*, but is only found growing in North Africa, not at all in Europe. Very rare in cultivation and is most likely a form *A. europaeum*, but with narrower foliage. This and its geographic location have earned it the separate species name.

A. leucanthum. This plant is listed in the Royal Horticultural Society's *Encyclopedia* as being native to South Africa. It is listed as having been first introduced in 1874, a time when the Cape Province area was much explored, so it would be logical to assume that this is either *A. eucomoides* or a variation. The plant is described as having 4 leaves, 5–6 in. long, lanceolate, slender, and pointed, with whitish flowers. It is not certain if this is a reference to *A. eucomoides* or to a separate species.

A. melanthioides. This is another species described in the RHS *Encyclopedia* as being from South Africa, Swaziland, and Botswana; introduced 1823. Described as having lower leaves that are long and narrow and bracts 2 in. long, some white turning pinkish with green veins, it would appear that this is close to *A. ciliolatum*. The brief description does not give reason to suppose that this is another or separate species.

A. pulchrum. Native to the same areas as *A. ciliolatum* and like it, except the bracts are purplish-red and the flowers pink. It is treated as a distinct species because of the color of the flowers and the lack of the minute, fringed margin on the leaves. As the two species are found in the same region of South Africa, we can assume that they are very closely related, and the merit of separate species names is perhaps questionable. Introduced circa 1921.

A. rechingeri. What makes this species unique is that it is found growing only in one place—Elaphonisi Island—about 1 mi. off the coast of Greece, not far from Peloponnesus. The island is only 4 mi. long and 3 mi. wide, with a population of some 500 people. The plant is close in form and habit to *A. europaeum*, the main difference between the two being that the tepals of *A. rechingeri* are sharply pointed. Additionally, the seed capsule does not split open but breaks up unevenly and irregularly. Flowering in winter and early spring.

A. roseum see *A. gramineum*

ANEMONE — RANUNCULACEA

Anemos, Greek "wind." Theophrastus was the first to use the name "anemone"; however, some authorities believe the word comes from the Syrian *nama'an*, the cry for the dead Adonis whose blood is described in legend as returning to life in the scarlet anemones. There are many species (more than 50) not all tuberous or rhizomateous and thus are not in this book. The majority of those worthy of cultivation come from the temperate regions of the Northern Hemisphere. The tuberous-rooted species are easy to grow, preferring sandy soil that is also rich in humus, preferably leaf mold, as found in established woodlands, where they receive both sun and partial shade.

The foliage is quite fernlike, finely divided in most species, frequently with a whorl of leaves produced on the flower stalk just below the flowers. The flowers are actually sepals which resemble petals and have numerous free stamens. Colors vary by species and range from white to pink to red to blue. Many fine cultivars are offered by nurseries.

CULTURE Almost all tuberous types grow well in good garden soil, preferring a sandier type. Though moisture is appreciated, they do not like very wet soil. Best planted early fall. If soil is not moist at planting time, tubers must receive water to start into growth. Once started, plants must be given adequate moisture until flowering is past and foliage begins to die down. Mid to late summer, lift and divide tubers if propagation is desired. Most can be left in ground and lifted and divided only when plantings become overcrowded. See special cultural requirements given under "Species and Cultivars" as great diversity exists among species. Easiest method is to lift and divide tubers. Allow foliage to die down and then transplant small tubers that have formed. Tubers can be kept almost indefinitely as long as kept dry and held at moderate temperatures. Where soil is not moist, soaking tubers in lukewarm water for about 24 hrs. will speed start of growth. Where possible, should be left *in situ* to form good-sized plantings. Also easily raised from seed, which should be fresh as possible. Sow in sandy mix. Take care in transplanting as they do not establish easily.

PESTS & DISEASES Few problems, other than perhaps slugs and snails. Check plants grown in greenhouses and out-of-season for aphids, and take appropriate control measures.

USES Both 'St. Brigid' and 'de Caen' are excellent cut flowers, as well as garden and container plants. Plant *en masse* for best effect in garden. Few better early flowering plants as woodland species when allowed to naturalize. Plant preferably on shady bank beneath high-crowned trees so these low-growing plants can be appreciated, but away from foot traffic where dormant plants may be trampled. Avoid compaction of soil.

SPECIES & CULTIVARS

A. **appenina.** Native to southern Europe and quite common so the date of introduction not recorded. Now found growing in many open woodlands throughout Europe. Excellent plant for naturalizing in rough grass and woodland settings receiving part sun and part shade. Grows well in good garden soil, but add organic matter if soil is sandy. Do not disturb once planted. Thick, almost tuberous, and elongated, very dark-colored rhizomes should be planted 2 in. deep and 6 in. apart. Single flowers, carried on 6 in. stems, are a good blue. White form, var. *alba*, offered by nurseries. Has tinge of light blue, noticeable on reverse of petals. Double flowers produced occasionally. *A. apennina* flowering in late winter/early spring, late February in milder climates, March in colder areas.

A. **blanda.** Native to Greece; introduced 1898. Should be grown in light shade where summers are hot; in full sun in cooler climates. Plant tubers 2 in. deep, spaced about 6 in. apart. 6 in. in height. Tubers rounded and globose. Spring-flowering; blossoms very early in warmer climates. Single flowers, generally deep blue, but pale blue, white, and pink forms have been noted. Many cultivars or selections available. Among the best are:
> *A. b. alba.* White form; slightly more dwarf than species; flowers a little larger; excellent rock garden plant.
> *A. b. atrocoerulea.* Flower color varies; generally deeper than species.
> *A. b. rosea.* Older cultivar still grown frequently and listed in catalogs. 'Charmer' preferred if dark pink is desired, rather than *rosea*'s clear pink.
> 'Blue Star.' Light blue.
> 'Bridesmaid.' Large, pure white flowers; possibly same as var. *alba*.
> 'Charmer.' Deep pink flowers; selection of var. *rosea* introduced by Van Tubergen.
> 'Pink Star.' Larger flowers than species; deep pink.
> 'Radar.' One of finest for garden; bright red with white center, sometimes with pinkish hue.
> 'White Splendour.' One of largest flowered; pure white; flowers last a little longer than species.

A. **coronaria** (syn. *A. bucharica*). Widely distributed from southern Europe into Asia; introduced 1596. Does not grow well in gardens in cooler climates, prefers warmth. Must be given protection of frames or greenhouses in colder areas. Plant late fall for spring-flowering; March/April for midsummer flowering; May/June for late summer/early fall-flowering. Responds well to light forcing for spring-flowering, after which bulbs generally discarded. Early forcing can be done by home gardener with a small greenhouse, providing good cut flowers for home. Roots misshapen, knobby, brown tubers. Flowers single, varying greatly in color from red to yellow, 8–12 in. in height. Popular with florists. Has been extensively hybridized. Main varieties are the poppy-flowered singles 'de Caen' and the double form 'St. Brigid.'

> *'de Caen'* cultivars
> 'His Excellency'. Bright scarlet; very large flowers on good stems.
> 'Hollandia'. Strain of 'His Excellency'.
> 'Mr. Fokker'. Blue.
> 'Sylphide'. Violet.
> 'The Bride'. Pure White.

> *'St. Brigid'* cultivars.
> 'Lord Lieutenant'. Bright blue.
> 'Mount Everest'. Pure white.
> 'The Admiral'. Cyclamen-violet.
> 'The Governor'. Vermilion-scarlet.

A. × **fulgens.** Found growing in southern France and regarded as naturally occurring hybrid between *A. pavonina* and *A. hortensis*. Late spring/early summer-flowering. Scarlet flowers on stems that reach 12–15 in. in height. Prefers sun, with a little shade in hottest climates; good, rich soil.
> 'Annulata grandiflora'. Flowers a little larger than species with yellow center.
> 'Multipetala'. Same as species but with double row of sepals making it semidouble flower.
> 'St. Bavo Hybrids.' Raised by Van Tubergen and listed by them under this name as a hybrid between *A.* × *fulgens* and *A. coronaria*; however, some authorities regard them as correctly listed under *A. pavonina*. Native of Greece and western Asia Minor. Also known as Peacock Eye. Excellent garden plant, grow in rich soil and full sun. 10–15 in. high. Early flowering. Lovely colors vary from white through pink to salmon, brick and dark red to violet-blue. Flowers over long period, from mid-spring to early summer.

A. **nemorosa** (Wood Anemone). Native to Europe, including Britain, thus date of introduction not known but has been long in cultivation. Found growing in woodlands. Needs moisture, shade, and rich humus. Should be allowed to naturalize, with creeping rhizomes set 2 in. deep and 6–8 in. apart. 6–8 in. in height. Spring-flowering. Flowers are white, often tinged with pink, 1 in. in diameter, although selected forms have larger flowers. One of the best is *A. n. alleni* ('Allen's Form'), with 1½ in. flowers, rose-lilac on the outside, bluish interior. Other cultivars offered are:
> 'Alba Plena'. Double white flowers.
> 'Leed's Variety'. Large, single white flowers.
> 'Robinsoniana'. Delicate lavender-blue.
> 'Royal Blue'. Violet-blue, tinged with rose-purple on reverse side of petals; very dark-green foliage tinged with purple. Somewhat more dwarf than type.

A. **ranunculoides** (Buttercup Anemone). Native to southern Europe; date of first introduction into cultivation lost in the mists of time. Common name refers to yellow color. Slender rhizome spreads horizontally and rapidly if given peaty soil and abundant moisture. 6 in. in height. Full sun, except in hottest locations. Var. *superba* has larger flowers than species and foliage is bronze-colored. Hybrids between *A. ranunculoides* and *A. nemorosa* called *A. lipsiensis*. Best form of this cross has sulfur-yellow flowers and is probably the *A. ranunculoides* offered by some nurseries. If purchasing rhizomes from nurseries, give plants light shade, as recommended.

ANEMONE

A. altaica Native to E. Russia, Asia. White, (Rhizome slender). 4–6 in. in height. April flowering.
A. americana Native to N. America. White. April flowering.
A. baicalensis Native to E. Asia. White (Rhizome). 6–15 in. in height. Spring flowering.

A. baldensis (syn. *A. fragifera*) Native to France, Switzerland & N. America. Introduced 1792. White (Rhizome). 6 in. in height. May flowering.
A. biflora Native to Iran. Introduced 1936. Red turning yellow orange, (Tuberous). 3–6 in. in height. March–April flowering.
A. bucharica see *A. coronaria*.
A. caroliniana Native to E. U.S.A. Introduced 1824. Purple or whitish, (Tuberous). 9 in. in height. Spring flowering.
A. caucasica Native to Caucasus, N.W. Iran, N.E. Turkey. Blue or white, (Rhizome). 2–3 in. in height. April–June flowering.
A. drummondi (form of *A. baldensis*) Native to N.W. America. White tinged blue. 4–12 in. in height. May–Aug. flowering.
A. fragifera syn. of *A. baldensis*.
A. gortschakovii Native to Central Asia. Yellow (Tuberous). 5 in. in height. April–May flowering.
A. graeca form of *A. pavonina*.
A. heldreichii (similar to *A. blanda*) Native to Crete. Blue (Rounded tuber). Dwarf. Feb.–March flowering.
A. hortensis (syn. *A. stellata*) Native to Central Mediterranean. Violet or rose (Tuber). 12–15 in. in height. March–April flowering.
A. × intermedia Native to Denmark, Sweden, Russia. Pale sulphur (Rhizome). Dwarf. Feb. flowering.
A. palmata Native to S.W. Europe. Introduced 1597. Golden yellow (Tuberous). 6 in. in height. May flowering.
A. pavonina Native to Balearic Is., Italy, Balkans. Purple to whitish (Tubers brown). 8–16 in. in height. March–April flowering.
A. rupicola Native to the Himalayas, W. China. Introduced 1913. White (Long woody rhizome). 4–10 in. in height.
A. stellata syn. of *A. hortensis*.
A. tschernjaewi Native to N. Afghanistan & Central Asia. Whitish to pale purple. 3–8 in. in height. March–May flowering.
A. tuberosa Native to S.W. USA. Rose (Tuberous). 4–12 in. in height. April–May flowering.

ANEMONELLA — RANUNCULACEAE

There is but one species of this charming little plant, *Anemonella thalictroides*. The name is derived from "Anemone", as there is a great deal of similarity between the two. Its common name is Rue Anemone, because of the foliage, which is like that of the meadow rue (*Thalictrum* sp.); and because of the flowers, which are not unlike those of the anemone. It is native to the eastern United States from Maine to northwestern Florida and west to Alabama, Mississippi, and on into Arkansas and Oklahoma. It flourishes in open woodlands in a wide range of soils.

The tuberous roots, clustered close to the base of the stems and quite close to the surface of the soil, are brownish, ovoid, and rarely more than 1 in. or so in length and ½ in. in diameter. The glabrous, light green leaves are about 1 in. wide and a little longer than wide, and are ovate with 3 rounded lobes. The flowers range in color from white to pink, with white predominating. Double forms are found occasionally. The color comes from petaloid sepals, there being no actual petals, which vary in number from 5–10. Stamens and pistils are numerous. The flowers, which reach a diameter of 1 in., are carried in an umbel on the very slender pedicels. This is the main distinguishing feature between the *Anemonella* and the *Anemone*, as well as the fact that the flowers arise from the fragile-looking stems at the same point as the leaves, which are either in a whorl or in pairs. The overall height of *A. thalictroides* seldom exceeds 10 in. and is more commonly 6–8 in. It flowers in springtime, April–May. Undoubtedly one of the plants sent to England by early settlers in the United States.

CULTURE These plants, growing in a wide range of climatic conditions, are fully hardy. They prefer light and high shade, and should not be exposed to direct sunlight throughout the day. Soil should be rich in organic matter, as would be found in woodlands. Give adequate moisture in spring, when the plants start into growth after their dormant period during the winter months, and frequent, if irregular, moisture during the summer months. Good drainage also must be provided. The plants are not suited for hot, dry areas.

The tubers are best planted in early autumn or spring. Place them just an inch or so below the soil level, some 8–10 in. apart. Once planted, they should not be disturbed for a number of years. In soil where the humus content is high, no feeding is necessary; in poor soils, however, apply a top dressing of well-decayed organic matter in late fall, after growth has died down, or in spring, prior to growth commencing. No additional feeding is necessary but light feedings of liquid organic matter can be given once the plants are established. If there is a preference of soils, it would be toward the acid side, but the plants are quite adaptable as long as organic matter is present.

PROPAGATION In the garden, the best means of propagation is the removal of tubers from established plants without actually lifting the plants out of the ground. In early fall or in the spring, remove the outer tubers from the outside of the clumps, leaving the clumps undisturbed, and plant them back as soon as possible. If a large increase is desired, the entire plant can be lifted and the tubers separated and planted back. While this will rapidly increase the numbers, flowering after this type of procedure will be delayed until the plants have again established themselves, usually at least one complete growing season.

Raising plants from seed is not difficult. Seed should be harvested as soon as ripe and stored for sowing in the spring. No protection from the weather (cold and rain) is needed. Sowing can be done in warmer areas as soon as temperatures rise and in colder areas when the danger of severe frost is past. Scatter seed in organically rich compost, barely cover, and give moderate temperatures in the 45°–50°F range at night and protection from frost during the day. Keep moist throughout the growing season. Sowing the seed *in situ* also can be carried out and is best done in the late summer, as soon as the seed is ripe, duplicating the schedule of the plant in the wild. Seedlings can be grown on in a shaded bed, and transplanted as soon as large enough to handle. That fall or the following spring the young plants can be set out into their landscape positions, and flowering will occur in the second season.

For smaller quantities, sowing can be done in containers, using a light soil mix high in organic matter. A mixture of leaf mold, peat moss, and some sand to provide drainage is suitable. Placed outside where there is protection from bright sun, the seed will germinate quite rapidly and the seedlings can be either transplanted to small pots as soon as large enough to handle or planted out in a suitably prepared bed. They can be planted out that fall or the following spring.

PESTS & DISEASES Few pests or diseases bother these plants. Slugs and snails can be a problem among germinating seedlings.

USES An ideal and lovely plant for woodland areas where it can enjoy organic soil and shade. Best planted in drifts and allowed to naturalize. Do not plant underneath shrubs which create dense shade but rather in areas with high, open shade and occasional sun. *Anemonella* is best seen when a goodly number are planted together, so that plantings should be of at least 15–25 plants in an area. If such space is not available, group them close to a path in as large numbers as possible. Although this delightful little perennial is common throughout the eastern part of the United States, it deserves wider recognition and should be considered in any garden with a suitable woodland setting.

ANOIGANTHUS — AMARYLLIDACEAE

The name is derived from the Greek *anoigo* ("to open, to expand") and *anthos* ("flowers"). The name refers to the habit of the flowers which at first seem to open reluctantly (when in bud they appear to be "pinched") then do open wide. Native to South Africa, with quite a wide distribution from the Cape to Natal, Swaziland, and north into Zimbabwe; introduced 1888.

There is little to distinguish the two species, *A. breviflorus* (syn. *A. luteus*) and *A. gracilis*. It should be noted that in some literature the plant is given the incorrect name *A. brevifolius*.

For a great number of years this species (I feel there is only one) was included in the genus *Cyrtanthus*, but was separated from that genus because of the yellow flower color and relatively short perianth. Now often considered again as *C. breviflorus*.

The bulb is white with a brown, membranous tunic. In its native habitat it is found in swampy areas, but also is found high in the mountains, and has been noted as occurring at 5,500 ft. The short flowering stalk is stocky and only some 6–8 in. in height. Often the 3–4 long leaves are produced after the flower. The leaves are over an inch in width, 12 in. in length.

The flowers are carried in an umbel with up to 10 flowers (more commonly only 5–6), bright yellow with a "pinched" look midway on the petals just before opening fully; petals a little over an inch in length with only the very lower portion forming a tube. While in a tight bud the flowerhead is covered by brownish, papery, persistent bracts. The pedicels are quite sturdy and upright; however, when open the flowers are more likely to be outward facing, the upper petals recurving more than the lower.

In the wild it flowers in September–October, and reports are that it flowers at a similar time in the Northern Hemisphere, but it is more likely to adapt and flower in early summer. The common name given to the plant in South Africa is Mock Fire Lily.

CULTURE Well-drained soil with adequate moisture is required. Setting the bulbs some 3–4 in. deep, spacing some 8–10 in. apart. Requires sun but will appreciate some shade in the hotter, drier areas during the heat of the day. Should be left undisturbed once established. Requires winter protection in areas of severe frost. In warmer climates can be mixed with other plants requiring moisture during spring and early summer.

PROPAGATION Seed provides the best means of propagation. Sow in spring in sandy-peaty soil mix in moderate temperatures; just cover the seed. Small plants should be potted up the second season and grown on. Flowering size is reached in some 2–3 seasons; sometimes more time is needed, depending on length of growing season. In warmer climates shorter time required.

PESTS & DISEASES No special problems.

USES Sunny borders among perennials, too straggly for rock gardens.

SPECIES

A. breviflorus (syn. *A. brevifolius, A. luteus, Cyrtanthus breviflorus*). Described above.

A. gracilis is noted in the literature. Native to Tanzania but would appear to be the same species—small differences possibly due to geographic location.

A. luteus: see *A. breviflorus*.

COMMENTS It should be noted that the bulb is suspected of being poisonous. Worthy of consideration by gardeners wishing something unusual yet of easy culture.

ANOMALESIA — IRIDACEAE

The name is derived from the Latin *anomalus* ("irregular"), referring to the irregular shape of the flowers. The lobes of the perianth segments are quite distinctive, the central segment of the upper 3 being shaped like an inverted spoon shielding the anthers and stigma. The other 2 lobes then form "wings" to this upper segment, while the lower segments are reflexed into the sack formed at the base of the upper segment. At times all 3 lower segments are reflexed, sometimes only 2. Seems to vary within the species.

The literature mentions only 2 species, however, reference can be found to 3. It being likely that the species *A. splendens* is a form of *A. cunonia*, as *A. cunonia* is described much like *A. splendens* but is bigger with a larger lower lip, which one would expect if it is indeed just a larger form. The other species, *A. saccata*, shows a greater difference between it and *A. cunonia*: its stem is much darker in color, as are the flowers. It is also twice the height, reaching 36 in., *A. cunonia* only 18 in. The flowers are always in the red range of colors, flowering in September in their native habitat of the Cape Province of South Africa.

For a number of years these plants were included in the genus *Antholyza*, from which they differ mostly in the unbranched stems, which do not zigzag, and in the lower segments (5 in *Antholyza* of more or less equal length with the upper much longer, being not nearly so recurved.

The plants have narrow, linear leaves. The flowering stem is quite two-sided, the flowers emerging from pointed buds, the lower flowers well developed before the upper open, the sepals remaining tight around the base of the flowers. The rootstock is a corm.

CULTURE The plants are not hardy; however, if planted deeply in the soil, they will withstand light frosts, and with protection such as heavy mulch in a warm location they can be grown in areas where the soil does not freeze. In colder areas they can be lifted in the fall and replanted in the spring. The corms should be set at least 3 in. deep and some 8–12 in. apart in a sunny location. They should be kept as dry as possible during the winter months and provided with moisture in the spring, being kept moist until the foliage starts to die down in late summer.

They grow well in containers where they should receive light feedings of liquid fertilizer during the period of emergence from the soil until they begin to flower. In average garden soil in the open border they do not require any feeding.

PROPAGATION The corms produce many offsets that can be separated in the fall and grown on in containers in cooler areas or in open borders in warm areas, the little cormels just being covered. Seed can be sown in the fall in warm areas or in the spring. The seed should be grown in a sandy soil mix and barely covered, requiring a temperature of 55°F at night and good light. The seed germinates quite quickly, and flowering-sized corms will be produced in 2 seasons. The seedlings should be transplanted when large enough to handle; space them some 4–5 in. apart. They can then be grown on, lifted in the fall (in cold areas), and replanted in the spring.

PESTS & DISEASES No special problems.

USES Good plants for pots and for sunny locations in a rock garden or flower border. Their bright colors are outstanding. Do not plant too close to other plants that flower at the same time; it is preferred that they be seen against green foliage. A rare plant.

SPECIES

A. cunonia (syn. *Petamenes cunonia*). Native to the Cape Peninsula; introduced 1756. 12–18 in. in height. The leaves are long, up to 12 in. and narrow. The flower spike will carry 8–10 flowers up to 1½ in. in length, the upper perianth segment being bordered by the 2 wings formed by the other of the 3 upper segments. The color is coral-red; foliage and stem somewhat glaucous. In its native habitat flowering in early spring, Sept.; April–May in the Northern Hemisphere. Rare in cultivation.

A. saccata ssp. *saccata*. Native to Cape Province of South Africa in Namaqualand (syn. *Kentrosiphon saccatus, Petamenes saccatus*). Reaches up to 36 in. in height. Foliage is narrow and linear, 12–16 in. in length with 3 prominent ribs. The number of flowers varies but seldom more than 10–12 per flowering spike, deep red, with the sack found at the lower part of the upper perianth segment being darker with hint of green. The wings are not so prominent as those of *A. cunonia*, making the pinching of the corolla tube more noticeable. The 'wings' tend to shield the entry to the sack rather than rise above the perianth tube. Flowering in spring or very early summer. Rare, doubtful if in cultivation.

A. splendens (syn. *Petamenes splendens*). Similar to *A. cunonia* but taller with the mid-lobe of the lower lip being longer. It seems to me that this is not so much a different species as a geographic form of *A. cunonia*. The other possibility is that it is a meager description of *A. saccata*. In my opinion there are only 2 species, *A. cunonia* and *A. saccata*.

COMMENTS Showy plants. Not in cultivation except in specialized collections, and then their identity is doubtful. Deserves greater recognition but might never receive that attention.

ANOMATHECA — IRIDACEAE

Name derived from Greek *anomos* ("singular and irregular") and *theca* ("capsule") due to form of the seed pod. Species now included in *Anomatheca* were formerly in *Lapeirousia* and in the literature will generally be found there today. The separation, however, was made in the early 1970's by Dr. Goldblatt of the Missouri Botanic Garden. The appearance of the plants, apart from the shape of the corm, is the same.

The distinctions are discussed in the entry for *Lapeirousia*. Suffice to mention here that the differences between the two genera are: *Anomatheca* corms are fibrous with rounded bases; in *Lapeirousia* the corms are woody and the bases are flat. In *Anomatheca* the chromosome count is 11; in *Lapeirousia* the number of chromosome count varies but is never 11.

There are 5 species in this genus, 1 having 2 subspecies. Of the species, only 3 are in cultivation and that rarely, the other species being very rare even in the wild.

CULTURE See *Lapeirousia*.

PROPAGATION See *Lapeirousia*.

PESTS & DISEASES See *Lapeirousia*.

USES See *Lapeirousia*.

SPECIES

A. fistulosa (syn. *Lapeirousia fistulosa*). Native to the Cape Peninsula in South Africa. Found growing in cracks in the rocks near the seacoast at an elevation of 600–1000 ft. Rootstock is a small corm covered with netted fibers. Plants are dwarf, only 3–5 in. in height. Foliage is basal; the 2 leaves are prostrate and some 5–7 in. in length. Flowers are slightly fragrant; white to creamy white, with the reverse of the tepals purplish-mauve; perianth tube is narrow, up to 1½ in. in length. Each flower is carried on a separate stem; each corm sometimes produces more than one stem. Summer-flowering.

A. grandiflora (syn. *Lapeirousia grandiflora, L. laxa* ssp. *grandiflora*). Native to the warmer subtropical areas of South Africa in the Transvaal; introduced 1887. Corm cylindrical; plant height 12 in., sometimes more. Foliage erect, ½ in. in width, 12–14 in. in length; up to 8 leaves, most commonly 5–7, carried on the usually unbranched stems. Flowers are bright red with a diameter of 2 in.; perianth tube narrow, some 1¼ in. in length, lower perianth segments have a darker blotch of color toward the base. Flowering mid- to late summer.

A. laxa (syn. *Lapeirousia cruenta, L. laxa*). Native to the same areas as *A. grandiflora* and quite similar in appearance; introduced 1830. This species is, despite its native habitat, one of the hardiest species in *Anomatheca* or *Lapeirousia*. Leaves are basal, linear, 12–14 in. long, narrow and sharply pointed, held in a flat fan shape; there are no stem leaves or bracts. Leaf bases cover the lower part of the flowering stem. Up to 10 flowers are carried on the 10–12 in. high stem in an unusual way: the topmost flower, which is the first to open, continues the upright direction of the stem, but below this flower the stem bends and along it, but always upright, the other flowers are carried. The diameter of the flowers is 1 in. Flower color is red with darker markings on the lower tepals; perianth tube is over 1 in. in length but seldom over 1½ in. Flowering period is over several weeks, beginning in August. An unusual feature of this plant is that the seeds are sometimes red in color, not black as in many of the other species.

Also listed in the literature are:

- *A. verrucosa*. 3–8 in. in height; pink flowers appear Aug.–Oct. in native habitat of Cape Province.

- *A. viridis* ssp. *crispifolia*. Foliage crisped, otherwise similar to ssp. *viridis*.

- *A. viridis* ssp. *viridis* (syn. *Lapeirousia viridis*). 4–14 in. in height; green flowers appear July–Sept. in native habitat of Namaqualand to southern Namibia.

ANTHERICUM — LILIACEAE

From *antherikos*, the name used by Theophrastus, because the flower stalks resemble a wheat stalk. Introduced into cultivation in the latter part of the 16th century. A native of the alpine meadows of Europe, its popular name, St. Bernard's Lily, was given to honor St. Bernard of Montjoux, whose association with the Alps is commemorated in the St. Bernard breed of dogs. While there are a goodly number of species in widely scattered geographic locations, ranging from the Tropics and subtropical regions of Africa to Europe and to the Americas, not many species are found in cultivation. It is found in mountain meadows, appearing after the snows melt, flowering in the summer in temperate regions. It is, therefore, fully hardy and, indeed, to do well should have a period of low temperatures during the winter months.

The rootstocks are short, with large, thick, fleshy roots. The leaves, usually grasslike and 12 in. in length, are narrow and found in a rosette at the base of the flower stalks. The flowers are carried on stems varying from 12–36 in. in height. While the flower colors can vary from white to yellow according to the species, that most commonly grown, *A. liliago*, is white.

CULTURE *A. liliago* needs a cold winter to perform well, while tropical and subtropical species need warmth. Plants need well-drained soil and should not be without moisture during the growing season. Less water needed after flowering. Plant tubers 4–6 in. deep in good soil, slightly deeper in sandy soil, not quite so deep in clay soil. In areas where winters are severe and snowfall scant, ground should be mulched so roots do not experience fluctuating temperatures. Give full sun. In spring, as soon as growth is observed, add dressing of 5-10-10 fertilizer. Flower spikes emerge in March/April and flower in late May or June. Can be grown in containers in mix of good topsoil, organic matter, and liberal amount of sharp sand to ensure good drainage. Because of large root size, good-sized containers, at least 10 in. deep, should be used. Water sparingly after flowering but do not allow container to dry out completely. Move to semi-shady location which duplicates growing conditions in native habitats, where sun may be very warm but ground slow to warm, never getting hot.

PROPAGATION Seed sown as soon as ripe in tropical areas; if not sown immediately in subtropical areas, may need cold winter in order to germinate. For this reason seed should be sown and placed in cold frame. Transplant as soon as large enough to handle.

Tubers lifted and divided in spring or early fall, only when necessary. Roots object to disturbance, so few if any flower spikes will be produced the next season.

PESTS & DISEASES Usual aphids and slugs and snails. If care is not taken when tubers are lifted, bruising can cause rotting.

USES Good rock garden plant. Unusual plant for front of perennial border, despite ultimate height, as foliage is not tall-growing. Good in containers; leave until quite pot-bound before dividing.

SPECIES

A. liliago (St. Bernard's Lily; syn. *Paradisea liliago*). Native to alpine meadows of Europe; introduced 1596. Leaves grasslike, narrow, up to 12 in. long. Flowers white, 1½–2 in. in diameter, open wide, reaching 24 in. in height. Early summer-flowering. Var. *major* somewhat taller with larger flowers, also native to alpine meadows of Europe and introduced a little later than the species.

ANTHERICUM

A. algeriense Native to Mts. of Algiers. Introduced 1900. White. 12–36 in. in height. April–May flowering.
A. angulicaule Native to Transvaal. White with green-brown stripe. 12–20 in. in height. Spring flowering.
A. boeticum Native to S.E. Spain. White. 24 in. in height. June–July flowering.
A. cooperii Native to Transvaal. White with green keel. 12–36 in. in height. Flowering May–July Northern Hemisphere; Nov.–Jan. South Africa.
A. echeandioides Native to Mexico. Introduced 1883. Orange-yellow. November flowering.
A. graminifolium syn. of *A. ramosum*.
A. liliastrum now *Paradisea liliastrum*—which see.
A. ramosum (syn. *A. graminifolium*) Native to W. & S. Europe. Introduced 1570. White. 12–36 in. in height. June–Aug. flowering.

ANTHOLYZA — IRIDACEAE

From Greek *anthos* ("flower") and *lyssa* ("rage"), the open flower resembling the open mouth of an angry animal. These unusual plants are commonly known as Rat's Tail in their native habitat of South Africa, where they grow in the sandy soil near the coastline just north of Cape Town. There are only 2 species in the genus, *A. ringens* and *A. plicata*. The corm is very large and subglobose. *A. ringens* reaches about 8–9 in. and *A. plicata* 12–18 in. in height. The leaves, which come right out of the ground, are rigid and ribbed. The flower stem shoots out at an angle and is sessile. Many brilliant scarlet, funnel-shaped flowers are produced, from which the stamens project. The plants flower in early spring.

CULTURE Corms planted 6 in. deep in good soil, preferably on sandy side, 8–9 in. apart. Moisture required during winter months. Should be grown only where winter temperatures do not fall below 25°F, unless protected by thick mulch so that ground temperature does not get too low. Best to lift corms in late fall and overwinter indoors in colder climates. Take time to clean and divide first. Start into growth again in early spring, using well-draining compost of equal parts of topsoil, peat moss, and sharp sand. When active growth seen, weak feedings of balanced liquid fertilizer given.

If grown in pots or containers, corms can be planted closer to surface, 2–3 in. deep, but need twice as much soil underneath for good root development.

PROPAGATION Corms will produce offsets, which can be grown on. Set outside as soon as large enough. Seed should be sown as soon as ripe. Will germinate in spring, and by end of that summer will be large enough to set out then or the following spring after overwintering.

PESTS & DISEASES No special problems, but be careful not to overwater.

USES Grown more because they are unusual than for their beauty. Good as pot plants or in warm locations in rock garden. Should be treated much like gladiolus.

SPECIES

A. merianella. See *Homoglossum merianellum*.

A. plicata (syn. *Anaclanthe namaquensis, Anaclanthe plicata*). Native to Namaqualand in the southwestern Cape Province of South Africa. The plants can reach a height of 16–18 in. but more commonly are shorter. The stem has 2–4, horizontal-to-ascending, hairy branches. The foliage is narrow, sword-shaped, and covered with minute hairs. Flowers are set close together on the stem and are up to 3 in. in length; color is bright red but the smaller segments are greenish. Bracts are covered with silver hairs 1 in. in length. Early spring-flowering. In its native habitat it grows in white sand and in culture must be given a sandy soil with excellent drainage.

A. revoluta. See *Homoglossum watsonium*.

A. ringens. Native of coast just north of Cape Town, South Africa. Grows in sandy soil. Rigid, ribbed leaves 8–10 in. in length. Sessile flowers, bright crimson-scarlet. Height of leaves and flowers seldom more than 9 in. Flowers in August in native habitat.

A. watsonioides. See *Homoglossum watsonioides*.

APODOLIRION — AMARYLLIDACEAE

The name denotes concealed peduncles: *a* ("without'"), *podos* ("foot'"), and *lirion* ("lily'"). There is a scarcity of information regarding these South African plants. They are similar to *Gethyllis* (which see for cultivation and propagation). Reading the descriptions of these plants it would appear that they are much like *Gethyllis*, the difference being that the stamens are in two ranks near the throat.

There are reportedly 4 species. The brief descriptions of those species I have been able to check read much like the descriptions of *Gethyllis* in height, color, habitat, etc. They are so much like this genus that one questions the validity of the name—perhaps research will show they are all one.

The species mentioned, all from South Africa, are *A. bolusii*, which also produces 2–3 leaves with pinkish-white flowers; *A. buchananii* (syn. *A. ettae*), introduced in 1896, which reaches a height of 3–4 in., with perianth segments white tinged red, flowering in June (summer), with leaves produced after the flowers and 7–9 in. in length; *A. lanceolatum*, which produces a solitary leaf some 7–9 in. in length; and *A. macowanii*, introduced in 1896, which produces 2–3 coiled leaves.

You will note that these sketchy descriptions are very similar to *Gethyllis*. It is certain that this genus soon will be revised and no doubt will be included with *Gethyllis*.

ARGYROPSIS — AMARYLLIDACEAE

This genus is sometimes listed in the literature; the plants *Zephyranthes candida* being given the name *Argyropsis candida* (which see).

The reason for the division of the genus from *Zephyranthes* is quite possibly that *Z. candida* is evergreen and thus deserving of being separated.

ARISAEMA — ARACEAE

From the Latin *aris* and Greek *aron* ("arum'"), and *haima* ("blood red"). About the only permanently red feature of these plants is the red berry produced in late summer. While the botanic name may be strange to many, one of the species is well known—*A. triphyllum*, the Jack-in-the-Pulpit. There are many species, more than 100, but only a very few are in cultivation. They are at home mostly in Asia; a few are found in tropical Africa, while Jack-in-the-Pulpit is native to the eastern coast of North America. *A. triphyllum* is recorded by Cutler as being used by the Indians when cooking venison. The roots are shredded and, together with the berries, boiled. It also is recorded that the starch of the root, which must be separated from the acrid parts by boiling, is delicate and nutritious. Two species consumed in the Himalayas, *A. costatum* and *A. curvatum*, also have to be prepared before eating the roots. Some are buried in a pit until fermentation takes place, dug up, washed, and cooked. The root of *A. tortuosum* is considered succulent by the mountaineers of Nepal.

For the most part the roots are tuberous, fleshy, and quite large. The leaves are long and divided into lobes, most commonly 3. The flowers on the spike, or spadix as it is properly known, are monoecious, either male or female; rarely are male and female flowers found together. This is the principal difference between the *Arisaema* and the *Arum*. A spathe, which covers the flowers, forms the "pulpit," wrapping around the spadix and shielding it, often ending in a long tail. The spathe is generally green or green and purplish in color.

CULTURE Tubers are planted 3–4 in. deep and spaced 10–12 in. apart. Once established, they should be left in the ground, lifting only when it is necessary to increase stock. Tropical species will grow in warmest places where no frost is expected. Plants emerge late and die down toward the end of summer, becoming dormant before onset of winter. In coldest areas, plants will benefit from winter mulch. As these plants emerge in late spring, mark location if any cultivation is to be carried out in area so developing shoots will not be damaged. Unless soil is poor, no feeding is necessary; however, organic matter should be incorporated into soil at planting time if lacking. More sun should be given if planted alongside a stream, but should receive some shade during most of day, especially in warmer climates where late morning and afternoon sun is strong.

PROPAGATION Lift tubers and separate young tubers produced from parents. Should be done in early fall. Replant young tubers as soon as possible so some root growth can take place before winter. If tubers are to be kept overwinter, best buried in damp peat and protected from frost.

PESTS & DISEASES No special problems.

USES Plants belong in woodland settings or alongside streams where there is adequate moisture and much organic matter in soil—consider for small gardens only if such conditions are present. Since flowers are more unusual than attractive as well as inconspicuous, overall merit rests with foliage. Plants look best among other foliage plants, such as *Hosta* and *Rodgersia*, which also like moisture and shade.

SPECIES

A. anomalum. Native to Malaya; introduced 1890. Flowering about May/June and, unlike some other species, produce leaves and flower spikes at same time from fleshy rootstock. Spathe has white stripes on dark brown to greenish-purple background; flowers reach some 12–18 in. in height. Plant has merit for warmer climates, where moisture is plentiful, provided some shade is given.

A. candidissimum. Native to western China; introduced 1924. Quite hardy, except in very cold climates with no snow cover. Mulch must be used in such areas. Flattish, round tuber. Flowers appear first, snaking out of ground, in early summer reach a height of 15–18 in. Spathe is white with pinkish stripes on inside; outside pale apple-green. Spadix yellowish-green. Orange seeds frequently produced in late summer, which can be scattered in suitably boggy, marshy locations to germinate.

A. dracontium (Dragon Root; Green Dragon; syn. *Arum dracontium*). Native to North America. Oblong tuber, about ¾ in. thick. Common names given because spadix often is more than 10 in. long sticking out from spathe and both are green in color. Spadix sometimes slightly yellowish. Leaves 18 in. long, divided, slender, and pointed. Flowers nestled under them, reach 18–30 in. in height. Leaf stalk mottled white at base. Late spring/early summer-flowering. Good plant for moist, shady areas. Orange-red berries produced late summer.

A. speciosum. Native to the Himalayas; introduced 1872. Jointed tuber, somewhat like a rhizome. Leaves perhaps most attractive of the genus, typically trilobed with pale rose margins. Leaf stalk mottled with deep purple. More than 36 in. high from bottom of leaf stalk to tops of leaves. Spathe greenish-purple on outside; inside more violet. Unusual spadix is variable in color, greenish-purple to creamy, but most distinctive characteristic is a prolongation from tip that will often reach 15–20 in. in length. Flowers lie below foliage. Early spring-flowering. Grows well in climates such as found along mild coastal regions and in temperate sheltered locations where frost is not too frequent or prolonged. Requires damp, moist conditions.

A. triphyllum (Jack-in-the-Pulpit; Indian Turnip; syn. *A. atrorubens*). Native to the North American Atlantic coast as far north as Quebec Province, Canada, and as far south as Florida and Louisiana in the United States; introduced 1664. Found growing in damp woodlands and in almost swampy ground, conditions necessary in garden if plants are to perform well. American Indians cooked roots, which eliminated the strong, burning taste experienced if eaten raw. Tuber nearly round with lateral rhizomes. Leaves are large with typical 3 lobes of genus. Spadix is brown; spathe is green or purplish-brown, streaked or mottled with purple on inside. Spadix carries bright red berries in late summer/early fall. Overall height 12–24 in. Flowers nestled below foliage. Flowering in April in warmer areas; June in northern, colder climates.

ARISAEMA

A. atrorubens syn. of *A. triphyllum*.

A. bakerianum var. of *A. fimbriatum*.

A. concinnum Native to the Himalayas. Female spathe—green or purple & white spotted; male bluish-purple & striped white (tuber globose). 12–24 in. in height. April–May flowering.

A. consanguineum Native from E. Asia to Yunnan. Introduced 1893. Green, brownish-purple striped (tuber flattish-round about 4" across). 18–36 in. in height. Early Summer flowering.

A. curvatum syn. of *A. tortuosum* var. *helleborifolium*.

A. elepnas (syn. *A. wilsonii*) Native to Yunnan. Introduced 1886. Spathe—maroon-purple with whitish stripes (tuber flattish 2" across). 12 in. in height. May flowering.

A. fargesii similar to *A. speciosum* Native to Szechwan. Introduced 1917. March flowering.

A. fimbriatum Native to Malaya. Introduced 1884. Spathe—brownish-purple, striped whitish; spadix—tip covered with purplish threads. 18 in. in height. Summer flowering.

A. f. var. *bakerianum* Native to Malaya. Introduced 1897. Not striped. Summer flowering.

A. flavum Native from Mediterranean region to the Himalayas. Introduced 1896. Spathe—deep purple within with green veins (tuber nearly round, small). 15 in. in height.

A. franchetianum syn. of *A. purpureogaleatum*.

A. galeatum Native to Sikkim. Introduced 1879. Spathe—pale green lower part purple, striped white. Tube—purple within & peduncle—pale green. 12 in. in height.

A. griffithii Native to Sikkim. Introduced 1879. Spathe—lined & spotted violet (tuber 3–5", flattish). 24 in. in height. Spring flowering.

A. helleborifolium var. of *A. tortuosum*.

A. hookeri similar to *A. griffithii* but from higher elevations in the Himalayas.

A. jacquemontii Native to the Himalayas. Yellow-green. 10–14 in. in height. June flowering.

A. japonicum Native to Japan. Introduced 1899. Spathe—green with white veins & stripes (tuber globose). 24–36 in. in height. March–April flowering.

A. leschenaultii (syn. *A. papillosum*) Native to India. Introduced 1864. Spathe—green with wide purple stripes (tuber globose 2" across). 24 in. in height. Summer flowering.

A. murrayi Native to India. Introduced 1847. Tube green, blade white (tuber large). 15 in. in height.

A. neglectum (syn. *A. wightii*) Native to Ceylon, India. Spathe—pale green (tuber globose). 8–16 in. in height.

A. nepenthoides Native to Himalayas. Introduced 1879. Spathe—variegated, blade—brownish (tuber flattish 2½" wide). 8–16 in. in height. Spring flowering.

A. papillosum syn. of *A. leschenaultii*.

A. praecox variety of *A. ringens*.

A. pradhanii Native to Sikkim. Introduced 1934. Spathe—chocolate-purple netted green, blade with alternate chocolate purple stripes alternating with green veins. 24 in. in height. May–June flowering.

A. propinquum Native to Himalayas. Purplish-striped greenish. 12 in. in height. May–June flowering.

A. pulchrum Native to India. Introduced 1879. Spathe—purplish striped green or white (tuber globose).

A. purpureogaleatum (syn. *A. franchetianum*) Native to Himalayas, W. China. Spathe—deep purple peduncle—purple (tuber globose). 15 in. in height.

A. ringens Native to Japan. Spathe—striped green & white, auricles—purple within, blade—deep-purple (tuber flat). 10–12 in. in height. Spring flowering.

A. r. var. *praecox* Spathe—striped green & grey without, brown & white on hood within.

A. r. var. *sieboldii* Spathe—plain purplish.

A. sazensoo syn. of *A. sikokianum*.

A. sieboldii variety of *A. ringens*.

A. sikokianum (syn. *A. sazensoo*) Native to Japan. Introduced 1938. Spathe—deep purple-brown spadix—white. 18 in. in height. May–June flowering.

A. tortuosum Native to N. India. Spathe—green, purple within (tuber flat-round up to over 4" across). 24 in. in height. May–June flowering.

A. t. var. *helleborifolium* (syn. *A. curvatum*) Same as species but with smaller leaflets & blade of spathe purple.

A. utile Native to Sikkim. Introduced 1880. Spathe—striped below netted brownish purple above (tuber flattish). 10 in. in height. May–June flowering.

A. wightii syn. of *A. neglectum*.

A. wrayi Native to Malaya. Introduced 1889. Spathe—pale yellowish-green or lilac with darker stripes (tuber). 12–18 in. in height.

ARISARUM — ARACEAE

Genus name was originally used by Dioscorides. This genus is small—some authorities recognize 2 species, while others regard these as merely geographic variations of the same species. Plants are native to the Mediterranean region.

Arisarum are dwarf plants, seldom taller than 8 in.; *A. vulgare* is the taller of the two. The roots are tuberous, varying according to species. Generally, only 1–2 heart-shaped or ovate, 2 in. wide leaves are produced; the leaf stalks are longer then the leaf blades. The flowers are monoecious and held on a straight tube. The feature distinguishing *Arisarum* from *Arisaema* is that the former has but 1 stamen, while the latter has 2 or more. Both species are easy to grow, although each prefers a different type of soil. Their most interesting feature is not the color but the size of the flower spikes, both among the smallest in the Araceae. Common names are Friar's Cowl and Mouse Plant, the latter given because the flowers resemble long-tailed mice.

CULTURE Tubers for both species should be set 4–6 in. apart and 1–2 in. deep. *A. proboscideum* prefers soil richer in humus than *A. vulgare*. Moisture should be available to *A. proboscideum* during the spring and summer, and it prefers light shade. *A. vulgare* prefers drier conditions in late summer and likes the sun but will grow well in some shade. Both species will withstand cold temperatures, but some protection should be given if the temperature of the soil drops below 32°F for long periods of time. A layer of mulch will provide such protection.

PROPAGATION Lifting and dividing the tubers in the spring is the most common method of propagation. Seed can be sown in the fall in milder areas or held over and sown in the spring. A sandy-humus soil mix accommodates both species.

PESTS & DISEASES None.

USES Unusual ground covers for shady areas or rocky places with some shade, depending on the species grown. More unusual than interesting, they deserve consideration if a widely varied collection of plants is desired.

SPECIES

A. proboscideum (Mouse Plant; syn. *Arum proboscideum*). Native to Italy and Spain; long in cultivation. Found growing in woodland areas where there is a fair amount of summer moisture and shade. The tubers are quite long and rhizomatous. The plants are seldom more than 6–8 in. tall, often much less and, when in a colony, form a mat of vegetation. The dark green leaves have distinct, pointed lobes. The flowers, produced in a spathe a little more than 1 in. long, are deep maroon with a whitish base and a long, slender "tail" often more than 4 in., which accounts for its common name, Mouse Plant. The spadix is small, whitish in color, and entirely exceeds the spathe. Early spring, March/April flowering. The berries are green.

A. vulgare. A common plant to the Mediterranean region; introduced in 1596. Found growing in sunny, dry locations, often among rocks. A little taller than *A. proboscideum*, sometimes reaching more than 10 in. in height. Leaves are ovate or heart-shaped, green or sometimes blotched, veined, or marbled with silver. The spathes, up to 2½ in. long with a short point at the apex, vary in color from green to chocolate brown, with many shades in between. The spadix is greenish, protrudes slightly from the spathe, and tends to curve downward. Flowers are produced from late winter to early spring. The berries are greenish. Ssp. *simorrhinum* is regarded by some authorities as a separate species, *A. simorrhinum*. Others consider it as only a geographic variation of the species, not deserving full species rank. Most commonly found growing in North Africa. Flowers are purplish-green in color, with a blunt spathe.

ARUM — ARACEAE

From the Greek *aron* ("wake-robin"), the name used by Theophrastus. A great number of plants now classified in other genera were formerly assigned to the genus, however, arum is still the common name for some of them. The genus now is comprised of only 12 species, all from Europe and around the Mediterranean.

The roots of the *Arum* contain a great quantity of farina, unfortunately combined with acrid properties which can be separated only by heat and water. *A. dioscoridis* is recorded by Theophrastus as having been eaten in ancient Greece. The leaves and roots were first soaked in vinegar. *A. italicum* was used by the inhabitants of the Balearic Islands who cooked the roots and mixed them with honey to make cakes. For several years the plants were grown in Guernsey for the commercial production of arrowroot from the roots. *A. maculatum*, with its thick and tuberous roots, was grown in the southern part of England for the production of arrowroot. There also are records of its being cooked and eaten by the English when other food was scarce. This is probably true as it is a common plant of hedgerows, liking the moisture and shade found in such locations. Accounts also show that it was cooked and eaten in Albania and the leaves eaten by the Greeks. It should be noted, however, that this was done only after they had first been dried and then boiled. Taro, a staple food of many Pacific islands, was included at one time in the genus *Arum*. (See *Colocasia*.)

The flowering spike is unusual as there are no flowers that would be recognized as such. Instead, there is a spadix, as with the *Arisaema* which, in turn, is surrounded by a spathe, or cover. In the *Arum*, however, the female and male flowers are separated by a distinct space between them. The female flowers are found on the bottom portion of the spadix. Above them are neutral and sterile flowers, with the males on the uppermost portion. The female flowers have 1 celled ovary. The tip is sterile. The leaves are not lobed. These two features—unlobed leaves and the male and female flowers on the spadix with sterile flowers between—are the characteristics distinguishing between these two otherwise similar genera. The leaves are produced in the fall or winter, followed by the flowers in early spring.

While these plants are not of great importance to the flower bed gardener, they are appropriate for the woodland garden or the wild garden. In former years they were of significant economic importance, as noted above.

CULTURE Tubers should be planted 3–4 in. deep and 12–18 in. apart; they prefer warm locations and a little sun. Where summers are hot, will perform well only if given shade and adequate moisture throughout growing season. Organic matter essential if plants are to thrive. Species native to the Mediterranean region will perform better in warmer climates and, because of their origin, can be given full sun in areas where summer temperatures are moderate, i.e., where 90°F days are few and far between. Do not mind being dry in late summer and will come into growth again with fall rains. No feeding necessary in average soils but top dressing with well-rotted compost given each year will help retain moisture and increase organic content of soil. Leave undisturbed once established. Lift for propagation purposes only.

PROPAGATION Easiest method is to lift and divide tubers from the quite large root structure. Small pieces of tuber can be cut or broken away from parent and will grow readily. Best time to do this is late summer/early fall. Plant tubers back as soon as possible. Berries containing seeds also produced. Seeds can be sown as soon as ripe and grown either in permanent location or in flats and transplanted when large enough to be handled. Plants raised in flats should be kept in shade at temperatures under 50°F until planted out.

PESTS & DISEASES No special problems.

USES Can be regarded only as woodland plants—grow where there is ample room and leave undisturbed.

SPECIES

A. creticum. Native to Crete and other islands in the area; introduced 1928. Found growing on rocky hillsides. Perhaps one of the best species for warmer climates. Needs protection where winters are severe. Attractive as it has a pleasant scent. Tuber is subglobose. The leaves with stalk are about 12 in. long, appear in the fall and are good deep, shiny green. Leaf stalk 8–9 in., blade 4–5 in. long and 3 in. wide. Spadix is yellow sometimes inclined to be more white in color. The only *Arum* with a spathe which has a swollen tube. Flowers in April–May, stalks being 10–15 in. in height. The yellow form makes a good garden plant, especially for its fragrance.

A. dioscoridis. Native to southern Turkey and south to Israel as are its varieties. The tuber is large and rounded, often over 2 in. in diameter. The leaves can reach a length of over 12 in. with the stalk being twice or more the length of the blade. The leaves emerge in the winter and are followed by the spathe, which has an unpleasant odor, in April or May, reaching 12 in. in height. Spathe color is quite variable, sometimes almost black with yellowish markings, or ranging from light yellow to pale green. The result is a number of distinguishable varieties: var. *liepoldtii* has a green spathe with purple markings; var. *philistaeum* has a spathe of deep purple which is almost unspotted except at the apex where the color can be greener; var. *smithii* has a spathe which is paler on the interior; var. *spectabile* which has a spathe pale within and yellow-green above with purplish spots; and var. *syriacum* has a yellowish-green spathe spotted dark purple on the lower part. Indeed, the species varies greatly, as many intermediate forms also exist.

A. italicum (syn. *A. modcense, A. numidicum*). Native to Italy, spreading westward into southern France and Spain; introduced 1693. Now found throughout much of the south and west of Europe growing in hedges and rocky places. The leaves reach 18 in. in height appear in the fall of the year followed by the flowers in spring, which reach 18 in. in height. A variable species which has been divided into several subspecies. One of these, ssp. *italicum*, has beautiful foliage with white markings and has long been cultivated in gardens; known in some cases as "var. *pictum*," a syn. of *A. i.* var. *marmoratum*, however, it should now be called *A. i.* ssp. *italicum*. The other subspecies do not have such markings in the leaves. Ssp. *neglectum*, often found in the southern parts of Great Britain and western Europe, has a spathe that is green with a hint of yellow, this spathe being much shorter than the leaves; ssp. *albispathum* is aptly named as the spathe is white inside, it is found from the Crimea all the way to southwestern Asia; ssp. *byzantinum* is from Greece and parts of Turkey, distinguished by a taller flowering spike, greenish with purplish markings. The type species has a pale yellow spathe and a spadix of brighter yellow. Some authorities rank the subspecies as species. *A. italicum* is indeed variable; it is not uncommon to find some markings in the leaves of the subspecies, but such markings are more prominent on *A. i.* ssp. *italicum*.

A. maculatum (Lords-and-Ladies; Cuckoopint). Very common in Britain and other parts of Europe. Leaves 12–18 in. long, often spotted with purple. Spathe is pale yellowish-green on the inside, deeper colored and spotted purple on the outside. Spadix darker yellow. Flower in spring with berries produced in late summer; berries not so attractive as those of *A. italicum*. Common to hedgerows and woodlands. Rampant grower, so plant where there is space to spread, reaches 12 in. in height.

A. nigrum. Native to middle and southern Europe. Tuber flattish, round, about 2 in. wide. While name implies very dark spathe, it is actually a deep purple, a little lighter in color inside. Leaves 18 in. long; flower spikes some 6–8 in. shorter. Has merit in gardens when planted with other dark-colored plants.

A. pictum (syn. *A. corsicum*). Native to Corsica, Sardenia, and Spain. Tuber sub-globose. Like *A. maculatum*, one of hardiest of species. Spathe purple on inside, lighter color outside, reaches 15–18 in. in height. Produced before leaves, which are 18–20 in. long and 3–4 in. wide, and are a good deep green with yellowish veins. Principal merit is that it flowers in fall.

ARUM

A. conophalloides (syn. *A. detruncatum*) Native to Turkey, Syria, Lebanon. Introduced 1889. Spathe—white within margin purplish; blade—green without mauve within (tuber large, flattish round). 8–12 in. in height. April–June flowering.

A. corsicum see *A. pictum*.

A. crinitum see *Helicodiceros muscivorus*.

A. detruncatum syn. of *A. conophalloides*.

A. dracontium see *Arisaema dracontium*.

A. dracunculus see *Dracunculus vulgaris*.

A. elongatum Native to E. Europe. Spathe—brownish purple. 10 in. in height. April flowering.

A. hygrophilum Native to Syria. Introduced 1860. Spathe—purplish or base whitish; blade—pale green margin purple. 12 in. in height. June flowering.

A. magdalenae syn. of *A. palaestinum*.

A. melanoleucum syn. of *Zantedeschia melanoleuca*.

A. orientale Native to Orient. Spathes—maroon-greenish cream spadices (tuber flat-round). 12 in. in height. April–June flowering.

A. palaestinum (syn. *A. magdalenae*) Native to Syria, Lebanon, Israel. Introduced 1864. Spathe—purplish-green outside; inside—purplish-black; spadix—bluish-black (tuber flat-round). 8–12 in. in height. April flowering.

A. pentlandii see *Zantedeschia angustiloba*.

A. philistaeum variety of *A. dioscoridis*.

A. proboscideum see *Arisarum proboscideum*.

A. rehmannii see *Zantedeschia rehmannii*.

A. spirale see *Cryptocoryne spiralis*.

A. tenuifolium see *Biarum tenuifolium*.

Achimenes **'Paul Arnold'**. A delightful plant, at home in containers or hanging baskets. I.B.

Acidanthera bicolor. Confusion over the validity of this genus continues. The other name for this plant is *Gladiolus callianthus.* The long tubular flowers make this plant outstanding, and it is frequently listed in catalogs. AUTHOR

Acidanthera bicolor var. *murielae.* While there may be a discussion over the exact nomenclature of *Acidanthera,* their beauty commands attention. These flowers add much to flower arrangements. WARD

Aconitum napellus. The stems of this group are not generally branched. This photo shows their typical appearance. B.A.G.S.

Plate 17

Acidanthera capensis

Botanical drawings reproduced by courtesy of the Bentham-Moxon Trust.

Plate 18

Aconitum napellus. If the stems branch, and they do occasionally, the lateral branches carry smaller racemes than the terminal branch. As another guide, the majority of this group are lighter in color, most usually in the white/light yellow tones, as seen here. B.A.G.S.

Aconitum noveboracense. The rootstock of this Monkshood is like a turnip. It is native to the east coast of the United States and has the common name of New York Monkshood. It comes into flower in mid-summer and remains attractive for a long period. KITTY KOHOUT

Agapanthus africanus. This photograph, taken in the wild near the Cape of Good Hope, shows the true form of the plant. In cultivation the foliage becomes more lush, and the flowers are not then shown to advantage. AUTHOR

Agapanthus africanus. Whether this is a distance species or not has not been fully determined. Often this form is sold under such names as *Agapanthus africanus* 'White', etc. In cultivation, the flower heads become a little more lax than in the wild. DE HERTOGH

Plate 19

Albuca acuminata. A rather rare species, and because it is not pretty, likely to remain rare. B.R.I.

Albuca altissima. While not an outstanding plant for the garden, in the wild it is quite striking. The reason for the common name, Sentry Box, is not as easy to understand when the flowers are not open, but could be applied to the entire plant, as they stand straight and tall, like soldiers. AUTHOR

Albuca altissima. The same plant showing the flowers carried on drooping pedicels. AUTHOR

Albuca batteniana. Rare, not very attractive, a plant for the collector, but quite easy to grow in warmer climates. AUTHOR

Albuca cooperi. This plant, photographed in the wild, deserves more attention. Quite an interesting looking plant, and planted "en masse", would be an attractive addition to the rock garden. AUTHOR

Allium. In botanic gardens, plants of families are grouped together, to show the varied forms and different genera in a family. Here *Allium* are grouped in the 'Order Beds' at the Oxford Botanic Garden. Such plantings are now rare, but fascinating. AUTHOR

Allium aflatunense. This species, from Central China, is much prized for its color, and is often used with striking effect in flower arrangements. It is a good garden plant. The plants shown are 3 years old. DE HERTOGH

Allium caeruleum. Planted in bold groups, this ornamental onion from Siberia puts on a good show of color in early to mid-summer. Despite flowering during the warmer months, the flowers are quite long lasting. DE HERTOGH

Plate 21

Allium amethystinum

Plate 22

Allium caeruleum

Plate 23

Allium callimischon. Not frequently found in cultivation, but perhaps deserving of more attention. WARD

Allium cernuum. Not an uncommon plant in European gardens, it is not as popular in American gardens, perhaps because it is native to North America, being found in the Alleghany Mountains. Bold plantings can be very effective. WARD

Allium christophii. Though a recent introduction, this species has become quite popular. An excellent cut flower. WARD

Allium cowanii. Regarded as a form of *A. neapolitanum,* this plant has the purest white flowers of any species. I.B.

Allium cyaneum. This species is unusual in that it has bell shaped flowers. Not a common plant in cultivation, but likely to become more popular. A low growing species, it should be planted near the front of a border. WARD

Allium dichlamydeum. Native to the coastal ranges of California, this *Allium* has good color, and while low growing, deserves consideration. It might well prove to be an interesting parent in hybridizing. J.H., by CHARLES S. WEBBER

Allium flavum. The lovely yellow flowers of this species, which is native to southern Europe, is a welcome addition to any garden. WARD

Allium giganteum. This ornamental onion is one of the most popular species, both for planting in the open ground, and in containers. It is always an eye catcher due to its impressive size and color. I.B.

Plate 25

Allium karataviense. Apart from the rather striking flowerheads, which seem out of proportion to the height, the foliage of this species is also quite ornamental. I.B.

Allium moly. This species, frequently listed in catalogs, is a truly dual purpose plant. It will naturalize in the garden and also provides good cut flowers for the home. It is native to the Mediterranean area of Europe. I.B.

Allium narcissiflorum. Not a commonly cultivated species, but the bright color of the flowers are showy. Even though a dwarf species, it is a good plant for the rock garden. WARD

Allium neapolitanum var. *grandiflorum.* Found frequently around the Mediterranean region of Europe, this plant would appear to have spread from northern Italy, as its name indicates that it is from the region of Naples. The species shown has flowers that are a little larger than the type, and is the one most commonly listed in catalogs. DE HERTOGH

Allium moly

Plate 27

Allium paniculatum

Plate 28

Allium oreophilum. It is not surprising that the great bulb growers have selected good forms of this species. This selection, known as 'Zwanenburgh', is offered by Van Turbegen and, with its deep colored flowers, is a good garden plant. DE HERTOGH

Allium pulchellum. There are several species that are native to the Mediterranean region of Europe, but this one is a little different as it flowers in late summer. WARD

Allium pulchellum var. *album.* This plant, to be effective, should be planted closer together than shown in the photograph, however within a few years, this planting will become quite striking. WARD

Allium roseum var. *grandiflorum.* This species, another from the Mediterranean area, is perhaps more at home in a botanic garden than in private gardens. AUTHOR

Plate 29

Allium roseum

Plate 30

Allium siculum. An unusual species that is placed in the genus *Nectaroscordum* by some authorities. The plants are not, upon first glance, thought to belong to the genus *Allium*. A plant that is well regarded, and often made into a feature plant, at exhibits in European flower shows. WARD

Allium siculum. Here in this close up, the rather graceful flowers are seen to advantage, as is the unusual color combination. WARD

Allium sphaerocephalon. Rather an unusually shaped flowerhead, makes this rather common and widespread European species, a good garden plant. Unlike many species, this plant likes summer moisture, perhaps one of the reasons why it is found in the British Isles, as well as most of Europe. DE HERTOGH

Allium wallichii. This plant, native to the Himalayas, is not commonly grown. A robust plant, and of comparatively easy culture. Its beauty is not outstanding, and many other species are more commanding of space in our gardens. WARD

Allium triquetrum

Alophia drummondii. Dr. Boussard, who took this photograph, regards this plant as being the most 'whimsical' of the irids, and remarks that it is his preferred species in the entire family. The glorious colors and form certainly merit such admiration. It is not an easy plant to obtain, despite being a native of Arkansas, Louisiana and Texas. BOUSSARD

Alstroemeria pelegrina. The lovely markings of the flowers are seen to advantage, however today, there are many cultivars on the market to which the species take a backseat, but such species provide a good gene bank for breeders. BOUSSARD

Alstroemeria. **Various cultivars.** In recent years many new and colorful hybrids have been introduced. These plants have the advantage of being long lasting cut flowers, as well as excellent garden plants. As will be seen from these flowers, the color range is wide and, without a doubt, more and even more exciting colors, will be forthcoming from the worlds leading hybridizers. AUTHOR

Plate 33

Alstroemeria aurantiaca

Plate 34

Alstroemeria haemantha

Plate 35

Alstroemeria ligtu

Plate 36

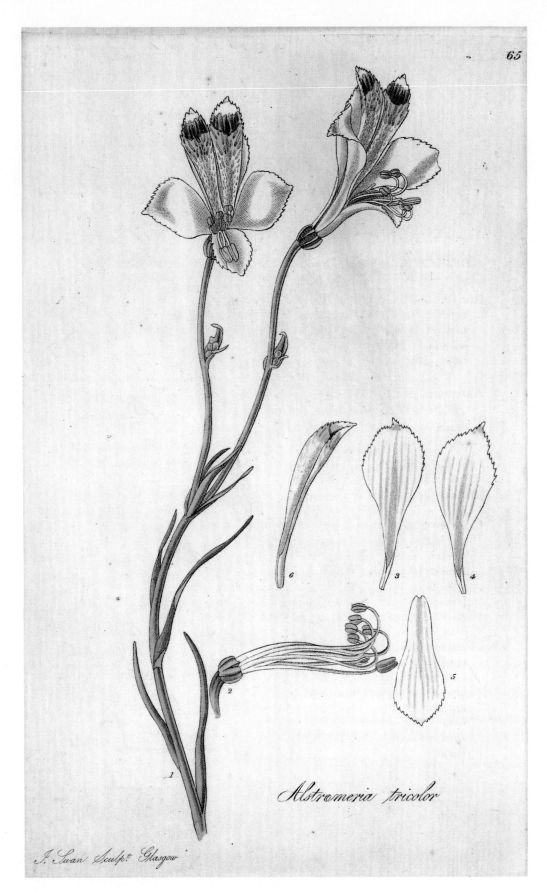

Alstroemeria pulchra

Plate 37

Alstroemeria pulchella

Plate 38

Amaryllis belladonna. Today there are many selections listed in catalogs of this lovely plant. While they each have merit, the species seen here is as beautiful as any selection. MATHEW

Ammocharis coranica. This lovely plant, which is the principle species that remains in the genus *Ammocharis,* (if not the only one), is well worth growing in any garden enjoying a warm climate. Remaining in flower over a long period, these plants do not like being disturbed. BARNHOORN

Anapalina caffra. This lovely plant is a common sight alongside the roads in the north-eastern parts of Cape Province. The flowers on this plant are not arranged on the spike in typical fashion. As can be seen, the flower spikes tower over the foliage. AUTHOR

Anapalina caffra. When this plant grows in subdued light, the color varies. This photo was taken in Tsitsikama National Park in September, which demonstrates the long flowering period of this plant, that deserves more recognition. BOUSSARD

Plate 39

Androcymbium melanthoides. The picture shows the plants in the wild at the end of their flowering period, when the pinkish tones are noticeable. This picture, supplied by the Botanic Research Institute of Pretoria, gives credit to the belief that indeed this is a distinct species. As these plants are very rare, even in their native habitat, I am not convinced that this is not a plant of *A. ciliolatum,* perhaps a regional form. More work on this genus is needed. B.R.I.

Androcymbium **sp.** I am grateful to the Kirstenbosch Botanic Garden for this slide, but even this great botanic institution was unable to give the species of this plant. It does show the unusual form and is 'typical' of the genus. Native of Cape Province, flowering in June–August in the wild. K.N.B.G.

Anemone blanda **'Bridesmaid'.** Amongst the various cultivars of *A. blanda* listed, 'Bridesmaid' is not only easy to grow, but remains a long time in flower. Left undisturbed, the plantings will increase in size each year. AUTHOR

Anemone blanda. **Mixed.** In my opinion these cultivars of *A. blanda* are seen at their best when planted in separate colors, not mixed as here. DE HERTOGH

Plate 40

Anemone coronaria. The single forms of this popular species are in great demand, not only as plants for the border, but also as cut flowers. AUTHOR

Anemone coronaria **'Lord Lieutenant'**. One of the cultivars of the St. Brigid types, an excellent cut flower and great garden plant. DE HERTOGH

Anemone nemorosa. The Wood Anemone is a delightful plant, and loves to grow in areas where the soil is rich in organic matter. While fragile looking, it is a hardy plant that will increase in size and should be planted in any woodland area, as few plants rival it for attractiveness in the spring. AUTHOR

Anemone ranunculoides. Given the right conditions, this plant can become almost invasive in the garden, but it is such a pretty plant, that this can be regarded an attribute, not a disadvantage. WARD

Anemonella thalictroides. A native to the eastern parts of the the United States, this lovely little plant is much hardier, and tougher, than its looks. KITTY KOHOUT

Anomalesia cunonia. The flowers are a delight, and the glaucous color of the foliage and stems is seen here to advantage. When full grown, and in full flower, this plant is outstanding. TRAGER

Anomalesia saccata. While rare in cultivation, this species is deserving of consideration for rock gardens and sunny borders in areas where the winters are mild. K.N.B.G.

Anomatheca fistulosa. It is doubtful if this plant will ever become popular, but has merit, as it looks very pretty growing amongst rocks in its native habitat. BOUSSARD

Anomatheca laxa

Plate 43

Anthericum liliago. For many years this plant was known as *Paradisea liliago* but now is placed under the genus *Anthericum;* they are very closely related. In certain catalogs it may well be still listed under its former name. WARD

Antholyza plicata. While this species is now syn. with *Anaclanthe namaquensis,* it would appear that there is reason for this being separate. It is quite distinct from the other species. B.R.I.

Antholyza ringens. No one could ever call this plant anything but unusual. A dwarf plant, it does deserve being more widely grown, but has to be placed in an area where the flowers can easily be seen, such as near a path. Photographed in its native habitat, which is sandy soil, near the coast. ORNDUFF

Argyropsis candida. For many, this species is *Zephyranthes candida,* but is perhaps deserving of generic rank as it is evergreen. A plant more for the collector than the average gardener. WARD

Plate 44

Arisaema anomalum. This illustration, in the Botanical Magazine of 1891, was made just one year after the plant was introduced into culture from Malaya. J.H.

Arisaema candidissimum. Of easy culture, this species, from western China, is quite attractive and worth a place in boggy areas of the garden. WARD

Arisaema flavum. Not very conspicuous, as the flowerhead is hidden in the foliage. Unusual, and not a common plant in cultivation. WARD

Arisaema helleborifolium. Regarded by some authorities as a var. of *A. tortuosum*, the spadix is certainly following a tortuous route from the spathe! A distinctive looking plant. WARD

Plate 45

Arisaema jaquemontii. Like so many of the genus, an unusual plant, and with it, it is difficult to become enraptured. WARD

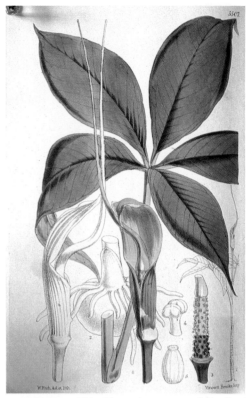

Arisaema neglectum. This illustration appeared in the Botanic Magazine in 1865. J.H.

Arisaema nepenthoides. This must be one of the more attractive species in the genus. Introduced way back in 1879, from the Himalayas, it has never become a very popular plant; perhaps it deserves to be more widely known. WARD

Arisaema pradhanii. This illustration appeared in 1936. A rather unusual plant, because the coloring is quite attractive. J.H.

Plate 46

Arisaema sikokianum. Perhaps worthy of greater use in the garden. The plants are quite attractive and make a rather bold statement. WARD

Arisaema speciosum. The tip of the spadix is not much elongated in this picture, but this species enjoys a fair popularity, perhaps correctly so. WARD

Arisaema tortuosum. A plant that is not commonly grown, a native of India, which will reach to 24 in. in cultivation. WARD

Arisaema triphyllum. Found in the wild along the Atlantic coastline of the United States, this plant enjoys the common name of Jack-in-the-pulpit. Of easy culture, providing that a moist woodsy area is its home. WARD

Plate 47

Arisaema wrayi. This illustration appeared in the Botanical Magazine in 1890, just one year after the plant had been introduced from Malaya. Rather stunning, and quite an unusual plant, it is not commonly grown. J.H., by M. BIRDSEY

Arisarum vulgare. This is a common plant around the Mediterranean area of Europe. While small, with spathes seldom over 2½ in. in length, it flowers over a long period, from the end of winter into spring. Given a good location, with sun and dryer conditions towards the end of summer, this plant can become well established. It is not as demanding as *A. proboscideum*. WARD

Arum creticum. A good species for warmer climates, this plant can be quite distinctive in almost any setting, but especially on the edge of woodlands. WARD

Arum dioscoridis. The colors found in this species vary considerably, all of them are interesting. WARD

Plate 48

Arum italicum **ssp.** *italicum.* While the flowering spike in not unattractive, the red berries carried in the summer and into the fall are of even greater interest. WARD

Arum maculatum. The upstanding spathe is quite a common sight in the hedgerows of Europe. This is the plant to grow, if you wish to establish a good colony of easy to grow plants in woodland settings. It is easy of culture and can become invasive. WARD

Babiana stricta. In this species, there is great variation in the color of the flowers. For this reason, many nurseries offer selections. These plants were photographed in the wild, near Cape Town. AUTHOR

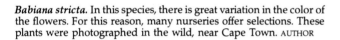

Babiana disticha. On Table Mountain, which rises above Cape Town in South Africa, the open areas are covered with this species in midspring. Due to the plants not being hardy, they are not as frequently grown in the northern hemisphere. This species should be considered by those living in areas with little or no frost, it is of easy culture. AUTHOR

Babiana pulchra. Not a common species, but the intense color makes this plant attractive. No doubt it has merit as a parent in the production of cultivars. AUTHOR

Plate 49

Babiana sinuata. A rare, but aptly named species, 'sinuata' meaning strongly waved; the margin alternately uneven with concavities and convexities. An apt description for this rather untidy flower, native to the South Western Cape area of South Africa. BOUSSARD

***Babiana stricta* 'White King'.** One of the cultivars offered by nurseries. I am of the opinion that these are selections made from many seedlings of the species. Planting, under protection or in spring after danger of frost is past, is a worthwhile operation as these flowers are so pretty. I.B.

Babiana vanzyliae. Sometimes spelt *vanzylii*, this unusual species is very rare in the wild, and even more so in cultivation. The unusual color makes it a prime subject for breeding. Like other species, it flowers in spring, reaches a height of some 6–8 in. B.R.I.

Babiana villosa. These easy-to-grow plants look well in mass plantings. They will increase naturally, and can be counted on to give a good display each year. AUTHOR

Babiana villosa

Plate 51

Bellevalia ciliata

Plate 52

Baeometra uniflora. These striking plants are quite common in the Cape of Good Hope Area, an area rich in bulbous species. The reverse of the petals is a rich orange, but the rather thick texture of the petals does not allow the exterior color to change the pure yellow of the interior. Not fully hardy, but well worth consideration in warmer climates, but, unfortunately, quite a rare plant in cultivation. ORNDUFF

Tuberous begonias are amongst the most showy of summer flowering plants, being at home in the border, or in hanging baskets. They will take a little sun, but much prefer to be in high shade with filtered sunlight. The catalogs are filled with introductions, some have been around for years and are propagated vegetatively; others, raised from seed, will give a wide range of color and flower form. AUTHOR

Belamcanda chinensis. The flowers are short lived, but quite attractive, as many are produced. It is the seeds, that follow the flowers, that last a long time. The dark purple color is quite attractive. BLISS

Biarum bovei. This illustration appeared in the Rumphia in 1835, which would suggest that this plant had been introduced just prior to this date. The plant shows considerable variation in both size and color. J.H.

Biarum tennuifolium* var. *abbreviatum. The spadix bending, the lower part of the spathe being underground, would seem to justify this being a distinct species. This is under discussion. These plants seem to be very variable. The spathe in this species is often bent over the spadix. An added reason why it bends is to get into position so pollination can be facilitated. WARD

Bloomeria crocea. The flowers of this California native last several weeks. The type has just a tinge of orange in the petals, and the perianth segments form more of a cup at their base. There would seem to be too much attention given to minor differences, in the color and petal form, to merit varietal differences in this species, such being perhaps more attributable to geographic modifications. H.C.R.L.

Plate 54

Blandfordia nobilis

Plate 55

***Bloomeria crocea* var. *montana*.** Very similar to the type, but the petals open to give a flatter flower, with the petals recurving directly from their base. J.H., by DONALD MYRICK

***Bongardia chrysogonum*.** This species flowers for a long time, and while the flowers are not large, the plant puts on quite a good show of color. MATHEW

***Boophane disticha*.** The size of the inflorescence matches the size of the bulb, as both will reach over 12 in. in diameter. While introduced into cultivation in the 18th century, it is a rare plant despite being of easy culture, provided warm winter temperatures are given. K.N.B.G.

***Bravoa geminiflora*.** An attractive species, with quite long lasting flowers. A great species for a cool greenhouse. The typical habit of the flowers, being in pairs, can be distinctly seen. LOVELL

***Bowiea gariepensis*.** This genus was, for a long time, considered to be monotypic. Recently a new species, seen in this photograph, was discovered and described. The flowers are somewhat larger and, being white, are more attractive than *B. volubilis*. The habit, time of flowering are the same or nearly so for both species, the principle difference being the color of the flowers, and the perianth tube is longer in the illustrated species. K.N.B.G.

Plate 56

BABIANA — IRIDACEAE

Babiaantje is the Dutch diminutive for "baboon," an animal native to South Africa, as are the *Babiana*. The baboons dig up the corms for food. It is recorded that the colonists at the Cape also ate the corms after boiling. The fibrous-covered corms are quite small.

The leaves are ribbed, tapered, plaited, stout and, along with stems, rather hairy. There are 30 species in the genus and plants seldom surpass 18 in. in height. The flowers are brightly colored, often two-toned. The number produced on a stem varies from 3 to a great many. They have 6 petals and open to a diameter of 2 in. with a hint of a funnel at the base. The flowers often have a heavy fragrance, reportedly intoxicating.

CULTURE Considered tender, thus ideal for areas enjoying Mediterranean climate. Able to withstand colder climates if planted against warm wall in sandy soil. Do not grow outdoors where temperatures fall below 20°F. Planting best done in late summer/early fall, as soon as corms become available. Set 6 in. deep in average and sandy soils, a little shallower in heavy clay soils, 6 in. apart, in full sun. Appreciate moisture during growing season but soil should be well-drained. Add organic matter if soil is very light in texture. If planted in good location corms should be left undisturbed for years.

PROPAGATION Plants can be lifted after several seasons. There will be a number of smaller corms that can be separated out and grown on to flowering size. Raising large stocks in this manner is a slow process as natural production of small corms is not great.

Seed can be sown in areas with milder climates in late summer. In colder areas store seed overwinter and sow in spring in light, sandy soil mix. Barely cover seed. Germination quite rapid. After growing season, lift seedlings and grow on in nursery -rows. Keep moist but not too wet during active growing period. Most plants will take three full seasons to reach flowering size, although a few will take only two. Regular sowing each year necessary for succession of plants. When growing various species in small area, take precautions to avoid cross-pollination by insects. Pollination best done by hand, protecting fertilized flowers from undesired pollen being deposited on stigmas.

PESTS & DISEASES No special problems.

USES Delightful flowers that do not occupy much space. Plant in small, sunny garden where variety is desired. Good container plants; striking flowers; some have a fragrance.

SPECIES & CULTIVARS

B. disticha (Blue Babiana; syn. *B. caerulescens, B. plicata, B. reflexa*). Native to Cape Province, South Africa; introduced circa 1774. Regarded by some authorities as variety of *B. plicata*. Dwarf type, seldom more than 12 in. in height. Pretty yellow-throated flowers with sweet hyacinth fragrance. Flowering May–June. Foliage held in fan, sword-shaped, to 5 in. long, ¾" wide.

B. hypogea. Native to South Africa and parts of Zimbabwe. Leaves 10 in. long. Mauve flowers, close to ground, reaching 2–4 in. only. Late summer-flowering.

B. plicata. Native to Cape Province; introduced 1774. Dwarf type, only about 4 in. high. Fragrant flowers, vary in color from mauve, to cyclamen, to purple, often with white markings at base of petals. Early spring-flowering. Foliage same as *B. disticha*, regarded now as a dwarf form, thus *B. plicata* not regarded as a valid name, but often listed.

B. rubrocyanea (Wine Cup Babiana). Native to Cape Province of South Africa; introduced circa 1795. Regarded by some as subspecies of *B. stricta* but due to its small size, 6 in. in height, deserves own specific name. Common name denotes flowers whose petals have two distinct colors—royal blue at top and rich crimson at base. One of the best species for heavier soils as it needs moisture throughout winter. Early spring-flowering. Foliage lanceolate to sword-shaped, 4 in. long ¾ in. wide, pubescent.

B. stricta. Native to South Africa; introduced 1795. One of taller-growing species, reaching 18 in. in height. Foliage lanceolate to sword-shaped-5 in. long, ½ in. wide. Colors very diverse, ranging from cream to crimson to lilac to blue. Flowering March/April and into May in cooler climates. Many selections offered by commercial nurseries:

- 'Blue Gem'. Violet with purple flakes; 8 in. in height.
- 'Purple Sensation'. Bright purple with white; 16 in. in height.
- 'Purple Star'. Dark cyclamen with white-striped throat; 12–14 in. in height.
- 'Tubergen's Blue'. Large lavender-violet, darker blotched flowers; 12–16 in. in height.
- 'White King'. Snow-white flowers, with pale blue stripes on reverse of petals; lobelia-blue anthers; 12–16 in. in height.
- 'Zwanenburg Glory'. Van Tubergen of Holland, which introduced it, regards this as most beautiful in its collection; outer petals dark lavender-violet, inner same but with large, white blotches; flowers 1¼ in. diameter; plant 18 in. in height.

B. tubata. Native to Cape Province; introduced 1774. Leaves 10–12 in. long, flower stem a little shorter. Flower tube longest of species; colors purple and creamy yellow, often with three red spots at base of outer petals. Flowering late spring/early summer. Now regarded as *B. tubulosa* var. *tubiflora*.

BABIANA

- *B. ambigua* Native to South Africa. 4–5 flowers blue to mauve; lower lobes white or pale yellow. 8 in. in height. Spring flowering.
- *B. caerulescens* see *B. disticha*.
- *B. curviscapa* (possibly closely related to *B. velutina* var. *nana*). Native to Namaqualand, South Africa. Magenta to purple; lower lateral lobes have white markings. 10 in. in height. Spring flowering.
- *B. gawleri* Not a valid name. Described as: Native to Cape Province. Introduced 1802. Lilac-mauve dark purple & white at base. 4 in. in height. Spring flowering.
- *B. hiemalis* Not a valid name. Described as: Native to Cape of Good Hope. Violet-blue. A few inches in height. Feb.–March flowering.
- *B. marcantha* Not a valid name. Described as: Native to South Africa. Cream with a maroon blotch at bottom of cup.
- *B. patersoniae* Native to Eastern Cape Province. Blue. ≅ 12 in. in height. April flowering.
- *B. pulchra* (syn. *B. angustifolia*) Native to South Africa. Bright, deep blue with wine-colored suffusions. 10 in. in height.
- *B. pygmea* var. *blanda* Native to Cape of Good Hope. Red with crimson markings. A few inches in height. Feb.–April flowering.
- *B. reflexa* see *B. disticha*.
- *B. sambucina* Native to South Africa. Introduced 1799. Bluish-purple. 6–9 in. in height. April–May flowering.
- *B. sinuata* Native to S.W. Cape Province. White, greenish brown at base of petals. 8–10 in. in height. Spring flowering.
- *B. socotrana* Native to Socotra. Introduced 1880. Pale violet-blue. 3–4 in. in height. September flowering.

B. stricta var. *grandiflora* Native to South Africa. Introduced 1757. Bright blue with pink. 12 in. in height. May, June flowering.
B. stricta var. *sulphurea* Native to South Africa. Introduced 1795. Cream or pale yellow, anthers blue, stigmas yellow. 9 in. in height. April–May flowering.
B. stricta var. *villosa* Native to South Africa. Introduced 1778. Brilliant crimson, anthers violet blue. 6 in. in height. Aug.–Oct. flowering.
B. tubulosa var. *tubiflora* Native to Coastal South Africa. Creamy white with red markings. 6–12 in. in height. March–April flowering.
B. vanzyliae Native to S.W. Cape Province. Yellow. 1½–5 in. in height. Late Spring–early Summer flowering.
B. velutina var. *nana* Native to Namaqualand. Magenta with white markings on two of the petals. 4–5 in. in height. Spring flowering.

BAEOMETRA — LILIACEAE

The genus gets its name from Greek *baesos* ("small") and the Greek *metron* ("measure"), a reference to the small size of the plant. A monotypic genus, *B. uniflora*, native to the Cape of Good Hope. The rootstock is a small bulb. In the wild it is quite a common plant and would appear to spread by seed quite rapidly. It was introduced in 1787.

Plants reach a height of 8–12 in. There are several leaves, each of them clasping the stem, the lower leaves being up to 9 in. in length and the upper shorter in length. The topmost leaves are only 2–3 in. long. The flowers are on a simple raceme, and there are 4–5 flowers per stem. The color is orange-red on the exterior of the petals and pure yellow on the inside with a black blotch at the base of each petal. The petals are free to the base, being narrower at base to appear almost stalked. The flowers are held erect, and when in flower the plant is quite striking. In the wild it flowers in August–September; April–May in the Northern Hemisphere. The flowers reach a diameter of 1 in. when fully open, and the perianth segments are ½ in. in length, lying almost flat when opened.

CULTURE The plants require sun, well-drained soil, and are suitable for outside culture only in warmer climates. They will withstand some frost, but not severe winters. The bulbs should be planted 2–3 in. deep, spaced 6–8 in. apart. Give adequate moisture early in the spring as soon as growth commences and little moisture after the flowers have passed. In colder climates it requires the protection of a greenhouse. It might be worth trying this plant as a spring-planted bulb and lifting in the late summer for overwintering in a frost-free area.

PROPAGATION Offsets removed while the plants are dormant. Or propagate from seed which should be sown in the spring, given night temperatures in the 50°F range and kept moist with bright, indirect light. They should be transplanted as soon as they are large enough to handle and then grown on to flowering size which will take 2, sometimes 3, seasons.

PESTS & DISEASES No special problems.

USES Good plants for the sunny border, but more for the collector than the home gardener.

BEGONIA — BEGONIACEAE

This genus was named in honor of Michel Bëgon (1638–1710), a French botanist. There are more than 350 species, found in almost every region having a warm, moist climate. With so many different species, it is possible to have *Begonia* in flower all months of the year.

Among *Begonia* are found plants with abundant growth, striking foliage, and beautiful flowers. The flowers, usually showy, can be very large, in a wide range of colors. They are unisexual; the staminate (male) is the showiest. The ovary is inferior, the number of stamens many, and the filaments are free or united at the base. There are 2–4 styles and the stigmas are either branched or twisted. The fruit is a capsule and often is winged, containing many minute seeds.

While the only type of *Begonia* relevant to this book is the tuberous, it is interesting to take note of the many different types in the genus. A good horticultural classification, grouping together those types that require the same culture, has been worked out by Mildred and Edward Thompson and published in the excellent book *Begonias, The Complete Reference Guide* (Times Books, 1981). It was based on the *American Begonia Society Classifications for Show Purposes* published in 1969. Much earlier, in 1914, Liberty Hyde Bailey had put the *Begonia* into four main groups in *The Standard Cyclopedia of Horticulture*. Since then, many horticulturalists have contributed to similar classifications. The Thompsons have refined this prior work in an effort to simplify this rather complex genus.

The horticultural classifications or divisions are: canelike, shrublike, thick-stemmed, semperflorens, rhizomatous, rec cultorum, tuberous, and trailing-scandent.

With one notable exception, species of tuberous *Begonia* are found today only in collections of plants in botanic gardens and those of serious collectors and hybridizers. Few are available to the home gardener as those now offered are cultivars. The exception is *B. grandis* ssp. *evansiana*, known as the Hardy Begonia. It will survive in areas with quite severe winters provided it receives protection, such as a thick mulch. Introduced in 1804, this native of China, Japan, and Malaysia reaches a height of 24 in. The leaves are toothed and deep green on the upper surface while the underside is deep red, as is the leaf stalk. Summer-flowering, the blossoms are very numerous and pink. It has a rather unique habit of producing bulblets in abundance, formed in the axils of the leaves.

The ancestry of the tuberous *Begonia* we grow in our gardens is complex. The first such hybrid was *B.* × *sedeni* in 1870. It was named for John Seden, who worked for James Veitch of Chelsea, London, England. The firm had introduced into culture several tuberous species, the result of numerous plant-hunting expeditions it had sponsored. On one such exploration, between 1865 and 1868, Richard Pearce, who was employed by Veitch, sent back three important bulbs—*B. boliviensis*, *B. pearcei*, and *B. veitchii*. The exact parentage of *B.* × *sedeni* is unknown, being a cross between an Andean species and *B. boliviensis*. This was the first hybrid of many the firm introduced over the next 18 years. The importance of this work and the direct influence on today's hybrids can be seen in the foliage of the tuberous *Begonia*, as the yellows and many oranges show the marbled and red undermarkings inherited from the leaves of *B. pearcei*, named in honor of the man who brought the plant into cultivation.

BULBOUS BEGONIA

The one truly bulbous *Begonia* is *B. socotrana*, which has given rise to numerous hybrids. It comes from the island of Socotra, which came under British jurisdiction in the latter part of the 19th century. The species was found by the Scottish botanist Sir Isaac Bayley Balfour of the Royal Botanic Garden, Edinburgh, and was introduced into culture in 1880. It has a unique flowering habit. It starts into growth in September and continues throughout the winter until March. It then goes dormant, but little bulbs, resting at the base of the stems, survive to give rise to new growth in September. This unusual growth habit has been exploited by various breeders resulting in many fine winter-flowering hybrids. Unfortunately, the species is not commonly grown today due to the difficulty of overwintering. Lemoine, the famous 19th-century nurseryman of Nancy, France, introduced one of the best hybrids, 'Gloire de Lorraine'. Several cultivars are derived from this line. Large specimens grown in hanging baskets can be kept in flower for 12 months or more. They do require careful cultivation but are worth the effort.

CULTURE   In March, hold back water on those plants that have flowered during the winter, but do not allow to dry out completely. During May through August, keep on the dry to very dry side. In late August/early September, buds at the base of older plants, from which dead foliage has been removed, will begin to swell. Raise temperatures, with 65°F minimum at night. Light feedings of liquid fertilizer can be given as soon as growth becomes active. Pot in loose soil mix, such as good topsoil or potting mix, peat moss, and enough sand to keep mix porous. Plants can be grown in cooler temperatures when new growth has been achieved and flowers produced. Continue this culture through the winter months until foliage begins to die down. Then, gradually reduce water and stop feeding so plants enter dormant state, in which they will remain until the following August/September. Plants should receive as much sun as possible during the growing period, except in the most southerly latitudes, where a little shade is needed. In warm climates, such as southern California, place plants outside when in flower.

PROPAGATION   Cuttings of small, over-summering, dormant shoots can be taken in fall when in growth and placed in a frame with sandy soil mix. They root readily and can be potted and grown on to flower.

Seed can be sown. As it is very fine, barely cover, if at all. Seedlings are treated in the same way as the tuberous *Begonia* (see below).

TUBEROUS BEGONIA

CULTURE   Most retail outlets have tuberous-rooted *Begonia* for sale at the beginning of the year, with the exact time varying from region to region. In southerly latitudes, where January temperatures are quite mild (50°F at night), plants can be purchased and planted at that time, in cooler regions later. Make sure tubers are sound and firm to the touch, without blemishes. Discard those showing signs of rotting. Tubers need night temperatures of 50°–55°F to grow well. In northern climes, where days are shorter, additional light may be needed. Gro-Lux or other lamps with full spectrum light for indoor plant growing should be used to give 12-hour days.

Starting tubers into growth must be done only when warmth can be provided. If early flowering is required, early starting is necessary, as tubers take 12–15 weeks to come into flower. Cold frames and unheated greenhouses can be used at later dates when sun will give needed warmth. Tubers can be transfered outdoors after all danger of frost has passed and nights are warm.

As a general guide, tubers should be started in January for early summer-flowering and in March for midsummer-flowering (July). A variety of containers can be used — make sure they are at least 2–3 in. deep, have good drainage, and are clean. Good, sterilized top soil mixed with equal parts peat moss and sand or perlite provides a good mix that is quick-draining but able to retain moisture.

Place tubers in mix and barely cover, with convex side on surface of the soil, concave side below. Roots develop from many different points on the surface of the tuber, hence need for covering top lightly. Space tubers 2–3 in. apart, slightly more for larger ones. Because growth proceeds at different rates and some may need repotting before others, spacing is important so roots of those that have not yet reached sufficient size will not be disturbed when removing others.

Soil mix must be kept moist but not wet; never allow it to dry out completely. Buds on top of the tuber might well be visible when planting; if not, they will soon appear but growth will be slow. When shoots are some 2–3 in. long, remove from container in which they were started and plant in individual pots. These should not be too large, usually a 4 in. pot is large enough, but a 6 in. pot is required for larger tubers. As these plants are choice and only a limited number can generally be grown, I prefer to use clay pots to allow roots to breathe. Plastic pots retain more moisture, so additional sand or perlite in soil mix should be used. With plants in active growth, the amount of topsoil or good potting mix — one part peat and one part sharp sand or perlite — can be increased and a feeding program of weak liquid fertilizer begun.

When shoots have reached 6 in., move into final locations in garden or containers. Container need not be too large as plants do not appreciate much root space. Few tubers will require a pot larger than 8 in. unless more than one tuber is planted. Soil mix is same as that for first repotting, although additional amounts can be added without causing a problem.

Water is important during the succeeding months of growth after potting. While never allowed to become completely dry, plants should ideally become almost dry before being given a good watering. Keep water off crown of plants, especially if temperatures are cool. Amount of light is important. Filtered sunlight is ideal; morning or late afternoon direct sun is all right; no direct sunlight, however, during the heat of the day.

More than one shoot is usually produced. Short, compact types should not be touched unless propagating from cuttings (which see). Those of the large-flowered, picotee, and grandiflora types are best restricted to the 2 or 3 strongest stems by pinching off weaker ones. The number left is determined by the actual size of the tuber — small ones allowed to produce 1 stem, the largest, 3 or perhaps even 4.

Most home gardeners find it traumatic to pinch out the first buds, as they are anxious to see the plants bloom. But this is an important cultural operation that allows the upright-growing and large-flowered single, double, and picotee types to produce more and better flowers. Compact types can be pinched if plants become leggy; pinching of terminal-growth buds in pendulous types will induce branching as well. The number of buds removed depends on the use to be made of the plant. For summer bedding, only the first few buds are pinched off since size in bedding is not so important as mass display; individual plants grown in pots to adorn a deck or patio will look better with fewer but larger flowers. First flowers should be allowed to develop only after plant has reached 10 in. or so in height.

Most plants, except compact growers and pendant types, will need some support. Insert a stake very carefully, so as not to injure the tuber. Tie stem to stake with soft twine, but avoid bruising stem.

Plant tubers in beds or containers that receive good sunlight at all times, but with filtered sunlight at midday, and out of the wind. Wind can cause containers, especially moss-filled hanging baskets, to dry out quickly.

Toward summer's end, when flower production dwindles, stop fertilizer program. Growth of entire plant will slow down; stems and leaves will turn yellow. Cut foliage to within 4–6 in. of the tuber and withhold water completely. Lift the tuber out of the soil and allow to dry. Stems will separate easily from tuber in short time. Clean tubers thoroughly and place in moist, but not wet, bed of peat or perlite. Check them from time to time while in storage to ensure they are moist and healthy; do not allow to become completely dry or they will shrivel. Store at 50°F, with good air circulation until growing time comes around again.

PROPAGATION

Asexual.   Division, the simplest form of propagation with most bulbs, is not possible with tuberous begonias. However, the tuber can be cut into sections and each part grown on to maturity. Only a few cuttings can be made from a single plant.

Cutting Tubers. Start tuber into growth in the manner described above. Larger ones frequently will produce more than 1 bud. Cut tuber with 2 or more buds with sharp knife so each section has its own bud. Grow on the cut sections as with regular tubers. The main disadvantage is that the scar tissue that forms to heal the cut will not produce roots. Thus, the root system is less than from a full-sized, uncut tuber. It is advisable not to divide tubers into too many sections for this reason. It is advantageous to dust cut surfaces with a fungicide to prevent infection.

Despite this drawback, cutting is a commonly practiced method, and many fine plants are propagated in this manner.

Cuttings. Those taken during the growing season will form small tubers by end of the season. Start the tuber in the normal way. When buds are growing well and some 3–4 in. long, remove from tuber by cutting with sharp knife, with a small section of the parent the tuber.

Remove any leaves or tissue other than the stem that will be covered by soil mix. The cutting with small "heel" of parent is placed in rooting medium of equal parts peat moss and perlite. Insert cutting to depth of 1 in. Keep moist at a temperature of 65°F. If possible, keep in a closed case where humidity remains high. Cuttings can be put in individual pots or in a bed on a greenhouse bench. For just a few, construct a plastic mini-greenhouse of polyethylene, or cover with a large glass jar.

Keep propagation area clean to minimize the possibility of infection. Examine cuttings regularly. Remove any foliage that may drop to the surface. A month after insertion, check cuttings for root production.

As soon as a good number of roots is observed, pot into stronger, more nutritious mix and grow on. Removal from high-humidity area, however, should be gradual so cutting is "weaned" from initial environment and becomes accustomed to regular greenhouse conditions. After several weeks in the greenhouse, provided outside conditions are favorable, cuttings can be moved and grown outdoors in the usual way.

Stem cuttings can also be taken later in the plant's life. Side shoots will be produced from central stem(s). When 3–4 in. long, separate from main stem, insert in soil mix, and grow on in the manner already stated.

Seed. The most satisfactory method of raising tuberous begonias to increase these fine plants is from seed. As with any plants grown from seed, slight variations will be seen; i.e., if red forms are purchased, all seedlings will be red, but with minor variations in shade. Interestingly the white will show less variation in strains. Expect slight differences in form as well. While some plants raised from seed might be disappointing, chances are equally good that spectacular plants will be produced.

Begonia seed is very fine—some 2 million to the ounce. Size should be kept in mind as it affects method of sowing. Sow in February or March at latest. Seed must receive 14 hours of light per day at 65–75°F for good germination.

Because seeds are so tiny, sowing surface must be level and firm. A mixture of sifted peat moss and fine grade of perlite works well. Mix should be moist. As no great depth is required, small pans or other shallow containers can be used. One inch of material is sufficient.

When surface is firm, level, and moist, scatter seed. Take care not to sow too thickly. Let seeds drop from packet one by one if at all possible, moving packet over surface to achieve even distribution. When sown, soak container in water until surface is moist; be careful not to bury fragile, tiny seeds. After moistening, cover container with a sheet of clear glass at temperature of 65°–70°F, with full spectrum light for 14 hours photoperiod out of each 24. The glass will accumulate moisture droplets, so wipe each day.

Most vigorous seeds will germinate in approximately a week. Remove glass and aerate during day, but replace during hours of darkness. Remove completely when most seeds have germinated. Some 3 weeks after first seeds have germinated, true leaves will form and seedlings can be transplanted to another container. Space about 2 in. apart; continue giving 14 hours of light daily (such as Gro-Lux) until natural sunlight becomes available for that period of time. Once seedlings sprout, feed with a weak liquid fertilizer. Lower temperatures by about 10°, 55°F at night and 65°F during the day.

Transfer seedlings to individual pots, about 4 in. in size, when leaves start to touch each other. Slowly modify temperature to that of the out-of-doors. When danger of frost has passed and night temperatures are in the 50s, plants can be set out. Nip out all flower buds produced in early stages; those produced from end of May/middle of June on should be allowed to develop. Plants to be grown indoors should be placed in 6–8-in. pots.

By fall, small tubers will have been produced, which can be stored and planted next year, as are tubers purchased in nurseries. Note that tubers formed by seedlings are much more regular in appearance than those propagated from cuttings.

Always keep in mind that sufficient moisture must be provided throughout lifetime of plant. At no time, however, should the medium become waterlogged. Give excellent drainage to ensure that roots are getting enough air.

PESTS & DISEASES

Aphids. A common problem but easily controlled with insectide. Many products do not kill eggs, which are laid in great numbers, so several applications at weekly intervals need to be given. Follow directions on labels. Aphids stick feeding parts into cells of plants, weakening them. Presence of ants often indicates aphid infestation. Examine foliage and growing points regularly and carefully as aphids come in a variety of colors and are difficult to see.

Spider Mites. Mites often go unnoticed until damage is already done. They are very small so easily escape detection. The red spider mite forms webbing, which is more apparent, but the first evidence usually is seen as speckles on the foliage caused by insects sucking on the underside of leaf, thus killing plant tissue. There are a number of miticides on the market. Combination products that control both aphids and mites are available. These broader spectrum products are very good provided they have been cleared for use on begonias. If only one or two plants are affected, isolate or discard them, but spray all others as a precaution.

Scale. Scale is most commonly found on indoor plants but occasionally occurs on outdoor plants. Small, circular or oval, brown-black spots, shaped like a turtle shell, will be found on leaves, mainly along the veins. Insects under the shell move around when young but adults stay put and die, leaving numerous eggs underneath their bodies. These hatch, march around, find their favorite spots, and go through the same life cycle. The young feed from time to time but the most damage is done by adults. There are a number of insecticides available that will kill scale.

Whiteflies. Here, it is the young, not adults, that do the damage. They devour the tender surface of leaves, as well as sucking the sap of the plants. A regular spraying program to kill egg-laying adults (which are highly visible as they fly around), eggs, and all intermediate stages is essential for control.

Mildew. Recognized by small spots on leaves which slowly enlarge, covering tissues with gray mold and causing tissue to die. Mildew is most evident on leaves but stems also can be attacked. High temperatures and poor air circulation contribute to proliferation of fungal problems.

USES Tuberous begonias, in all their different forms, can be used many ways in the garden and for short periods indoors. Pendant types are best grown where characteristics can be seen to advantage, such as an elevated position so foliage hangs over the edge of the container. Very decorative in hanging baskets or baskets attached to a wall. Because shade is desirable in hotter climates or during hottest part of day during summer, pendant types are fine hung from tree branches or in an arbor.

Most forms are effective in beds or containers outdoors. It is preferable to plant in beds with a mulch of small-grade bark chips to avoid splattering leaves with soil. Pots can be brought into the home when their plants are in flower. It is best to circulate containers often as plants will become too leggy if grown indoors for long periods. Spectacular flowers add a touch of elegance to the garden. Just a few plants add interest and color. Texture of foliage in itself is elegant, making flowers just that much better.

A very unusual and effective way to grow these plants is in a hollowed-out log. Provide good drainage. The contrast of foliage and flowers against the texture of wood is eye-catching and a sure conversation piece.

CULTIVARS The popular hybrid tuberous begonias come in a variety of forms. In each case the male, or staminate, flowers are the showiest. On the basis of the characteristics of the flowers and the habits of the plants, cultivars have been classified into 17 categories (see below). In each group many varieties are offered by mail-order firms, nurseries, and other retail outlets. While most are produced in Europe—mainly Belgium, the largest single producer in the world—there are several firms in North America producing new varieties.

Categories

(In many cases, experts will refer only to the numbers, e.g., a flower in category #4 = Cristata)

1. 'Large-flowered Single'—all same color; diameter usually more than 4 in.; tepals entire; cluster of stamens easily seen in center of flower.
2. 'Crispa'—flowers larger than those of 'Large-flowered Single' except that edges of tepals are ruffled.
3. 'Crispa Marginata'—same as 'Crispa' except border of tepals is different color than rest of tepals.
4. 'Cristata'—large flowers; portions of tepals are crested.
5. 'Large-flowered Double'—all same color; not complicated by any variation from 'Large-flowered Single' type except flowers are double; arrangement of tepals varies: those with center like a rosebud called "rose form"; those with flowers not unlike a camellia known as "camellia-like."
6. 'Picotee'—large flowers, more than 4-in. diameter; border of tepal different color from rest of tepal; border color can vary, appearing only at very edge of tepal or as a definite border of little width or diffusing into principal color of flowers.
7. 'Ruffled Picotee'—same as 'Picotee' except tepal edges are ruffled.
8. 'Fimbriata'—similar to 'Ruffled Picotee' except tepals are fringed in addition to being ruffled; overall appearance that of a carnation.
9. 'Grandiflora-Compacta'—as name implies, low-growing and sturdy; this is main difference between it and 'Large-flowered Doubles'; rarely more than 4 in.
10. 'Maxima'—smaller double flowers, 2–3 in. diameter; produced in great quantities; all of one color with no additional characteristics; low-growing and compact.
11. 'Bertinii'—single, pendulous flowers with pointed tepals, produced in goodly number; stems thinner and longer rather than pendulous.
12. 'Bertinii Compacta'—produces both single and double flowers with no special characteristics; differ form 'Maxima' types in that leaves a little larger and are pointed; flower stalks and stems upright.
13. 'Multiflora'—many double flowers produced; smaller than 'Maxima' type, rarely more than 2½ in. diameter; well-rounded, compact plants, shorter than 'Maxima'.
14. 'Pendula'—pendulous habit both in foliage and flowers; grown mainly in shade but will tolerate more sun than other forms; group broken down into "small-flowered," "large-flowered," and "picotee."
15. 'Duplex'—best described as having flowers that resemble a poppy; flowers semidouble rather than fully double; edges of petals ruffled.
16. 'Marmorata'—large, double flowers of more than one color; color appears as variation in tepals with distinct pattern; edges often of darker color.
17. 'Narcissiflora'—so called because they resemble flat-faced daffodils.

BEGONIA

B. acaulis Native to New Guinea. Introduced 1943. Rose-pink (tuberous). To 10 in. in height. Ever blooming under fluorescent lights.
B. andersonii Native to India. Introduced circa 1949. Pink (tuberous). To 10 in. in height. Summer flowering.
B. baumannii Native to Bolivia. Introduced 1890. Fragrant rose red (tuberous). To 10 in. in height. Summer flowering.
B. biserrata (syn. *B. palmaris*) Native to Guatemala. Introduced 1847. Rose white (tuberous). 12 + in. in height. Summer flowering.
B. bogneri Native to Madagascar. Introduced 1973. Pale pink (tuberous, semituberous). 8–12 in. in height. Late Winter to Fall flowering.
B. boliviensis Native to Bolivia. Introduced 1859. Red (tuberous). 24–36 in. in height. Summer flowering.
B. bulbillifera Native to Mexico. Introduced 1831. Rose (tuberous). 12 + in. in height. Summer flowering.
B. caffra syn. of *B. homonyma*.
B. cavum Native to Mexico. Introduced 1948. White (tuberous). To 10 in. in height. Summer flowering.
B. cinnabarina Native to Bolivia. Introduced 1849. Fragrant, orange red (tuberous). To 10 in. in height. Summer to Fall flowering.
B. clarkei Native to Bolivia & Peru. Introduced 1867. Deep rose (tuberous). To 10 in. in height. Summer to Fall flowering.
B. crenata Native to India. Introduced 1791. Pink (tuberous). To 10 in. in height. Summer flowering.
B. davisii Native to Peru & Bolivia. Introduced 1876. Orange scarlet (tuberous). To 10 in. in height. Summer flowering.
B. dregei Native to South Africa. Introduced 1836. Bluish white (tuberous). 12–36 in. in height. Summer to Fall flowering.
B. evansiana syn. of *B. grandis* ssp. *evansiana*.
B. evansiana var. *alba* Of garden origin. Introduced 1940. White, pink tinted (tuberous). 12 in. in height. Summer flowering.
B. fimbristipula Native to China. Introduced 1883. Fragrant, pink (tuberous). To 10 in. in height. Summer flowering.
B. flanaganii syn. of *B. partita*.
B. froebelii Native to Ecuador. Introduced 1874. Scarlet (tuberous). To 10 in. in height. Late Summer to Winter flowering.
B. gemmipara Native to Himalayas. Introduced 1855. Fragrant white, tinted rose (tuberous). To 10 in. in height. Sept.–Oct. flowering.
B. gracilis Native to Mexico. Introduced 1825. Pink (tuberous). 12 + in. in height. Summer flowering.
B. gracilis var. *martiana* (syn. *B. martiana*) Native to Mexico. Introduced 1864. Fragrant rose pink (tuberous). 12 + in. in height. Summer flowering.
B. grandis ssp. *evansiana* Native to China. Introduced 1939. Fragrant pink (tuberous). 12 + in. in height. Summer to Fall flowering.
B. homonyma (syn. *B. caffra*) Native to South Africa. Introduced 1840. White (tuberous, semituberous). Spring to Fall flowering.
B. ignea Native to Guatemala. Introduced 1864. Pink (tuberous). 12 + in. in height. Summer flowering.
B. josephii Native to India. Introduced 1859. Rose red (tuberous). To 10 in. in height. Summer flowering.
B. macbethii syn. of *B. dregei* 'Macbethii'

B. martiana syn. of *B. gracilis* var. *martiana*.
B. micranthera Native to Argentina. Introduced 1874. Orange red (tuberous). 12 + in. in height. Summer flowering.
B. micranthera var. *fimbriata* Native to Argentina. Introduced 1941. Fragrant orange red (tuberous). 12 + in. in height. Summer flowering.
B. micranthera var. *foliosa* Native to Argentina. Introduced 1941. Red (tuberous). 12 + in. in height. Summer flowering.
B. micranthera var. *venturii* Native to Argentina. Introduced 1941. Apricot (tuberous). 12 + in. in height. Summer flowering.
B. monophylla Native to Mexico. Introduced 1859. (Tuberous). To 10 in. in height. Summer flowering.
B. morelii Native to Tropical Asia. Introduced 1975. White very pale pink tinted (tuberous, semituberous). Everblooming.
B. natalensis Native to Natal. Introduced 1855. Pink tinted yellowish white (semituberous, tuberous). 18 in. in height. Winter flowering.
B. octopetala Native to Peru. Introduced 1788. White (tuberous). To 10 in. in height. Oct.–Nov. flowering.
B. ovatifolia var. *cretacea* Native to Himalayas. Introduced 1879. White (tuberous). To 10 in. in height. Everblooming under fluorescent lights.
B. palmaris syn. of *B. biserrata*.
B. partita (syn. *B. flanaganii*) Native to South Africa. Introduced 1961. White (semituberous, tuberous). Spring to Fall flowering.
B. pearcei Native to Bolivia. Introduced 1865. Orange yellow (tuberous). To 10 in. in height. Summer to Fall flowering.
B. picta Native to India. Introduced 1807. Fragrant pale rose (tuberous). To 10 in. in height. Everblooming under fluorescent lights.
B. princeae Native to Tropical Africa. Introduced 1902. White (tuberous). 12 + in. in height. Summer flowering.
B. pygmaea Native to Zambia. Introduced 1961. (Semituberous, tuberous).
B. richardsiana syn. of *B. suffruticosa*.
B. rosiflora syn. of *B. veitchii*.
B. sikkimensis Native to Sikkim. Introduced 1859. Bright red (tuberous). To 10 in. in height. Summer flowering.
B. suffruticosa (syn. *B. richardsiana*) Native to Natal. Introduced 1840. White (semituberous, tuberous). Summer to Fall flowering.
B. sutherlandii Native to South Africa. Introduced 1868. Yellow orange (tuberous). To 10 in. in height. Summer to Fall flowering.
B. sutherlandii var. *latior* Introduced 1961. (tuberous). To 10 in. in height.
B. veitchii (syn. *B. rosiflora*) Native to Peru. Introduced 1867. Fragrant vermillion (tuberous). To 10 in. in height. Summer flowering.
B. viscida Native to Mexico. Introduced 1969. Pale green (tuberous). 12 + in. in height. Summer flowering.
B. williamsii syn. of *B. wollnyi*.
B. wollnyi (syn. *B. williamsii*) Native to Bolivia. Introduced 1909. Greenish white (tuberous, semituberous). Winter flowering.

BELAMCANDA — IRIDACEAE

Genus name is derived from the name used in India for the plants. The genus has two species, both native to eastern Asia. The roots are said to be used as an antidote for the bite of cobras.

Belamcanda are tuberous-rooted, herbaceous perennials. The foliage is deep green, iris-like, and encircles the stem. The lower basal leaves are larger than those above. The flowers, which individually do not last very long, are produced in loose heads bearing up to 12 blossoms. Flowers are always spotted, up to 2 in. in diameter, narrow at the base, and, when fully open, almost flat; the tube is very short. Colors vary but they are always in the yellow, orange-red range with deeper red-brown spots. Both species flower in mid- to late summer and produce dark purple-black seeds which are visible as the seed pods open. The attractive seeds, which resemble the fruit of the blackberry, give this plant its common name, Blackberry Lily.

CULTURE Not fully hardy—needs protection against cold during winter in regions where temperatures drop below 15°F. In such areas lifted plants store overwinter, and replant in spring. Preferably planted in sandy, well-drained soil with humus, possibly against a wall in cooler climates but in warmest areas in high shade. Set in place with only an inch of soil over tubers, in fall and spring in warmer areas (preferably fall) and spring in cooler areas. Plant 6–8 in. apart. Keep moist during growing season.

PROPAGATION Gather seed in fall when easily separated from pod. Seed can be sown as soon as ripe if given temperatures around 65°F; however, more practical to sow in spring when temperatures remain warm. Sow outdoors in warmer areas, barely covering seed. Can be started indoors in flats in sandy soil mix, sowing some 30 days before last frost. Young plants can then be grown individually in small pots to be planted out after all danger of frost is past.

Tuberous roots can be divided in fall and planted back in milder climates or overwintered for replanting in spring. Store divided roots in very sandy soil mix, keeping mix on dry side but not so dry that tubers shrivel.

PESTS & DISEASES No special problems.

USES Plant to give added interest to shrub and herbaceous borders in mid-to late summer even though beautiful flowers are short-lived. Opening seed pods prolong interest. Unusual material for dried flower arrangements.

SPECIES

B. chinensis. Native to eastern Asia; introduced 1823. Grows to height of 24–36 in. Leaves iris-like, rich green, enclasp stem; lowest leaves are longest. Flower stalk tends to zigzag as it grows. Variable-colored flowers, generally in deep orange-red tones with darker spots of red-brown flowers in loose head, 3–12 in a head. When open, flowers are flat, 2 in. in diameter. Seed pods display as many as 15 seeds resembling a loosely formed blackberry.

B. flabellata. Native to eastern Asia. Differs from *B. chinensis* only in color and height. Color is good yellow spotted with orange. Height generally under 24 in. Despite the two variations, some botanists believe there is but one species, *B. chinensis*. Introduced a little later than *B. chinensis*.

BELLEVALIA — LILIACEAE

While there is a number of *Bellevalia*, the majority are of little or no horticultural interest. *B. paradoxa*, native to eastern Turkey, is described under *Muscari paradoxum* (which see), where it is now more generally found in the literature. Some authorities, however, still list it as a separate species under *Bellevalia*, as some slight differences do exist, i.e., the flowers are not constricted at the mouth and the anthers lie beneath the lobes, carried in the mouth. It is doubtful if any of the species will find their way into cultivation; however, the following list of species should be on record.

BELLEVALIA

B. atroviolacea Native to N.E. Afghanistan & adjacent U.S.S.R., Central Asia. Introduced circa 1973. Deep indigo blue to purple. 6–8 in. in height. April flowering.

B. aucheri Native to Turkey. Introduced 1939. Brownish. 6–8 in. in height. April–July flowering.

B. brevipedicellata Native to S.W. Crete. Whitish. To 6 in. in height. Feb.–March flowering.

B. ciliata (very similar to *B. sarmatica*) Native to Greece, Southern Italy, Turkey. Pale-purple turning greenish-brown. 6–14 in. in height. March–April flowering.

B. clusiana Native to Turkey. Introduced 1843. Top of raceme violet, lower flowers brownish. 6–8 in. in height. April–July flowering.

B. crassa Native to N.E. Turkey, Black Sea. Introduced 1980. Pale color. Dwarf. April–July flowering.

B. dubia Native to Italy, Yugoslavia, Greece & Turkey, Austria, Crete, Sicily, Portugal. Introduced 1939. Bright steely blue or violet in bud & turning greenish-brownish. 12–16 in. in height. March–April flowering.

B. fomini Native to Transcaucasia, S.E. Anatolia, Turkey. Introduced 1927. Violet in bud, opening grayish-lilac with blackish lobes. 6–8 in. in height. April–July flowering.

B. forniculata (syn. *Muscari forniculatum*) Native to N.E. Turkey, U.S.S.R. Introduced 1922–1923. Bright-blue. Dwarf. April–July flowering. [A species with possibility of garden merit].

B. gracilis Native to Turkey. Introduced 1939. Yellow or whitish in bud, turning brown. Dwarf. April–July flowering.

B. hyacinthoides see *Strangweia spicata*.

B. kurdistanica Native to S.E. Turkey, N. Iraq. Introduced 1939. Pale lilac turning whitish. 6–8 in. in height. April–July flowering.

B. latifolia Native to S.E. Turkey, N. Iraq. Introduced 1939. Greenish in bud, opening white with green veins, turning brownish. 6–8 in. in height. April–July flowering.

B. lipskyi Native to Crimea. Deep dull purple turning yellowish-brown. 6–14 in. in height. April–May flowering.

B. longipes Native to Syria, Iran, Iraq. Brownish. To 15 in. in height. March–April flowering.

B. longistyla Native to E. Turkey, Iran, Caucasus. Introduced 1928. Dull purple in bud, turning brownish or dirty white. 6–8 in. in height. April–July flowering.

B. modesta Native to Turkey. Introduced 1980. Creamy with purple-brown shading. 6–8 in. in height. April–July flowering.

B. paradoxa (syn. *Muscari paradoxum*—Which see for a more complete description) Native to Eastern Turkey. Introduced 1884. Dull blue, anthers yellow. 6 in. in height. April–July flowering.

B. pycnantha (syn. *Muscari pycnanthum*) Native to E. Turkey, N. Iraq, N.W. Iran, Soviet Armenia. Introduced 1935. Dull blue or blackish blue with yellow margins. 12 in. in height. May–June flowering.

B. rixii Native to Eastern Turkey. Introduced 1980. Deep violet-blue, anthers violet. Dwarf. April–July flowering.

B. romana (syn. *Hyacinthus romanus*) Native to S.W. France, Corsica, Greece, Italy, Sicily, Yugoslavia. Dirty white. 6–14 in. in height. April–May flowering.

B. sarmatica (syn. of *B. trojana*, closely related to, or syn. with *B. ciliata*.) Native to Turkey & E. Mediterranean. Introduced 1927. Purplish, turning brownish-purple. 6–14 in. in height. April–July flowering.

B. tauri Native to Cicilian Taurus. Introduced 1939. Brownish-purple. 6–14 in. in height. April–July flowering.

B. trifoliata Native to Crete, Italy to Greece, Turkey, Egypt & has become naturalized in Southern France. Introduced 1843. Dull violet or purplish, margin greenish. 6 in. in height. March–May flowering.

B. trojana syn. of *B. sarmatica*.

B. warburgii Native to Syria & Israel. Brownish-purple. 6–14 in. in height. April–July flowering.

B. webbiana Native to Northern Italy. Deep purplish turning brown. 6–14 in. in height. April–May flowering.

BESSERA — AMARYLLIDACEAE

The genus is named in honor of Dr. W. S. J. G. von Besser (1784–1842) who was Professor of Botany at Brody in the U.S.S.R. A genus of some two or three species but only one of which, *B. elegans*, is mentioned in the literature. Native to Texas and Mexico.

This genus will be found placed in various families. The flowers have superior ovaries and thus some place them in Liliaceae. However, the flowers are carried in umbels, hence their placement in the Amaryllidaceae. In recent time such flowers have been placed in Alliaceae, adopted for use for those plants with superior ovaries but with flowers in an umbel. *Bessera* have a tunicated corm which when mature is 1 in. in width. The plants have narrow and linear basal leaves and are often furrowed. The flowers form an umbel on a hollow flower stalk, are bell-shaped with a short tube, and then the perianth segments spread widely, not unlike the flowers of garlic. Flower color is coral or lilac and thus *B. elegans* gets its common name of 'Coral Drops'; some authorities say they resemble the flowers of snowdrops.

CULTURE The plants are not hardy and will not withstand any frosts. In areas where such occur they should be grown in containers and protected during the colder months. They like full sun, soil mix that has good drainage with a high organic content. The corms should be planted 4–5 in. deep in the spring, spacing them some 12 in. apart. Ample moisture should be given while the plants are in active growth, reducing the amount given as the foliage matures. During the winter months they should be kept dry. The species *B. elegans* flowers in mid- to late summer, July–Sept.

PROPAGATION Cormels are produced and these can be removed in the fall, after the foliage has died down. They should be stored overwinter and sown in a sandy, well-drained compost in the spring. Grown on they can be transfered to nursery rows the following spring; those of over ¼ in. in diameter can be planted out in a warm, sunny border. Seed can be sown in the spring, seedlings transplanted as soon as large enough to handle and then grown as per cormels. A light sandy soil mix should be used. At all times the plants must be kept moist during their growing period and then kept dry until growth recommences.

PESTS & DISEASES No special problems.

USES Collector's item. In warm climates plant in sunny borders.

SPECIES

B. elegans. Native to Mexico; introduced 1830. The only *Bessera* that is cultivated. The foliage is basal, up to 30 in. in length, green, linear, with a furrow on the upper surface. The flowering spike reaches some 2 ft in length, with up to 30 flowers in an umbel. The color of the flowers is variable, from scarlet to purple with green stripings on the outside and whitish on the inside. The stamens are longer than the perianth segments and the filaments are joined in a tube for about half of their length.

BIARUM — ARACEAE

The name was used by Dioscorides for a related plant. This is a small genus containing only a few species native to regions around the Mediterranean and the Middle East. In their native habitat, the plants are found in rocky places where there is an accumulation of soil. Being native to the Mediterranean area, they like the sun; they like to receive their rainfall during the winter months, with a long, warm, dry period in the summer. As they are frequently found at quite high elevations, they can withstand some frost, but in areas where temperatures fall below 25°F they should be protected.

Biarum tubers are whitish and fleshy and are found just below the surface of the soil. The spathes, which appear before the leaves, are distinctive in that they are stemless and are produced at or just below ground level. Their colors will vary according to species but are mainly in the white to green-white range, frequently covered with spots of maroon, which may be so dark as to appear black. The flowers on the spadix are maroon or dark purple, with the male and female flowers separated by a series of sterile, hairlike flowers. The fruits are white or greenish, looking for all the world like a basket of small, spherical eggs nestling on the ground. The majority of the species are unpleasant to smell, but this attracts their pollinators—flies and crawling insects.

CULTURE *Biarum* should be grown in moderately good soil in areas with perfect drainage and full sun. Little or no moisture is needed during the summer months; winter moisture is essential. The plants must be allowed to be warm and dry during the summer for them to flower well. They should be planted with the tubers just below the surface of the soil—a depth of about 2 in. is sufficient. Spacing can be 4–6 in. apart. Once established they should go undisturbed. Feeding is not necessary. Although they can withstand some frost, it is a good idea to protect them with a mulch in those regions where an unexpectedly heavy frost might occur.

PROPAGATION The tubers can be lifted and divided, which is best done in the summer while the plants are dormant. However, seed is the best means of propagation, except in *B. tenuifolium* (which see). The seed should be harvested as soon as ripe and sown in the early spring in well-drained, sandy soil, barely covered. They will soon germinate if placed in a warm location with temperature in the 55°F range at night. As soon as they are large enough to handle, place the young plants in individual containers, preferably peat pots so there is little disturbance to the roots when planted out in their final spot in late summer or early fall.

PESTS & DISEASES There are no particular problems; however, if there are slugs and snails around, protect the plants by using snail/slug bait.

USES These small plants should be considered only for a rock garden, and then planted close to the path so they can be appreciated. They do not make suitable container plants, and, while they can be grown in a warm, sunny border, they look their best in locations that duplicate their natural habitat—rocky, sandy locations in full sun.

SPECIES

B. bovei. Native to the Mediterranean; introduced 1860. The subglobose tubers often are found among sparse scrub or in bare areas. The 5–10 leaves produced are almost an inch wide and some 2–4 in. long, the stalk up to 6 in. long. The spathes are produced at ground level, with the apex varying from quite blunt to erect and long-tapering; thus, the size also varies from 3–4 in. to almost twice that length. The color also varies from dark green to dark brown. The spathe appears quite inflated toward the base and is produced before the leaves emerge, sometimes at the same time, or even after. No sterile flowers are found above the male. While principally in flower in late summer/early fall, on occasion it can be spring-flowering.

B. carratracense. Native to the southern part of Spain in Malaga Province; introduced in 1860. While some authorities regard it as a separate species, others regard it only as a geographic variation of *B. bovei*.

B. davisii. Native to Crete; introduced 1854. Found growing on arid hillsides and in rocky crevices. Considered one of the most attractive of this genus as it does not have the unpleasant smell common to most of the other species. The spathes are produced at ground level reaching 3–4 in. in height, the bottoms often just below the surface of the soil. Light cream in color, covered with deep maroon spots; spathes are swollen into a bladder shape, with the cup about 2–3 in. across. The pointed apex, which forms a hood over the spathe, is about 4–5 in. high. They are quite open at the top. The spadix is maroon at the top, yellowish below, and barely protrudes from the spathe. The 3–8 leaves that appear after the spathes are short, only about an inch long, and barely ½ in. wide. Flowering in November, just before the leaves.

B. eximium. Native to central Turkey; introduced 1854. Found growing in stony ground. It is thought by some to be the best species because the spathe is quite large, taller than it is wide, and often up to 4 in. high. The lower part is cylindrical and the upper part quite flat. The exterior is variable in color, from whitish-green to pure white, spotted with maroon dots; the interior is very dark maroon. The spadix is thick and long, often as long as the spathe, and upright. The ovate-oblong leaves, 3–4 in. long, narrowing to the stalk, appear after the spathe, in the fall. Total height of plant is under 12 in.

B. kotschyi. From the same areas of Turkey as *B. eximium*, which it resembles greatly. *B. kotschyi*, however, has a narrower spathe, which has a tendency to curl over at the tip. The lower parts are bottle-shaped and closed to a greater degree, expanding to a distinct blade after narrowing at the neck. Flowering in September, before the leaves emerge, under 10 in. in height.

B. spruneri. Native to Greece; introduced 1894. Found growing in rocky ground in exposed, sunny spots. This species is quite similar to *B. tenuifolium* in appearance, but some of the differences are that the spadix does not grow beyond the spathe, there are no sterile flowers above the male, and it is in flower much earlier, April–June. The leaves are spatulate to lanceolate, and the spathe is purple and greenish. Total height only 6–8 in.

B. tenuifolium (syn. *Arum tenuifolium*). Native to the central and eastern Mediterranean regions; introduced 1570. Found growing among rocks in full sun and is one species of this genus that increases readily by offsets. The spathe is erect and narrow, often twisting, up to 6–8 in. high, pale greenish-brown to almost chocolate in color on the inside and greenish exterior. The spadix is slender and erect, up to 8–10 in. in length, and a deep blackish-purple. It carries male flowers, with sterile flowers both above and below. The flowers are produced in July-November, before the leaves. The leaves vary considerably and can be only a few inches in length or up to 6–8 in., sometimes with wavy edges, but generally narrow, seldom over an inch wide, and often quite grasslike. They may appear before or after the spathe. Reaches 10 in. in height.

B. tenuifolium var. ***abbreviatum.*** Native to northern Greece and Italy. A number of variations exist between this plant and *B. tenuifolium*, so *B. t.* var. *abbreviatum* could be a distinct species. Its principal distinguishing characteristics are: the lower part of the spathe is white and found below ground, with the remainder only about 4–5 in. in length, hooded over and causing the spadix to bend in order to appear outside the spadix cup; the upper part of the spathe is bright green, with a blackish-purple interior; the spadix is blackish. Flowers in September, at about the same time as the leaves are produced. A little shorter in height than *B. tenuifolium*.

COMMENTS   While these are unusual flowers and, therefore, well worth growing by the collector, only one species—*B. davisii*—is truly beautiful and also lacks the rather disgusting smell describable as rotting mutton. However, since *B. tenuifolium* is the species that increases the most rapidly, you should try growing this species first if you are considering this genus at all.

BLANDFORDIA — LILIACEAE

This genus is named in honor of George Spencer-Churchill (1766–1840), an ancestor of Sir Winston Churchill. He became the Marquis of Blandford and later the fifth Duke of Marlborough. The genus was first introduced and described early in the 19th century.

The plants, of which there are about 4–5 species, are native to Australia and Tasmania, found growing in sandy places or moorland up to heights of 4000 ft. In their native habitat they are known as 'Christmas Bells', as they are in flower at that time of the year there; in the Northern Hemisphere they flower in midsummer.

Plants are rhizomatous. The leaves are linear, long, and pointed. The lower leaves sheathe the stem, and, depending on the species, there are several stem leaves that are upright in habit and some 4–6 in. long and up to an inch wide. The basal leaves, which sheathe the lower part of the stem, are up to 36 in. long and are plicate at the base, similar to the Dutch iris.

The flowers are most attractive, subtended by bracts that are narrow, pointed, and generally brownish in color. The flowers grow in umbels or racemes on recurved pedicels and are tubular, as the perianth is funnel-shaped. They are mostly 1½–2 in. long and in red or yellow tones, some being of both colors. Often as many as 15 are found in a flower head. The width of the open flowers varies, some opening like a bell, others remaining tubular. The seed pods are held vertically, often with the remains of the withered flowers at their bases.

CULTURE   When planting, rhizomes should be set some 3–4 in. deep and spaced some 8–10 in. apart in well-drained soil. *Blandfordia* do not like temperatures below 35°F at night during the winter, their dormant season. In climates where winters are mild and temperatures remain above the 40°F mark, they can be planted in sunny borders. In other climates, winter protection is needed, such as that provided by a greenhouse or a cold frame. In these areas heat is needed to maintain minimum temperatures above 35°F at night, preferably in the 45°F range.

Provide moisture at the first sign of growth and continue right through until the foliage starts to mature in late summer/early autumn. Then, the plants should be kept barely moist, watered only when the pots are really dry. Repotting, if plants are grown in containers, should be done after the foliage ripens.

Weak feedings of fertilizer, preferably organic liquids, can be given once growth is active. The plants should receive good light or full sun, except in the hottest regions. After the foliage has died back, it should be removed, whether the plants are being grown in a border outdoors or in containers.

PROPAGATION   Division of the rhizomes after the growing season has finished is a good method of propagation. Offsets also will be produced, and these can be separated from the parents. Seeds can be sown in the spring, barely covered with a sandy soil mix. The young plants can be transplanted to small pots and grown on or, in warm climates, lined out in nursery rows. Protect them from strong sunlight. Will take 2–3 seasons to reach flowering size from seed.

PESTS & DISEASES   No special problems. Should aphids be found, control them with a product cleared for use on lilies.

USES   While for years regarded as great greenhouse plants and so cultivated in many countries in display greenhouses, there is a good case for these plants being used in sunny borders in warmer climates. In colder regions they are good container plants if potted in the fall, kept barely moist until growth commences when water can be increased, kept warm at all times, and given good light once the foliage appears.

SPECIES

B. flammea. Native to New South Wales; introduced 1849. Grows up to 24 in. in height, with bell-shaped, drooping flowers in an umbel rising above the foliage. The species color is a matte yellow. Var. *aurea* has brighter yellow flowers; var. *elegans*, reddish flowers tipped with yellow; and var. *princeps*, orange-red flowers on the exterior, yellow on the interior. All are June-flowering.

B. grandiflora (syn. *B. cunninghamii*). Native to Australia; introduced 1812. The leaves are rough along the edges. The 2 in. long flowers reach 18–24 in. in height. They are produced in racemes and are in a narrow tube and then flare out. The yellow flowers have a red base; the flowers of var. *intermedia* are completely yellow. July-flowering.

B. marginata. Native to Tasmania; introduced 1842. The leaves are almost erect and rough along the margins. The orange-red flowers are narrowly tubular, gradually tapering at the base, and are found in a crowded raceme which has many blossoms; July-flowering. This may be a variation of the other Tasmanian species, *B. punicea*, though both are valid names.

B. nobilis. Native to New South Wales; introduced 1803. The plant reaches a height of 35 in. The leaves are toothed along the margins. About an inch long, the orange flowers with yellow margins are found on long pedicels, are very narrow at the base, and flare to an almost straight tube. The number of flowers on the loose raceme is restricted; July-flowering.

B. punicea. Native to Tasmania; introduced circa 1840. Perhaps the most beautiful of the genus. The leaves are 18–24 in. long, rather coarse and narrow, with rough edges. Flower heads reach 36 in. in height, carrying as many as 20 1½ in. flowers. They are crimson with yellow interiors and tips, tubular, narrow at the base, pendant on short pedicels, and subtended by the brownish spathes. The compact racemes are carried on a greenish-brown stalk with few stem leaves. This species is the most free-flowering and most deserving of culture. July-flowering; December in its native habitat.

B. aurea. See *B. flammea* var. *aurea*.

B. cunninghamii. See *B. grandiflora*.

B. intermedia. See *B. grandiflora* var. *intermedia*—sometimes raised to specific rank.

B. princeps. See *B. flammea* var. *princeps*—sometimes raised to specific rank.

BLOOMERIA — AMARYLLIDACEAE

These native California plants, of which there are only two species, are named for H. G. Bloomer (1821–1874), a San Francisco botanist. The corms are covered with a fibrous coat. The few leaves produced are long (12 in. or more) and narrow, carried at the base of the stems. The flowers are yellow and similar to *Brodiaea*, except that the perianth segments are free to the base. There are 6 stamens, which form a sheath around the ovary. The flowers are carried in umbels and blossom in late spring/early summer. The height of the plant can vary from 12–18 in. *Bloomeria* are found growing in various locations, from sea level up to 5000 ft. *B. clevelandii* is found in warmer areas then *B. crocea* and its variety *B. c. aurea*. The common name of the plant is Golden Stars.

CULTURE Best planted in spring unless grown in a warm climate with few or no frosts. Protect well or lift in late summer and store overwinter where frosts are severe. Provide adequate moisture in spring and early summer, after which allowing soil to dry out. At no time during culture should ground be waterlogged. Plant in a sunny location.

When planting in pots, use sandy soil mix. Allow pots to dry out in protected location over the winter months.

PROPAGATION Easiest method is to lift and divide corms at end of summer or early fall. Corms propagate readily. Seed can be sown in shallow pans in early spring. Use sandy soil mix and barely cover seeds. Keep at 65°F. As soon as seedlings are large enough to handle, transplant into individual small pots, about 3 in., and grow on during summer. Plant out in fall or hold over until following spring, depending on climate. Seedlings will flower in second full season.

PEST & DISEASES No special problems.

USES Adds interest to flower borders and when planted among shrubs. Best if left undisturbed as clumps will multiply and become of greater importance. Looks good in rock gardens, but position with care so rather tall flower spikes will not overpower neighboring plants. For all intents and purposes *Bloomeria* should be regarded as yellow *Brodiaea*, and combined plantings of both are effective, especially if you select species that flower at same time. Also grow well in pots and make excellent cut flowers.

SPECIES & VARIETIES

B. clevelandii. Native to California; introduced 1896. Shorter than *B. crocea* and flowers not so brilliant. Shorter flowering time.

B. crocea and var. *aurea.* Most common of genera. Natives to southern California. Spring flowering, lasting several weeks. *B. crocea* has orange-yellow flowers and darker colors on outer petals; var. *aurea* (syn. *Nothoscordum aureum*) has attractive umbels of bright yellow flowers. Both produce a single leaf, which must be protected against slugs and snails or entire plant will be lost. Stems reach 12–18 in. in height. When well-grown and established, flower spikes may reach 24 in., with corresponding increase in number of flowers produced. Var. *aurea* was first introduced in 1869; its corm is about size of a hazelnut. Var. *montana* is very similar, but the petals recurve more directly and flowers do not have a semblance of a 'cup' being formed by the bases of perianth segments. Such differences are perhaps overstressed and as such, consideration is being given to this being a geographic variance.

BOMAREA — ALSTROEMERIACEAE

The genus is named in honor of Jacques Christophe Valmont de Bomare, of Paris, a person much interested in plants who lived 1731–1807. *Bomarea* is native to the tropical highlands of tropical America and is a genus of some 100 species.

The rootstock is sometimes tuberous; indeed, in one species, *B. edulis*, the tubers are eaten and has the common name of White Jerusalem Artichoke. From the literature it is difficult to determine just how many of the species are possessed of tuberous roots. Apart from *B. edulis*, references can be found to *B. shuttleworthii* as having tubers but it would appear that the majority of the other species have fleshy roots but not distinct tubers.

The flowers are carried on the vines in clusters; they are tubular with the perianth segments all being equal or divided into 2 unequal series. The foliage is oblong and pointed and is carried on quite long leaf stalks. The plants can become quite large over a period of time.

CULTURE The plants are not hardy, preferring temperatures that do not drop below 45°F at any time. They are thus only for the warmest climates or for the heated greenhouse. The tubers should be planted so that they are just at soil level; soil should be high in organic matter. Spacing will vary—in containers 1 per 10 in. pot or in a greenhouse border 3 can be grouped so the resulting growth can be trained around suitable supports. They must receive good moisture during the growing season and should never be allowed to become dry, but the amount of moisture should be restricted when plants are not in active growth. Regular feedings of liquid organic fertilizer should be given during the warm months. The enjoy bright light but prefer some shading from bright, direct sunlight.

PROPAGATION Careful division of the rootstock provides the best means of propagation. Seed can be sown in the spring in a soil mix high in organic matter and the temperature should be kept in the 65°–75°F range. Seeds germinate in about 20 days. After germination the plants can be transplanted to individual pots and repotted as growth dictates.

PESTS & DISEASES No special problems.

USES In warm climates these plants make attractive climbers. In other climates they should be given protection but can be grown outdoors in summer. They are thus best grown in large containers with a trellis so the plants can climb.

SPECIES

B. edulis. Widely distributed from Mexico to Peru and also found in Cuba; introduced in the latter part of the 19th century. A climbing vine often reaching many feet. The flowers are in clusters, up to 1½ in. in length with the outer segments rose or yellow tipped with green and the inner yellow or greenish and spotted with rose; summer-flowering. The leaves are lanceolate, up to 5 in. long and 1 in. in width. The tubers are edible. Var. *chontalensis* is more vigorous and has larger flowers of bright rose, the inner parts being greenish-yellow; var. *ovata* has leaves that are wider and more rounded at the base, and hairs on the undersides; the flowers of var. *punctata* have bright reddish-orange outer segments with the inner golden yellow, spotted with crimson, and is a particularly attractive plant.

B. shuttleworthii. Native to Bogota; introduced 1881. The flowers on the twining vine which can reach to many feet are up to 3 in. in length, funnel- or long bell-shaped; the outer segments orange-vermilion tinged with green and the inner bright yellow with red midrib, with the tips green with dark spots. Summer-flowering. Leaves 5–6 in. long, ovate-lanceolate, 2 in. wide glabrous.

COMMENTS These lovely vines deserve greater recognition and should be grown in warm climates and in display greenhouses where they will become very popular plants with visitors.

BONGARDIA — BERBERIDACEAE

A monotypic genus named in honor of Heinrich Gustav Bongard, a German botanist; A tuberous plant native of western Asia, Syria, and Iran; introduced in 1740. Found growing on stony hillsides and in cultivated fields. The leaves arise directly from the tuber, thus differing from *Leontice*, (which see), and are radical and pinnate. The leaflets are sessile and cleft at their tips, with either 3, 4, or 5 clefts. The yellow flowers are in branched panicles, opening to a flat flower. These are produced over a period of several weeks from March–May.

The tuber is dark brown, rounded with a definite neck, some 2 in. in diameter and about 1½ in. in width. It is reportedly quite hardy, but in colder climates it should be protected by a mulch during the winter months or grown in a cold frame. It reaches a height of some 6 in. The flowers are ½ in. in diameter and the leaves are of the same length. It is not a commonly grown plant.

CULTURE   Set the tubers some 1–3 in. deep in a very light peaty-leafy soil. Provide some moisture in the spring, but keep plants on the dry side, expecially during the winter. Plant in a sunny border and allow to grow undisturbed.

PROPAGATION   The tubers can be divided, but seed is the best method of propagation. This should be sown in the spring in a light soil mix high in organic matter and kept on the dry side. The plants should be given protection from excessive moisture and set out at the end of their second season.

PESTS & DISEASES   Rotting from being too wet is the only serious problem.

USES   Interesting plant for the perennial border or rock garden. More unusual than of great beauty.

SPECIES

B. chrysogonum (syn. *B. rauwolfii*, *Leontice altaica*, *L. chrysogonum*, *L. odessana*). Described above.

COMMENTS   Will never be widely grown.

BOOPHANE — AMARYLLIDACEAE

The name is derived from the Greek *bous* ("an ox") and *phonos* ("slaughter") and it should be noted that, while the more common spelling of the genus is *Boophane*, other spellings are used: *Boophone* and *Buphane*. A genus of 5 species, all native to South Africa and found over much of that country growing in dry grasslands. The rootstock is a large bulb, as much as 6 in. in diameter and often a little longer, covered with dry scales and sets two-thirds out of the ground. The leaves are basal, linear, often with a fluted margin, up to 14 in. in length and almost 2 in. in width, spreading in the shape of a fan. They often appear after the flowers have opened.

The plants produce a single flower head, which is large—over 8 in. is not uncommon. The flower stalks are thick, and the individual flowers are on long pedicels, 2 in. or more in length at flowering time, which lengthen as the seed pods develop. Flowers are produced in the spring.

The Xhosa people use the dry, outer covering of the bulb as a dressing after circumcision and also as an application for abscesses and boils and decoratively as a head mask. Europeans moisten the dry scales to use as a dressing for boils or use the fresh leaves on cuts to check bleeding. The bulb was used by the Bushmen as one of the components of their arrow poisons. While all the species are used as described above, the more common *B. disticha* is the most frequently used. The common names are Century Plant, Sore Eye Flower, and Cape Poison Bulb. The sweet fragrance of the flowers and the inhalation of the pollen will cause a severe headache, drowsiness, and sore eyes.

CULTURE   The bulbs should be planted so that the neck and about half of the bulb are aboveground. Space them about 12 in. apart in a sunny location. The type of soil is of no great importance as these bulbs seem to do well in all types, from sandy to quite heavy clay. While the bulbs prefer moisture during both the flowering season and the growth period of the leaves, they are able to withstand periods of drought. The plants cannot, withstand cold weather ,however, and should not be planted where they will receive any frost. For this reason they are best grown in warmer climates or in containers in frosty areas, being brought indoors during periods when temperatures drop below 40°F.

Once planted they should be left undisturbed as the bulbs resent being moved—they require at least a season, sometimes 2, before they will flower after being transplanted.

PROPAGATION   Some smaller bulbs are produced by the larger ones, and these can be removed without disturbing the larger. Seed is quite freely produced and should be sown in the spring in a sandy soil mix, given night temperatures of 45°–50°F, and kept on the dry side. After germination the small plants can be transplanted to small individual containers and grown on for 2–3 seasons, when they can be planted out. Give good light at all times, but protect the seedlings from strong, bright sunlight.

PEST & DISEASES   No special problems.

USES   Bulbs for the sunny border or in a corner of the garden where they will receive sun and good drainage. In colder areas they can be grown in deep containers.

SPECIES

B. disticha (syn. *Boophane longipedicellata*, *Buphane disticha*; originally described as *Brunsvigia disticha*, syn. *B. toxicana*). Occurs in all provinces of South Africa in grasslands and is also found in Angola; introduced into culture in Europe 1774. Leaves arranged in two ranks as a fan, up to 14 in. in length, about 2 in. wide, tapering, and often with fluted edges. Leaves are produced after the flowers, which appear in their native habitat Oct.–Dec.; in the Northern Hemisphere in late spring/early summer. The large heads of pink flowers can be over 12 in. in diameter, held upright on a flowering stem that is often only just a few inches out of the bulb reaching 12 in. in total height. The pink perianth segments curl and twist but are not much reflexed. The individual flowers are ½ in. or more in diameter, segments narrow, with quite prominent orange stamens. Frequently there will be over 50 flowers in a head. This plant has an unusual seed distribution habit. The umbel of ripening seed pods will separate from the stalk and become a ball, which will be blown around the veldt by the wind, scattering the seeds as it rolls. This species will perform well even with little moisture during the winter months.

B. flava. Native to South Africa. Pale yellow; 18 in. in height; flowering in March–May in Namaqualand. Rare, if indeed in cultivation.

B. guttata (syn. *B. ciliaris*, *B. ciliata*). There seems to be some question as to the correct name: some authorities prefer *B. guttata*. Originally described as *Brunsvigia ciliaris*. This plant is found growing in flatlands and on low hills, and flowers most frequently after fire has scorched the area. This is possibly due to the plant then receiving more sunlight and not being shielded from the sun by grasses. Flowers in an umbel, often with over 100 in the flower head. Height 8–12 in.; flower head up to 10 in. in diameter. Individual flowers on long pedicels, which lengthen after the flowers have bloomed; flowers in fall before the leaves. Flowers are small, with greenish-brown, reflexed perianth segments and prominent stamens. Leaves are up to 15 in. long and often over 3 in. wide, varying in number from 4–6. Flowering time in the wild is March.

B. haemanthoides. Native to South Africa. 12–20 in. in height; cream with reddish tips; flowering Nov.–Dec. (late spring/early summer) in Namaqualand. Rare, doubtful if introduced into cultivation.

COMMENTS While needing warmer climates, *B. disticha* does deserve consideration because of its size. Its flowering over a considerable period of time has certain merit not only for bulb collectors but for keener gardeners as well.

BOWIEA — LILIACEAE

The genus is named in honor of J. Bowie (1789–1869), a collector for Kew Gardens who worked mostly in Brazil. A small genus of only 2 species, *Bowiea volubilis* and *B. gariepensis*. The rootstock is a somewhat flattened bulb that reaches 4–5 in. in diameter and is produced at or near the surface of the soil. The plants are perennial climbers, scrambling over vegetation and found growing under scrub and in the protection of rocks in many parts of South Africa, Malawi, Zambia, and Zimbabwe. Reference is found in the literature to a *B. gariepensis*, having more attractive white flowers(see illustration), than *B. volubilis*. They produce 1 or 2 rudimentary leaves at the base, then the climbing shoot which is many branched; this stem is rounded and flowers are produced at the ends of the various branches. The tepals are short, quite thickened, and recurved, and the flowers face in all directions as the plant climbs through the scrub. The unattractive flowers, green with a hint of yellowish-white, are produced in July. Stems can reach up to 15 ft in length.

The bulbs are much used in native medicine. Roasted and powdered they are used as a purgative in water. Many remedies are made from the uncooked bulbs, including headache powders and an alleged cure for barrenness in women. The bulbs have much the same effect as digitalis.

CULTURE Not hardy, but able to withstand temperatures in the lower 30°F range. Needs a sunny, well-drained border preferably with the bulb, which is planted at the soil level or just below, being protected from the sun by a rock. Water is needed in the winter and spring, with a dry period in late summer. No fertilizer is needed. The bulbs should be left undisturbed. In colder climates they must receive winter protection.

PROPAGATION By offsets removed in the fall. The seed can be sown on sandy soil in the spring and kept moist but never wet. After germination the seedlings should be given good light and temperatures in the 50°F range at night. The bulblets should be transplanted the following spring into their own containers and grown on until the bulb is some 1 in. in diameter, then planted out in a sheltered position.

PESTS & DISEASES No special problems.

USES For the collector only; plant among shrubs where the plant can climb and wander.

SPECIES

B. gariepensis. Described above.

B. volubilis. Described above.

BRAVOA — AGAVACEAE

The derivation of this genus name is sometimes attributed to the brothers Leonardo and Miguel Bravo, who were Mexican botanists; others attribute the genus to Nicholas Bravo, a hero of the War of Mexican Independence. One thing is certain: the species, of which there are 3, perhaps 4, are native to Mexico. Some authorities regard these plants as being in the genus *Polianthes*, but others separate them because of their colored flowers: the *Polianthes* are white or greenish-white; the *Bravoa* are red or orange.

The majority of these plants produce their flowers in pairs, in long racemes, and they flower in the summer. The foliage is mostly basal, varying from being grasslike to having quite wide leaves. There are also some leaves on the stems, but these are small and sparse.

The flower tubes are cylindrical, the tube terminating to give very short lobes; these are nearly equal in length. The rootstocks are tubers that are fleshy, as are the roots produced.

CULTURE The plants are not hardy, and where frost is experienced they should be grown in a cool greenhouse or the tubers lifted in the fall, stored overwinter in a frost-free area, and planted out in the spring as soon as the ground warms and there is no danger of frost. In warmer climates they can be grown in sunny, well-drained borders, but they must have good moisture during the growing season and kept on the dry side during the winter months.

The tubers should be set 3–4 in. deep, spaced some 8–10 in. apart. They like a rich soil with good organic content and will benefit from frequent feedings of liquid organic fertilizer during their growing season. They are good container plants but should not be too crowded. Only in the very hot areas should these plants be given any shade and then only during the heat of the day.

PROPAGATION When the plants are lifted in the early fall the offsets can be removed from the parent plants and grown on. Division of the mature plants also should be done at this time of the year. Seed is quite freely produced and can be sown in spring in a rich organic soil mix and just covered. Germination is good and the plants can become quite large the first season. In the second spring they should be individually potted—by the following spring the tubers will be of flowering size and ready to plant.

PEST & DISEASES No special problems.

USES Lovely plants for the cool greenhouse in cooler climates and for growing outdoors when treated like dahlias or other tender plants. In warm climates, pleasant garden plants for the sunny border.

SPECIES

B. geminiflora (syn. *Polianthes geminiflora*). Native to central Mexico; introduced 1841. Is at home growing in rather damp areas.

Reaches some 24 in. in height, with leaves that are mostly basal, 1/2 in. wide and some 18–20 in. long. The flowers are produced in pairs on a stem that is almost leafless. Color will vary a little between rich reddish-orange to yellowish hues. The tubes are 1 in. long, carried on very short pedicels, the flowers being almost horizontal but more likely facing slightly downward. The plants flower in early to midsummer.

B. graminifolia (syn. *Polianthes graminifolia*). Found growing in central Mexico in grassy areas; date of introduction into culture not known but possibly circa 1840. The flowers are produced in pairs but are a truer red than *B. geminiflora*. Another difference is the foliage which is, as the name implies, grasslike and produced at the base of the flowering stalk. The height is about 20–24 in. and the flowering time midsummer.

B. platyphylla (syn. *Polianthes platyphylla*). Native to central Mexico; quite possibly introduced circa 1840. Leaves are about or just more than an inch in width and not long, reaching only some 4–5 in., lying close to or on the ground. This species flowers as the others, however, the flowers are somewhat shorter in the tube than the other species. The great difference among the 3 species mentioned is that in this species the stamens are not attached toward the bottom of the perianth segments but closer to the mouth of the flower. The color is not so attractive, being a duller and rather pale red. Flowering in midsummer.

COMMENTS With quite a variation in color within the various species, it is interesting to conjecture what hybrids could be produced. It is also interesting to consider the result of crosses between these species and *Polianthes* as the tuberose (*Polianthes tuberosa*) is such a widely grown cut flower.

BRIMEURA — LILIACEAE

Named in honor of Marie de Brimeur, an ardent French flower gardener of the 16th century.

For many years the species in this genus, which is native to areas around the Mediterranean, were included with *Hyacinthus*. In point of fact, they bear little resemblance to the hyacinth, appearing more like a bluebell (*Scilla*), but are different from either. *Brimeura* has a long, tapering bract that subtends each flower stalk; in *Hyacinthus* this bract is insignificant. In *Brimeura* the tepals are joined into a short tube; in *Scilla* they are free.

There are but two species in this genus, *B. amethystina* and *B. fastigiata*, produced from small bulbs some 3/4 in. in diameter. The linear leaves, narrow and grasslike, appear in early spring, followed by the flowers, which are produced April–May. The individual flower stalks are longer than the bracts. The flowers are bell-like, carried in racemes. The number varies—sometimes solitary in young plants or up to 15 on older plants and also varying according to species. Also according to species, height will vary from 2–10 in.

CULTURE This depends on the species; *B. amethystina* prefers the sun, while *B. fastigiata* prefers some shade. The bulbs should be planted in the fall, placed 1–2 in. deep and 3–5 in. apart. Good drainage is essential, especially during the summer months. They are quite hardy, withstanding temperatures down to the 20°F range, as well as considerable heat in the summer, with little or no moisture. During the growing season and throughout the winter months, however, moisture should be given. Unless severe conditions are experienced during the winter (for example, temperatures below 15°F), little or no protection is required. In cold areas a covering of leaf mulch is helpful. Placement near rocks also seems beneficial, as the reflected heat is appreciated. It is not necessary to fertilize, very little if at all—the leaf-mold mulch is sufficient.

PROPAGATION Ripe seed can be sown in shallow containers in a sandy soil mix. It should be barely covered. After germination, seedlings can be kept growing until the foliage dies back naturally. Small plants then can be allowed to become quite dry, removed from the soil, stored in a frost-free area to overwinter, and planted early the following year, as soon as the days start to lengthen. When the seedlings have formed bulbs 1/4 in. in diameter, they can be planted out into their permanent locations.

If given good growing conditions the bulbs will multiply and form small colonies, a natural increase. This occurs after they have been growing in their permanent locations for a number of years. When the foliage has died down, in late spring or early summer, the bulbs can be lifted and divided according to size. The larger bulbs can be planted back in the fall; the smaller ones also can be planted back then or, if preferred, held until early spring and treated like plants raised from seed, as described above.

PEST & DISEASES No particular problems.

USES These low-growing bulbs have a place in a sunny border or in a rock garden. Raised in small containers, they provide interest due to their rareness rather than their beauty. While not spectacular, they deserve a place in the garden if bulbs are favorite plants.

SPECIES

B. amethystina (syn. *Hyacinthus amethystinus*). Native to the Pyrenees; introduced 1759. Found growing in rocky places and on thin, grassy slopes. The bulb is ovate, up to 3/4 in. in diameter. The 6–8 leaves are narrow and bright green, up to 12 in. long. The flower spike is up to 10 in. high, carrying 5–15 bright blue flowers in a one-sided, rather loose raceme. The flowers, which appear in April/May, are almost 1/2 in. long, pendant, and have tepal lobes that are shorter than the tube. A white form has been found in gardens but it is extremely rare in the wild.

B. fastigiata (syn. *Hyacinthus fastigiatus, H. pouzolzii*). Native to Corsica and Sardinia; introduced 1882. A low-growing plant, rarely more than 4 in. in height. Generally about 5 leaves appear, sometimes more but seldom fewer. They are up to 6 in. long, dark green, linear, and narrow. The pale pink or white flowers, held horizontally (rarely slightly erect), are carried all around the 4 in. stem and are clustered together at the top. The number varies from a solitary blossom to 10 per stalk. They are bell-shaped, 1/3 in. long, and the flaring tepals give a starry appearance. Appear in April/May. It is the less worthy of the 2 species in this genus.

BRODIAEA — LILIACEA

The genus is named in honor of James Brodie (1744–1824), a Scottish horticulturalist. This extensive genus has some 30 species, most of which usually are not obtainable from commercial sources, and has been the center of much change. Over the years botanists have divided the genus, and many species formerly listed under *Brodiaea* are now placed in other genera; i.e., *Triteleia, Bloomeria, Dichelostemma, Ipheion,* and *Muilla.* (The latter is quite appropriate as it is "Allium" spelled backwards.) As many are still listed under *Brodiaea* in catalogs, they are so listed here; however, in this chapter I also have listed the various species of the genus with their new names. In so doing I feel I am helping rather than hindering the reader, because one is more likely to find the *"Brodiaea"* in the listings, and also because one is more apt to think of them by that name.

There has been much discussion as to which family *Brodiaea* belong. This also applies to the genera into which many of the *Brodiaea* are now placed. The controversy lies in the fact that the ovary in these genera are superior, as in the Liliaceae, while the flowers are in umbels, as in the Amaryllidaceae. There are some who place the genera in their own family, the Alliaceae. As discussion will no doubt continue for some time to come, I list them in Liliaceae. See Appendix, "Families of Bulbous Plants," for further discussion.

For the most part, *Brodiaea,* commonly called Fool's Onion, are native to the west coast of North America with the majority found in California. Two species come from South America. All have corms, with the exception of the Argentinian species, *B. uniflora,* which is a bulb. This species is also known as *Triteleia uniflora,* but has now been given the name *Ipheion uniflorum,* a small genus. After being classified into many different genera, it would seem to have found a home at last (we hope).

Brodiaea have three fertile stamens, as do *Dichelostemma.* The distinctive difference between these two genera is that the leaves of the former are rounded underneath (as are those of *Muilla),* while those of the latter are keeled. Another slight difference is that the stigma of *Brodiaea* is distinctly three-lobed, a characteristic not so distinct in the *Dichelostemma. Triteleia, Bloomeria,* and *Muilla* all have 6 fertile stamens. *Bloomeria* differs from *Triteleia* in that the stamen filaments are cuplike at the base instead of straight.

The majority of the species produce grasslike foliage, and all of the species have more or less linear leaves. Not infrequently the leaves have died away prior to the flowers being at their best. The onionlike flower heads usually rise 12–24 in. over the foliage. The individual flowers in the umbels are usually blue when wide open and funnel- or bell-shaped. The fruit is a capsule, and seeds are black.

CULTURE The required culture varies only a little from species to species. The climate on the west coast of North America is ideal: the plants should receive moisture during the winter months, some in the spring, and occasional moisture during the summer. Late summer and early fall should be a dry time. At no time should the plants be allowed to become waterlogged. They should be given full sun for at least part of the day and never grown in deep shade. In areas where the winter temperatures fall below 20°F, protect the plants, either by covering them with a thick, heavy mulch or by lifting them entirely.

The rounded and slightly *Gladious*-like corms should be planted some 5 in. deep, a little deeper in sandy soil. As the flower stalks often are produced after the leaves, it is best to plant many *Brodiaea* together to make a good showing. The plants are happy if left undisturbed for years. They should be lifted and divided only when the groupings become overcrowded. This should be done in late summer, after flowering and after the stems have died down. They then can be planted back immediately.

PROPAGATION The corms produce a great number of cormels, or offsets, which should be sorted according to size and planted back, spaced 4–5 in. apart. The smaller corms can be planted in nursery rows and grown on for a season prior to planting in their final locations. This should be done in the fall of the year but early enough so that the corms can settle in before the onset of winter.

PESTS & DISEASES The corms are subject to rotting if the ground is poorly drained. Small rodents have a special liking for them, so protect with small-meshed wire. And, of course, watch for the usual slugs and snails, but these can be taken care of very easily.

USES *Brodiaea* can be used in a number of places in the garden. Loamy, well-drained yet moisture-retentive soil, as would be found on the edge of a woodland garden with some shade, makes an ideal home. While sometimes suggested as plants for the rock garden, care must be taken to ensure that adequate moisture is given during the growing season. Thus they should be grown alongside a stream or beside a pool, but never in a formal setting.

In grasslands they can be naturalized but moisture must be provided, duplicating the frequent summer rains as in their native habitats.

SPECIES & CULTIVARS

B. bridgesii (syn. *Triteleia bridgesii*). Native to the United States from Oregon to central California; introduced in 1888. The leaves are grasslike, up to 20 in. long and a little less than ½ in. wide. The flower head is quite large, sometimes exceeding 4–5 in. in diameter, and reaches a height of 24 in. The total number of individual flowers in the umbel can be as high as 25, sometimes more. They open into violet-blue blossoms, which become bluer as they mature and measure an inch in diameter with an inch-long tube. They open gradually, thus giving an attractive display for many weeks during the summer months. Makes a good cut flower.

B. californica. Native to northern California in the United States; introduced in 1896. Found growing in grasslands at lower elevations on well-drained soils. It is the largest and tallest of the *Brodiaea* that can be purchased commercially, often reaching a height of more than 24 in. The flower head will contain some 12–15 individual flowers. The perianth is about an inch long, and the diameter of each flower is just over an inch. The color varies from pale blue, almost white, to intense blue, to pinkish. Those offered in catalogs are often the selections of the pinkish shade. Flowering early summer. The leaves are narrow, linear, and 12–24 in. in length.

B. coccinea (syn. *B. ida-maia;* also known as *Dichelostemma ida-maia*). Native of California; introduced 1870. The common name, California firecracker, is appropriate for this outstanding species. The umbels are large and a bright red, the petals tipped with green. The flower head itself can reach several inches in diameter but the individual flowers are quite small, about ½ in. Height to 24 in., usually less. Usually has 3 grasslike leaves, about 24 in. long and ½ in. wide. This plant needs open spaces to grow well and must have dry summer conditions after flowering for the corm to ripen. Flowers early summer.

B. coronaria. Native to British Columbia to California; introduced 1806. Found growing in clay, rather alkaline, soil. The corm is small and roundish. This is one of the shorter species, rarely more than 12 in. in height, with foliage about the same length. There are seldom more than 6–7 individual flowers in an umbel, each only an inch wide. Flowers are violet to lilac in color, but white and pinkish forms are known. Var. *macropodon* is a very good dwarf form, 4 in. tall; var. *rosea,* found in Lake County, California (introduced 1896), is pale lavender, turning pinkish. Good plants for the rock garden but must have a dry period at the end of summer/beginning of fall. 2–3 leaves are produced which are linear, pointed and grooved on the upper surface, 7–10 in. in length.

B. grandiflora (correctly known as *Triteleia grandiflora*). Native to the western United States. Commonly found growing in grassy areas and on quite rocky hillsides. The compact flower head, rarely more than 10 in. high, carries the short-stemmed, short-tubed flowers, which are an inch long, an inch in diameter, and dark to medium blue in color. The plant can be used effectively in the rock garden but it should be grown where it has good drainage and can be left undisturbed. Flowering early summer. Regarded by many as a form of *B. coronaria*.

B. ixioides (Golden Star; also known as *B. lutea* and *Triteleia ixioides*). Native to Oregon and northern California; introduced 1831. Mature plants reach a height of 18 in. The flower head sometimes carries only a few of the small flowers or sometimes as many as 18–20. The flowers are about an inch in diameter and are of a good yellow color with a darker stripe down the center, which can be purplish, earning it its common name. Flower has a short tube, rarely more than ¼ in. long. Flowers late May–early June. The foliage is fleshy, linear, and 3–8 in. long.

B. lactea (also known as *B. hyacinthina* and *Triteleia hyacinthina*). Native to British Columbia south to California; introduced 1833. The plant grows to 18 in. in height, with 3 in. diameter umbels. Individual flowers are 1 in. in diameter, milk-white, on strong, wiry stems. They prefer moisture during the spring until flowering in June. This species is easy to cultivate in the garden, and the flowers can be dried for winter use. *Brodiaea lactea lilacina*, also known as *Triteleia hyacinthina lilacina*, is slightly taller than the species, with a lilac tinge to the flowers. Leaves are linear, pointed, and 12–18 in. in length.

B. laxa (Ithuriel's Spear; also known as *Triteleia laxa*). Native to Oregon and California; introduced 1832. The common name refers to the angel in Milton's *Paradise Lost*. This species has the largest umbels of any in the genus, often up to 6 in. in diameter. The flowers also are large, more than an inch across. The umbels are not so tight as in other species, and the pedicels are often more than 2 in. long. The deep blue flowers have anthers that also are blue; they are carried on strong, wiry stems, often up to 24 in. in height. This is one of the most attractive of the genus for the garden and has potential as a cut flower. A selected form, 'Queen Fabiola', a little taller and with stronger stems, has been introduced by Van Tubergen of The Netherlands. Flowering late June/early July. The leaves are 12–24 in. in length, linear, and pointed.

B. peduncularis (also known as *Triteleia peduncularis*). Native to California; introduced 1896. The very long pedicels, often more than 3 in., can mean the flower heads are more than 10 in. across. The white flowers are similar in size to *B. laxa*, and often are more than 24 in. high. This species makes a good cut flower. Late-June-flowering. Leaves 12–30 in. in length, linear, only ¼ in. wide.

B. × tubergenii (also known as *Triteleia × tubergenii*). A hybrid raised by Van Tubergen in The Netherlands; the parents are *B. laxa* and *B. peduncularis*; introduced in the 20th century. This is a strong grower, reaching 24 in. in height. The large umbels carry pale blue flowers, whose exterior is a darker hue. A good cut flower, well worth growing in the garden. Flowering late June/early July. Leaves stout, linear, and 20–30 in. in length.

B. volubilis (also known as *Dichelostemma volubile*). Native of California; introduced 1874. This is an unusual species in that the flower stem is twining and needs the support of low-growing shrubs, such as a *Hebe* or a *Daphne*. It is not so hardy as many of the other species, requiring protection where the winter temperatures are below 20°F. The umbels are small and pinkish-mauve in color, reaching a height of some 30 in. Flowering June/July. Native of California. The leaves are narrow, linear, and up to 12 in. in length.

BRODIAEA

B. appendiculata Native to California. Deep violet-purple. 4–18 in. in height. April–May flowering.
B. aurea (syn. *Nothoscordum aureum*) Native to Patagonia, Uruguay, Southern Argentina & Chile. Introduced 1838. Yellow with a green stripe (corm whitish). 2–4 in. in height. April flowering.
B. capitata (syn. *Dichelostemma pulchellum*) Native to California. Introduced 1871. Violet-blue. 12–24 in. in height. May flowering.
B. clementina Native to California. Light blue. 12–18 in. in height. March–April flowering.
B. crocea (syn. *Triteleia crocea*) Native to California. Yellow. 6–12 in. in height. May–June flowering.
B. douglasii (syn. *Triteleia grandiflora*) Native to Oregon, Washington. Introduced 1876. Violet-blue (bulb small globose). 12–18 in. in height. May flowering.
B. eastwoodiana Native to California. Introduced 1936. White. 20–24 in. in height. June flowering.
B. gracilis (syn. *Triteleia montana*) Native to California. Introduced 1876. Deep yellow with brown nerves. 3–4 in. in height. July flowering.
B. hendersonii (syn. *Triteleia hendersonii*) Native to Oregon, California. Salmon with purplish midribs. 8–20 in. in height. May–July flowering.
B. howellii (syn. *Triteleia grandiflora* var. *howellii*) Native to California. Introduced 1880. White. 18–24 in. in height. July–August flowering.
B. hyacinthina syn. of *B. lactea*.
B. ida-maia See *Brodiaea coccinea*.
B. jolonensis [Probably *Triteleia*] Native to California. Violet. 1–8 in. in height. April–June flowering.
B. leichtlinii of uncertain standing in *Brodiaea*. Native to Chile. Introduced 1873. White face. No longer in cultivation. Early Summer flowering.
B. minor (syn. *B. purdyii*) Native to California. Pinkish or violet-blue. 4–8 in. in height. June–July flowering.
B. multiflora (syn. *Dichelostemma multiflorum*) Native to California. Introduced 1892. Blue-purple. 12–18 in. in height. May flowering.
B. orcuttii Native to Southern California. Introduced 1896. Purple-violet. 4–12 in. in height. April–July flowering.
B. porrifolia Should be placed in *Ipheion* because it's bulbous & from S. America. Native to Chile. Introduced 1874. White with violet streaks & green below. 6 in. in height. Early Summer flowering.
B. pulchella (syn. *Dichelostemma congesta*) Native to California. Introduced 1871. Deep violet-blue. 12–24 in. in height. May flowering.
B. purdyii syn. of *B. minor*.
B. rosea syn. of *B. coronaria rosea*.
B. stellaris Native to California. Bright purple with white center. 4–6 in. in height. July flowering.
B. uniflora (syn. *Ipheion uniflorum*) Native to Argentina, Uruguay. Introduced 1836. White to deep purple-blue. 6–8 in. in height. April–May flowering.
B. venusta (syn. *Dichelostemma venusta*) Native to Northern California especially foothills of Sierra Nevada range. Rose. 16–36 in. in height. Early Summer flowering.

BRUNSVIGIA — AMARYLLIDACEAE

In the 18th century, when the great houses in Europe vied in collecting and flowering unusual and different plants, Ryk Tulbagh (1699–1771), then Governor of the Cape Province, sent bulbs of the plant to the Duke of Braunschweig in Germany. The genus is named in honor of the House of Brunswick. There are 16 species, all native to South Africa. The common name for this plant is the Candelabra Flower.

The bulbs of *Brunsvigia* are among the largest produced by any plant. They frequently weigh well over a pound and can be as much as 24 in. in length, with a diameter of more than 6 in. Leaves are broad, produced in pairs, and most commonly lie close to or flat on the ground. The flowers are produced before the leaves and are carried aloft on strong, quite fleshy stems. Up to 30 flowers grow in umbels. They are always in the red/pink/purple tones. The zygomorphic flowers have a distinct tube, and the tips of the petals curl back a little. The appearance is not unlike a *Nerine*, except in the *Brunsvigia* the pedicels lengthen after the flowers have passed. The fruit is a capsule.

CULTURE Bulbs do not like frost. *B. radulosa* is perhaps hardiest of all, being able to withstand mild frost, but no temperatures below 25°F unless protected. Sunny locations, good moisture, and well-drained soil required during growing season. Plant bulbs so shoulders are at soil level. Give water in moderate quantities as soon as flower stalk appears. Increase moisture when leaves show and keep moist until leaves start to die back. Reduce moisture at this point so bulbs can enjoy resting season.

If growing in containers, they should be large enough so bulbs can remain undisturbed for a number of years, as bulbs take a season or two to return to flower after being moved. This is the same for bulbs grown outdoors. Weak feedings of liquid fertilizer can be given, but only when starting into growth.

PROPAGATION Mature bulbs will produce a number of offsets which can be separated from the parent and grown on to flowering size. Offsets take 2 or 3 years to reach flowering size.

Seed can be sown as soon as ripe. Obtaining flowering-sized bulb from seed takes several years, so offsets are the quicker way to propagate.

PESTS & DISEASES Bulbs subject to rot if excellent drainage not given. Susceptible to aphids, snails, and slugs.

USES Excellent container plants; must be moved indoors in winter. Some protection needed in warmer climes when weather turns chilly. Can be used effectively against walls of house where additional heat is appreciated. Can be grown in slight shade. Should be cultivated only if necessary dry resting period can be given.

SPECIES

B. gregaria. Native to eastern Cape Province; introduced 1822. Bulb ovoid, 2–3 in. thick; outer sheath frequently dry and brown, held loosely from bulb by membranous tunics. In common with many other *Brunsvigia*, leaves lie flat on surface of soil, but, unlike others, only 4 leaves produced. These are quite short, about 5 in. in length, and appear after flowers. Pinkish-red flowers appear late summer on stalks 10–12 in. high.

B. josephinae. Native to eastern Cape Province; introduced 1814. Bulb 5–6 in. thick. Leaves 3–4 in. wide and up to 36 in. long. One of most attractive of species. Coral-red flowers carried on 18–20 in. high stems. Late summer-flowering, often with more than 25 blossoms in each head of 6–8 in. long pedicels. Color sometimes more purplish. *B. josephinae* is parent of intergeneric hybrid × *Brunsdonna tubergenii;* the other parent is *Amaryllis belladonna*, which has violet-pink flowers.

B. orientalis (syn. *B. gigantea; B. multiflora*). A native of the Cape Peninsula to Knysna; introduced 1752. Leaves of moderate length produced after flowers; lie flat on ground; will often persist throughout winter. Largest-flowered of species, often more than 24-in. diameter umbel. Red flowers carried on 12–15 in. high sturdy stems. Bulb should be allowed to dry out in spring; will flower again in midsummer. Species has unusual and interesting method of seed distribution in the wild. As flowers fade, ovaries enlarge and become papery. Finally, complete flower head breaks away from stalk and is carried by wind, scattering seeds over considerable area.

B. radulosa. Native to open, rocky, grassland of Transvaal. Hardiest of species. Regarded by some as most decorative. Leaves rounded at tip; lie flat on soil. Leaves and flowers produced at about same time, with flower stalk emerging a little before leaves. Flowers are good pink, 10 in. across, with 12 in. high stems. Should be protected in all areas except those that are practically frost-free, although can withstand some frost. Very rare in the wild, and, to my knowledge, has not been introduced into cultivation.

BRUNSVIGIA

Key: SH = Southern Hemisphere

B. appendiculata Native to Namaqualand. Pink with darker veins. 7–12 in. in height. Flowering Late Summer, March–May SH.
B. bosmaniae Native to Karoo, Namaqualand. Pink with darker veins. To 8 in. in height. Flowering Late Summer, March–May SH.
B. ciliaris See *Boophane guttata*.
B. comptonii Native to Little Karoo. Pink. To 5 in. in height. Flowering Summer, Feb.–March SH.
B. falcata syn. of *Cybistetes longifolia*.
B. gigantea syn. of *B. orientalis*.
B. litoralis Native to Eastern Cape Province. Red. To 28 in. in height. Flowering Summer, Feb.–March SH.
B. marginata Native to Southern Cape Province. Bright scarlet. To 8 in. in height. Summer flowering.
B. minor similar to *B. gregaria* Native to Namaqualand & W. Karoo. Introduced 1822. Flowers smaller, bright pink. This may well be a geographic variant of *B. gregaria*. 6–9 in. in height. Flowering Late Summer, March–April SH.
B. multiflora syn. of *B. orientalis*.
B. natalensis Native to Transvaal. Rare. Crimson. 20 in. in height. Flowering Summer, Nov.–Jan. SH.
B. striata Native to Mossel Bay, Humansdrop, South Africa. Pinks. 6–10 in. in height. Flowering Summer, March–April SH.
B. toxicaria See *Boophane guttata*.
B. undulata Native to South Africa, Natal. Red. 18 in. in height. Flowering Summer, Feb.–March SH (dormant in Winter).

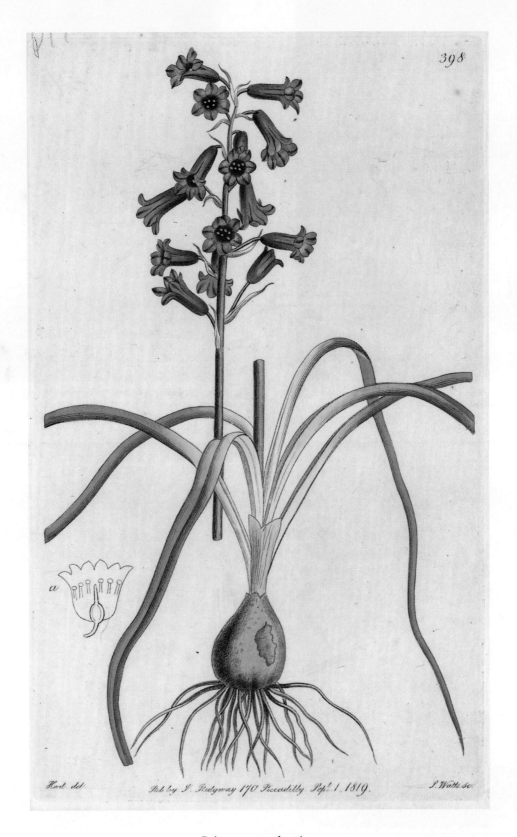

Brimeura amethystina

Plate 57

Brodiaea capitata. Often classed as *Dichelostemma pulchellum*, this California native plant is most attractive, putting on a good show of color, each spring. J.H., by WARREN COOLBAUGH

Brodiaea coccinea. It is easy to appreciate why this plant has the common name of California Firecracker. This solitary flower gives only an indication of the glorious color and stunning display, of a full inflorescence. TRAGER

Brodiaea ixioides. The 'Golden Star', the common name for this rather small plant, well describes its form. Native to Oregon and northern California, it is quite hardy, and a plant comparatively easy to cultivate. J.H. by WARREN COOLBAUGH

Brodiaea laxa. I can not help wonder about the common name of this plant, 'Ithuriels Spear'. It is uncommon to have a common name based on a character in literature, in this case the angel from Milton's Paradise Lost. This species has the largest flowerheads of the genus. J.H. by CHARLES WEBBER

Plate 58

Brodiaea laxa **'Queen Fabiola'**. This selection from the species is a popular garden plant, especially in Europe. An introduction by the Dutch firm Van Tubergen, for so many years a firm in the forefront of bulb introductions. DE HERTOGH

Brunsvigia josephinae. The bulbs of the genus reach an enormous size. This species is found in the Karoo, but is rare in the wild. Photograph was taken at Kirstenbosch Botanic Garden. AUTHOR

Brunsvigia marginata. A lovely species that, unfortunately, is rare in cultivation. The size of the umbel often equals, or nearly so, the height of the plant. While not hardy, these bulbs are worth considering as plants for spring planting, and then lifting and storing over winter in a frost-free location. B.R.I.

Brunsvigia radulosa. This photograph, taken in the wild, shows the beauty of this species which is perhaps the hardiest of the genus. It is well worth growing in areas enjoying mild winters. It is not demanding with regards to the soil conditions, but like the majority of bulbous species from South Africa, does not like to have wet feet, at any time. B.R.I.

Bulbine. This genus is very similar to *Bulbinella,* the distinguishing feature being that in *Bulbine* the filaments of the anthers are hairy, as seen in this picture. ORNDUFF

Bulbinella cauda-felis. These white flowerheads stick well above the lower grasses, and are quite a distinctive sight in early spring. Individually the flowers are not very attractive, but combined in this thin poker shape they are quite noticable. ORNDUFF

Bulbinella floribunda. While the small individual flower are not long lasting, put them in a many flowered spike, and you have a plant that is worthy of a place in the garden. AUTHOR

Bulbocodium vernum. This lovely little plant looks much like a *Crocus* when it first emerges from the ground, but when it matures the perianth segments separate. Variable in color, it makes a fine addition to the alpine house. B.A.G.S.

Plate 60

Caladium. It is quite understandable that these plants are often called Fancy Leaved Caladiums. Only in one's fancy could such patterns be devised, yet here they are, albeit aided a little by man in selection, and in various breeding programs. The color range, striations and patterns in the foliage is surprising. DE HERTOGH

Caladium. Besides being unusual, the foliage of Caladiums lends itself to many decorative uses, as seen here, being displayed with other plants in a cool greenhouse. The foliage is quite thin in texture, and heavy winds or other harsh treatment can damage the foliage in a hurry. AUTHOR

***Colocasia antiquorum* var. *esculenta*.** This plant, which is of great economic importance in tropical areas, is also an attractive ornamental plant with its bold, large, and attractive foliage. COURTRIGHT

Calochortus albus

Plate 62

Calochortus barbatus

Plate 63

Calochortus luteus

Plate 64

BULBINE — LILIACEAE

The name is derived from the Greek *bulbos* "bulb." This is a genus of some 25 species, very few of which have definite bulbous rootstocks. However, many of the species have fleshy, tuberous-like roots that allow the plants to withstand drought in their native habitats of South Africa and eastern Australia.

The plants have linear leaves held at the base of the plant. The flowers are small and the segments spreading, held in a teminal spike on fleshy stems. The genus is closely related to *Bulbinella* but distinguished from that genus by having hairy stamen filaments and by having 4 or more seeds in each cell of the seed capsule instead of the 2 in *Bulbinella*. The rootstock of the species described below are bulbous tubers or tuberous rhizomes.

CULTURE Cannot be regarded as hardy plants but are suitable for culture in areas where frost is light and does not linger long into the day. In other regions the plants should be given protection against frost, such as a cool greenhouse. The plants will grow quite well in poor soil, and while they will grow well in sun they also will stand some shade. The rootstocks should be set 2–3 in. deep. Moisture should be given during the growing season and then reduced during the dormant season. Little or no feeding is needed, but if given should be in the spring.

PROPAGATION The plants are easy to raise from seed, which can be sown in the spring either indoors or outside after the danger of frost is past. The young plants can be set out the first season, either growing in pots in cold climates or in the border in warmer areas. As these plants are easily raised from seed it is not necessary to increase the stock by division of the rootstocks, although this is another valid method of propagation.

PESTS & DISEASES No special problems.

USES In the rock garden or in the border these plants put on quite a display. As the plants are succulent in appearance, they also can be grouped with other succulents.

SPECIES

B. alooides. Native to the Cape Peninsula of South Africa; introduced 1732. The rootstock is a tuberous rhizome. The foliage is fleshy, the leaves being rounded with a definite furrow on the side held next to the flowering stem. The flowers are yellow, held in a loose raceme. Petals recurve and flowers are upright facing, becoming pendant as they age, when the seedpods are then held upright. Has 30 or more flowers per stalk, flowering in August in the wild, very early spring in the Northern Hemisphere. The flowering spikes tower above the foliage, reaching a height of up to 18 in., the foliage only half as tall.

B. bulbosa. Native to eastern Australia. Produces a bulb-like tuber. The foliage is basal, awl-shaped, and up to 12 in. in length, with a short, sheathing base. The flowering spike reaches up to 24 in. in height, the last 6 in. being the flowering part. The flowers are yellow; the perianth segments are ½ in. in length and held upright. The pedicels lengthen after the flowers are past. The leaves are fleshy and thick, up to 12 in. in length.

COMMENTS These bulbous members of the genus are rare in cultivation.

BULBINE

B. asphodelioides Native to Cape Peninsula, Cape Province, South Africa. Pale Yellow. 12–15 in. in height. Mid to late Summer flowering.
B. favosa Native to Cape Peninsula, Cape Province, South Africa. Yellow with orange stripe. 3–4 in. in height. Winter/Spring flowering.
B. lagopus Native to Cape Peninsula–Port Elizabeth area. Yellow. 24 in. in height. March–June flowering.

BULBINELLA — LILIACEAE

The name is a diminutive of *Bulbine*. It is a small genus of perhaps 10–12 species, native to South Africa and New Zealand, not all of which are bulbous. In the bulbous types the rootstock is an erect rhizome covered with fibers. The leaves are linear and basal. The flowers are carried in a simple raceme, looking like a small Red Hot Poker; however, the colors are white to yellow-orange, and the flowers are ⅓ in. in diameter. There are often well over 100 flowers in a spike. They flower in late winter/early spring. *Bulbine* differs from *Bulbinella* in that its filaments are clothed with fluff or are bearded.

CULTURE Plant in the late summer, setting the rootstocks just below the surface of the soil, in sun or light shade. They must receive good moisture in the fall and winter but can withstand dry conditions in summer. They seem to prefer slightly acid, well-drained soil. They should not be tried in areas where there is more than a little frost; in colder areas they need the protection of a greenhouse. They can be planted in the spring for late-summer-flowering and lifted and stored overwinter.

PROPAGATION Division of the rootstocks in late summer is best. Sow seed in the spring in a soil mix with good humus, barely if at all covering the seeds. Transplant as soon as large enough to handle and plant out after the second season. Night temperatures must be in the 50°F range, and, when this temperature is reached at night outdoors, the plants can be set outside.

PESTS & DISEASES No special problems.

USES Good plants for the sunny border with other perennials, or in front of shrubs in the shrub border.

SPECIES

B. floribunda (syn. *B. robusta, B. setosa*). Native to the western area of Cape Province. Leaves are basal, linear, narrow, have many veins, and are light green in color. The flowers can be yellow to deep orange, very small but many in a spike. The lower flowers are long gone before the entire spike has developed, thus giving a long flowering period. The stamens extend well clear of the flat, opening flowers. The height can be as much as 36 in. *B. f.* var. *latifolia*, also from the Cape Province, has broader, deeply grooved leaves. It is doubtful that either of these plants is in cultivation.

B. cauda-felis (syn. *B. caudata*). Native to the southwestern Cape Province and Namaqualand. It is not so tall growing as *B. floribunda*, but is otherwise quite similar. The flowers, which can vary from white to yellow, are carried on a slimmer spike. While present at flowering time, the leaves are not always fully developed. The flowers are small, ⅓ in. in diameter. Previously the species name, *B. cauda-felis*, was given only to the yellow form found growing in stream beds that become dry in the summer. Recent work in South Africa has determined that there is but one species and because of historical preference *B. cauda-felis* is the correct name. In the literature reference is often made to *B. cauda-felis*, referring only to the yellow form. Lovely plants, but rare, if indeed they are in cultivation at all.

COMMENTS Of the rarer and lesser known plants these deserve consideration as an attractive ornamental for warmer regions, especially those with wet winters and springs.

BULBINELLA

B. triquetra Native to Cape Province. White with bright yellow stamens. 6–18 in. in height. Aug.–Nov. flowering.

BULBOCODIUM — LILIACEAE

The name is derived from the Greek *bulbos* "bulb" and *koduon* "wool." The reason for the genus being given this name is unclear. There are only two species in *Bulbocodium*, both native to Spain, the Alps, and the Caucasus. For many years the genus was regarded as monotypic; however, in 1778, Thomas Mawe, in his dictionary of gardening, listed two species: *B. serotinum* (probably the species known today as *B. versicolor*), which he described as a late-flowering mountain plant with rush-like leaves; and *B. vernum*, which he said was native to Spain and had spear-shaped leaves.

Bulbocodium is related to *Colchicum* and *Merendera*. The common mane, Mountain Saffron, is very apt because the plants resemble the crocus. The genera differ in that in *Bulbocodium* and *Merendera* the tepals are free, whereas in *Colchicum* they are joined and form a tube; however, *Bulbocodium* has only 1 style, while *Merandera* has 3.

CULTURE Hardy genus; corms can be planted in any good garden soil in sun or light shade. If season is mild, will flower in January; if cold, will flower March/April. Leaves appear after flowers. Plant with 3–4 in. of soil over tops of corms in Sept./Oct., or as soon as available. Best left undisturbed so plantings can increase. If possible, allow to ripen in summer by withholding moisture until natural rains come in fall.

PROPAGATION After plants have become established and are of good size, lift corms, remove offsets, and grow on. Lift corms after foliage has started to die down. Hold corms in storage until fall or replant immediately after separation, but only if location will not be too wet during summer. Plants seem to benefit from being lifted and divided every 3–4 years.

PEST & DISEASES No special problems.

USES Flowers not very showy but do provide interest in borders due to early flowering. Excellent container plants where early flowering is desired and where protection can be given.

SPECIES

B. vernum (syn. *Colchicum vernum*). Native to Spain, the Alps, and the Caucasus; introduced in circa 1753. Corm elongate with brown tunic. Leaves, while present at time of flowering, continue growing as flowers fade, reaching length of 5–8 in. Flowers rarely more than 4 in. high, looking like the crocus when first produced but inclined to become loose and lose this form after a short time. Color varying shades of purple, with occasional white form. Flowering March–May.

B. versicolor (syn. *B. ruthenicum*). Native to Spain, the Alps, and the Caucasus. Very similar to *B. vernum* but not so large, 3–5 in. in height. Leaves strap shaped, produced after flowers. According to Brian Mathew and Christopher Grey-Wilson in their book, *Bulbs*, flowers have rounded lobes at base of the blade of tepals instead of teeth, as in *B. vernum*. Flowering March–April.

B. autumnale. See *Merendera montana* (syn. *M. bulbocodium*).

B. trigynum. See *Merendera trigyna* (syn. *M. caucasica*).

CALADIUM — ARACEAE

From the South American Indian term *kelady* "unknown." There are about 16 species according to some authorities. Specific rank given to varieties of *C. picturatum*, *C. schomburgkii*, etc.; all native to South America, mainly Brazil. The *Caladium* were introduced in Western culture in the 19th century. At that time they were grown in greenhouses only, their brightly colored leaves being easy to produce from the tuberous rhizomes. They have become popular plants for gardens in warmer climes. The majority grown at present are of hybrid origin, raised for *C. bicolor* and, to a lesser degree, *C. picturatum*. In South America, the rhizomes of *C. bicolor* were eaten after being roasted or boiled; in the West Indies the leaves were boiled and eaten as a vegetable.

The leaves are sagittate, peltate, ovate, etc., according to species, and vary in color. The leaf stalks often are variegated as well. The spadix is divided into three zones: the top is male and gray-yellow in color; the center sterile and whitish-gray; and the lower portion produces the female flowers and is light yellow. The spathe covering the flowers is hooded, and the lower portion is rolled.

CULTURE Start into growth early in the year. In a greenhouse, plant in soil mix containing a liberal amount of peat moss, with adequate moisture, and at a temperature of 70°–75°F. As soon as shoots emerge, bright, indirect light must be provided 8–10 hours a day. Transplant into individual pots using soil mix that will retain moisture but not impede drainage—roots do not like standing in water. Humidity must be kept high once growth has commenced. Once a goodly number of roots has developed, feed liquid fertilizer. The plants can be hardened off and placed outside if protection is given from wind and nighttime temperatures remain in the upper 50s. Protect them, however, from direct sunlight. Some late evening or early morning direct sun can be tolerated by mature, well-grown plants.

In cooler climates plants are best started indoors in late April and planted outdoors when temperatures at night are 55°F+. In warm climates, where there is no frost, they can be started into growth some 3 weeks before outside temperatures stay in the upper 50s. Planting prior to this will harm them and result in small, misshapen leaves.

Prepare the bed with abundant organic matter and keep moist. High humidity is necessary, as drying winds also harm the decorative leaves. Grown in protected areas free from direct sunlight but where light intensity is high.

The plants will begin to mature in late summer/early fall. Reduce moisture at that time. When plants are quite dry, lift and store overwinter in warm (55°F+) location, in damp peat. At no time should they completely dry out or they will shrivel.

PROPAGATION The roots can be divided, provided each portion has a bud; larger roots can be cut into sections but each also must have a bud. The plants also can be raised from seed, but variation of the progeny is considerable; therefore, vegetative propagation is recommended. This is best carried out in spring, prior to starting them into growth.

PESTS & DISEASES If the tubers are kept in water they will rot, so drainage is of great importance, both in containers and in the garden. Slugs and snails like developing leaves, which must be protected.

USES Caladiums are ideal for getting color into shady areas of the garden where summer temperatures are high and humidity is adequate. They can be raised as house plants provided there is no exposure to direct sunlight or wind. They are good plants for summer culture in greenhouses with high humidity, and are becoming more popular as a greenhouse crop, being marketed from very early spring through summer as indoor plants.

SPECIES & CULTIVARS

C. bicolor. Native to Brazil, Trinidad, and Guyana; introduced 1864 from Brazil. All true varieties introduced circa 1864. There are innumerable variations in the leaf colorations of this species. It is this characteristic that has led to its being the most commonly used species in the raising of new cultivars. The rhizome is flattish but round. The leaves are arrow-shaped or ovate, with the basal lobes joined up to one-third of their length. Under ideal conditions, as in their natural habitats, the leaves can reach, or even exceed, a length of 30 in.; however, they are more frequently in the 8–10 in. range. The leaf stalk will vary in length, often many times that of the blade. The average height, however, is under 24 in. The spathe tube is green and whitish-green on the interior, fading to white at the top and often purplish at the base. It must be remembered that this species is extremely varied—and variable—in the colors of the leaves and in the locations of the colors on the leaves. Hence, varieties from the wild that have been collected and introduced have produced a plethora of forms. Among them are the following, all of which, unless otherwise stated, will have the overall habits and growth patterns described.

> Var. *agryrospilum*. Leaves are green with a crimson center; many white spots between veins; stalk streaked white-rose. (From South America.)
> Var. *baranquinii*. Large leaves, often more than 20 in.; crimson center; green margin with scarlet veins. (From Brazil.).
> Var. *brongniartii*. Green leaves with reddish veins on the lower part; stalk variegated with lilac and green. (From Brazil.)
> Var. *chantinii*. Leaves mostly reddish along the veins, with hues of the same but paler color on the blade; whitish spots between the veins; violet stalk. (From Brazil.)
> Var. *devosianum*. Green leaves spotted with white, with a quite pronounced but narrow crimson border; violet stalk. (From Brazil.)
> Var. *eckhartii*. Green leaves with a few marginal spots of rose and a few white spots in the middle; purple stalk at base fading to green.
> Var. *hendersonii*. More correctly a cultivar as it is of garden origin and, as such, not found in the wild; hence it is not a variety. Green leaves with red colorations along the veins and small but distinct scarlet spots; stalk is variegated green and violet.
> Var. *kochii*. As the preceding entry, is of garden origin; large, more rounded leaves of deep green with whitish spots and margin.
> Var. *leopoldii*. Leaves have large, rounded, reddish disk; margin and veins deep crimson to scarlet; stalk is violet at the base turning reddish on upper part. (From Brazil; introduced 1864.)
> Var. *macrophyllum*. Deep green leaves with many white spots, which are sometimes rose-colored; green stalk. (From Brazil.)
> Var. *neumannii*. Leaves deep green spotted with white and rose; green stalk. (From Brazil.)
> Var. *regale*. More pronounced sagittate leaves; bright green with purple margin; white spots on leaves vary in size.
> Var. *rubivenium*. Leaves are a smaller size; midrib varies in color from pale gray to reddish-green; veins reddish or flesh-colored; like many varieties has not been found in the wild but in the garden, and thus should be regarded as a cultivar.
> Var. *verschaffeltii*. Green leaves with bright red spots scattered over the blade. (From Brazil.)

One can appreciate that with so many variations the exact description of the species type is difficult to make. At various times many of the above varieties and those that are perhaps cultivars have been given specific rank. The complexity of the nomenclature given to such a variable plant over the years makes it almost impossible to determine the exact nomenclature.

C. humboldtii (syn. *C. argyrites*). Native to Brazil; introduced 1858. One of the smaller species, seldom taller than 8–10 in., making it useful for table decoration. The leaves are more oblong, with a short but slender point; light green, the center and margins white. Spathe color same as *C. bicolor*. Leaf stalk slender and variegated. Var. *myriostigma* has leaves with small, white dots.

C. marmoratum. Native to Ecuador. Approximately 18–24 in. in height; the leaves are 6–8 in. long, half as wide; dark green with spots of white, gray, or yellowish-green scattered irregularly over the surface. Spathe is pale green on the exterior and even paler on the inside. Leaf stalk varies in length from 12–18 in.

C. medio-radiatum. Native to Colombia. Reaches up to 12 in. in height. Oblong leaves are 6–10 in. long, dark green with a silvery midrib and wavy margin. Green leaf stalk is marbled brown.

C. picturatum. Native to Peru; introduced 1854. Like *C. bicolor* this is a variable species from which many varieties have been described; often given specific rank. The rootstock is spherical in shape. It can be distinguished from *C. bicolor* by its narrower leaves, which are variously colored above while paler below. The spathe tube is ovoid, yellowish-green on the inside, dark purple at the mouth, with a white or pale yellow-green blade. Much used in hybridizing. Height 20–30 in.

> Var. *adamatinum*. Leaves densely spotted with white along the veins; leaf stalk striped with rose and brownish-green.
> Var. *belleyenei*. Leaves more slender than type; white except for veins, which are green, as is the margin, which also is spotted white.
> Var. *lemaireanum*. Deep green leaves with pale veins that are almost white.
> Var. *sagittatum*. Long, sagittate leaves; midlobe 6 in. long and 1½ in. wide; veins spotted with red.
> Var. *troubetzkoyi*. Much like the previous species, but leaves are deep green with pale red bands along the veins and small white or rose spots.

Many of these varieties of *C. picturatum* have been given specific rank in certain books. As with *C. bicolor*, the nomenclature is hazy, understandable for a plant with such wide differences of color. As noted, however, it is distinguished from *C. bicolor* by its narrow and greener leaves.

C. pubescens. Native to Peru; introduced 1908. The tuber is globose, with a large number of buds. As the name implies, the plant is hairy all over. Height varies from 14–24 in. The bright green leaves are more heart-shaped and 8–12 in. long. The stalk varies in length from 6–12 in. The spathe is 6 in. long and glabrous on the inside; the tube is globose; the spadix is up to 4 in. long; and the pale green blade is some 4 in. long, with a long, slender, white and greenish-white point.

C. schomburgkii. Native to Brazil and Guyana; introduced 1858. A variable species, although generally the leaves are elliptic-ovate, not peltate, up to 6 in. long, and green spotted with white with paler veins that are silvery or sometimes reddish. The stalk is a distinguishing point as it is up to 16 in. long, sheathing the lower third thus total height 20–24 in. The spathe tube is 1½ in. long; the blade is 4½ in. long and whiter than the light green spathe.

> var. *argyroneurum.* Same color as type but veins are silvery, as is the midrib.
> Var. *erythraeum.* Red midrib and veins.
> Var. *pictum.* White-spotted leaves; sometimes reddish between the red veins.
> Var. *rubescens.* Slender leaves only 2 in. wide and 6 in. long; reddish; leaf stalk also dark red.
> Var. *subrotundum.* Leaves are rounded at base; spotted white or red.
> Var. *venosum.* Leaves are almost deltoid, up to 10 in. long; dark green with red margin and yellowish lines; pale green stalk, sometimes spotted or lined with black.

Hybrids

C. hortulanum. These are the "Fancy-leaved Caladiums." There are many hybrids raised in the United States, varying from 12–24 in. in height, mostly in Florida. Over the years the many crosses producing the cultivars offered have been of mixed parentage. As a general guide, those with heart-shaped leaves are derived from *C. bicolor;* those with narrower leaves, *C. picturatum.* This partial listing will give some idea of the many available.

> 'Ace of Hearts'. Transparent leaf, rose to a good crimson with soft green margins and blood-red ribs.
> 'Bleeding Heart'. Heart-shaped leaves, rose-colored, surrounded by white with green margins and crimson veins.
> 'Candidum'. White leaf with green border and dark green veins.
> 'Cleo'. Overall background of leaf is green with red center and veins.
> 'Debutante'. Leaf has red center with fingers of red extending toward margin with the rest white on green, green border.
> 'Elizabeth Dixon'. Silvery white leaves, silver color breaking down toward margins, flecking the green border; crimson veins.
> 'E. O. Orpet'. Narrow leaves, bright red flowing into narrow green border; not so tall-growing as some.
> 'Frieda Hemple'. Bright red leaves with deeper red veins and green border; sturdy, can stand a little wind.
> 'Jody'. Narrow, quite arrow-shaped leaves, red between the white veins, green border.
> 'Keystone'. Large leaves of good green with white spotting and border of green-crimson veins with lighter, almost white, streak in center.
> 'Lord Derby'. Almost transparent leaf of a good rose with green veins and border.
> 'Mrs. Arno Nehrling'. Fine, dark green leaf with network of lighter green veins and crimson major veins.
> 'Pothos'. Thick leaves, white, marbled with green, veins with hint of purple.
> 'Rio'. Tall-growing; many transparent leaves, rose with splashes of green in center, green margin dotted with pink.
> 'Roehrs Dawn'. Large leaves, creamy white, red veins, green border.
> 'White Princess'. Pure white leaves with dark green veins and border.

C. adamantinum. See *C. picturatum* var. *adamantinum.*

C. argyrites. See *C. humboldtii.*

C. sagittatum. See *C. picturatum* var. *sagittatum.*

C. venosum. See *C. schomburgkii* var. *venosum.*

CALIPHRURIA — AMARYLLIDACEAE

The name is given because the flowers are beautiful (*karos,* Greek) and they are enclosed in a spathe (*phroura,* Greek, meaning a prison). A small genus of perhaps 2 species, only 1, however, *C. hartwegiana,* is in cultivation. The genus is native to the Andes of South America.

Closely related to the *Eucharis* but differ in two respects: the spathe encloses the flowers and the bulbs produce stolons. The perianth tube is narrow and funnel-shaped and the stamens have a bristle on each side. The foliage is bright green, some 6–8 in. in length and up to 2 in. in width. The white flowers are carried on a strong stem that reaches 12 in. in height and as many as 8 flowers are produced. The bulbs are ovoid in shape and covered with a dark papery tunic, much like a tulip bulb.

CULTURE The bulbs cannot withstand temperatures below 45°F and thus are for a warm climate or greenhouse growing. They require a well-draining soil mix that has a good organic content and should be planted just below the surface of the soil. Planting is done in the fall. They should be spaced 6–8 in. apart in the border or 3 to a 12 in. container. Water is sparingly given until good growth is observed when more water and light feedings of liquid organic fertilizer should be given every 7–10 days. They require bright light. The flower stalk will appear in April and the plants flower in May-June. After flowering the plants should be allowed to dry and ripen in a warm location while water is withheld.

PROPAGATION After growth has died down, the bulbs can be lifted and offsets or pieces of the stolons removed and grown on. Seed also is set and should be sown in the spring in a sandy compost and given temperatures at night of some 60°F. The seedlings should be transplanted into individual pots and grown on as soon as large enough to handle.

PESTS & DISEASES No special problems.

USES As container plants or plants in sunny borders in warmer climates. They are rare and grown only by collectors.

SPECIES

C. hartwegiana sometimes listed as *C. hartwegii.* Native to Colombia; introduced in 1843. Described above.

CALLIPSYCHE — AMARYLLIDACEAE

The name is derived from *kallos* ("beauty") and *psyche* ("butterfly"), given because the flowers are pendant and have very long filaments which extend as much as 6 in. beyond the perianth tube.

Authorities differ on the number of species in the genus. Some say 3, others 4: the species in dispute is *C. bicolor* (now *Eucrosia bicolor*, which see), which at one time was placed in this genus. However, *C. bicolor* produces many leaves, not 2 as in *Callipsyche* and, in addition, the base of the filaments form a rudimentary corona. All, however, are bulbous plants and are native to Peru and Ecuador.

Callipsyche produces few leaves, which are often as much as 4–6 in. in width and appear after the flowers in mid- to late summer (one species may flower twice, see below). The flowers appear in early summer and are carried in umbels on a hollow flower stalk. The flowers are pendant and funnel-shaped with a very short tube from which the anthers protrude for several inches.

CULTURE The bulbs are not hardy, requiring temperatures that do not drop below 50°F and therefore are suitable only for greenhouse growing or for warm climates. The bulbs should be set with the top of the bulb just at or barely below soil level. They need bright light, moisture during their active growth cycle, and afterward a drier resting period, with just enough moisture provided to prevent the bulbs from shriveling. They must have good drainage and prefer a sandy soil mix but with good organic matter present. They should be lightly fertilized with liquid organic matter during their growing period. They are best planted in the fall or in the case of *C. aurantiaca* as soon as the foliage has died down and bulbs are dormant.

PROPAGATION Offsets from the bulbs are the best means of propagation. These should be separated from the parent bulbs and grown on in individual containers or the smaller ones placed in a flat and grown on until of a size to put into individual containers. Seeds should be sown in early spring, given temperatures in the 65°F range at night, using a sandy soil mix and barely covering the seed. The seedlings should be transplanted after the foliage has died down and grown on as for small offsets.

PESTS & DISEASES No special problems.

USES Good container plants provided warmth can be given. Can be planted out in warm climates in a sunny border.

SPECIES

C. aurantiaca. Native to Ecuador; introduced 1868. The bulbs are ovoid and some 2 in. in diameter. Leaves are few; oblong with sharp point and are 6 in. long and almost 4 in. wide. The flowers are deep golden yellow, with 6–8 in an umbel; produced in late winter with another flowering often taking place in mid- or late summer. The height of the flower stalk is 24 in. and the green stamens extend beyond the flowers as much as 3 in.

C. bicolor. See *Eucrois bicolor*.

C. eucroisioides. Native to Ecuador with reference in the literature that it also is found growing in Mexico; introduced 1843. The bulbs are about the same size as *C. aurantiaca* but more rounded. Produces few leaves, 4 in. in width and 12 in. in length, of a good green, which contrasts well with the flowers that are scarlet and green sometimes tending to be more yellowish. Up to 10 flowers in an umbel on a flowering stem reaching 24 in. in height. Flowers in May, sometimes earlier, and often a second batch of flowers will be produced in late summer.

C. mirabilis. Native to Peru; introduced 1868. Often described as one of the most spectacular flowers in cultivation. Usually 2 leaves are produced, which are 6 in. wide and twice as long. A great number of flowers are produced in the umbels, as many as 30 not being unusual, arranged in very orderly fashion on flowering stems that reach up to 36 in. in height. Flowers are 2–3 in. in length, pale greenish-yellow; white stamens protrude for as much as 5 in., thus making an extremely attractive flower head.

COMMENTS As with all tropical plants, the time of flowering varies greatly, and often more than one batch of flowers is produced in a year, depending on the cultivation given. If warm temperatures and bright light are combined with careful attention to feeding and moisture, more than one batch of flowers is almost certain to be produced.

CALOCHORTUS — LILIACEAE

From the Greek *kalos* ("beautiful"), and *chortos* ("grass"). An apt description of some of the most underrated bulbs grown in our gardens; the foliage is indeed grasslike. All *Calochortus* are native to the western United States. The majority are found in California, with other species in Utah, Oregon, Washington, and Colorado. It is interesting to note that the Mormons, when they first arrived in Utah, ate the bulbs of *C. luteus*, which were described as very palatable and nutritious. These bulbs also were eaten by the Indians of that area, while the Indians of the Pacific Northwest ate the bulbs of *C. elegans*.

There are some 60 species in the genus, with a great deal of variation. An in-depth study of these plants was published in 1940 by Ownbey in the *Annals of the Missouri Botanical Garden* (Vol. 27). He designated 3 divisions or types for *Calochortus*:

Cyclobothra — bulbs covered with reticulated fibers; flowers usually pendulous, except *C. weedii*, with erect flowers; fruit not winged.

Eucalochortus — shiny green leaves; nodding flowers, hairy on the inside, giving rise to one common name, Cats' Ears, also known as Fairy Lanterns; seed capsules commonly winged.

Mariposa — bulb covered with thin, paperlike membrane; leaves shiny green, less noticeable than those of Eucalochortus; mostly upright flowers.

The common names for the various species illustrate the great diversity to be found in this genus: Fairy Lantern, Globe Lily, Butterfly Tulip, Star Tulip, Cats' Ears, Mariposa Tulip. The great variation in colors is said to be caused by the type of soil, and temperatures as well. While very attractive in flower, they have a reputation for being difficult to grow under cultivation. Many writers in the last few years have pointed out that this reputation is unwarranted as these plants will perform well provided they receive the same conditions in cultivation as in the wild. It is surprising that breeding has not been done in view of the great variations shown by the genus. These plants truly deserve greater attention by breeders and gardeners. Few commercial firms offer the bulbs for sale. Until recently, some collectors offered bulbs collected in the wild, but this practice has been largely discontinued, perhaps rightly so.

The bulbs of *Calochortus* are small and tunicated and often, after flowering, divide and do not flower again the following year but miss one or two seasons. This may be one of the reasons for it being said they are difficult to grow. The plants have swordlike leaves, with the flower spikes arising from the center of the plant. The flowers are held on thin but sturdy stems. The individual flowers are not large; the inner segments are broader and larger than the outer and the majority are bearded on the inside. The outer segments generally have much less color than the inner and are greenish toward the base. There are 6 stamens.

CULTURE Most important aspect is the need for quick drainage. Adequate moisture required during winter and early spring, but none during summer. Care should be taken, therefore, in placement in garden. Can withstand a lot of warmth during summer. When well-ripened by summer dry periods, can withstand cold in winter. Ideal location is in sunny, well-drained soil with moderate amount of organic matter where bulbs can be left dry during summer. Cover with about 3–4 in. of soil when planting. If area where the bulbs are planted does not dry out, lift them after foliage has died down and replant in fall.

In containers, soil mix should not be rich but in proportions that will provide good drainage, such as 2 parts good topsoil and equal parts of peat moss and sharp sand.

PROPAGATION Small bulbs can be divided or offsets separated and grown on to flowering size, which will take at least 2 seasons. As indicated previously, certain species produce bulbils that can be grown on to flowering size.

Seed is best sown as soon as ripe or in early Fall. Sandy mix as described above is suitable. Covering with sand is recommended. Moist but not wet conditions are best. Transplant after second year. As the number of species offered is limited, obtaining seed from other gardeners unfortunately is the only way to acquire additional species.

PESTS & DISEASES Incorrect cultural conditions present greatest hazard. Moisture during summer will cause rotting of bulbs. Rarely attacked by pests or other diseases.

USES Should find a home in gardens of those who love to grow unusual and attractive plants. Plant in containers in full sun, in a protected frame in colder areas, or at the base of a wall to shield them from too much winter rain.

SPECIES & CULTIVARS Presented according to Ownbey's 3 divisions.

Cyclobothra

C. barbatus (syn. *C. flavus, C. lutea, C. pallidus, C. barbata*). Native to Mexico. Nodding flowers of good, deep yellow, often with purplish markings on outer surface; fringed; interior lined with hairs. Height 1–4 feet. Unusual in that bulbils sometimes will be produced in leaf axils and can be treated as seeds. Leaves linear, pointed, 9–12 in. long. Late-summer-flowering.

C. weedii (syn. *C. citrinus, C. luteus weedii*). Native to southern California. Found growing on rocky hillsides with dry summer conditions. Has one of largest flowers of species, often more than 3 in. diameter. Upright, yellow flowers carried on a branched flower spike to height of 18 in. Purple hairs on inside of flower; deeper yellow blotch surrounded by thin purple line often found at base of inner petals. Flowers from May to July, 24–36 in. in height, basal leaf up to 16 in. long, stem leaves much reduced.

Eucalochortus

C. albus. Native to the coast ranges of California; introduced 1832. Seldom more than 12 in. in height. Stems carry 4–8 flowers, sometimes more. Basal leaf 12–24 in. in length. Stem leaves lanceolate to linear up to 10 in. long. Flowers pendulous, inner petals 1 in. long, outer petals shorter. Inner petals white fringed and hairy; color can vary to a good pink (this form often called *C. a.* var. *rubellus*). April–May flowering.

C. amabilis (syn. *Cyclobothra amoena*). Native to Sierra Nevada range; introduced 1892. 10–12 in. in height. Lilac-pink. Gland found on lower part of inner petals extends across petal, distinguishing it from gland found in var. *rubellus,* which is confined to center of petal. Flowering May/June.

C. caeruleus (syn. *C. maweanus*). Native to the Sierra Nevada. Found growing at about 5000 ft among pines and oaks. Dwarf; average about 4–5 in. in height but may grow to 8 in. Often listed as *C. maweanus* in catalogs. Color varies but always in blue-purple-white shades. Flowers have a great number of long hairs on inside. Flowering early May/June. Basal leaf 4–8 in. long.

C. pulchellus. Native to California; introduced 1832. Several nodding flowers on each stem carried 12–15 in. high. Flowers just over an inch in diameter, yellow petals fringed with hairs which are also found on inside of inner petals. Flowering April/May. The foliage on the stem is linear to lanceolate, 2–10 in. long. Basal leaf 4–16 in. in length.

C. uniflorus. Native to moister regions of southern Oregon and California; introduced 1868. Regarded as one of the easiest of the genus to grow. About 12 in. in height. Several flowers produced per stem. Flowers erect, up to 2 in. in diameter, pale pink or lilac, with darker spot at base of petals and a few hairs on inner segments. Flowering early summer, April–June. Produces bulbils in axils of stem leaves, which are short, basal leaf 4–16 in. in length.

Mariposa

C. clavatus. Native to the Pacific coastal range and the Sierra Nevada, although mostly southern ranges of both; introduced 1897. Upright flowers carried on strong stems reaching 24 in., sometimes more. Flowers are a good, deep yellow with brownish anthers, which color sometimes noted in veins and in outer petals. Early summer-flowering. Lower leaves 4–8 in. long, linear upper leaves smaller.

C. kennedyi. Native to Arizona and drier regions of southern California; introduced 1892. Vividly colored flowers of deep scarlet; color may vary to almost orange. Height varies from 6–18 in. or more. Lower leaves 4–8 in. height, upper smaller. Flowers March–May in native habitat.

C. luteus (syn. *C. citrinus*). Native to California; introduced 1831. 12–18 in. high. Often several flowers per stem but seldom more than 5. Lower leaves linear 4–8 in. long upper reduced. Good yellow flowers with brown blotch at base of inner petals; hairs found on lower part of inner petals. Late spring/early summer-flowering.

C. luteus var. *citrinus;* syn. of *C. superbus.*

C. l. var. *oculatus;* syn. of *C. superbus.*

C. l. var. *vesta;* syn. of *C. vestae.*

C. venustus. Native to Pacific coastal range and the Sierra Nevada; introduced 1836. Height varies from 8–24 in. One of most colorful species; colors range from white to yellow to purple to red. Erect flower, more than 2 in. in diameter, inner segments hairy, often with 2 blotches of color on petals, the lower petals being much darker in color

    'El Dorado'. Large, yellow flowers; offered by Kapiteijn of The Netherlands.

    'Eldorado'. White with red blotches; offered by Russell Graham of Oregon.

CALOCHORTUS

C = Cyclobothra.
E = Eucalochortus.
M = Mariposa.

C. apiculatus (C) Native to Idaho. Straw-colored. 12–18 in. in height. Summer flowering.
C. benthamii (C) syn. of *C. monophyllus*.
C. catalinae (M) (syn. *C. lyonii*) Native to Southern California coast. White to lilac-purple. 12–24 in. in height. June flowering.
C. citrinus syn. of *C. luteus* & *C. weedii*.
C. elegans (E) Native to Oregon. Introduced 1826. Greenish-white. 8 in. in height. June flowering.
C. elegans luteus (C) See *C. monophyllus*.
C. eurycarpus (M) Native to Washington, N.E. Oregon & Montana. Introduced 1826. Creamy-white or lilac with indigo marks. 18–24 in. in height. Aug. flowering.
C. flavus (C) syn. of *C. barbatus*.
C. flexuosus (M) Native to Southern Utah. White with lilac tinge or deep purple. April–May flowering.
C. ghiesbreghtii (C) Native to Mexico & Guatemala. Purplish. Dwarf. Summer flowering.
C. greenei (C) Native to Northern California & Oregon. Introduced 1876. Clear lilac. 12+ in. in height. Summer flowering.
C. gunnisoni (M) Native from S. Dakota to New Mexico. Light lilac (endures 6 mos. frozen). 6–15 in. in height. June–July flowering.
C. howellii (M) Native to Oregon. Introduced 1890. Creamy-white. 12–18 in. in height. Summer flowering.
C. invenustus (M) Native to Mts. of Colorado. Clear lilac. 8–12 in. in height. July flowering.
C. liechtlinii (M) Native to Sierra Nevada. Creamy-yellow or white with brighter yellow hairs, deep reddish purple blotch. 2–24 in. in height. May–June flowering.
C. lilacinus (E) syn. of *C. uniflorus*.
C. longebarbatus (C) Native to Washington. Introduced 1890. Pale or purple-lilac with purple stripe. 12 in. in height. July flowering.
C. lyallii (C) Native from Washington to Mts. of California. Cream, tinged blue. 6–8 in. in height. June flowering.
C. lyonii (M) syn. of *C. catalinae*.
C. macrocarpus (M) Native to California. Introduced 1826. Lavender. 12 in. in height. June–Aug. flowering.
C. madrensis (M) syn. of *C. venustus*.
C. maweanus (C) syn. of *C. caeruleus* which see.
C. m. var. *majus* syn. of *C. tolmiei*.
C. monophyllus (C) (syn. *C. benthamii, C. elegans luteus*) Native to Sierra Nevada. Introduced 1877. Yellow brownish spot, yellow hairs. 4–12 in. in height. July–Aug. flowering.
C. nanus (C) variety of *C. elegans*.
C. nitidus (M) very like or syn. of *C. eurycarpus*.
C. nuttallii (M) Native to Eastern California, Oregon, Montana to Colorado & New Mexico. Introduced 1869. White to deep yellow. 6–18 in. in height. May–Aug. flowering.
C. obispoensis (M) Native to S. California. Introduced 1889. Petals Orange to yellow, tipped reddish brown, long dark hairs, anthers orange. 2–24 in. in height. May–June flowering.
C. oculatus (M) syn. of *C. vestae*.
C. pallidus (C) syn. of *C. barbatus*.
C. palmeri (M) Native to California. White to lavender, anthers white, brownish about glands. 4–12 in. in height. May–July flowering.
C. persistans (M) Native to California. Introduced 1940. Purplish, yellow haired above gland. 4 in. in height. June–July flowering.
C. plummerae (M) Native to Southern California. Pink to dark rose, orange hairs. 12–24 in. in height. May–July flowering.
C. purdyi (C) syn. of *C. tolmiei*.
C. purpureus (C) Native to Mexico. Introduced 1827. Outer segments: green & purple without, yellow within, inner segments: purple without, yellow within. 36 in. in height. Aug. flowering.
C. splendens (M) Native to Southern California. Introduced 1832. Pale lavender or pinkish-lilac. 12–18 in. in height. May–June flowering.
C. superbus (M) (syn. *C. luteus citrinus, C. luteus oculatus, C. venustus* var. *citrinus*) Native to California. Introduced 1893. White, cream or yellow sometimes lavender, with a central brown or deep maroon spot with bright yellow occulation surrounding. May–July flowering.
C. tiburonensis Native to California. Introduced 1971. Yellow-green with many fine hairs, bronze foliage. 18–24 in. in height. June flowering.
C. tolmiei (C) (syn. *C. maweanus majus, C. purdyi*) Native to Oregon. Introduced 1898. Creamy white tinged lilac with purple or white hairs. 5–12 in. in height. April–July flowering.
C. umbellatus (C) Native to California. White, tinged lilac or greenish white. 4–12 in. in height. March–April flowering.
C. uniflorus syn. *C. lilacinus* which see.
C. vestae (M) (syn. *C. luteus vesta, C. oculatus, C. venustus pictus*) Native to California. Introduced 1895. White with rosy spots at base & brown blotch on each segment. May–July flowering.

CALOSCORDUM — AMARYLLIDACEAE

The exact derivation of the genus name is unknown, but undoubtedly comes from *kalos* ("beautiful") and possibly *cordum* ("piping" or "edging"). In this genus only one species is of interest and found in the literature: *C. neriniflorum*, a native of eastern Siberia, Mongolia, and China. It is a hardy species, flowering in July and August; reaches to 16 in. in height. It formerly was known as *Nothoscordum neriniflorum* and once was given the name *Allium neriniflorum*. This plant was first described as *Caloscordum* and has been returned to that genus.

The bulb is small, just over an inch in diameter, rounded, and with a white tunic. Flower stalk will reach up to 16 in. and carries an umbel of bright pink flowers, 30 or more not being uncommon. The flower head measures some 2½ in. in diameter. Each flower is a little less than ½ in. in diameter. While the overall effect is of pink flowers, the base of the segments in the flower is whitish. The stamens are attached to the perianth segments and thus they differ from *Allium*. The flowers are on pedicels about an inch long. The foliage is flat and onionlike, sparse, and often withers before the flowers are produced.

CULTURE Tolerates a wide range of soils but prefers well-drained soils in full sun. Needs moisture in the spring and early summer, but will withstand drought in summer through fall. Set bulbs 2 in. deep and 6–8 in. apart. Plant in fall or early spring. Will need protection in areas where ground freezes.

PROPAGATION While some bulbils are produced, seed is freely produced and germinates well. Sow in spring using a sandy soil mix, barely covering the seed; keep moist, with night temperatures in the 45°F range. Transplant as soon as large enough to handle or after the foliage has died down. Lift and store in an airy place for replanting in early spring. During the resting period, it is possible to leave the seedling bulbs in the container where they have grown, but this means keeping the soil mix on the dry side.

PESTS & DISEASES No special problems.

USES A great little plant for a sunny border in drier areas.

SPECIES

C. **neriniflorum** (syn. *Allium neriniflorum; Nothoscordum neriniflorum*). Described above.

COMMENTS Not widely grown in cultivation but worthy of consideration, especially in gardens in warmer climates where the bulbs can be left undisturbed.

CALOSTEMMA — AMARYLLIDACEAE

The name is derived from the Greek *kalos* ("beautiful") and *stemma* ("crown"), the flowering spikes being crowned with many individual flowers. Natives of Australia, these plants with tunicated bulbs flower during the winter months. The straplike leaves die down each year. Leaves are produced after the flowers, or may appear at the same time. There are 3 species. Being subtropical plants, seldom if ever exposed to frost in their native habitats, they are plants for the greenhouse in all but frost-free areas. They appreciate good moisture in early fall as the bulbs awake from their summer resting period, continued moisture duing the growing season and then drier conditions as the foliage begins to turn brown and die. The bulbs are of medium size, somewhat bottle-shaped, with quite a long, narrow, and flattened neck.

The funnel-shaped flowers are attractive. The 6 tepals form a narrow tube above the ovary, then open out to form flowers 1 in. in width with the 6 anthers well displayed, the filaments being flattened and forming a corolla in the center of the flower. The stigmas are thin, pointed, of greater length than the stamens, and protrude above the corona. As many as 20 flowers are produced in an irregularly shaped flower head. The green bracts that cover the flower buds persist, gradually turning brown as the age of the flowers advances.

CULTURE Will not tolerate frost, and in areas where frost is experienced should be grown in a greenhouse where the coldest night temperatures are in the 45°F range. They like a soil mix that is sandy, with humus. Keep moist but never wet, allowing the spring-flowering species *C. album* to become a little dry toward the end of summer.

Bulbs should be planted with the neck of the bulb level with the soil and spaced some 8–10 in. apart if being grown in a border. They also do well in containers but drainage should be good. They like sun, and in frost-free areas can be planted in a sunny, well-drained border with some shade during the hottest time of the day. Established plantings appreciate some fertilizer during the spring and early summer, but feedings should be light.

PROPAGATION Offsets provide the best means of propagation. These should be separated from the bulbs after the bulb has finished flowering. Seed can be sown in the spring. Sow thinly over a sandy soil mix and barely cover. Seedlings should be given good light and temperatures in the 50°F range at night. Keep the soil mix moist.

PESTS & DISEASES No special problems, but watch for aphids when greenhouse grown.

USES Greenhouse culture in all but warmest areas. In warm areas a good bulbous plant best displayed toward the front of the border.

SPECIES

C. **album.** Native to Australia; introduced 1824. Reaches a height of some 12–18 in. with pure white flowers produced in spring or early summer. The leaves are short, only 4–6 in. in length and 2–4 in. in width.

C. **luteum.** Native to Australia; introduced 1819. Lovely yellow flowers produced in the fall on stems 12–18 in. in height. The leaves are strap-shaped and narrow, up to 10–12 in. in length. The foliage will die down after the flowers are produced, then the plants should be kept on the dry side. Water again more freely when the foliage reappears in late summer.

C. **purpureum.** Native to New South Wales and southern Australia; introduced 1819. Flowers are deep purplish-pink; exterior whitish at the base in the funnel part of the flower, contrasting with the dark green ovary reaching 12–18 in. in height. Flowering in November. Strap-shaped, narrow foliage dies down after flowering. The variety *C. p. carneum* from same habitat has paler, sometimes almost whitish flowers.

COMMENTS These plants deserve a place in borders in warm climates. They will increase in number if clumps are undisturbed.

CALYDOREA — IRIDACEAE

The name is derived from the Greek *kalos* ("beautiful") and *dorea* ("gift"). This genus is at home in South America, and, while it is noted in the literature that there are some 10 species, little is written about them. Only 2 species would appear to be at all known. While this is not an uncommon occurrence with genera from South America, it is certain that the genera from this part of the world deserve evaluation and study. Leaves of *Calydorea* are not unlike the more familiar *Sisyrinchium* species; indeed, one species was so named when first introduced, namely *S. speciosum* = *Calydorea speciosa*.

The species are dwarf, the flowers fleeting, carried on leafless stems or nearly so. The colors are blue, sometimes with yellow toward the base of the petals. Flower heads are compact, often branched, and the individual flowers can be over an inch in diameter with the perianth segments of equal length. The leaves are narrow, linear, and longer than the flower spikes. The globose corms are a little less than an inch in diameter and reportedly are edible.

CULTURE Corms should be planted about 1 in. deep in well-drained soil. Moisture is required in the spring and during the growing season, but the corms should be allowed to become dry during the winter months. These plants are native to warm regions and thus should be grown under protection in the coldest climates or planted in the spring and lifted in the fall. If grown in containers they should be allowed to dry during the winter months and then repotted in the spring. In climates where there is little or no frost they belong in the rock garden or in a border at the base of a wall. As they are not tall growing they must be carefully positioned, to be easily seen. They require good light, but as they are mostly in the blue shades some protection from the hottest sun is suggested. Such protection is further indicated by their grassy native habitat.

PROPAGATION While the corms are reputed to form numerous cormels, the rarity of the plant suggests that raising stock from seed would be best. Sow in the spring in moderate temperatures and in a sandy, well-draining soil mix, barely covered. They should be kept moist but not wet and the young plants transplanted to larger containers as needed.

PESTS & DISEASES No special problems.

USES For the collector only, as these plants are very rare. They look fine in a rock garden but are best cultivated in containers until sufficient stock is built up for planting.

SPECIES

C. nuda. Native to Uruguay. Found growing in grassy fields. Flowers blue and up to 1 in. in diameter, sometimes a little larger. Several flowers in a branched but compact flower head. 4–6 in. in height. Foliage narrow and linear. Rare if indeed it is in cultivation.

C. speciosa (syn. *Sisyrinchium speciosum*). Native to Chile; introduced 1836. Foliage up to 9 in. in length but often less. Flowers are blue with yellow at the base of the petals, height of the flower head reaches some 3–9 in. with some reduced leaves. Like the others, this species is rarely grown.

CAMASSIA — LILIACEAE

From *quamash*, the name given to the plants by the Indians of the Pacific Northwest. Its common names are camass or quamash, and sometimes it is called the wild hyacinth. There are 6 species, all from the Americas. One South American species is quite widespread, being found in Peru, Bolivia, Chile, and Argentina. *C. scilloides* is native to the midwestern and southern United States. The other species are from North America, principally the areas along the Pacific Coast. These plants are closely allied to the *Scilla*, a genus not found in the Americas. Although they are found in many areas, including the plains and foothills of mountain ranges, the majority of *Camassia* are at home near streams and lower meadows where summer moisture is available. They are very attractive plants, especially when seen as a mass in their native habitats.

For many years the plant roots provided a major portion of the food supply of the Indians, who boiled and roasted them. When boiled in water, roots yield a good molasses, which was used on festive occasions by a number of the tribes.

The leaves are long and narrow. The flowers are in long racemes with either upright or horizontal stalks. In some species, the withered flowers may cover the developing seed capsule. Together with the color of the flowers, these features are used for differentiating among members of the genus.

CULTURE *Camassia* is at home in most garden soils, especially if there is good summer moisture. It will exist in drier areas but will not perform well. Plant bulbs in fall, about 4–5 in. deep. If placed alongside a stream or pond, where they can obtain the moisture they like, the bulbs should be set above the waterline. Grow well in full sun but if abundant moisture is not available should be grown in slight shade, especially during hottest parts of the day. Quite hardy and will withstand frost and cold winters. However, *C. biflora*, the South American species, needs the protection of a frame in colder climates. Once established, do not disturb.

PROPAGATION The preferred method is starting from seed, from which plants come easily. Ripe seed should be sown as soon as possible in soil mix that provides good moisture and drainage. Bulbs grow to considerable size and will produce a few offsets that can be grown on to flowering size. Selections of *Camassia* offered by nurseries must be reproduced by offsets.

PESTS & DISEASES No special problems.

USES South American species, which are not so attractive as North American varieties, are suitable for growing only in frost-free areas or in a greenhouse. North American species make excellent plants for growing in grassland, providing a summer alternative to b. Ideal plants for extending flowering of border alongside stream or pond. Should be planted *en masse*, as one or two bulbs are not effective, except perhaps for the connoisseur.

SPECIES & CULTIVARS

C. biflora (syn. *Fortunatia triflora*). Native to South America; recently introduced into Britain. Found growing in dry, rocky places. Not hardy. Named after habit of producing 2 (sometimes 1 or 3) flowers per bract. Narrow basal leaves some 24 in. long. Flattish flowers rarely more than ½ in. in diameter. Loose racemes of white or pale pink flowers carried on stems 18–24 in. long; flowers between fall and spring in a temperate greenhouse.

C. cusickii. Native to Oregon; introduced 1888. Very large bulb, weighing about 4–8 oz. Plant reaches 30 in. in height, sometimes more. Leaves up to 2 in. wide and 20 in. long. Numerous flowers, usually pale blue. Deeper, intense blue can be found in *C. c.* 'Nimmerdor', which makes a good cut flower, and *C. c.* 'Zwanenburg', introduced circa 1900 by Van Tubergen of The Netherlands. Stalks horizontal, turning up at ends. When flowers fade do not cover developing seed capsule. Flowering early summer (May–June).

C. esculenta: See *C. quamash*.

C. howellii. Native to southern Oregon; introduced circa 1938. Bulb rather small. Will reach more than 24 in. in height. Flower stalks horizontal. The deep blue-violet flowers cover developing seed capsule as they wither. Flowering May.

C. leichtlinii. Found wild over much of western United States and into British Columbia; introduced 1853. Bulb ½–1½ in. thick. One of the tallest growing, sometimes reaching more than 3 ft. Leaves up to an inch wide and seldom more than 24 in. long. Flowers vary from white to deep purplish-blue; twist around developing seed capsule as they wither. Stalk intermediate, between horizontal and upright. Regarded as one of best garden species. Several forms have been given subspecies rank: ssp.

typica, with white/creamy white flowers, often sold under name 'alba'; and ssp. *suksdorfii*, introduced circa 1942, with blue to violet flowers. Great variation, some flowers semidouble, offered by nurseries as *plena* or *semiplena*, all in the white range. Var. *atroviolacea*: deep purple, regarded as one of the finest of all Camassias.

C. **quamash** (syn. *C. esculenta*). Widespread habitat, ranging from British Columbia down the Pacific Coast to California and eastward to Montana and Utah; introduced 1837. Popular food plant of the Indians. Bulbs reach good size when growing in areas with adequate moisture, but can also exist on hillsides where summer moisture is scant. Height varies from 12–24 in. Great variation in flower color, from deep blue to pale blue to white. When they fade, flowers sometimes drop away or cover seed capsule. Stalks vary from horizontal to erect. Flowering early April through May. Hardy species, of easy culture. Botanists have divided species into 8 subspecies based on flower and foliage color:

Ssp. *azurea*. Leaves grayish; flowers light blue-violet.
Ssp. *breviflora*. Leaves gray-green; flowers blue to deep blue violet.
Ssp. *intermedia*. Leaves green; flowers pale blue-violet.
Ssp. *linearis*. Leaves green; flowers deep violet-blue.
Ssp. *maxima*. Leaves grayish; flowers deep blue-violet.
Ssp. *quamash* (ssp. *typica*). Leaves grayish-green; flowers pale to deep blue.
Ssp. *utahensis*. Leaves gray; flowers pale blue-violet.
Ssp. *walpolei*. Leaves green; flowers pale blue or blue-violet.

Two selections are offered by nurseries: 'Orion', deep blue; and 'San Juan', even deeper blue.

C. **scilloides** (syn. *C. angusta, C. fraseri; Quamasia hyacinthina*). Native to the midwestern and southern states of the United States. Seldom grows higher than 24 in.; leaves about same length, up to an inch wide. Flowers vary from deep blue to white; fall away, leaving capsule uncovered. Stalks more upright than horizontal. May/June flowering.

CANNA — CANNACEAE

From the Greek *kanna* ("reedlike plant"). There are some 50 species, native to the tropical parts of South America and Asia, with one species native to Florida. The common name Indian Shot is given to the many garden hybrids because the seeds are black and very hard, resembling the shot or pellets in shotgun cartridges. Most of the plants grown in gardens today are hybrids. The principal species used is *C. flaccida*, from Florida, but many others also have been used.

The seeds are used as beads in some countries. The roots are used to make a form of arrowroot. The particular starch made from the roots of indigenous species is known as *tous-les-mois* in the West Indies.

Not all species produce rhizomes, as some are bifrous-rooted, but those rhizomes produced are fleshy and covered with a dark brown skin. The stocky, often stubby, rhizomes have many projections, and the leading buds are rounded. For the most part, the flowers are not made up of petals but of stamens, which are petaloid. Flowers have 3 sepals, which are most commonly green; 3 long petals, which are also colored and not very wide; and up to 5 stamens, which look like petals and are broad and also colored. One of these stamens is carried to form the lower lip of the flower. Generally, only one stamen is fertile and will often be petaloid on one side. There is a single, long style leading to an inferior ovary, with the fruit being a 3-celled capsule bearing the many hard, black seeds.

CULTURE They grow in the wild in warm, humid climates, so must receive warmth wherever grown in order to thrive. Must be lifted and stored overwinter where there is severe frost and cold. In more moderate areas, can be left in the ground but must be given adequate winter protection. Do not grow where shade cools the soil or where there is much winter moisture, as rhizomes must not lie in cold, damp soil.

Ideal location is in good soil, with full sun and humidity during warm months. Plant as soon as all danger of frost is passed or in March in frost-free areas. Set with 4–6 in. of soil over roots. Plants appreciate rich soil so should be fed as soon as growth is observed with a complete fertilizer, once a month until flowering. This is necessary for lush growth and especially if in sandy or poor soil. Adequate moisture needed throughout growing season. Reduce water toward end of summer as foliage starts to die back. Store rhizomes overwinter in moist, frost-free place; if allowed to dry out they will wither and rot will set in.

Grow well in containers if number of shoots to a rhizome is restricted to only one. Remove weaker growths. More than one rhizome can be placed in large container. Regular feedings of liquid fertilizer should be given until plants are well into flower.

PROPAGATION Can be propagated from seed but, to obtain same plant as parent, asexual propagation is necessary. Seeds are hard so soak them in warm water for 24 hours before sowing, or notch seed coat with a knife prior to sowing. Sow during first months of year in sandy peat mix and transplant as soon as germinated. Temperature should be in 65°–75°F range. Take care when transplanting as roots are quite brittle and break easily. Can be sown in individual pots using richer soil mix, such as half good topsoil and half peat and sand. Grow seedlings in minimum temperature of 60°F until hardened off and planted outdoors.

Rhizomes can be lifted and stored overwinter, then cut into sections in spring. Each section should have at least one prominent bud. Best to start roots pieces in good soil mix, water a little until growth is made, and then increase water given. Place in containers as soon as they sprout; plant out when of good size. If porous soil mix is used, liquid feedings of fertilizer can be given as soon as new roots produced. Temperature in 65°–70°F range induces best growth. Plants like to be fed constantly but not until growth is evident.

PESTS & DISEASES Plants can be attacked by a number of pests and diseases but generally damage is not great. Mosaic can cause leaves to become mottled with yellow in no organized pattern; plants affected by this virus, which is spread by sucking insects, should be discarded. Slugs and snails like the fast-growing shoots so take precautions when first setting out plants in garden. Also protect established plants as soon as shoots emerge in spring.

USES With their bright flowers and foliage patterns these are striking plants for the summer border. Remain attractive throughout season. Also can be used in containers to highlight deck, patio, or entrance to home.

SPECIES & CULTIVARS

C. **discolor**. Native to Central America; introduced 1872. Leaves up to 36 in. long and 12 in. wide; green with purple margin, undersides purplish. Flowers red, yellow, and purple, standing as high as 6 ft. Summer-flowering.

C. edulis (syn. *C. esculenta*). Native to the West Indies and South America; introduced 1820. Source of *tous-les-mois* (an arrowroot). Rhizome stout, tuberlike, edible. Tall-growing, often reaching 8–10 ft. Leaves green with purplish-red undersides, 24 in. long and 6–8 in. wide. Flowers red and yellow. Summer-flowering.

C. flaccida. Native to Florida; introduced 1788. A principal parent of modern garden hybrids. Leaves 10 in. or more in length, 4–6 in. wide. Flowers in summer and are in shades of yellow, with stems that reach up to 6 ft. in height. Lip of flower is large and wavy, giving a floppy look.

C. indica (Indian Shot; syn. *C. patens*). Not native to India, however, but rather Central and South America and the West Indies; introduced 1570. One of the oldest species in cultivation. Not so tall-growing as others in species, most often in the 3–5 ft. range. Green leaves 18 in. long and 6–8 in. wide. Flowers are in loose racemes in red, rose, or shades of pink.

C. iridiflora (syn. *C. gigantea*). Native to Peru; introduced 1816. One of the earliest to flower, blooming in early summer. Tall, often more than 8 ft. Leaves 8–10 in. wide, 24 in. long. Rose-colored flowers unusual in that petals form slender tube while petal-like stamens have definite notch at tips. Leaves 24 in. long—wide and oblong.

C. latifolia (syn. *C. gigantea*). One of the tallest growing, often more than 15 ft. Leaves 36 in. long, up to 10 in. wide. Flowers in shades of red, in the summer.

CANNA

C. achiras syn. of *C. lanuginosa*.
C. aurantica variety of *C. lutea*.
C. esculenta see *C. edulis*.
C. gigantea syn. of *C. iridiflora* & *C. latifolia*.
C. glauca Native to W. Indies, South America. Pale yellow, rhizome long, slender. 48–72 in. in height. Summer flowering.
C. lanuginosa (syn. *C. achiras*) Native to Brazil, Peru. Yellowish-red. 36–48 in. in height. Summer flowering.
C. limbata Native to Brazil. Introduced 1818. Greenish yellow. 36 in. in height. Summer flowering.
C. lutea Native to Tropical & Subtropical America. Introduced 1824. Pale yellowish white. 30 in. in height. Late Summer flowering.
C. patens see *C. indica*.
C. pedunculata Native to South America, West Indies. Yellow-green. 5–6 in. in height. Summer flowering.
C. sanguinea syn. of *C. warscewiczii*.
C. speciosa Native to India, East Indies. Introduced 1820. Pale purple. 72 in. in height. August flowering.
C. warscewiczii (syn. *C. sanguinea*) Native to Costa Rica, Brazil. Introduced 1849. Reddish, varigated foliage. 72 in. in height. Summer flowering.

Hybrids

A number of hybrids are offered by nurseries. The species that play the major roles are *C. flaccida* for flower size and *C. warscewiczii* for variegated foliage. Height ranges from 18 in. to more than 48 in.

Very dwarf (18–20 in.).
  'Nirvana'. Green leaves striped with yellow; red buds; deep yellow flowers.
  'Seven Dwarfs'. Flower colors of yellow, salmon, pink, red, crimson.

Dwarf (30 in.).
  Most dwarfs offered are hybrids raised by German hybridizer Wilhelm Pfitzer and frequently are listed as 'Pfitzer's Primrose Yellow' or whatever color is available; all have green foliage. Among them are: 'Chinese Coral', 'Primrose Yellow', 'Salmon Pink', and 'Scarlet Beauty'.

Standard (48 in.).
  'Black Knight'. Bronze foliage; red flowers.
  'President'. Green foliage; bright red flowers.
  'Richard Wallace'. Green foliage; bright yellow flowers.
  'Stadt Fellbach'. Green foliage; orange flowers.
  'Wyoming'. Bronze foliage; orange-red flowers.

Very tall (more than 48 in.).
  'Eureka'. 4–5 ft. high; dark green leaves; pure white flowers.
  'Red King Humbert'. Often more than 6 ft. high; bronze foliage; red flowers.

COMMENTS It is interesting to note that in the literature Memoirs of the Botanical Survey of South Africa Vol. No 51 published in 1985 and entitled *Species of Southern African Plants*, both *C. edulis* and *C. indica* are listed, the only reference I have found of their being native to southern Africa.

CARDIOCRINUM — LILIACEAE

From Greek *kardia* ("heart"), and *krinon*, a kind of lily. For many years this genus was combined with *Lilium*. This is quite understandable when viewing them. The flowers have the same shape and are very fragrant. Closer examination reveals distinct differences; however, the major difference is the shape of the leaves. These are heart-shaped in the *Cardiocrinum*. In the Chinese and Japanese species leaves are arranged in a whorl halfway up the stem; in *C. giganteum* they are arranged along the length of the stem. The leaves are often a foot long and as wide. The other main difference is in the bulb. Those of *Cardiocrinum* have a few overlapping scales and are not tunicated. New scales are added each year but all die when the plant flowers. Young offsets are formed over the period of time the parent is preparing to flower. Another difference is in the seed capsules, which have distinct teeth along the edges. While there are 3 species, only one, *C. giganteum*, is commonly found.

Cardiocrinum is the only bulbous plant that provides a musical instrument. The hill people of Nepal, where *C. giganteum* is native, make musical pipes of the hollow stems.

CULTURE They are woodland plants that thrive in rich, moist soil (will not thrive in dry soil) and like filtered sunlight. As they take up to 5 years (or more) to reach flowering size, it is recommended that bulbs of different sizes be planted together. Do not plant too deeply; top of bulb should just break the soil. Plants send up non-flowering shoots each year, often reaching considerable height, and then the shoots die back. After flowering the plant dies, allowing offsets to continue. Because offsets often become quite crowded, it is advisable to lift dead plant, separate young plants from parent, and replant, to avoid overcrowding. Some protection might be needed in very cold winter areas, not so much to keep cold from plants as to avoid thawing and freezing that can lift and damage bulbs.

PROPAGATION   Offsets provide best means; separate them from parent when it dies in fall and replant offsets at once. Seed can be sown in fall but resulting plants will take as many as 7 years to flower. Plants raised from seedlings are tallest.

PESTS & DISEASES   Subject to same problems as *Lilium*, which see.

USES   Best in woodlands. Do not like to be grown in containers. Lovely when seen against a solid background of evergreens, such as rhododendrons.

SPECIES

C. **cathayanum** (syn. *Lilium cathayanum*, *"Lilium mirabile"*). Chinese species from eastern and central China; native habitat is dense woodland; introduced 1939. Leaves produced in whorl about halfway up stem, with a few smaller leaves in a random pattern above. Leaves are large, about 6–8 in. long and up to 6 in. wide, with stalks 4–6 in. long. Plant produces 1–5 funnel-shaped flowers, held horizontally; not large, seldom more than 4 in. in diameter; greenish white outside, creamy white inside. Height of plant from 24–48 in. Summer-flowering.

C. **cordatum** (syn. *Lilium cordifolium*, *L. cordatum*). Japanese species widely distributed in the islands; introduced 1876. Specific name describes leaves, which are heart-shaped and produced in a whorl but sometimes scattered over middle portion of stem; up to a foot long, often as broad as long. Summer-flowering, the strong 48–60 in. stem bears as many as 15, sometimes more, flowers. Flowers creamy white with lower 3 tepals having distinct yellow band toward base, often ornamented with brown spots. Disadvantage is that leaves are produced early in year and can be harmed by a late frost.

C. **giganteum** (syn. *Lilium giganteum*). Native to Himalayas and found over large area stretching from Nepal and Upper Assam to northern reaches of Burma and into southeastern Tibet; introduced 1852. By far finest species. Basal leaves form a rosette; others scattered up stem. Leaves 18 in. long and almost as wide. Strong stem, often 4–6 in. in diameter at base of strongest flowering stems, often surpassing 8–9 ft. in height. Fragrant flowers open late summer; as many as 20 produced on each half pendant; pure white outside, pure white interior, purplish marking toward base; 6 in. in diameter. *C. g.* var. *yunnanense*, native to western China, not so tall, only 4–6 ft.; brown stems, rich bronze-tinted young foliage.

CARPOLYZA — AMARYLLIDACEAE

The name is derived from the Greek *karpos*)"fruit") and *lyssa* ("rage"), from the peculiar method of dehiscense.

A monotypic genus, which, while first described and introduced in 1791, is seldom grown. The one species is *C. spiralis*, a native of the Cape Peninsula of South Africa, growing on sandy soil in coastal areas. Seldom more than 3–4 in. high, the starry white flowers are ½ in. in diameter, the reverse of the petals retaining just a hint of the pinkish color the flower buds exhibit. Seldom more than 2 or 3 flowers are produced on the thin, wiry stems.

It flowers in winter and very early spring, May–June in its native habitat, and is inconspicuous so one has to look carefully to find the plants. Rootstock is a bulb, small, the 3–4 thread-like leaves that are produced are held in a transparent sheath at their base then emerge twisting and turning, their length some 3–4 in.

CULTURE   Full sun in sandy soil, bulbs set just below surface of the soil, good moisture in the winter months, dry conditions at other times. Not hardy and must be protected from temperatures below 35°F. No feeding is required.

PROPAGATION   A certain number of bulblets are produced which should be removed from established plants during the dormant period, in late summer. Seed also can be sown in a sandy soil mix in early spring, the plants being transplanted when large enough to handle, grown on in a sandy soil mix.

PESTS & DISEASES   No special problems.

USES   For the collector only.

SPECIES

C. **spiralis** (syn. *C. tenella*). Described above.

CEROPEGIA — ASCLEPIADACEAE

The name is derived from *keros* ("wax") and *pege* ("a fountain"), as the flowers are sometimes waxy in appearance. There are many species, some authorities putting the number as high as 200, but only a few have bulbous rootstocks, either in the shape of a corm or fleshy tuber. The species is distributed worldwide, but the bulbous species would seem to be confined to Africa: *C. debilis* is found in tropical East Africa, the others in South Africa.

The plants are either climbers or scramblers. They have opposite leaves, ovate or nearly heart-shaped, and unusual flowers. The corolla tube is very inflated at the base, then becomes slimmer, and then widens again at the top. The perianth segments then divide into 5 lobes which most commonly are again united at their tips. Thus the flowers form a small, cage-like structure, the edges of the perianth often recurving and showing the color of the interior.

In certain species, the leaves will drop off at an early age, leaving the stem leafless at the base. In southern Africa the common name of Bushman's Pipe often is given to these plants.

CULTURE   Tropical species need heat and night temperatures of 50°F in the winter months. Those from South Africa are hardier, but still need winter protection in the colder climates. Suitable for growing outside in frost-free climates or with a little protection in climates where frosts are mild or rarely experienced.

They need a well-drained soil in sun, much moisture during their growing season, and then a dry period during the winter months. The rootstocks should be planted 2–3 in. deep, where the stems will be able to wander or twine their way among other plants. Some feeding can be given in the spring as soon as growth commences.

PROPAGATION   Cuttings provide the best way to increase the stock. These should be rooted in pure sand or sandy soil mix with slight bottom heat. They should be potted up as soon as rooted. Seed is sometimes produced but is not to be relied upon due to its scarcity. The plants can be laid on the surface of a sandy soil mix and roots often will be produced at the nodes; this root production can be increased by nicking the plants with a knife at the point just below a node. These rooted sections then can be potted into containers and grown on. Seed should be sown with the seed only just covered and kept in a warm temperature at 50°F at night; keep moist and pot the individual plants as soon as they are large enough to handle.

PESTS & DISEASES   No special problems.

USES   Good and unusual climbing and scrambling plants for warm climates and cool greenhouses.

SPECIES

C. **amplicata.** Widely distributed in the southern part of Africa from eastern Cape Province up into the Tropics. A perennial climber that grows on dry hillsides and scrambles through the scrub often reaching many feet in length. The rootstock is a fleshy tuber, producing a succulent stem and small, dark green leaves that are found only at the tips of the shoots—they fall off quickly from the stems. The flowers are carried in the nodes of the plants, quite commonly more than 2 per node, appearing in summer; swollen or inflated at the base, then narrowing and the lobes of the 5 perianth segments meeting at the tips. Flower is white sometimes the outside is marked by green veins, sometimes a clear white; the interior is green and this color is seen along the edges of the segments as they recurve. It is not possible to see the reproductive parts of the flower without opening the corolla tube. In the wild the stem of the plants frequently will root at the nodes if they come into contact with the earth. This plant is comparatively hardy, but not sufficiently hardy to be grown in areas with frost.

COMMENTS   The unusual flowers make this an interesting plant. It is rare in cultivation and it is doubtful if it will ever be a common garden plant, no matter that it is deserving.

CEROPEGIA

C. africana Native to Cape Province. Green & dark purple. Root, tuberous. Twining. Summer flowering.
C. barklyi Native to Cape Province. Purple with darker lobes. Root, corm-like. Twining stems. Summer flowering.
C. debilis Native to Tropical E. Africa. Purple with darker lobes, lobes fringed with dark hairs. Root, corm-like. Twining. Summer flowering.
C. rendallii Native to Transvaal. Tube white, with purple base. Root, large tuber. Stems twining. Summer flowering.
C. woodii Native to Natal. Purple with dark lobes. Trailing. Summer flowering.

CHASMANTHE — IRIDACEAE

From the Greek *chasme* ("gaping"), *anthe* ("flower"), an apt description of the flowers, which are curved and hooded. Native of South Africa, only half of the 8–9 species are generally cultivated. Some authorities divide *C. aethiopica* into subspecies, i.e., *C. vittigera*, which they say merits subspecies status because of the wider leaves; others say this is a minor difference only. Produced from sub-globose corms, the plants are vigorous and need room to spread. They are unable to withstand much cold and should be grown in areas with little or no winter frost or in protected frames.

The flowers are abundantly produced; in some species the flower spike will be branched. The positioning of the flowers on the stem is also unusual as the flower head gives the impression of being flattened, the flowers either facing all one way, as in *C. aethiopica*, or on two distinct sides of the stem. The flowers themselves are tubular with the lower parts being pinched and much shorter than the upper. The upper tepals form a hood over the lower and also to some extent sheathe the long stamens that protrude just beyond the tip of the upper tepals. The lower tepals will often angle away from the upper tepals, often by as much as 90°. In all species the flowers have a tube and such angling only affects the last ¼ in. of the lower tepals. Lower flowers are open long before the upper; often the upper buds are still very small and carried closely together on the flattened flower spike as the lower flowers are fully opened. These first-opening flowers will be long past before the entire flower spike has fully developed.

The foliage is stiff, quite broad, up to 2 in. in width, and the leaves sword-shaped, tapering to a quite sharp point. To the touch they are coarse. The number of leaves per plant varies but usually more than 6 and sometimes as many as 8–10 per corm.

CULTURE   Plant corms in fall or spring 3 in. deep and 10 in. apart in full sun or light shade. Not too fussy about soil but must receive moisture during winter months and early spring. Plants will die down after flowering. Corms need not be lifted in warm areas unless overcrowded.

PROPAGATION   Quickest method is to lift and separate smaller corms and plant back. Seed quite easy to raise. Sow as soon as ripe. Seedlings will take at least one season before reaching flowering size.

PESTS & DISEASES   Not usually bothered, but in colder climates corms can become damaged and rot sets in. If in doubt, lift and store after flowering, keeping them on dry side.

USES   Corms will multiply if they like their location, so plant where there is no restriction on spreading. Good cut flowers, lasting for days. May become more popular in warmer parts of the United States because plants are resistant to summer drought.

SPECIES

C. **aethiopica** (syn. *Petamenes aethiopica*). Native to South Africa; introduced 1759. Large, sub-globose corm. Leaves arranged in fan shape; 12–18 in. long. Unusual flowering habit: all flowers on 48 in. high stem turn to one side. Top part of flower is reddish; lower part greenish-orange. Stamens are prominent. Flowers have distinct curve; just over 2 in. in length; 12–15 produced in late summer (July–early August).

C. **caffra** (syn. *Anapalina caffra, Petamenes caffra*). Native to South Africa; introduced 1928. Corm about three-fifths in. thick. Flower stem about half height (22–24 in.) of *C. aethiopica* but more flowers produced per stem. Leaves also fan-shaped, narrow, 12–14 in. long. Bright red flowers arranged on stem on each side as though flattened; upper petals almost twice length of lower, hiding stamens; lowest petals curve downward. Flowering early summer.

C. **floribunda** (syn. *Petamenes floribunda*). Native to South Africa. Most prolific flower producer, often as many as 30 per stem. Branching flower spike, 48 ins in height, each one producing 12–15 flowers. Color closer to orange than red and occasionally yellow. Leaves wider than other species (2 in.) but showing same fan-shaped arrangement; about 14 in. long. Flowering early summer. *C. fucata* is similar but has narrower leaves and the perianth tube is gradually constricted at the base. In *C. floribunda* the tube is sharply constricted. *C. fucata*, however, is not a valid name and should be grouped under *C. floribunda*.

CHASMANTHE

C. bicolor (syn. *Petamenes bicolor*) Native to South Africa. Upper perianth red, lower part of tube yellow to greenish. 18 in. in height Late Summer flowering.
C. fucata (see *C. floribunda*) Native to South Africa. Bright yellow & red, corm large, subglobose. 36 in. in height. Summer flowering.
C. intermedia (sometimes placed in *Anapalina*) Native to South Africa. Dark blotch at the base of the claws, shorter & wider foliage than *C. caffra*. Early Summer flowering.
C. peglerae (syn. *Petamenes peglerae*) Native to Eastern Cape, South Africa. Bright red-orange. 10–24 in. in height. Late Summer flowering.
C. vittigera (syn. *Petamenes vittigera*) Native to South Africa. Reddish yellow with red stripes, similar to *C. aethiopica*, but has wider leaves. 36 in. in height. Early Summer flowering.

CHIONODOXA — LILIACEAE

From Greek *chion* ("snow"), *doxa* ("glory"), hence their common name, Glory of the Snow, which describes their habit of flowering—peeking above the melting snow.

One of the finest of the early flowering bulbs, the some 9 species in *Chionodoxa* are native to the mountains of Crete and Turkey. They are related to the *Scilla* but differ in that the petals are united at the base, forming a short tube, and the flattened filaments form a cup in the center of the flower.

The leaves are basal, not numerous; often only 2 are produced. They are narrow, quite thick and stiff, and dark green. The flowers are borne in loose racemes, and, depending on the species, the number per stem varies from 1 to more than 10. While pink and white forms are known, the majority of the flowers are blue, with various splashes of white in the petals.

CULTURE Plant in fall, 3 in. deep, in well-drained soil. Must have winter and spring moisture so humus should be added to sandy soil. Like sun. Large bulbs will produce more than one flower spike; must be left in ground for number of seasons before many spikes grown. In order to reach good size and increase should not have competition from other plants; set in location where they can be left. Lift and divide only when overcrowded.

PROPAGATION Many offsets produced which can be separated from parents and planted in fall. Seeds germinate readily; sow as soon as ripe either in flats or in small rows if many are wanted.

PESTS & DISEASES No special problems.

USES Lovely early spring flowers for rock garden or for growing in containers to be brought indoors.

SPECIES & CULTIVARS

C. cretica. Native to Crete. Generally quite short but can reach height up to 8 in. Leaves almost grasslike. Flowers lilac-blue with white center; 1–3 per stem. Flowering early spring. Long in cultivation.

C. gigantea (syn. *C. allenii*). Native to Turkey; introduced 1878. Large, light to medium blue flowers, up to 8 in. in height. Some nurseries offer white form, *C. g. alba*.

C. luciliae. Introduced 1877. Most commonly grown and generally considered finest species. Fleshy, globose bulb. As many as 10 blossoms on 6–8 in. high stems, each up to 1 in. diameter. Color good blue with white center; however, several selections offered of other colors: 'Alba', pure white; 'Pink Giant', little taller than species, bright pink; 'Rosea', pink. *C. luciliae* in commerce is most likely *C. forbesii* and the original *C. luciliae* same as *C. gigantea*. There is need for distinct identification of species in this genus.

C. sardensis. Native to Turkey, introduced 1877. Rare in culture. Dwarf, 6 in. high, good gentian blue with small white eye. Flowering late spring.

C. siehei. Native to eastern Asia Minor; introduced 1904. Largest growing of genus, reaching 12 in. in height. As many as 15 flowers per stem. Flowers almost purple when first open, turning to deep blue with white center. Late-spring-flowering. Species that should become more popular.

C. tmoli (syn. *C. boissieri*). Small bulb. Grows to height of 3–6 in. Flowers blue with white center. Late-spring-flowering. A form of *C. luciliae* of gardens.

CHIONODOXA

C. albescens (syn. *Scilla albescens*) Native to Crete. Flowers smallest of genus, whitish, pinkish or pale blue. 3–12 in. in height. April–June flowering.
C. allenii see *C. gigantea*.
C. boissieri see *C. tmoli*.
C. forbesii Native to Turkey, Lycia. Introduced 1871. Blue. 6 in. in height. Spring flowering.
C. lochiae Native to Mt. Troodos, Cyprus. Introduced 1953. Uniformly blue. 6 in. in height. Spring flowering.
C. nana Native to Crete. Introduced 1879. White, pale blue or pale lilac. 4 in. in height. May–July flowering.

× CHIONOSCILLA — LILIACEAE

In the wild, bigeneric hybrids often will occur between related species and, occasionally, between genera. The resulting hybrids are sterile and propagation can be accomplished only by asexual means. The cross can occur many times but the number of such crosses is few. *Chionoscilla* is a cross between *Chionodoxa luciliae* and *Scilla bifolia*. The only species that has been described is *Chionoscilla × allenii*, introduced in 1889. Both parents flower as soon as the snows melt. The appearance is like a very fine form of *Chionodoxa luciliae*, except that the perianth segments are joined at the base. Plants grow to a height of 6–8 in., with 4–7 flowers per stem, rather like a large-flowered *Scilla bifolia*. The leaves are dark green, the flowers a good deep blue. Flowering in March.

CULTURE Same as *Chionodoxa*.

PROPAGATION By offsets. Lift plants after foliage starts to die down. Separate young bulbs. Store until fall, then plant 3 in. deep.

PESTS & DISEASES Can be attacked by fungus that causes anthers to turn black. Smut fungus, *Ustilago vaillantii*, can be transmitted to seedlings of related genera. Plants so infected must be discarded.

USES In rock gardens or sunny borders where plants can be protected from excessive summer moisture. Excellent plant for collectors.

SPECIES

C. × **allenii**. Described above.

CHLIDANTHUS — AMARYLLIDACEAE

From the Greek *clideio* ("delicate"), *anthos* ("flower"). This not-too-common genus is composed of but 2 described species—one from Mexico and the other from the Andes of South America, in Peru and Chile. These plants are not hardy and must be given protection in areas where temperatures fall below 26°F during the winter. They can be grown outside in frost-free areas, in full sun or light shade.

The tunicate bulb is small. The flowers are yellow in both species and both are fragrant, with trumpet-shaped blossoms produced in a few-flowered umbel. The main difference between the 2 species is that in *C. fragrans,* the most commonly grown, the flowers are almost sessile, while in *C. ehrenbergii* they have a flower stalk. The flowers are produced in midsummer and appear before the leaves.

CULTURE Plant in spring or early fall; spring is better. Keep barely moist until flower spike emerges, then increase amount of water. Soil should be well-drained but moisture-retentive. After leaves appear, give liquid feedings. When leaves start to die back in fall, reduce water. Overwinter in frost-free location. Start into growth again in spring.

If planted in containers, repot each spring in soil mix of equal parts of good topsoil, peat moss, and sand.

PROPAGATION Separate offsets from parent bulbs if increase of stock is required. This is done in fall or spring, prior to growth being started.

PESTS & DISEASES The usual slugs and snails. Bulbs will rot in storage if not given adequate aeration and kept dry.

USES Unusual plant that does well in containers. Conversation plant for sunny, well-drained border or rock garden in moderate but moisture-retentive soil. Excellent cut flowers.

SPECIES

C. **ehrenbergii**. Native to Mexico. Flower stems some 10 in. high; yellow flowers on 1–2 in. long stalks. Midsummer-flowering. Linear, dark green leaves.

C. **fragrans**. Native to Peru and Chile; introduced 1820. Most popular of genus. Some authorities think *C. ehrenbergii* is form of *C. fragrans*. Bulb ovoid and tunicated. Linear, dark green leaves. Sessile, yellow flowers, on stems 8–10 in. tall, delightful lemon scent. Few flowers per stem. Midsummer-flowering.

CIPURA — IRIDACEAE

The derivation of the name is not known. It is doubtful if any of the species are in cultivation; there are reportedly 2, possibly 3 known. They are native to the warm areas of South America. The rootstock is a bulb or corm. The literature is not clear and in fact contradictory on this.

The flowers are carried in terminal spikes, perianth tube short and the 3 inner perianth segments are shorter than the outer. The colors are white to pale blue, height up to 10 in., spring-flowering to early summer. Leaves are basal, 2 or 3 per plant, but for their size quite wide, as much as ½ in. There is generally one leaf on the flowering spike located just below the flower head. In the wild the plants are found in grassland and scrub.

CULTURE Need warm conditions and only suitable for growing outdoors where the temperature does not drop into the 40°F range at night. In most areas should be grown in a greenhouse. It might be possible to grow these plants in a sunny border, planting the bulbs some 1–2 in. deep in well-drained, friable soil. Moisture will be needed during the growing season with the plants allowed to become drier during their resting (winter) stage.

PROPAGATION A great number of offsets are produced which can be removed in the early spring prior to growth commencing. Seed is also freely produced and this can be sown in a sandy soil mix, the seeds being barely covered, in moderate temperatures in the spring.

PESTS & DISEASES No special problems.

USES Collectors' item and should be grown in special locations.

SPECIES

C. **martinicensis**. See *Trimezia martinicensis*.

C. **paludosa**. Native to the West Indies, Guiana; first described in 1752. Corm is covered with a thick, brown, papery tunic. Flowering stem up to 8–10 in. high but often less, only reaching 2–4 in. The flowers are short-lived but several are produced. The outer segments are longer than the inner and are horizontally held or nearly so with a slight bend upward at the tip; the inner tend to curve over the style and anthers. The color is variable from white to a hint of blue with a touch of yellow at the base of the inner segments. Flowers an inch or a little more in diameter, lasting but a short time. Foliage linear and up to ½ in. wide with the veins being distinct.

COMMENTS This genus is one among several that is native to South America. They all should be looked at more closely for their potential as garden plants.

CLIVIA — AMARYLLIDACEAE

The genus was named in honor of the Duchess of Northumberland (?–1866) whose maiden name was Clive. *Clivia miniata* sometimes is referred to as *Imantophyllum miniatum*, (although this name has no standing), also sometimes used to cover all the *Clivias*.

This is a genus of some 4 species of South African evergreen plants which are, in fact, imperfect bulbs, and frequently regarded as belonging to the bulb category when loosely applied. The rootstock consists mostly of the bases of leaves not much if at all modified for storage. In their native habitats they are found in shady areas and mostly where there is abundant winter moisture. The plants form thick clumps and, in the wild, sizable areas of ground are occupied by these plants where few other species interfere with their spread.

The flowers of the species are quite striking. They are carried in umbels and are in the red, orange, and yellow color range. Flowers are tubular or funnel-shaped, with the 6 stamens attached to the throat of the tube. The dark green strap-shaped leaves are often quite wide and thick. The bases of the leaves are sheathed.

CULTURE The plants are not hardy but require protection only against frost. Grow in shade as strong sunlight will scorch the foliage. Best grown in organically rich soils with good drainage. They should be left undisturbed as these plants seem to thrive when crowded, and when grown in containers can be left undisturbed for a good number of years. The plants should be planted so the lighter part of the lower leaves is in the soil and plants set some 10–12 in. apart. In containers they can be closer together. In border plantings, top dress with good top soil every other season and fertilize with weak feedings of organic fertilizer given in late summer until the flower spikes begin to appear. While these plants prefer shade, if too dense the plants will not produce flowers. In areas where frost occurs they should be grown in a cool greenhouse where the temperature can be kept above 35°–40°F in winter. Moisture always should be given but these plants do not seem to mind being neglected a little during early summer.

PROPAGATION Lifting and dividing the plants in late spring provides the best means of propagation. This can be difficult as the roots often intertwine and a knife may have to be used. Seed can be sown in the spring as soon as ripe, placing them in a soil mix with good humus content and keeping moist and in a temperature range of some 55°F at night. The young seedlings should be transplanted as soon as they are 3–4 in. in height and grown on in individual pots until some 8–10 in. in height when they can be planted out.

PESTS & DISEASES No special problems but slugs and snails can be a problem.

USES In mild, frost-free climates these are great plants for the shade and for planting under trees where they can be left undisturbed. They make superb container plants and can be used indoors when in flower. Being of a hardy constitution with regard to the actual care they need, they should be considered when plantings are needed indoors in office complexes. They seem to have no objection to drafts as long as these do not reduce the temperature to the low 30's.

SPECIES

C. caulescens. Native to eastern Transvaal. Produces umbels of waxy looking, reddish-orange flowers that curve downward with the ends of the tubes being dentated and colored dark green, fading to lighter green, and tipped with yellowish-white. As many as 15 flowers are produced on a sturdy, flattened spike that reaches 18 in. or so in height. The leaves are dark green, 24 in. in length, sometimes more, and 2 in. or more in width. The plants flower in October in their native habitat; spring in the Northern Hemisphere. Rare in cultivation, if at all.

C. × cyrtanthiflora. Hybrid of *C. miniata* and *C. nobilis*; introduced 1877. The flowers are pendulous and salmon or orange with inner perianth segments being much broader than the outer; flowering in late winter/early spring. Foliage wider than either parent. Height 18–24 in.

C. gardenii. Native to Natal and Transvaal; introduced 1862. Narrow foliage up to 24 in. length with tendency to arch; a deep green in color. Flowers in an umbel, downward curving, reddish-orange to yellow, 2–3 in. long, and carried on a strong stem which reaches up to 24 in. in height. Winter-flowering.

C. miniata. Native to Natal; introduced 1854. The most commonly grown species and differing as the flowers in the umbels are more trumpet-shaped and face outward. The inner segments are broader than the outer and the throat of the flowers often are lighter in color, sometimes quite yellow. There are a number of color variations, some forms being bright orange, others redder. The flowers are carried on strong stems which reach 24 in. or more in height, and flowers are produced during the winter months, often lasting many weeks. Many fine seedlings have been raised, some having foliage that is over 3 in. in width but always strap-shaped and of good green. The form known as *striata* has variegated leaves and var. *aurea* describes a yellow form.

C. nobilis. Native to South Africa; introduced 1823. Produces more flowers per stem than any other species—as many as 50 or even more not being unusual. The flowers hang downward and cover each other, forming, as it were, an umbrella. They are reddish-orange fading a little toward the mouth of the cylindrical flowers, the orange being replaced by green, which is more intense, at the mouth. Height 24–30 ins. The flowers are produced in October in the native habitat; spring in the Northern Hemisphere. This plant was first known as *Imantophyllum altonii*. The foliage is fleshy, strap-shaped, dark green, and often up to 36 ins. in length.

COMMENTS All of the *Clivias* make great container plants and deserve to be more commonly grown. In milder climates they have potential as shade plants and it is unfortunate that only *C. miniata* is commonly grown.

COLCHICUM — LILIACEAE

The name is derived from Colchis, a location in Asia Minor, from which the bulbs are distributed. *Colchicum* are found not only in the eastern Mediterranean but also as far east as Iran and Turkestan and in much of Europe, including Britain. There are between 50–60 species, and a goodly number of selections have been introduced by bulb firms over the years.

There is much confusion over the nomenclature of the *Colchicum*. This is due in part to the leaves being produced after the flowers and to the fact that, when the flowers fade, they are much the same color as the leaves, thus making herbarium specimens difficult to differentiate. There is also confusion over the difference between the *Crocus* and the *Colchicum*, due in no

Calochortus macrocarpus

Plate 65

Calochortus pulchellus

Plate 66

Calochortus albus × *C. pulchellus*. The colors of both parents are combined to give a most attractive flower. Many of the genus deserve to be better known and more frequently grown in gardens around the world. J.H., by CHARLES WEBBER

Calochortus catalinae. Native to the southern part of California, this species, which varies a little in color, is another that is worthy of being more widely grown. While a certain amount of selection has been done, there would seem to be an opportunity for even greater work and hybridizing. ORNDUFF

Calochortus venustus. No wonder nurseries list selections of this species. It is quite a startling plant, and when one considers the great variation in this genus, it is difficult to understand why they are not widely grown. J.H., by CHARLES WEBBER

Caloscordum neriniflorum. It is understandable why this plant was once given the name *Allium.* An unusual and rare plant, but most certainly deserving of greater consideration by keen gardeners. LOVELL

Calostemma luteum

Plate 68

Calostemma purpureum

Plate 69

Calydorea nuda. Rare, the flowers lasting for only a brief period, means that this plant will never become very popular and widely grown. A good plant for the collector. MATHEW

Camassia cusickii. While rather an untidy grower, inclined to flop over, the blue flowers of this easy to grow species are quite attractive. COURTRIGHT

***Camassia leichtlinii* ssp. *typica*.** Despite this being a common plant in much of the western United States, this plant is deserving of a place in many gardens. This subspecies is a plant of great garden merit. WARD

Camassia quamash. No wonder nurserymen have made several selections of this species; it is a lovely plant, and if the need arises, one could dig the bulbs and eat them, as did the Indians in years gone by. AUTHOR

Canna. For warmer climates, the selections of Cannas that can be grown is surprising. They can be counted on to give a good display. In cooler climates, they should be planted out, as early as possible, in the spring. They should not be exposed to frost. They perform quite well in containers, but as they can get quite large, the shorter cultivars are better for such use. Many Cannas have leaves with various markings, to some an added attraction, to others, a variation not admired. AUTHOR

Ceropegia. The unusual structure and appearance of the flowers of this genus can be seen in this photograph, taken in northern Natal. There is no doubt that this genus is rare in cultivation; indeed, this may well be the only illustration in print. B.R.I.

Cardiocrinum giganteum. One of the most striking plants for woodland gardens. Will not flower for several seasons after planting, but once established, will perform year after year, as many young plants are produced at the base of the monocarpic stalks. WARD

Plate 71

Chasmanthe bicolor

Plate 72

Chionodoxa luciliae. While this is considered to be the finest species, and several selections are listed in catalogs, some difficulty can be experienced in growing this plant in areas that do not enjoy a cold winter. Coming from the mountains of Crete and Turkey, cold temperatures seem to be needed, as does a dry period in late summer. I.B.

Chasmanthe floribunda. While easy to grow and propagate, this plant is not as popular as it should be, perhaps because it becomes untidy with age and must be lifted and divided. AUTHOR

Chasmanthe floribunda (**yellow form**). The rather untidy habit of the plant can be seen, but the beauty of the flowers cannot be denied. AUTHOR

Chionodoxa sardensis. While this low growing species has many attributes, it has never become widely grown, perhaps being in the shadow of the better known *C. luciliae.* DE HERTOGH

× ***Chionoscilla allenii.*** This lovely plant, a bi-generic hybrid, puts on quite a display in the spring. A grand addition to any collection of bulbs. MATHEW

Chlidanthus fragrans. If you want a mid-summer flowering bulb with a good fragrance, then this species is worthy of consideration. While this species is the more common, it is not often available. H.C.R.L.

Cipura paludosa. Photograph taken in the garden of Dr. Boussard. The distinct veining of the foliage is quite apparent. This species is rare in cultivation, and no doubt will remain so. BOUSSARD

Clivia caulescens. The genus *Clivia* contains several species that are not as well known as *C. miniata*. This photo shows *C. caulescens*, an attractive plant which grows well in containers and is worthy of wider distribution. AUTHOR

Clivia miniata. This species has been long in cultivation and deservedly so. There are several selections available, the colors varying slightly, and foliage varying from wide, to quite narrow. This photo shows the clear colors typical of the species. AUTHOR

Clivia nobilis. Another species that is not well known but *C. nobilis* produces more flowers per plant than any other, and is as easy to grow as *C. miniata*. AUTHOR

Colchicum aggripinum. Regarded by some as a hybrid, it shows the typical veining of the tessellate group. A low growing plant, and thus not as likely to be damaged by rain, but it does attract slugs and snails. WARD

Colchicum autumanale. It is always a surprise to see these plants in the fall; they seem to pop out of the ground. Common names include 'Naked Boys' and 'Autumn Flowering Crocus', and injustice as *Colchicum* has such large flowers. Once planted, these bulbs should be left undisturbed for years. AUTHOR

Colchicum autumnale album plenum. Despite the long, and perhaps, correctly speaking, invalid name, this plant is often so listed in catalogs. When combined with the true *C. autumnale,* in bold and contiguous plantings, makes an eye catching display. WARD

Colchicum bornmuellerii. While some authorities say this is a form of *C. gratia,* this plant has been recently raised to specific rank. Despite its rather fragile appearance, this is quite a tough little plant. WARD

Colchicum corsicum. A fine dwarf species that is rare in cultivation. WARD

Colchicum fasciculare. Perhaps better known as *C.illyricum,* this spring flowering species has a good clean look. WARD

Plate 76

Colchicum laetum. While only a couple of inches in height, and thus should be planted close to a path, this species puts on quite a display in the fall. WARD

Colchicum **'Princess Astrid'**. The exact parentage is not known, but as the flowers show the criss-cross markings, no doubt there was a tessellated species involved, most likely *C. bowlesianum*. It is one of several found listed in catalogs. WARD

Colchicum bowlesianum. One of the tessellated group, it is rather rare and seldom seen outside of collections. WARD

Colchicum speciosum album. Worth growing if only because it will increase quite rapidly if left undisturbed for several seasons, the flowers are not unlike a tulip in form, and can reach almost 10–12 in. in height, and thus are striking. WARD

Colchicum **'Water Lily'**. While an outstanding introduction and a great plant, the weight of the flowers is often its undoing as they just are too heavy for the stems. It is, however, a plant well worth growing. WARD

Plate 77

Commelina coelestis. These plants are offered for sale in various catalogs. The blue flowers are often over 1 in. in diameter, and offer an unusual color for the early summer border. B.A.G.S.

Commelina dianthifolia. Pretty, dainty, and of unusual color. Unfortunately the flowering period lasts only a few weeks. One wonders what potential lies in such bulbs; certainly, the color is attractive, and the late flowering time, July–August, is a period when bulbous plants can add much interest to borders. LOVELL

Conanthera bifolia. This native of Chile is rare in cultivation. The distinct cone formed by the anthers can be seen clearly. A variable plant, sometimes only a few inches in height, but always seems to carry a good number of flowers per stem. Despite the rather fragile appearance, the stems are quite strong. MATHEW

Corydalis transsilvanica. A rare and unusual species, there being some confusion as to just where this plant is native, the interesting shape of flowers found in this genus can be appreciated. WARD

Crinum americanum. In its native habitat in Florida, this lovely plant is found in marshy ground and along the banks of streams. It is often grown in ponds and in display greenhouses, as shown in this picture. AUTHOR

Crinum bulbispermum. This plant, photographed in the wild shows the typical habit. Known as the 'Orange River Lily', it is not an uncommon plant in its native habitat, and deserves a place in gardens enjoying a warm climate. AUTHOR

Crinum campanulatum. Another South African species, which must have adequate moisture to grow well, it grows well even standing in water. B.R.I.

Crinum macowanii. This lovely plant is an easy one to grow and should be in gardens enjoying a warm climate. It is trouble free as well as attractive. AUTHOR

Crinum macowanii. A closer look at a flowerhead of this species. AUTHOR

Crinum macowanii. This plant was photographed in Zimbawbe, close to Victoria Falls. It was growing in compacted soil in very dry conditions. This shows the adaptability of the species and the need for a closer look at the genus. Authorities in Zimbawbe insist that it is indeed *C. macowanii,* but is more likely a variant. AUTHOR

Crinum moorei. The other parent of *C. × powellii,* from which it gets its more trumpet shape. Unfortunately, this species is not as hardy as one would wish and can only be grown under protection or in warmer, frost-free (or nearly so) areas. B.R.I.

Crinum × powellii. A hybrid between *C. bulbispermum* and *C. moorei.* This plant is a little hardier than either of its parents and is well worth growing in the garden. Being of such ease of culture, it should be the first of this genus to be tried. I.B.

Plate 80

small measure to their common names of Autumn Crocus and Meadow Saffron. They do look alike but, upon close examination, the basic differences between them are distinct and easy to recognize: In *Colchicum* there are 6 stamens, 3 in *Crocus;* in *Colchicum* there are 3 distinct styles, in *Crocus* there is just 1 which is divided into 3 just below the tip (visible when flowers are aboveground); when lifted, *Colchicum* will show a superior ovary, the flower parts joining the stem below the ovary, while the *Crocus'* ovary is inferior, joining the stem above the ovary.

The *Colchicum* have been known in medicine for centuries. All parts of *C. autumnale* contain colchicine, which both Theophrastus and Dioscorides described as poisonous. The Arabians, however, prescribed it for the treatment of gout, which is its only medical application to the present day. Tragus in 1552 warned against its use for this purpose. Some 25 years later (in 1577) Dodonaeus also warned about the use of this poisonous alkaloid, but Hermodactyl described it as being good for various ailments. It became more popular as a garden plant during the 17th and 18th centuries.

Colchicine also has found a useful purpose in horticulture. It can cause dwarfing and malformation in plants and has been used to change the chromosome count in pollen and egg cells. This can result in the production of larger plants, as well as deformed plants. The dilution needed will vary with the species of plants being treated.

Professor Rolf Nordhagen of Norway in 1933 wrote of the possibility that ants distribute the seeds of *C. autumnale*. He recorded that the ants removed 23 seeds in a 12-minute period. He found that these seeds, as well as those of *C. speciosum,* are covered with a sugary layer when ripe. It appeared that the ants devoured the edible sugar coating, but the embryo and cotyledon were protected due to the presence of colchicine. This is one of the rare but not unknown symbiotic relationships between insects and plants.

The corms of *Colchicum* can be quite large, up to 4 in. long. The tunic is brown and leathery when mature and is composed of the tubular portion of the first and outermost leaf. Corms are irregularly shaped, convex on one side and flat on the other, and extend downward, forming a footlike projection. There is a sharper projection from the upper side, which is sometimes hidden by the tunic; from this projection a groove runs down the corm, which is wider at the base. The bud forms at the bottom of this groove and rises to the surface through a tube, which also forms. The top of the tube is open, the bottom closed, but the roots soon push their way through. The roots generate from a definite point located on the outer side of the base of the new shoot.

The size of the foliage varies with a species and can be as much as 12 in. long and half as wide. This can be a problem as the leaves can smother other plants growing close by. In the winter- and spring-flowering species, the leaves and flowers are produced at the same time. The foliage of the autumn-flowering species is produced in the spring; however, sometimes it will follow the flowers and will remain growing for long periods of time, with only a limited dormant period. This foliage is produced by the same corm that has flowered. The spring foliage seems to exhaust the corm, but a new corm forms to flower the following autumn. The seed capsule ripens in early summer and is found in the cup formed by the innermost leaves.

CULTURE Plant in July. Autumn-flowering species should be planted an inch deeper than the winter- and spring-flowering kinds. Cover winter- and spring-flowering bulbs with 4 in. of soil set in average soil, slightly more in sandy soil. In hot, dry areas, give shade for hottest part of day. Will stand full sun in cooler climates. Prefer good depth of well-drained, moisture-retentive soil. When planting, take care that bulbs are set in area where foliage will not cover other plants. Continue watering as long as foliage remains green; withhold when foliage starts to die back, usually in July. Bulbs should be left undisturbed as long as possible, allowing foliage to die down first. Lift and divide just before foliage has matured so bulbs will be easier to find.

PROPAGATION Allow bulbs to grow at least two seasons before lifting and dividing. Can be grown on and will flower in second to third season, depending on size when separated. Can be raised from seed. Sow in July as soon as ripe in sandy soil, covering with sandy soil mix. In second year seedlings can be transplanted to rows spaced 3–4 in. apart and then grown on for another two seasons. Will have reached good size by this time and can be lifted in July and planted out.

PESTS & DISEASES Foliage sometimes attacked by fungal smut disease, which causes black streaks to appear. No known cure so discard bulbs attacked and replenish soil discarding that which is infested. Corms also subject to gray bulb rot like that of tulips. Slugs and snails quite partial to foliage when it first appears, as are flower buds of autumn-flowering species. Take precautions if these pests are present.

USES Although sometimes suggested as suitable for growing in grass, best planted along shrub borders to bring color at unusual times of year. Grown in this way, leaves can grow and nourish the plants without diminishing other plants. Not good as container plants; flowers of certain species, although they remain tight and hold their form for first week or so of opening, become floppy and untidy.

SPECIES & CULTIVARS There are three natural divisions of *Colchicum:*
1. Autumn-flowering species and hybrids.
2. Winter- and spring-flowering species. This group contains the only yellow-flowering species, which is a native of Kashmir.
3. Tessellated species, those marked with a crisscross pattern on the petals in colors of dark and light rosy mauve.

Autumn-flowering Species

C. autumnale (Naked Boys). Native to much of Europe, including Britain; cultivated since at least 1753. Common name possibly because color is pale pink. Large, ovate corm. Moisture required to get roots started; moisture should be given also toward end of summer. One of earliest species to flower, often late August or early September. If left unplanted on a shelf will often start into flower. Frequently as many as 6 flowers produced from 1 corm. Flowers reach height of only 4–6 in. and soon flop over. Leaves produced after flowers. Vary in number from 5–8, about 10 in. long and up to an inch wide. Nurseries offer both single and double forms. A white single form sometimes sold as *album;* the double white often listed as *album plenum*. Both of these were found in cultivation, selected and introduced after the species. They are the same height as the species.

C. byzantinum. Introduced in 1629. Large corm, often 2 in. or more long and as wide, will produce upward of 20 flowers. Flowers in September but leaves do not appear until winter. Generally 5 or 6 leaves produced, dark green, more than a foot long and up to 4 in. wide. Flowers are on stems reaching up to 6 in. high, pale lilac or bright mauve with white center. Species does not appear to set seed; E. A. Bowles suggests that plants now grown have been propagated vegetatively and over the years have lost ability to seed. Species sometimes listed as *C. autumnale major* and some authorities regard it as a form of *C. autumnale*.

C. cilicicum. From Asia Minor; introduced 1896. Very similar to *C. byzantinum* but flowers a little later in September and leaves produced soon after flowers have passed; flowers are deeper lilac color and fragrant. Var. *purpureum* from Turkey flowers at the same time and has an even deeper-hued flower.

C. giganteum. Native to Turkey; introduced by Max Leichtlin of Baden-Baden and appeared in his lists of 1890 and in various other lists up to 1903. Some authorities regard *C. giganteum* as another form of *C. speciosum;* however, the former flowers a little later than the latter, late October or November, and is described as stronger and larger. Flowers are soft purple, reaching height of 10–12 in.

C. speciosum. From Turkey and Caucasus; introduced about 1850. Regarded as one of the finest for the garden. Corm very large, 2 in., with neck. Tulip-like flowers up to 12 in. high; color varies from pale to deep reddish-violet, with paler throat, sometimes white. Also white form offered as var. *album*. These flowering Sept.–Oct., with foliage appearing in spring. Left in situ these plants form large clumps; 1 bulb covering 1 sq. ft in 2 seasons and doubling in size with the passage of a couple of good seasons.

Autumn-flowering Hybrids

A number of hybrids have been introduced over the years. The parentage is varied but involves *C. giganteum, C. speciosum, C. sibthorpii,* and *C. bowlesianum*. Many are quite similar to one another. The parentage may include some of the tessellated varieties and this characteristic will be found, to a greater or lesser degree, in several of the hybrids.

'Autumn Queen'. Lilac with white throat, silvery white tessellation; flowering Aug.—Sept.

'Disraeli'. Light purple flowers, quite large; flowers September.

'Gracia'. Light violet-purple, inner segments violet-purple with large, whitish base.

'Lilac Bedder'. Light violet-purple with darker veins.

'Lilac Wonder'. Large flowers; narrow, amethyst-violet segments with white lines in center.

'The Giant'. One of tallest hybrids offered, reaching 10–12 in.; very free-flowering; large flowers, violet with large, white throat.

'Violet Queen'. Flowers checkered with imperial purple and white lines in throat; distinctive orange anthers in September.

'Water Lily'. Large, double-flowered, often with 20 petals, lilac-rose; main disadvantage is that flowers often fall over due to weight; flowering Sept.–Oct.

Winter- and Spring-flowering Species

Few of these species are offered by nurseries, and it is only in plant collections that these bulbs are to be found. The exception is *C. luteum,* which, as mentioned, is itself an exception as it is the only yellow-flowered member of the genus.

C. catacuzenium. Native of Greece; introduced 1938. Dwarf, only 2–3 in. in height. Pale rosy lilac flowers carried on ground with 3 leaves just appearing with flowers and surpassing them in height by several inches. Flowering in March. Leaves sometimes appear before flowers.

C. luteum. Native to northern India and Afghanistan; introduced 1874. Only yellow species, sometimes tinged with lilac. Three or 4 small flowers per bulb, 3–4 in. in height. Leaves appear with flowers but continue to grow, as long as 10–12 in., even after flowers fade.

Tessellated Species

C. agrippinum (syn. *C. tessellatum*). Introduced circa 1874 but most likely a plant of garden origin. Ovoid corm, about 1 in. thick. Regarded by some as possible hybrid between *C. autumnale* and *C. variegatum*. Pale lilac flowers have tessellation of darker lilac-purple. Narrow leaves pointed like flowers; 3–4 in. in height. Flowering Sept.–Oct.

C. bowlesianum (syn. *C. algeriense, C. bivonae, C. latifolium, C. sibthorpii, C. visianii*). Native to Greece; introduced 1821. Flowers 7–9 in. high, very large, tessellated, rosy lilac with darker purplish-violet. Foliage grasslike, up to 10 in. long. Flowering Oct.–Nov. Offered by some nurseries but not generally found in gardens.

C. variegatum (syn. *C. parkinsonii*). Native to Turkey and Aegean islands; introduced 1753. 3–5 in. in height. Leaves, 3–4 and 5–7 in. long, spread along ground. Nicely checkered lilac-purple flowers, with white perianth tube. Flowering late Nov.–Dec.

COLCHICUM—Autumn Flowering Species

C. alpinum (syn. *C. parvulum*) Native to S.E. France, Italy, Sicily, Sardinia, Corsica. Introduced 1820. Rosy lilac. 3 in. in height. Aug.–Sept. flowering.

C. andrium Native to Greece (Isle of Andros). Introduced 1949. Rosy lilac. 3 in. in height. Oct.–Dec. flowering.

C. arenarium Native to Hungary. Introduced 1805. Pale rose, anthers yellow. 4–5 in. in height. Sept.–Oct. flowering.

C. atropurpureum Native to Europe. Introduced 1934. Crimson purple. 3–4 in. in height. Sept.–Dec. flowering.

C. autumnale var. *alboplenum* Native to Europe. Introduced circa 1753. White with pink flush. 4–6 in. in height. Sept.–Oct. flowering.

C. a. var. *pleniflorum* Native to Europe. Introduced circa 1753. Lilac. 4–6 in. in height. Nov.–Dec. flowering.

C. a. var. *striatum* Native to Europe. Introduced circa 1753. Striped pink & white. 4–6 in. in height. Oct. flowering.

C. balansae (syn. *C. candidum*) Native to S. Turkey, Rhodes. Introduced 1855. Rosy pink, anthers yellow. 4–6 in. in height. Oct. flowering.

C. baytopiorum Native to S. Turkey. Introduced 1983. Pinkish purple, anthers yellow. 2 in. in height. Oct. flowering.

C. bertolonii syn. of *C. cupani.*

C. boissieri (syn. *C. pinatziorum, C. procurrens*) Native to W. Turkey & S. Greece. Introduced 1876. Pinkish lilac, anthers yellow. 8 in. in height. Sept.–Dec. flowering.

C. borisii May be *C. micranthum* Native to Bulgaria.

C. bornmuelleri Native to N.W. Turkey. Introduced 1889. Pinkish purple. 4 in. in height. Sept.–Oct. flowering.

C. b. var. *magnificum* Native to N.W. Turkey. White throat, rosy lilac. 4–5 in. in height. Sept.–Oct. flowering.

C. callicymbium Native to Greece & Bulgaria. Introduced 1902. Lilac purple, purple throat. 16 in. in height. Sept. flowering.

C. candidum syn. of *C. balansae.*

C. caucasicum see *Merendera trigyna.*

C. corsicum Native to Corsica. Introduced 1879. Rosy lilac. 3–4 in. in height. Aug.–Sept. flowering.

C. creticum Native to Crete. Introduced 1939. Rosy lilac, dark purplish anthers. 3 in. in height. Oct.–Dec. flowering.

C. cupani (syn. *C. bertolonii*) Native from S. France to Crete, Sicily, Sardinia, Tunisia, Algeria. Introduced 1827. Pale pink, rosy lilac. 3 in. in height. Sept.–Dec. flowering.

C. decaisnei syn. of *C. troodii.*

C. glossophyllum Native to S. Greece. Pale pink, rosy lilac (perhaps a var. of *C. cupani*). 3 in. in height. Sept.–Dec. flowering.

C. guadarramense Native to Northern Spain. Introduced 1912. Warm pink. 6 in. in height. Sept. flowering.

C. haussknechtii Native to Iran. Pinkish lilac. 3–4 in. in height. Oct. flowering.

C. hiemale Native to Cyprus. Introduced 1897. Rosy. ½–¾ in. in height. Nov. flowering.
C. hierosolymitanum Native to S. Turkey, Syria, Lebanon, Israel. Slightly tessellated pinkish lilac. 5 in. in height. Sept.–Nov. flowering.
C. imperatoris-friderici syn. of *C. kotschyi*.
C. jesdianum Native to Iran. Introduced 1930. Whitish. ½ in. in height. Sept.–Nov. flowering.
C. kotschyi (syn. *C. imperatoris-friderici*) Native to S. Turkey, N. Iraq, Iran. Introduced 1853. White, pale pink, anthers yellow. 1½–2½ in. in height.
C. laetum Very narrow segments rose lilac. 2–3 in. in height. Sept. flowering.
C. lingulatum Native to Rhodes, Greece, S. Turkey. Introduced 1844. Pale pink, anthers yellow. 2–3 in. in height. Aug.–Sept. flowering.
C. lusitanum Native to Portugal, Spain, North Africa. Introduced 1827. Faintly tessellated deep purplish pink. 2 in. in height. Oct.–Nov. flowering.
C. micranthum Native to N.W. Turkey. Introduced 1884. Pale pink, anthers yellow. 2–3 in. in height. Sept.–Oct. flowering.
C. neapolitanum Native to S. France, Italy. Introduced 1826. Pale pink. 3 in. in height. Sept. flowering.
C. orientale syn. of *C. turcicum*.
C. parnassicum Native to Greece. Pale lilac. 2¼ in. in height. July–Aug. flowering.
C. parvulum syn. of *C. alpinum*.
C. peloponnesiacum Native to S. Peloponnese, S. Greece. Introduced 1949. Lilac, white. 2 in. in height. Oct.–Nov. flowering.
C. pinatziorum syn. of *C. boissieri*.
C. procurrens syn of *C. boissieri* (*C. procurrens* of catalogs, often *Merendera sobolifera*)
C. psaridis Native to S.W. Turkey, S. Greece. Introduced 1904. Pale pinky purple. 2 in. in height. Sept.–Dec. flowering.
C. pusillum Native to Greece, Crete. Introduced 1822. Lilac pink, white. 2 in. in height. Oct.–Dec. flowering.
C. speciosum album Native to Turkey, Iran, Caucasus. Introduced circa 1828. Pure white. 4 in. in height. October flowering.
C. s. atrorubens Native to Turkey, Iran, Caucasus. Introduced circa 1828. Dark purple. 4 in. in height. October flowering.
C. s. bornmuelleri Native to Turkey, Iran, Caucasus. Introduced 1889. Lilac, white throat. 4 in. in height. October flowering.
C. s. giganteum Native to Turkey, Iran, Caucasus. Introduced circa 1828. Rosy purple. 4 in. in height. October flowering.
C. s. maximum Native to Turkey, Iran, Caucasus. Introduced circa 1828. Rosy purple. 4 in. in height. October flowering.
C. s. rubrum Native to Turkey, Iran, Caucasus. Introduced circa 1828. Dark purple. 4 in. in height. October flowering.
C. tenorii Native to Southern Italy. Introduced 1858. Rosy to crimson, slightly tessellated. September flowering.
C. trapezuntinum syn. of *C. umbrosum*.
C. troodii (syn. *C. decaisnei*) Native to Cyprus, S. Turkey, Syria, Israel, Lebanon. Introduced 1862. Pale pink, white. 2–3 in. in height. October flowering.
C. turcicum (syn. *C. orientale*) Native to N.W. Turkey, E. Bulgaria, S. Yugoslavia, N. Greece. Introduced 1873. Pale to deep magenta. 2–3 in. in height. Aug.–Oct. flowering.
C. umbrosum (syn. *C. trapezuntinum*) Native to N. Turkey, Caucasus, Crimea. Introduced 1829. Pale pink, whitish, anthers yellow. 2–3 in. in height. Aug.–Sept. flowering.

COLCHICUM—Winter & Spring Flowering Species

C. acutifolium syn. of *C. szovitsii*.
C. aegyptiacum syn. of *C. ritchii*.
C. ancyrense syn. of *C. triphyllum*.
C. armenum syn. of *C. szovitsii*.
C. biebersteinii syn. of *C. triphyllum*.
C. bifolium syn. of *C. szovitsii*.
C. brachyphyllum (syn. *C. libanoticum*) Native to Lebanon. Introduced 1904. Pale rose, white. 4 in. in height. Jan.–Feb. flowering.
C. bulbocodioides syn. of *C. triphyllum*.
C. burttii Native to Turkey, Eastern Aegean Is. Introduced 1977. White, pinkish lilac. 3 in. in height. Feb.–March flowering.
C. croaticum syn. of *C. hungaricum*.
C. crociflorum syn. of *C. kesselringii*.
C. doerfleri Native to Macedonia. Introduced 1911. Rosy lilac. 4 in. in height. Feb. flowering.
C. falcifolium (syn. *C. hirsutum, C. lockmanii, C. serpentinum, C. tauri, C. varians*) Native to E. Turkey, Syria, Iran. Introduced 1885. White, pink purplish. 4 in. in height. Feb.–March flowering.
C. fasciculare (syn. *C. illyricum*) Native to Syria. Introduced 1826. Rose, white. 2 in. in height. Feb. flowering.
C. hirsutum syn. of *C. falcifolium*.
C. hungaricum (syn. *C. croaticum*) Native to S.E. Europe. Introduced 1921. White, pale lilac. 4 in. in height. Dec.–April flowering.
C. hydrophilum syn. of *C. szovitsii*.
C. hygrophilum syn. of *C. szovitsii*.
C. illyricum syn. of *C. fasciculare*.
C. kesselringii (syn. *C. crociflorum, C. regelii*) Native to Russian Turkistan, N.E. Afghanistan. Introduced 1880. Inside white, outside each segment centrally striped purple, rosy lilac. 3–4 in. in height. Jan.–Feb. flowering.
C. libanoticum syn. of *C. brachyphyllum*.
C. lockmanii syn. of *C. falcifolium*.
C. nivale syn. of *C. szovitsii*.
C. regelii syn. of *C. kesselringii*.
C. ritchii (syn. *C. aegyptiacum*) Native to Syria, Palestine, Egypt, Algeria. Introduced 1826. Pale pink to white. ½ in. in height. Jan. flowering.
C. serpentinum syn. of *C. falcifolium*.
C. stevenii Native to Syria, Lebanon, Israel, Southern Turkey, Cyprus. Introduced 1843. Deep pink, yellow anthers. 1–2 in. in height. January flowering.
C. syriacum syn. of *C. szovitsii*.
C. szovitsii (syn. *C. acutifolium, C. armenum, C. bifolium, C. hydrophilum, C. hygrophilum, C. nivale, C. syriacum*) Native to Turkey, Iran, Caucasus. Introduced 1834. White, pink, purplish. 2–3 in. in height. Jan.–Feb. flowering.
C. tauri syn. of *C. falcifolium*.
C. triphyllum (syn. *C. ancyrense, C. biebersteinii, C. bulbocodioides*) Native to Spain, Algeria, Turkey, Morocco. Introduced 1846. Almost white, rosy, purplish. 3 in. in height. Feb.–Mar. flowering.
C. varians syn. of *C. falcifolium*.

COLCHICUM—Tessellated Species

C. algeriense see *C. bowlesianum*.
C. bivonae see *C. bowlesianum*.
C. chalcedonicum Native to N.W. Turkey. Introduced 1897. Deep rosy purple, anthers yellow or brownish. 3–4 in. in height. Aug.–Oct. flowering.
C. latifolium see *C. bowlesianum*.
C. macrophyllum Native to Crete, Rhodes, W. Turkey. Introduced 1950. Pale lilac. 2–3 in. in height. Sept.–Nov. flowering.
C. parkinsonii see *C. variegatum*.
C. sibthorpii see *C. bowlesianum*.
C. sieheanum Native to Southern Turkey. Introduced 1926. Deep purplish pink. 1½–2 in. in height. Sept.–Oct. flowering.
C. tessellatum see *C. agrippinum*.
C. visianii see *C. bowlesianum*.

COLOCASIA — ARACEAE

The name is derived from the word used in Arabia for the plants, *kolkas* or *kulkas*. There are some 6 or 7 species. Certain species are widely grown in tropical areas as food crops. The species *C. esculenta*, yielding edible tubers, is of greater importance than the others. It is from the root of this plant that a paste called "poi" is made. In other warm areas the edible tubers take the place in the diet of the potato of colder climes; few tropical areas are without "taro", a common name for this versatile plant. The plants are native to tropical Asia. Certain plants of the genus also are grown as ornamentals, the large leaves having the common name of elephant's ear. Certain cultivars have been selected for unusual markings in the foliage. Even though the species has a definite ornamental value the foliage, on occasion, is blanched and eaten or often included in soups.

The flowers are carried on a spadix which is divided into 3 parts. The male flowers on the upper part, the female on the lower and between them flowers that are short, ovoid, and neuter. It is for the foliage, not the flowers, that the plants are used ornamentally.

CULTURE Tropical conditions are required with high temperatures and high humidity. They can be grown in areas where the summers are warm provided that ample moisture is given and a rich soil is available. The tubers should be planted 4 in. deep, spacing them some 48 in. apart. It takes about 7 months for the plants to reach maturity. They appreciate constantly being covered by 3–4 in. of water, which emphasizes their need for adequate moisture. No fertilizer is required but if the soil is very poor a good dressing with a balanced fertilizer should be worked into the soil at planting time. This should be of an organic nature. The tubers can be lifted at the end of the season, kept barely moist and warm during the colder months, and replanted when temperatures are again in the 55°F range at night, 65°–70°F during the day.

PROPAGATION Division and cutting of the tubers provide the best means of propagation and should be done at the end of the season or prior to planting.

PESTS & DISEASES No special problems.

USES As a crop of economic importance in warm climates and as an ornamental in areas where the summers are long and enable early planting to be done. The large leaves are striking, a good specimen plant for displays under glass and for the marshy borders of ponds and streams in warm climates.

SPECIES & CULTIVARS

C. **affinis**. Native to the tropical Himalayas. It has small, rounded tubers and a leaf-stalk up to 14 in. in length. Leaves have roundish blades, up to 8 in. in length; color is light green on the upper surface and glaucous on the underside. The spathe is a pale green and has a 2–3 in. long, slender tip. Var. *jenningsii*, also from the Himalayas, has a purplish leaf stalk and the leaves are spotted deep green or blackish-violet. Height of plant varies from 20–24 in. in culture.

C. **antiquorum** var. *esculenta* (Common names: Taro, Kalo, Dasheen; syn. *C. esculenta*). Native to the East Indies; introduced 1551. The most commonly grown species. This is a variable species, having a leaf stalk varying from 20–40 in. in length with the leaf blades heart-shaped at their base and up to 24 in. in length. The plants can vary in height from 3–7 ft. The spathe is pale yellow and up to 12–14 in. in length, the base greenish to purplish, the number of variations being considerable. The leaf color also will vary, the upper surface being dark green to a bluish-black. Among the many forms that are grown are the following:

    'Euchlora'. Dark green leaves and purplish margins; leaf stalks violet.
    'Fontanesia'. Purple veining to the foliage.
    'Illustris'. Leaves marked with bluish-black or purple.

Those selections most commonly grown for food are:
    'Globulifera'. Dark green leaves and leaf stalks marked with maroon lines; grows to some 80 in. or more in height and produces 20 or more good-sized tubers.
    'Ventura'. Redder leaf stalks; produces tubers of better quality but not so large as those of 'Globulifera'.

C. **fallax**. Native to the Himalayas. Differs from *C. esculenta* by producing only a solitary tuber. More commonly grown as an ornamental. Same height as *C. esculenta*.

C. **gigantea**. Native to Malaysia. Foliage is whitish on the undersides; spathe white, reaches the same height as *C. esculenta*.

C. × **marchallii**. Hybrid of *C. affinis* and *C. antiquorum* var. *esculenta*. Leaves 7–10 in. in length with undulated edges; green on the upper surface glaucous on the undersides with some spotting. Can reach over 60 in. in height.

COMMENTS Large, unusual, quick-growing plants when given tropical conditions, they deserve a place in warm display areas under glass or outdoors if possible. When of full size they are impressive plants.

COMMELINA — COMMELINACEAE

This genus, the type for the family, is named in honor of two Dutch botanists, Johann (1629–1698) and Kaspar (1667–1731) Commelin. There would appear to be some 100 species of this genus, with an extremely wide distribution, some found in South Africa and tropical Africa, South America, tropical Asiaz ; and Mexico and the United States. It is a genus that, in my opinion, needs to be closely studied, as with such a wide distribution, I feel differences will be found between the plants that will merit some species being raised to generic rank. As will be seen from the description of the species, certain plants offered in the United States are not correctly named, but here again closer study is needed to determine exactly the correct nomenclature.

While there are many species, not all of them are bulbous, in fact there are but 3, perhaps 4, that have tuberous roots. The tubers are fleshy, thick, and lightish-brown in color, becoming darker with age, and many roots are produced from them. The leaves are alternate, ovate, lanceolate or linear, and mostly sessile. The flowers, which are open for only a short period of time, are most commonly blue, sometimes yellow, and rarely white. Flowers emerge one at a time from a terminal spathe, the ovary is superior and 3-celled. Flower clusters exhibit alternate branching.

CULTURE   The tubers should be planted in the spring and lifted in the fall in areas where there is frequent frost. They will survive light frosts but not prolonged freezing. Planted at the base of a wall in a sheltered location and given the protection of a mulch, they will stand a better chance. In milder climates, however, they are quite safe outdoors. In this regard the guide should be the tubers of dahlias, with which most gardeners are familiar.

Tubers should be planted 2–3 in. deep in well-drained soil, spacing them some 6–10 in. apart. Moisture is required during the spring and summer but they should be allowed to become on the dry side during the winter. It should be noted that they should never be allowed to become completely dry or the tubers will suffer. This moisture requirement in the winter is quite tricky and, while it is suggested that it is better to be a little too dry than too wet, great care should be exercised.

The plants appreciate full sun, but in hot areas some shade during the hottest part of the day should be given.

PROPAGATION   Division of the tubers when lifting them or prior to planting is a good way to increase the stock, however, seed is an easy way to raise these plants. It should be sown in the spring in areas where some protection can be given, such as a cold frame or cool greenhouse, moving the seedlings outdoors as soon as night temperatures are in the upper 40°F's or low 50°F's. Just cover the seed, provide moisture, and the seedlings can be overwintered in a frame or cool greenhouse and then transplanted to give them room in the second spring once growth has commenced. Plants should flower in the second or third year from seed.

PESTS & DISEASES   No special problems.

USES   In a warm border or rock garden; good drainage essential, as is moisture.

SPECIES

C. coelestis (Blue Spiderwort). Native to Mexico; introduced prior to 1700. This plant sometimes listed in catalogs, but it is noted in the literature that plants under this name might well be *C. sikkimensis* or *C. tuberosa*. Note also that *C. sikkimensis* is a creeping perennial, not tuberous. The plant reaches 18 in. in height. Leaves are oblong-lanceolate, some 2 in. in width and 6–8 in. in length. The flowers are blue with hairy peduncles, the spathe is folded and pointed. Flowers often exceed 1 in. in diameter and bloom in early summer. Two varieties are known, var. *alba* has white flowers and var. *variegata* has blue and white flowers, both introduced after the species.

C. dianthifolia. Native to Mexico. Later in flower than the previous species but still flowers in summer, July–August. The flowers are a clear blue and while not long-lasting there are a great number on the flower head so the period of flowering continues several weeks. The flowers are not large, often less than ½ in. in diameter. The height is 5–10 in., with numerous lanceolate leaves. Thick, tuberous roots.

C. tuberosa. Native to Mexico; introduced 1732. Flowering in June–July with sky-blue flowers; spathes ovate-cordate, fringed with hairs. The foliage is oblong-lanceolate and the sheaths are hairy. Reaches 15–18 in. in height. Plants in culture under this name may be *C. coelestis*.

CONANTHERA — TECOPHILAEACEAE

The name is derived from Greek *konos* ("cone") and Latin *anthera* ("anther") as the 6 anthers form a quite prominent cone in the center of the flower. A small genus from South America, those known native to Chile. The rootstocks are of fair size; 1½ in. in height and some 1¼ in. in diameter in *C. campanulata*; the corm of *C. bifolia* has a very coarse netting extending above the ground and surrounding the lower parts of the flowering stem. Formerly placed in Liliaceae but now in Tecophilaeaceae.

CULTURE   The plants, coming from a warm area, will not withstand very cold weather but will withstand temperatures in the upper 30°F range. They are thus suitable for the cool greenhouse or planted out in areas where no or very little frost is experienced in a warm, sunny border. During their resting period, which is fall through the winter, they should be kept on the dry side.

The rootstocks should not be planted deeply, the necks being at or slightly below soil level. Sun is required and good drainage. The taller-growing species are spaced 10 in. apart, the shorter growing, 4–6 in. apart. Moisture should be given in the spring but, when the foliage starts to die back, allow the plants to become on the dry side. They also can be planted in early spring after frost has passed, and lifted in late summer and stored overwinter in a frost-free area.

PROPAGATION   Young bulbs should be removed from older plants and grown on. This is best done in spring prior to growth commencing. Seed should be sown in spring in a sandy soil mix, barely covering the seed; when large enough to handle, transplant into individual pots. However, if late in the season, young seedlings are best left in the containers where sown and overwintered and then transplanted when growth commences in the spring.

PESTS & DISEASES   Few problems except if not allowed to become dry during the resting period rotting will occur.

USES   In sunny, well-drained, and protected borders, but should be considered by the bulb fancier only.

SPECIES

C. bifolia. Native to Chile; introduced 1823. Rootstock a corm with elaborate tunic which is netted and much frayed in appearance. The flowering stem is leafless, often many-flowered, and can vary in height from a few inches to over 12. The flowers have petals which are reflexed and do not form a tube except at the very base. The color is deep blue with hint of purple, and from the center of the flowers the cone structure formed by the anthers protrudes, the yellow color being a good contrast to the blue. There are many flowers in a flower head, as many as 5 or more is not unusual. Foliage is linear and pointed, some 10–18 in. long.

C. campanulata (syn. *C. simsii*). Native to Chile; introduced 1823. There seems to be some discussion over which of these names is valid, some authorities preferring one, others the other. The principal difference between this species and *C. bifolia* is that the petals do not reflex and are joined in the lower third to form a campanulate flower. Also, unlike the previous species, the flowering stem carries 2–3 leaves which are sessile with the base of the leaf surrounding the stem for half of its diameter. Many flowers are produced in a branched inflorescence; the total number of flowers often being more than 10. They are of a good blue although authorities note that the color may vary from white to a darker purple-blue. Panicles remain attractive for a considerable period; flowering time is April to late May. The height varies a little but is generally about 12 in. The foliage is linear, 2–3 leaves per bulb, length 10–12 in., and their bases surround the base of the flowering stalk.

C. parvula. 2–2¾ in. in height. It is very similar to *C. campanulata* and may be only a dwarf form.

COOPERIA — AMARYLLIDACEAE

Name given in honor of Joseph Cooper, one-time gardener to Earl Fitzwilliam, Wentworth, Yorkshire, England. A small genus that some authorities regard as being placed in *Zephyranthes* but the separation is quite distinct. The differences between the 2 genera are that in *Cooperia* the anthers are erect, *not* versatile, and the flowers have a long perianth tube.

The flowers open at night and have a pleasant, primrose-like fragrance. The solitary flowers are produced in late summer and the color is white, sometimes flushed with pink. While flat when open, they are funnel-shaped in bud. The foliage has a glaucous tinge and is produced before the flowers appear. The bulb is quite large and has the appearance of a *Narcissus* bulb with its dark brown, papery tunic.

CULTURE Bright light, good moisture-retentive soil, and ample moisture during the growing season are required by these bulbs. They are not fully hardy, being able to take a little frost but not any prolonged cold period. For warmer climates they are good garden plants. The bulbs should be planted in the spring, spacing them 6–8 in. apart and 1–2 in. deep. In colder areas they should be grown only where good protection can be given against early fall frosts as these bulbs flower late in the summer and should enjoy pleasant temperatures at night until the foliage has died down.

PROPAGATION The best means is by offsets separated from the parent bulbs in the spring prior to growth commencing. These should be grown on in a protected area with temperatures similar to those given the parent bulbs.

Seed can be sown in the spring, given night temperatures in the 55°F range and a sandy but moisture-retentive soil mix should be used. Barely cover the seeds. Keep moist and growing through the first season and repot into individual containers in the second spring. The young plants should be of flowering size in the third season and then can be planted into their flowering positions.

PESTS & DISEASES No special problems.

USES In sunny borders in warmer climates but plant where their nighttime fragrance can be appreciated.

SPECIES

C. drummondii (syn. *Zephyranthes brazosensis*). Native to Texas. Erect flowers carried on stalks which are 6–9 in. in height. The foliage is glaucous, linear, twisted, and 10–15 in. in length. The perianth tube is long, often over 3 in., and the flowers open to a little over an inch in diameter. The color of the flowers is white, however, the tube will take on a reddish color as the flowers age. Flowering time July–August.

C. oberwettii. Also from Texas. Is very similar in height and size to the previous species but the foliage is not glaucous.

C. pedunculata (syn. *Zephyranthes drummondii*). Native to Texas and parts of Mexico. The linear and somewhat glaucous foliage is produced prior to the flowers appearing. Flowers are erect and carried on a stalk that is 3–6 in. long. The perianth tube is some 2 in. in length, greenish in color; the flowers open white, sometimes flushed pinkish-red. Diameter of the open flower is about 1 in. or a little more, and flowers are of a very fine, shiny texture. 9 in. in height. Leaves are 9 in. long.

COOPERIA

C. brasiliensis (syn. *Zephyranthes brasiliensis*) Native to Brazil. White, tinged red outside. 5 in. in height. Late Summer flowering.
C. smallii (syn. *Zephyranthes smallii*) Native to Texas. Yellow, tube green. 6 in. in height. Late Summer flowering.
C. traubii (syn. *Zephyranthes traubii*) Native to Texas. White, tinged pink outside. 6 in. in height. Late Summer flowering.

CORYDALIS — FUMARIACEAE

From the Latin *korydalis* ("crested lark"), as the spur of the flowers resembles the spur of the lark. This genus of more than 200 species is native to the temperate regions of the Northern Hemisphere, from China to North America to Europe to the eastern parts of the Mediterranean. Many are annuals; only a few species have tuberous roots. They are useful plants for the rock garden. The bulbous root of *C. bulbosa* was boiled by the Kalmuck Tartars, furnishing a starchy substance that was a mainstay of their diet.

The flowers are produced in racemes with a bract beneath each one. Flowers have 4 petals, the upper and lower ones being the largest, with the spur on the upper. The foliage is much divided and often glaucous. Even without the flowers, the plants are pleasant additions to the garden.

CULTURE Culture is varied according to locale. Seem to thrive everywhere in rich, well-drained soil with adequate supplies of organic matter. Where summer is very hot, should be given light shade; will even thrive in quite dense shade in warmer regions. Plant tubers 2–3 in. deep in the fall. Must be left in ground in order to increase, which does not take long. Lift and divide only when necessary to increase stock.

PROPAGATION Lifting and dividing rootstock is best method, but large amounts of seeds are produced which germinate readily. Sow seeds in fall and overwinter under protection. Plants reach flowering size in 2 or 3 seasons. Lifting and dividing rootstock is best done when foliage dies down in late summer/early fall. Offsets are planted back at that time.

PESTS & DISEASES No special problems.

USES Best grown in informal setting and in area where plants can spread. Good woodland and rock garden plants.

SPECIES

C. diphylla. Native to Kashmir. Low-growing, seldom more than 6 in. in height. Foliage gray and much dissected. Flowers white with purplish tips. Flowering early spring.

C. pumila (syn. *C. decipiens*) is a smaller form of *C. solida* and sometimes given specific rank by some authorities (as is *C. densiflora*). This plant is more likely a geographic variation of *C. solida*—found from Turkey to Scandinavia—it is very similar to *C. solida*.

C. solida. Native from the Balkans to Scandinavia. One of most popular of genus. Foliage much cut and grayish. Flowers on stems are 6 in. in height; light purple. Racemes are dense but quite short, seldom more than 2 in. Flowering April.

CORYDALIS

C. aitchisonii Native to Iran, Afghanistan & Central Asia. Golden yellow sometimes with mahogany on tips. 4 in. in height. March–May flowering.
C. allenii syn. of *C. scouleri*.
C. bracteata Native to Siberian Altai Mts. Yellowish or creamy white. 4–9 in. in height. May–June flowering.
C. bulbosa (syn. *C. cava*) Native from Scandinavia to Balkans. Purplish, more robust than *C. solida* & no scale-like bract on stem below flowers. 6–8 in. in height. Early Spring flowering.
C. cashmeriana Native to Himalayas, Kashmir to Bhutan & S. Tibet. Introduced circa 1935. Bright blue. 4–8 in. in height. May–Aug. flowering.
C. caucasica Native to Caucasus. Pinkish purple or sometimes creamy. 4–6 in. in height. April–May flowering.
C. cava syn. of *C. bulbosa*.
C. decipiens see *C. pumila*.
C. densiflora variety of *C. solida* from Turkey.
C. emanueli Native to Caucasus. Blue. 8–12 in. in height. July–Aug. flowering.
C. glaucescens Native to Central Asia. Pale white streaked purple. 8–12 in. in height. March–June flowering.
C. ledebouriana Native to Central & N.E. Afghanistan. Introduced circa 1875. Purple. 12 in. in height. March–June flowering.
C. longifolia syn. of *C. schanginii*.
C. macrocentra Native to Central Asia, especially Tadjikistan & N.E. Afghanistan. Yellow. 4–8 in. in height. March–April flowering.
C. marschalliana Native to Yugoslavia. Creamy yellow. 8 in. in height. Spring flowering.
C. nobilis Native to Siberia. Introduced 1783. Yellow. 12–18 in. in height. April–June flowering.
C. pauciflora Native to Alaska & E. Asia. Pinkish or bluish purple. 1½–3 in. in height. Spring flowering.
C. popovii Native to Central Asia, Pamir Alai. White with mauve petal tips. 6–8 in. in height. March–July flowering.
C. rutifolia Native to Crete, Turkey to Pakistan. Whitish, tinged purple or deep purple. 12 in. in height. April–May flowering.
C. schanginii (syn. *C. longifolia*) Native to Central Asia. Introduced 1832. Pale rose. 6–8 in. in height. Late Spring–Summer flowering.
C. scouleri (syn. *C. allenii*) Native to N.W. North America. Introduced circa 1895. Yellowish white tinged purple. 10 in. in height. Spring flowering.
C. sewerzowii Native to Central Asia. Introduced circa 1882. Yellow (tuber depressed globose). 8–12 in. in height. Late Spring–Summer flowering.
C. transsilvanica Origin obscure. Deep pink terra cotta. 4–6 in. in height. Spring flowering.
C. trifoliata Native to W. China, Himalayas. Introduced after 1954. Violet-blue (tubers small) 6–9 in. in height. Early Summer flowering.
C. verticillaris Native to Iran & Iraq. Introduced 1932. Pale pink, tipped maroon. 6 in. in height. Feb.–March flowering.

× CRINODONNA — AMARYLLIDACEAE

These bigeneric hybrids are the result of crossing *Amaryllis belladonna* with *Crinum moorei*, the seed being carried by *A. belladonna*. The cross took place in both Europe and America at about the same time in 1920, but it was first described and published in Italy by Dr. Ragioneri of Florence circa 1921. The American name, × *Amarcrinum howardii*, named for its first hybridizer in the United States, F. Howard of Los Angeles, has given way to × *Crinodonna* for this reason.

The results of such crosses are quite uniform and combine the characteristics of both parents. The plants flower in the fall. They have the persistent green leaves of the *Crinum* and pink, funnel-shaped flowers of the *Amaryllis*.

CULTURE Not very hardy; unable to withstand frost. Should be considered only as container plants in colder climates. In areas where frost is encountered only occasionally, can be planted in sheltered location in sun or high shade. Plant in September as soon as bulbs available. Set so top of bulb is just aboveground in good garden soil with good drainage. Provide adequate moisture except for a period at the end of summer/early fall when they can be allowed to ripen yet never completely dry out. As soon as growth is noticed, keep moist but not wet. Weak liquid feeding of organic fertilizer can be given in spring when the weather is warm but only if soil is poor; take care not to feed too heavily. A continuous supply of moisture must be given until the following summer after flower spikes appear. Plants should remain undisturbed unless they become overcrowded.

PROPAGATION Asexual propagation necessary, i.e., the separation of offsets in late summer. Best to leave bulbs in ground for at least 2 seasons after first planting so natural increase can occur. Lift and divide in late summer and replant at once.

PESTS & DISEASES Because of persistent foliage, protection against slugs and snails should be given.

USES Excellent garden plants to give color late in year. Grow in large containers in colder climates; make good fall color prior to first frosts.

SPECIES

C. × *corsii* (syn. *C. memoria—corsii*, × *Amarcrinum howardii*, × *A. memoria—corsii*). Long, arching, deep green leaves. Stems 30 in. in height bearing flower head of tightly packed pin flowers which open to a diameter of some 4–5 in. Not all flowers open at once so the fall-flowering season can be quite prolonged.

CRINUM — AMARYLLIDACEAE

Krinon is the Greek name for lily. While some 100 species of *Crinum* are known there are only a few in cultivation. They are found in South America, India, Southeast Asia, and both tropical and southern Africa. It is the species from South Africa that is commonly grown and these are good plants for mild climates. None are fully hardy and, in areas where frosts are common in the winter, need to have protection. In areas where the ground temperatures fall close to 32°F they should be considered only as container plants, being brought indoors during the winter months.

All of the species produce very large bulbs. Often the size of a large grapefruit, or even larger, the bulbs are rounded with a long neck that often can reach over 12 in. in length. The plants are most commonly found near water, alongside streams or ponds. The flowers also are large, the flower stalks short, and the base of the flower is a narrow tube widening to a blunderbuss shape. The petals are thick and most frequently have a ridge on the outer segments that is red or crimson.

The majority flower in late summer, the leaves being produced usually in one plane (distichous), broad and numerous, often as many as 20 or more. The flowers are carried on strong stalks and usually exceed 24 in. in height, 36–48 in. not being exceptional. There are many flowers in the umbel but seldom are more than 5–7 open at one time, the remainder often being in tight bud when the first flowers are open. Individual flowers often over 8 in. in diameter.

CULTURE The plants must have adequate moisture and the soil should contain generous amounts of humus. Plant in early summer, making sure the rounded part of the bulb is buried but that the neck is aboveground. Give full sun. In areas where winter frosts are likely but not severe, the plants should be covered during those months when frost can be expected.

After planting and when the leaves are being produced, the plants appreciate feedings of liquid organic fertilizer that can be given during the summer months until the plants send up their flower spikes. After they have flowered, less water is needed, especially if grown in areas where winter protection is required.

PROPAGATION The plants produce large seeds which will often germinate on the plant. Best sown as soon as possible, while still green, in a soil mix rich in organic matter. Sow 3 in. apart and an inch deep. Keep moist and, if possible, keep growing throughout the winter months. In this way the seedlings will reach flowering size in 3 seasons.

The quickest way to increase the stock is to separate the offsets from the parent bulbs. This should be done in May, replanting the parent bulbs and the offsets as soon as possible. The roots of all species are fleshy and they should not be broken. For this reason offsets are best removed while still quite small.

PESTS & DISEASES No special problems, but watch for basal stem rots on established plants and control with fungicides.

USES Good plants for the milder climates, they look well planted where the colors can be seen against a background of *Agapanthus* or other summer-flowering bulbs. In the herbaceous border they look well but should be considered only as additions and used as highlights. Good container plants that can be left undisturbed for a number of years but must be given winter protection.

SPECIES & CULTIVARS

C. **asiaticum**. Native to China; introduced 1732. One of the finest Crinums but must have warm conditions and cannot be grown outdoors where temperatures in the winter fall below 40°F. The fragrant flowers are produced during the summer months, white tinged a little pink near the edge in the center of the petals and have pink stamen filaments. Petals are recurved, 3–4 in. in length but not wide, often being only ½ in. Many flowers in the flower head, each on a flower stalk of up to 3 in. in length. When established there are many flowers produced, giving a long flowering period. Height 18–24 in. There is considerable variation in the flowers, some being greenish on the outer segments. The leaves are unlike other species described, as they are produced in a rosette. There are some 20 or more produced, up to 48 in. in length and 4 in. in width.

C. **bulbispermum** (Orange River Lily; syn. *C. capense, C. longifolium*). Native to South Africa; introduced 1752. The flowers vary from white to pale pink and have a rose-colored stripe through the center of each petal. The petals are 3–4 in. long and the flower stalks are of the same length. As many as 20 flowers in the flower head, which reaches up to 36 in. in height. Leaves are 24 in. in length, arc upward and then curve back to the ground, throwing the flower spike well above the foliage. Early summer-flowering. The leaves die right back to the bulb in the winter months and during this time the plant does not need much moisture.

C. **campanulatum** (Water Crinum; syn. *C. aquaticum, C. caffrum*). Native to South Africa. An unusual species as it can be grown in shallow water during the summer and taken out during the winter. 36 in. in height. Flowers are bright red to purple but smaller than other species, being only 2–3 in. long. Leaves are up to 48 in. long but narrow, up to an inch in width. An unusual species that has merit in warm climates where it can receive the moisture it likes during the growing season. Flowering in April/May.

C. **macowanii**. Widely distributed from Natal south to the eastern part of the Cape of Good Hope; introduced 1874. Easy to recognize in flower as it has black anthers, this lovely species grows up to 48 in. in height with large, trumpet-shaped flowers of white to pale pink with a crimson stripe down the middle of each petal. Leaves in a rosette up to 36 in. in length and 3–4 in. wide. Leaves die down in the winter and reappear early in the spring. Flowers in late summer and, while not popular, is a species that deserves greater recognition and cultivating in gardens in warmer areas. Fruit is round and knobby containing large, irregular seeds.

C. **moorei** (Cape Coast Lily; syn. *C. colensoi, C. mackenii, C. makoyanum, C. ornatum*). Native to Natal; introduced 1874. Has been grown for many years, being a favorite plant for the cool greenhouse where it is protected from frost during the winter months. The stems are 48 in. in height, and the flowers white or pale pink. The leaves are up to 36 in. in length and 4 in. wide. Unlike many other species this plant is found in forests, and should be given some shade in hotter areas. Needs a lot of moisture in the summer but prefers drier winter conditions.

C. × **powellii** (interspecific hybrid between *C. bulbispermum* and *C. moorei*). Large leaves, up to 48 in. in length and 4 in. wide but thinning toward the tip. Flower stalks 24–30 in. in height. Hardier than either of its parents, the bulbs will stand a little frost and cold weather but should be planted in the protection of a south wall in such climates. The flowers vary from pure white to a good pink. The flowers are large and open wide with up to 15 fragrant flowers on a stem, each over 4 in. in diameter. One of the best for garden use and should be the first *Crinum* to try in your garden. Some of the best cultivars include:

'Album'. Pure white; cultivated since 1893.
'Harlemense'. Pale shell-pink.
'Krelagei'. Deep pink, large flowers.

CRINUM

Species from tropical climates are in flower for many months of the year.

C. *abyssinicum* Native to Ethiopia. Introduced 1892. White, bulb ovoid, 3 in. thick. 12–20 in. in height. Summer flowering.
C. *acaule* Native to Zululand. White, keel red. 12–18 in. in height. Summer flowering.
C. *amabile* Native to Sumatra. Introduced 1810. Tube bright red, white with crimson central band, tinged purplish-red without, bulb small. 24–36 in. in height. Summer flowering.
C. *americanum* Native from Florida to Texas (marshes & swamps.) Introduced 1752. Creamy-white, bulb globose 3–4 in. thick, 18–36 in. in height. Flowers periodically Spring to Fall.
C. *amoenum* Native to India. Introduced 1807. Tube greenish, segments white. 12–24 in. in height. Summer flowering.

C. amoenum var. *mearsii* Native to Burma. Introduced 1907. White. Dwarf. Summer flowering.
C. angustifolium (syn. *C. arenarium*) Native to Northern Australia. Introduced 1824. 12 in. in height. Summer flowering.
C. aquaticum see *C. campanulatum*.
C. arenarium syn. of *C. angustifolium*.
C. augustum Native to Mauritius. Introduced 1818. Tinged red. 18 in. in height. Summer flowering.
C. bainesii see *Ammocharis tinneana*.
C. balfourii Native to Socotra. Introduced 1880. White with greenish tinge on tube. 18 in. in height. October flowering.
C. blandum variety of *C. angustifolium*.
C. brachynema Native to India. Introduced 1840. White, tube green. 12 in. in height. Summer flowering.
C. bracteatum Native to Seychelles, Mauritius. Introduced 1810. White, tube tinged green. 12 in. in height. July flowering.
C. braunii Native to Madagascar. Introduced 1894. White, tinged pink on margins, tube greenish.
C. caffrum see *C. campanulatum*.
C. capense see *C. bulbispermum*.
C. careyanum Native to Seychelles, Mauritius. Introduced 1821. White, tinged red. 12 in. in height. Flowering early & through Summer.
C. colensoi see *C. moorei*.
C. crassifolium syn. of *C. variabile*.
C. crassipes Native to Tropical & Subtropical Africa. Introduced 1887. White with pink keel. 9 in. in height. July flowering.
C. crispum Native to Transvaal, South Africa. Introduced 1932. White, pinkish without. 2–3½ in. in height.
C. cruentum Native to Mexico. Introduced 1810. Bright red. 24 in. in height. Late Summer flowering.
C. declinatum variety of *C. asiaticum*.
C. defixum Native to India. Introduced 1810. White, tube greenish or tinged red. 12–18 in. in height. October flowering.
C. distichum Native to Sierra Leone. White, keeled bright red. June flowering.
C. douglasii Native to Thursday Island. White. 30 in. in height. Summer flowering.
C. erubescens Native to Tropical America. Introduced 1780. White, tinted claret-purple. 18–24 in. in height. Summer flowering.
C. erythrophyllum Native to Burma. White. 10 in. in height. Summer flowering.
C. falcatum see *Cybistetes longifolia*.
C. flaccidum Native to New S. Wales, Southern Australia. Introduced 1819. White. 18–24 in. in height. July flowering.
C. forbesianum Native to Delagoa Bay. Introduced 1824. White, tinged reddish outside. 12 in. in height. October flowering.
C. forgetii Native to Peru. Tube green, segments white. 12 in. in height. Late Summer flowering.
C. giganteum Native to W. Tropical Africa. Introduced 1792. White. 24–36 in. in height. Summer flowering.
C. graminicola Native to Transvaal. Clear pink or white with rose-pink stripes. 12 in. in height. October flowering.
C. × *grandiflorum* (*C. bulbispermum* × *C. careyanum*).
C. × *haarlemen* form of *C.* × *powellii*.
C. heterostylum see *Ammocharis heterostyla*.
C. hildebrandtii Native to Comoro Is. Introduced 1886. White. 12 in. in height. September flowering.
C. humile Native to Tropical Asia. Introduced 1826. White, tube greenish. 12 in. in height. October flowering.
C. intermedium Native to Wai Weir Is. White, tipped yellow. 12–18 in. in height.
C. jemense syn. of *C. yemense*.
C. johnstonii Native to Central Africa. Introduced 1900. White, tube greenish, segments pinkish. 24 in. in height. Summer flowering.
C. kirkii Native to Zanzibar. Introduced 1879. White, striped red on back of segments. 12–18 in. in height. September flowering.
C. kunthianum Native to Colombia. Introduced 1890. White. 12 in. in height. Late Summer flowering.
C. lastii see *Ammocharis tinneana*.
C. latifolium Native to India. Introduced 1806. White, tinged red. 12–24 in. in height. Summer flowering.
C. leucophyllum Native to Damaraland, S.W. Africa. Introduced 1880. Pinkish. 12 in. in height. August flowering.
C. lineare Native to Eastern Cape, South Africa. Segments tinged red without. 12 in. in height. September flowering.
C. loddigesii var. of *C. cruentum*.
C. longiflorum Native to India. White. 18 in. in height. Summer flowering.
C. longifolium see *C. bulbispermum*.
C. mackenii see *C. moorei*.
C. makoyanum see *C. moorei*.
C. mauritianum Native to Mauritius. Introduced 1817. White segments tipped pink. 36–48 in. in height. April flowering.
C. mearsii variety of *C. amoenum*.
C. ornatum see *C. moorei*.
C. parvum Native to Tropical Africa. Introduced 1896. White, striped red. 6–9 in. in height. Summer flowering.
C. parvum of gardens is *Ammocharis heterostyla*.
C. pedunculatum Native to E. Australia. Introduced 1790. White. 24–36 in. in height. Spring flowering.
C. plicatum variety of *C. asiaticum*.
C. podophyllum Native to Old Calabar (W. Africa) Introduced 1879. White. 8–9 in. in height. November flowering.
C. × *prainianum* (*C. moorei* × *C. yemense*).
C. pratense Native to India. Introduced 1872. White, tube greenish. 12 in. in height. June–July flowering.
C. procerum variety of *C. asiaticum*.
C. purpurascens Native to W. Africa. Introduced 1826. White, tinged reddish purple. 12 in. in height. Summer flowering.
C. pusillum Native to Nicobar Is. Whitish. 12 in. in height. Summer flowering.
C. rattrayi Native to Albert Nyanza, S.W. Kenya. Introduced 1904. White. 36–48 in. in height. March flowering.
C. rhodanthum see *Ammocharis tinneana*.
C. roozenianum Native to Jamaica. White within, tube purple-crimson, segments crimson on back. 18–24 in. in height. Summer flowering.
C. samuelii Native to Central Africa. Introduced 1901. White. 24 in. in height or taller. Summer flowering.
C. sanderianum Native to Tropical Africa, Sierra Leone. Introduced 1884. White with crimson band down middle of segments. 20 in. in height. Summer flowering.
C. scabrum Native to Africa? Introduced 1810. White with bright red on back of segments. 12–24 in. in height. May flowering.
C. schimperi Native to Mts. of Abyssinia. Introduced 1894. White, tube reddish-green. 24 in. in height. July flowering.
C. schmidtii var. of *C. moorei*.
C. sinicum var. of *C. asiaticum*.
C. strictum Native to South America. Introduced circa 1872. White with pale green tube. 24 in. in height. September flowering.

C. suaveolens Native to Ivory Coast. Introduced 1912. White, tube greenish. 24–30 in. in height. Summer flowering.
C. submersum form of *C. scabrum*.
C. sumatranum Native to Sumatra. White, tube greenish. July flowering.
C. tinneanum see *Ammocharis tinneana*.
C. thruppii see *Ammocharis tinneana*.
C. undulatum Native to N. Brazil. Introduced circa 1875. White, tube greenish. 12 in. in height. November flowering.
C. variabile (syn. *C. crassifolium*) Native to South Africa. White, flushed red, tube green. 12–18 in. in height. April flowering.
C. wimbushii Native to Lake Nyasa, Central Africa. Introduced 1898. White, tinged pink. 2 in. in height. Summer flowering.
C. woodrowi Native to Bombay. Introduced 1897. White. 24 in. in height. Summer flowering.
C. × *worsleyi* (*C. moorei* × *C. scabrum*).
C. yemense (syn. *C. jemense*, very near *C. latifolium*, possibly a variety.) Native to Arabia. Introduced 1892. Pure white. Spring–early Summer flowering.
C. yuccaeflorum Native to Sierra Leone. Introduced 1785. White, banded red on back, tube greenish. 12 in. in height. June flowering.
C. zeylanicum Native to Tropical Asia & Africa. Introduced 1771. White with red band on back, tube reddish or greenish. 24–36 in. in height. Early Spring flowering.

CROCOSMIA — IRIDACEAE

Name derives from the Greek *krokos* or Latin *crocus* ("saffron") and *osme* ("smell"). The dried flowers, when immersed in warm water, smell strongly of saffron. The genus is well known, as it includes the montbretia hybrid of gardens. There are not many species in this genus, as many have been combined with *Tritonia*, to which it is closely related.

The montbretias resulted from the first cross between *C. aurea* and *C. pottsii*, made in France by Mr. Lemoine in 1880. Since that time many other crosses between these 2 species have been made, and it is due to their easy culture that the plants have become well known. Because they are of easy culture they are often regarded as less desirable plants and have fallen into disrepute among some gardeners.

Crocosmia have 6 perianth segments which open widely and, relative to the actual size of the flowers, a long perianth tube and 6 stamens that are often longer than the perianth segments. The leaves are sword-shaped in a flattish fan and are not shiny. The flower spike rises from the leaves and will carry several shorter leaves on the lower part of the spike. The color of the flowers varies but is always in the yellow, gold, orange, and red range. They are very easy to grow except that some species are not hardy and will require protection in colder areas. Where they grow well they can become a nuisance as they multiply quickly.

CULTURE Light soil with organic matter suits these plants. They will thrive in sun or light shade and should be planted where they have room to spread. They are ideal for the narrow border but should be lifted every 3 or 4 years to prevent overcrowding and loss of flower production.

The plants are fairly hardy, especially the hybrids, but will need some protection, such as straw or other loose material in areas where the temperature falls below 20°F for any period of time. They require moisture in spring and early summer but toward the end of summer do not mind drier conditions.

PROPAGATION The plants will produce many offsets that can be lifted and planted back in the same location, provided some additional organic matter is added. This should be done during the dormant period, the best time being before growth starts in the spring. Plant the corms 1–2 in. deep and 8–10 in. apart. Seed is easily raised but takes at least 2 seasons of growth before reaching flowering size. Sow as soon as the seed is ripe in small containers in a sandy soil mix. Transplant seedlings to individual pots as soon as they are large enough to handle.

PESTS & DISEASES Very clean plants; not subject to any special pests or diseases.

USES A great border, such as along driveways in confined areas, can be made of these plants but they are best planted in bold groupings. Once established, demand little care as they will crowd out other plants. Good in containers especially for late-summer-flowering and will provide good cut flowers.

SPECIES & CULTIVARS

C. aurea (Falling Stars). Native to South Africa; introduced 1846. Usually found in forest shade on the central and southern escarpments in Transvaal. The flowers are golden yellow and become reddish as they age. They are carried on stems 24–36 in. in height; spikes are branched, each bearing 8 or more flowers, which means that there are often 30 or more flowers per stalk. The flowers are 2 in. in diameter and, as not all open at once, the flowering time is considerable, beginning in late summer and continuing into fall. There are frequently 4–6 leaves on each flower spike, located toward the base. Corms globose with offsets from clefts in the side, covered with a dry tunic. Leaves are about a foot long, light green, and arranged in a fan shape. Though the hybrids are superior in many ways, *C. aurea* is still a pleasant plant to grow. Best grown where it can spread to form a mass.

C. × **crocosmiiflora**. An interspecific hybrid that has become one of the most widely grown plants in gardens. Result of a cross made in France in 1880 by Mr. Lemoine of Nancy, the parents are *C. aurea* (the seed parent) and *C. pottsii*. Leaves sword-shaped, more or less in a fan shape, quite coarse but not unattractive. The flowers are produced in a zigzag, upright panicle. The perianth is funnel-shaped and the slender tube is just over an inch in length. Flowers are orange-scarlet, with hybrids (see below) in varying hues. Flowers carried on spikes reaching up to, but rarely more than, 36 in. in height. Flowering period quite long and starts in August.

    'Citronella'. A very pleasant, light lemon-yellow but not quite so free-flowering as the type.

    'His Majesty'. More pronounced orange-scarlet with flowers of greater size, overall appearance more scarlet due to the outside of the petals being crimson.

    'Star of the East'. The flowers are best in the shade as the color is a pale orange-yellow which shows up well in the shadier areas.

C. masonorum. Native to the southeastern coastal regions of South Africa. Flowers are a brilliant orange-red. Not so hardy as the other species but has the advantage of coming into flower earlier in August. The flower spike reaches a little over 24 in. in height but the stem bends to become nearly horizontal. Needs lots of moisture from spring until flowering, and, in all but the mildest areas, should be lifted when the foliage turns color, stored overwinter, and planted again in the spring when the worst of the frosts have passed. An excellent seaside plant. Leaves are pleated, held in a fan 18–24 in. long.

C. pottsii. Native to an area a little farther up the coast than *C. masonorum,* coming from Natal; introduced circa 1880. Taller than others, flower spikes will often exceed 48 in. in height; flowers are produced in a one-sided fashion. Flower color is best described as flame, the yellow flowers being flushed with red and orange.

Flowering in midsummer (Aug.). Foliage linear, in 2 ranks 18–24 in. long, 1¾ in. wide.

COMMENTS These plants deserve greater use, but must receive moisture in the spring and early summer in order to perform well. A plant that should be considered where color is needed in late summer, especially along the coast.

CROCOSMIA

'Aurore' Garden Origin. Introduced 1890. Orange. 36 in. in height. Autumn flowering.
'Carmin Brilliant' Garden Origin. Introduced circa 1895. Orange-red. 24 in. in height. Summer flowering.
'Emily McKenzie' Garden Origin. Introduced 1954. Orange. 24 in. in height. Autumn flowering.
'Jackanapes' Garden Origin. Red & yellow. 24 in. in height. Summer flowering.
'James Coey' Garden Origin. Introduced 1921. Red. 24 in. in height. Summer flowering.
'Lady Hamilton' Garden Origin. Introduced 1911. Orange-yellow. 36 in. in height. Autumn flowering.
'Queen of Spain' Garden Origin. Introduced 1916. Orange-red. 36 in. in height. Summer flowering.
'Solfatare' Garden Origin. Yellow. 24 in. in height. Summer flowering.
'Vesuvius' Garden Origin. Introduced 1907. Red. 24 in. in height. Autumn flowering.

CROCUS — IRIDACEAE

Crocus is a Chaldean name given to the genus by Theophrastus. In Greek it means "saffron." In the wild the plants are found over much of Europe, especially around the Mediterranean, in North Africa, and in western Asia as far as Afghanistan. Although regarded by many as spring-flowering bulbs, a number of species flower in the autumn, including *C. sativus,* from which saffron is collected.

Saffron has long been regarded as one of the choicest of spices, as well as a brilliant dye, and for its medicinal powers. It is claimed to be the "Karcom" of the Hebrews (Song of Solomon, 4:14). The word also can be traced to the Arabic "Za-Feran." The spice was once cultivated in and around Saffron Walden in Essex, England. Today, the majority is imported from Spain. It is costly, for it takes more than 4,000 flowers to produce one ounce of saffron. The only part of the flower used is the stigma. It is recorded that the price was so high that the value of the crop often surpassed the value of the ground upon which it was grown.

No mention of this genus can be made without reference to the superb work that was carried out by horticulturist E. A. Bowles, who was with the Royal Horticultural Society. His book, *A Handbook of Crocus and Colchicum,* is as complete as could be desired. It was first published in 1924 and revised in 1954.

There are more than 80 species of *Crocus* and many more varieties and cultivars. A very popular genus, well over 40 species are listed in catalogs and offered by nurseries today. A very comprehensive collection can be had for your garden.

The styles of the *Crocus* vary greatly from species to species. They can be dissected into 3 lobes, as in *C. sieberi;* into 6, as in *C. olivieri;* or very finely dissected with numerous lobes. The character of the corm is another important identification feature. There

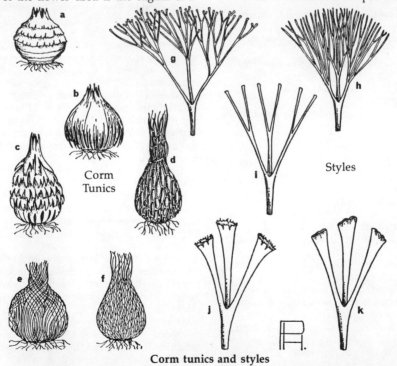

Corm tunics and styles

(a) Annulate (*C. chrysanthus*) (b) Parallel fibres (*C. aureus*) (c) Shell-like tunic (*C. laevigatus*) (d) Coarsely reticulate (*C. cancellatus*)
(e) Woven fibres (*C. fleischeri*) (f) Finely reticulate (*C. sativus*) (g) Much-dissected (*C. boryi*) (h) Much-dissected (*C. vitellinus*)
(i) 6-lobed (*C. balansae*) (j) Trilobed (*C. sieberi*) (k) Trilobed (*C. chrysanthus*)

are those with annulate tunics, i.e., circular rings around the entire corm; those with fibers that run parallel from top to bottom; shell-like tunics with overlapping scales; coarsely or finely reticulate, i.e., coarse or fine fibers around the corm; and those with fibers that form a close tunic, looking like the weave of cloth.

Other characteristics that aid in the identification of species are the appearance of the foliage before or after flowering, the number of leaves, and their growth habit. The obvious, of course, is the flower color, as well as the color of the anthers.

The classification of *Crocus* as proposed by George Maw in 1886 has been followed since that time by all major contributors to the literature of the genus. Brian Mathew, in his excellent book *The Crocus*, published in 1983 by Timber Press of Oregon, writes that, while Maw's classification, with only a few modifications, would accommodate all new species found since his time, it is possible to group all species in a more natural way.

Mathew's book is necessary reading for all those wishing to have extensive knowledge of the genus. His 'Informal Key to Crocus Species' will, it is sure, be of immense help to all interested. No hybrids, subspecies, varieties, or cultivars are included in *The Crocus*, but it is an extremely important work and no doubt, as time passes, will assume its rightful place in the literature of these plants, indeed ranking with E. A. Bowles' masterpiece.

The classification below was proposed in 1886 by Mawe and has been followed by most authorities since.

DIVISION I. INVOLUCRATI
Species with a basal spathe springing at the base of the scape from the summit of the corm.
  Section I. *Fibro-membranacei*, with a corm-tunic of membranous tissue, or of membranous tissue interspersed with nearly parallel fibers.
    Autumn-flowering: *C. asturicus, C. banaticus, C. cambessedesii, C. clusii, C. karduchorum, C. kotschyanus, C. nudiflorus, C. ochroleucus, C. salzmanni, C. scharojanii, C. serotinus, C. vallicola.*
    Spring-flowering: *C. imperati, C. malyi, C. minimus, C. suaveolens, C. versicolor.*
  Section II. *Reticulati*, with a corm-tunic of distinctly reticulated fibers.
    Spring-flowering: *C. corsicus, C. etruscus, C. heuffelianus, C. tommasinianus, C. vernus.*
    Autumn-flowering: *hadriaticus, longiflorus, medius, niveus, sativus.*

DIVISION II. NUDIFLORI
Species without a basal spathe.
  Section I. *Reticulati*, with a corm-tunic of distinctly reticulated fibers.
    Autumn-flowering: *C. cancellatus.*
    Spring-flowering: *C. ancyrensis, C. carpetanus, C. cvijicii, C. dalmaticus, C. gargaricus, C. reticulatus, C. sieberi, C. stellaris, C. susianus, C. veluchensis* (see also *C. Sieheanus*).
  Section II. *Fibro-membranacei*, with a corm-tunic of membranous tissue, or of membranous tissue interspersed with nearly parallel fibers.
    Spring-flowering: Lilac or White: *C. alatavicus, C. hyemalis, C. nevadensis.*
    Autumn-flowering: Lilac or White: *C. boryi, C. caspius, C. laevigatus, C. tournefortii.*
    Spring-flowering: Yellow: *C. aureus, C. balansae, C. biliottii, C. candidus, C. graveolens, C. korolkowi, C. olivieri, C. sieheanus, C. suterianus, C. vitellinus.*
  Section III. *Annulati*. Basal tunic of corm separating into annuli.
    Spring-flowering: *C. adami, C. aërius, C. biflorus, C. chrysanthus, C. crewei, C. cyprius, C. danfordiae, C. hartmannianus, C. isauricus, C. pestalozzae, C. tauri.*
    Autumn-flowering: *pulchellus, speciosus.*
  Section IV. *Intertexti*, with a corm-tunic of stranded or plaited fibers.
    Spring-flowering: *C. fleischeri.*

Series/Sections

Mathew writes that we are in a fortunate position today of having a great deal more information concerning the genus *Crocus*, and it is possible to group the species in a more natural way. He also states that a great deal more work is needed if evolutionary lines are to be traced with any degree of confidence. He also states that certain species in his classification may be found in more than one group, if their characters have a wide spread of variations. As examples he cites *C. cambessedesii* and *C. laevigatus*, the reasons being that they may flower in either fall or spring. Color variations in *C. reticulatus* allow it to fall in as many as four of the groups, sometimes more than once in a group, this happening as one uses his excellent 'Informal Key to Crocus Species.'

The great importance of Mathew's work cannot be denied. His 'Informal Key' is superb, as it takes into account practical features that can be seen easily. Such keys, in my opinion, are far superior to those that demand the dissecting of a plant. It is to be wished that more authorities would adopt such practical keys.

The various series into which Mathew divides the genus and the species in the various series are as follows:

Section (A) Species with a basal spathe. Type of section = *C. sativus.*

  Series (a) *Verni.* Corms with reticulated fibers, spring-flowering, flowers for the most part without conspicuous outer striping, bracts absent. Type of series: *C. vernus.* Species in series: *C. vernus, C. tommasinianus, C. etruscus, C. kosaninii, C. baytopiorum.*

  Series (b) *Scardici.* Spring-flowering, leaves have no pale stripe on the upper surface. Type of series: *C. scardicus.* Species in series: *C. pelistericus, C. scardicus.*

  Series (c) *Versicolores.* Spring-flowering, corms with tunics, which for the most part have parallel fibers, flowers with conspicuous exterior striping. Type of series: *C. versicolor.* Species in series: *C. malyi, C. versicolor, C. imperati, C. minimus, C. corsicus, C. cambessedesii.*

  Series (d) *Longiflori.* Autumn-flowering, yellow anthers, styles much divided. Type of series: *C. longiflorus.* Species in series: *C. nudiflorus, C. serotinus, C. longiflorus, C. medius, C. niveus, C. goulimyi.*

  Series (e) *Kotschyani.* Autumn-flowering, anthers white, styles for the most part three-forked. Type of series: *C. kotschyanus.* Species in series: *C. scharojanii, C. vallicola, C. autranii, C. kotschyanus, C. karduchorum, C. gilanicus, C. ochroleucus.*

  Series (f) *Crocus.* Autumn-flowering, anthers yellow, style distinctly three-branched. Type of series: *C. sativus.* Species in series: *C. pallasii, C. thomasii, C. cartwrightianus, C. sativus, C. moabiticus, C. oreocreticus, C. asumaniae, C. hadriaticus.*

Section (B): Nudiscapus. Species without a basal spathe. Type for Section = *C. reticulatus.*

  Series (g) *Reticulati.* Corm tunic for the most part decidedly covered with reticulated fibers, flower produced in winter or spring, style three-forked or much divided. Type of series: *C. reticulatus.* Species in series: *C. veluchensis, C. cvijicii, C. dalmaticus, C. sieberi, C. robertianus, C. cancellatus, C. hermoneus, C. reticulatus, C. abantensis, C. angustifolius, C. ancyrensis, C. gargaricus, C. sieheanus.*

  Series (h) *Biflori.* Tunics of corms split into rings at the base, either entire or with toothlike projections, leathery in texture, spring- or late-winter flowering, style three-forked. Type of series: *C. biflorus.* Species in series: *C. chrysanthus, C. almehensis, C. danfordiae, C. biflorus, C. pestalozzae, C. aerius, C. cyprius, C. hartmannianus, C. adanensis, C. leichtlinii, C. caspius.*

  Series (i) *Orientales.* Corm with parallel fibers or lightly reticulated, numerous leaves, spring-flowering, style three-forked. Type of series: *C. korolkowii.* Species in series: *C. michelsonii, C. alatavicus, C. korolkowii.*

  Series (j) *Flavi.* Tunics of the corms membranous, split into parallel fibers, spring-flowering, styles much divided. Type of series: *C. flavus*. Species in series: *C. flavus, C. antalyensis, C. olivieri, C. candidus, C. vitellinus, C. graveolens, C. hyemalis*.

Series (k) *Aleppici*. Tunics of the corms membranous, with split, parallel fibers, foliage produced at the same time as the flowers, fall- or winter-flowering. Type of series: *C. aleppicus*. Species in series: *C. aleppicus, C. veneris, C. boulosii*.

Series (l) *Carpetani*. Undersurface of leaves rounded with grooves, upper surface channeled, spring-flowering, style whitish, obscurely divided. Type of series: *C. carpetanus*. Species in series: *C. nevadensis, C. carpetanus*.

Series (m) *Intertexti*. Corm tunic fibrous with fibers interwoven, spring-flowering. Type of series: *C. fleischeri*. Species in series: *C. fleischeri*.

Series (n) *Speciosi*. Corm tunic splits into rings at the base, leathery or membranous, foliage after the flowers, autumn-flowering, style much divided. Type of series: *C. speciosus*. Species in series: *C. speciosus, C. pulchellus*.

Series (o) *Laevigati*. Corm tunic membranous or splitting into parallel fibers, sometimes leathery, foliage produced at same time as flowers, autumn-flowering, anthers white, style much divided. Type of series: *C. laevigatus*. Species in series: *C. laevigatus, C. boryi, C. tournefortii*.

Subgenus (2) *Crociris*. Anthers split open on side next to the style. Type of subgenus: *C. banaticus*. Species in subgenus: *C. banaticus*.

Little-known species, taxonomic position uncertain: *C. boissieri*.

CULTURE Crocuses are tolerant of a wide range of soils but they must have good drainage. The average garden soil is suitable, but, if on the heavy side, the incorporation of sand or small-grade gravel is advisable. This should be done in such a manner that the area beneath the bulbs is free-draining for a depth of 4–6 in. A good way to accomplish this is to remove the soil from the planting area to a depth of 6–8 in., work very sharp sand or gravel into the bottom of the area, and mix a little with the heavy soil to bring the depth up to 3–4 in. Then place the bulbs into position, spacing them some 6 in. apart, with 3–4 in. of soil over them. The bulbs have a way of moving away from the original planting position over a period of time as new corms are produced on the side of the parent. Some species like to be set a little deeper, but they will find their own depth.

The bulbs are best planted in the fall. In colder climates this should be done as early as possible. In warmer climates, where the temperatures remain in the 60°F range into October, plant toward the end of the month. Crocuses enjoy full sun but will perform well in shade if it is not completely shady all day. They should have at least 4 hours of sun per day.

The majority of the plants are dwarf, so the locations chosen should be where they can be seen. Thus, the edges of paths, drives, and small beds toward the front of borders are to be preferred.

They can be planted under grass as well. This method is to lift the turf, cultivate the soil, place the bulbs into position, and then lay the turf back over them. The grass should be mown short but all mowing must stop as soon as the flowers appear, as their beauty can be appreciated only when there is not much lawn growth. By the time the grass needs to be cut in the spring, the crocuses have made their leaf growth, and, while it is preferable to wait as long as possible before cutting their leaves, they can be cut before the lawn becomes unsightly. This method of growing *Crocus*, however, should not be attempted where the weather is warm in January and February.

Crocus can be grown in pots and perform well as long as the following procedures are maintained. Place 6 corms per specialized 6-inch *Crocus* pots, or use a regular container, covering them with about an inch of soil. Fern pots also can be used. Place in a cold location; if set outside, they can be covered with sand and soil. The ideal temperature is around 40°F. When they have made good root growth, they can be brought indoors, as described in the section under "Forcing" (see page 000).

The *Crocus* also can be grown in water, especially the stronger hybrids, but it is necessary to select the larger-sized corms. They should be placed either in a bowl with gravel or resting in the neck of a crocus vase so that the bottom of the corm is just above the level of the water. Place in a cool location until they have made root growth. They can be given a little more warmth but high temperatures should be avoided. Bulbs forced or grown in water should be discarded or they may be planted in the garden to regain their strength, but they should not be relied upon to give a display of flowers for one or two seasons.

If the plants are to remain for a long period in the same location, an application of fertilizer can be given each spring, as soon as the foliage appears. Incorporating bonemeal into the ground at planting time is beneficial.

PROPAGATION *Crocus* can be raised from seed, sown when ripe or saved until the following spring. It germinates quickly and easily. Using a sandy soil mix, sow thinly in a container, barely cover it, and keep moist but not wet. After germination, they can be plunged in the pots outdoors in a well-drained area with some shade. Leave them in the container until the second season. When the leaves die down, remove the newly formed corms and plant them out that fall into their correct locations. They may flower the following spring; if not, then the next spring.

The cormels produced by the parent corms offer the quickest way to obtain flowering-sized stock. The corms should be lifted as soon as the foliage starts to die down. The new corms, formed on the side or top of the old corms, are separated and then planted back in drills 2 in. deep and spaced 2–3 in. apart. These corms will produce more young offsets if they are lifted than if left *en masse* in the ground. Certain species do not produce rapidly, so these should be left in the ground until the clumps are of sufficient size to need thinning. This is true of some species but not of the cultivars.

PESTS & DISEASES The corms are subject to a number of rots while in storage and the time spent in checking them prior to planting is well-spent. The tunics can be stripped without harm to examine the corms. The various kinds are: hard rot (*Septoria*), which causes sunken patches on the corms; dry rot (*Sclerotinia*), which causes black patches with no shrinking; and gray bulb rot, which causes the corm to become covered with mold. Corms so attacked should be discarded at once. If in doubt, be on the safe side and discard. The plants also can be attacked by gray mold fungus (*Botrytis cinerea*), usually due to lack of air circulation when the plants are overcrowded but this can be kept under control by using a number of available fungicides. Mice like *Crocus*, so protect them from these rodents.

There are not many other problems to be expected when growing these lovely harbingers of spring.

USES These are ideal plants for early spring and late fall color. While the spring species and varieties are the best known, the autumn- and winter-flowering deserve more recognition and a place in the garden, adding another dimension to the fall landscape. Planted in large areas, *Crocus* can be eye-stoppers in the spring; planted under lawns (as described) or to complement other spring-flowering bulbs, they give added length of color to the early herbaceous and shrub borders. They look well in a rock garden and can be set out where other low-growing plants can be placed over and around them, such as plants with low rosettes of leaves that hug the ground.

They are great container plants and also do well grown in just water. As mentioned, there are crocus vases available that are designed for a single corm. Grown in this way they make unusual and attractive individual flower settings for the dinner table.

Crocuses are easy to grow and should be used wherever possible.

SPECIES & CULTIVARS
Autumn-flowering.

C. **cancellatus** ssp. **cancellatus**. (syn. *C. c.* var. *kotschianus, C. cilicicus*). Native to southern Turkey, Syria, Lebanon, and northern Israel; introduced before 1841. The corm has a coarsely reticulated covering. The 10–12 in. long leaves appear after the flowers. The flowers reach 2–3 in. in height and vary in color from pale to mid-lilac-blue. The perianth has a pale yellow throat, feathered at the base of the outer segments with violet. The stigmata are multi-branched and orange in color. Flowering time varies from the end of Sept.–early Dec. Ssp. *pamphylicus* is distinguished from the species only because the anthers are white instead of yellow.

C. **goulimyi**. Native to Greece; a comparatively recent introduction (1955). The tunic of the corm is hard and shell-like, splitting lengthwise. The leaves are 9–10 in. in length and appear with the flowers. The soft lilac flowers are 3–4 in. in height, and are produced Oct.–Nov. This species produces a large number of offsets.

C. **hadriaticus** (syn. *C. peloponnesiacus*). Native to Greece; introduced prior to 1843. The corm is finely netted and silky. The leaves are quite long, reaching 18 in. The flowers, 3–4 in. high, are white with purple markings on the outside segments toward the base; the throat is white or purple and xthe stigmata are orange. Flowering in Oct. Var. *chrysobelonicus* has the same characteristics as species but flowers are pure white with a yellow throat.

C. **kotschyanus** ssp. **kotschyanus** (syn. *C. zonatus*). Native to Lebanon; introduced before 1854. This is one of the finest of the autumn-flowering species and is often listed in catalogs as *C. zonatus*. The corm is irregular and flattened, with somewhat parallel fibers. The 12 in. long leaves have a white band in the center and appear after the flowers, persisting throughout the winter. The flowers are large and pale lilac with an orange band on the inside base of the petals; the throat is yellow reaching 2–3 in. in height. Flowering not infrequently at the end of Aug., but more commonly early in Sept. This is one of the most prolific offset producers in the genus, the many small cormlets spreading rapidly. It is an excellent species for naturalizing in open, woodland areas. Var. *leucopharynx*, sometimes listed in catalogs as *C. karduchorum*, differs from the species in that the flowers have a white throat and no orange ring at the base of the petals, but the petals are the same color and flowers reach the same height. This is a vigorous plant, not found in the wild, but it increases more rapidly than the common form.

C. **laevigatus**. Native to Greece; introduced before 1832. The unusual feature of the bulb is that the tunic completely covers it and is quite hard and smooth, like an eggshell. Leaves, which are produced before the flowers emerge, are 9–10 in. long. The outer petals of the flowers are feathered with deep lilac-mauve, which color appears only on the inside petals toward the base. The throat is pale yellow, the stigma is orange and featherlike in appearance, and the stamens are white. Flowers reach 2–3 in. in height. This can be regarded as a late-fall or early-winter species as it continues to flower from Nov.–Jan. Var. *fontenayi*, also from Greece, introduced circa 1938, has more globular-flowered form than the species, with ageratum-violet blossoms. Flowers in Dec. and of same height. The outside petals are feathered purple and the throat is inclined to be lighter. It flowers profusely, even when the temperature is hovering around 32°F. This is now considered by some to be included with the species.

C. **longiflorus** (syn. *C. odorus*). Native to southern Italy; introduced 1843. The corm is coarsely reticulated. The leaves are produced with or a little earlier than the flowers and are 8–9 in. long. This is one of the more interesting species, with its unusual flower coloring—the petals are light violet on the outside, more bluish on the inside, and darker veins and bronzing toward the base. As if this were not enough, the anthers are saffron-yellow and the stigma is scarlet and fragrant. The 5–6 in. high flowers appear in Oct. or Nov.

C. **medius**. Native to the southern part of France; introduced 1843. The corm has a finely reticulated tunic. The leaves are produced in the winter and are about 9 in. in length. The flowers, 3–4 in. in height, are light lilac in color with deeper veining, a white throat, and red stigma. Oct.–Nov. flowering.

C. **niveus**. Native to Greece; introduced prior to 1900. The large corms have a tunic that is covered with parallel and netted fibers. Usually, 3–4 of the 1–10 in. long leaves are produced after the flowers. This is regarded as being the best of the white-flowered *Crocus*, with an orange-yellow throat and scarlet stigma; about 3–4 in. in height, and appears in November.

C. **nudiflorus** (syn. *C. aphyllus, C. fimbriatus, C. multifidus, C. pyrenaeus*). Native to the Pyrenees and the mountain foothills in southwestern France; introduced before 1798. One of the most unusual of the genus in that the small corm, which has a tunic with parallel fibers, produces a stolon from which young corms are formed. It thus will spread quite quickly and will tolerate damper locations than many other species. Because of this stoloniferous root system, it can be established easily in open grassland and is best left undisturbed. The grasslike leaves are 6–8 in. long, ⅛ in. in width, and produced after the flowers. Very large, bright purple flowers with long petals allow the species to reach 6 in. in height. Sept.–Oct. flowering.

C. **ochroleucus**. Native to the eastern Mediterranean in Lebanon and surrounding areas; introduced prior to 1859. The corms are flattish with a thin tunic and produce many cormlets. The leaves are produced before the flowers and are 10–12 in. long, a contrast to the flowers which rarely grow taller than 2–3 in. in height. The flowers are small and white to creamy white with an ocher-yellow base to the petals. October-flowering.

C. **pulchellus**. Native to Greece and long in cultivation. The small corm has an annulate tunic. The leaves are produced at the time of flowering but continue to grow, reaching a length of some 10 in. The flowers, too, are quite tall, about 6 in. high, and are pale lilac with deeper veining, orange stigma, and white anthers. Early Sept.-flowering. *C. p.* 'Zephyr' is an introduction of Van Tubergen of The Netherlands. Shorter than species, but foliage same, as is the flowering time; white flowers with a golden-yellow throat.

C. **sativus**. Native to Italy and east to Turkey; introduced circa 1750. The silky tunic of the large corms has densely reticulated fibers. The leaves are produced with the flowers and are up to 18 in. in length. The petals of the flowers are lilac-purple with deeper color in the throat. This species is the one that produces the saffron of commerce. The unusual, orange-red stigma, from which the saffron is produced, is very large, often protruding beyond the closed flower and flopping toward the side. Sept.-flowering. There is a free-flowering form offered by some nurseries listed as *cashmerianus*, introduced from Kashmir. *C. s.* var. *thomasii*, introduced in 1900, has flowers that are slender and pointed segments.

C. **serotinus**. Native to Spain, Portugal, and North Africa. Autumn-flowering, and, with the exception of *C. nudiflorus*, Brian Mathew has proposed that all other autumn-flowering *Crocus* from these regions be placed in this species. Thus included are such established names as *C. asturicus, C. clusii*, and *C. salzmannii*. Mathew recognizes three subspecies, which are distinguished by the tunics of the corms: *C. s.* ssp. *salzmannii*, with a membranous tunic that splits into parallel fibers; *C. s.* ssp. *serotinus*, with a corm that has a coarsely netted tunic and stout fibers; *C. s.* ssp. *clusii*, with a corm that has a finely netted tunic and slender fibers.

Other differences occur, but, for practical purposes, it might be easier to distinguish these by the fact that *C. s.* ssp. *salzmannii* is not fragrant, while the others are, and *C. s.* ssp. *serotinus* has the leaves at the same time as it flowers or with the tips of them just visible while *C. s.* ssp. *clusii* often does not have the leaves at the time of flowering but they appear as the flowers wither.

C. s. ssp. *serotinus* reaches a height of 3–4 in. The flowers are pale to deep lilac, often with darker veining, the throat is white or very pale yellow, the style is orange and divided into many branches. Up to 4 leaves are produced, 3–4 in. in length at flowering time, elongating afterward. Flowering time in its native habitats is Oct.–Nov.

C. s. ssp. *clusii* (syn. *C. clusii*) is similar to *C. s.* ssp. *serotinus*, except for the tunic of the corm, noted above, and has up to 7 leaves.

C. s. ssp. *salzmannii* (syn. *C. asturicus, C. granatensis, C. salzmannii*) is similar to the above subspecies but the flowers are not fragrant and the corm, apart from the differences mentioned above, sometimes produces stolons. The flowering period in the wild is a little longer than the other subspecies, extending from Sept.–Dec. This plant also has a wider distribution in the wild, being the subspecies found in northern Morocco in North Africa.

C. speciosus ssp. **speciosus.** Native to Crimea, Caucasus, Turkey, and Iran; introduced before 1800. Along with its several forms and selections, this is regarded as one of the best of the autumn-flowering species and the easiest to grow. The large corms have annulate rings at the base and produce large numbers of small corms. If left alone, the species will sow itself and create large plantings. The leaves are only a few inches long when the plants are in flower, but grow to an eventual length of 12–18 in. The flowers are quite large, reaching a height of 5–6 in. Flower color is somewhat variable, as can be expected when there are a number of selections being made, but generally displays a deep violet-blue with deeper veining, yellow anthers, and deep orange stigmata. Sept.–Oct.-flowering. The following are some selections offered by various nurseries:

'Aitchesonii'. Pale blue, later into flower.

'Artabir'. Lavender-violet-blue with a hint of white under inner petals; creamy white throat; deep yellow stamens; deep-orange stigmata; slightly perfumed; 5–6 in. high.

'Cassiope'. One of the largest flowered; good aster-blue with yellowish base; opens wide to show orange stigmata; taller than others, reaching up to 7 in.

'Conqueror'. Clear, deep blue.

'Oxonian'. Also a good blue, but does not have the vigor of some offerings; large flowers carried on stems that are also blue.

Late-winter- and spring-flowering

C. alatavicus. Native to Turkestan; introduced 1877. The corm has a dark brown, fine-fibrous tunic. The leaves are a little more than 7 in. long. The flowers are white with a hint of cream with broad purple stripes on the exterior perianth segments, and are 3–4 in. high. On sunny days, when the flowers open fully, the bronze-yellow throat is seen at its best. Jan.–early Feb.-flowering.

C. ancyrensis. Native to Angora; introduced 1879. The tunic of the pear-shaped corm is coarsely netted. In comparison with the other species in this group, the leaves are long, reaching 12 in. This dwarf plant, only 2 in. in height, has orange flowers that are produced in January.

C. angustifolius (Cloth of Gold; syn. of *C. susianus*). Native to Crimea, Caucasus; known before 1597. The corm is coarsely reticulated. The leaves reach a length of 10 in. This dwarf plant, only a couple of inches high, is a deep golden-yellow tinged with mahogany on the outside. The stigma is orange-scarlet. An unusual feature is that the anthers are twice as long as the orange filaments. Early flowering in March. *C. s.* 'Minor'. A selection from the species, selected because the leaves are more upright and not quite so long. The flowers are deeper in color, as well, and the plant comes into flower later; height a little shorter.

C. biflorus. This plant has long been in cultivation. Even prior to its being included in Philip Miller's *Gardener's Dictionary* in 1768, it was being widely grown in various gardens. The plant most commonly grown and the one referred to by Miller is the same or very close to that known as 'Scotch Crocus.' This is a white flower with a yellow throat and prominent purple stripes on the exterior of the segments.

This plant may be equated with the Italian variant of *C. biflorus* and thus becomes *C. biflorus* ssp. *biflorus*. Until Brian Mathew undertook his close look at this genus, many of the variants of this species were given species rank. Mathew admits that this species constitutes one of the major taxonomic problems of the entire genus. There are many variations, so to lump all into *C. biflorus* would not be sensible, and if too critical a look was taken many taxa would result. Mathew has taken a sensible approach and divided the species into 14 subspecies, including many that previously had separate species rank.

All of these 14 subspecies are not of the same importance. Three are described here, the remainder in the chart that follows.

C. biflorus ssp. **biflorus.** Native to Italy, Sicily, Rhodes, and northwestern Turkey. Long in cultivation. Among the species now included are: *C. circumscissus, C. pusillus, C. argenteus, C. praecox,* and *C. annulatus* var. *biflorus*.

This subspecies is the variant that is the most westerly-occurring of those that have become naturalized in many parts of Europe. Its common name, 'Scotch Crocus,' is given to a sterile clone that has been many years in cultivation. Spring-flowering, fragrant, coming into flower in Feb.–March but often a little earlier or extending into April. The foliage is about the same height as the flowers at flowering time, 4 in., but extends to 10–12 in. once the flowers have withered. The background color is white or lilac-blue, with 3 purple or brownish-purple bands on the exterior of the outer perianth segments. There is frequently a fine feathering of this color in the segments. Filaments and anthers yellow, with the style divided into 3 orange or reddish, apically expanded branches. While the white or lilac-blue flowers normally have the 3 striking bands of color, unstriped flowers are not unknown, and these often have a silvery or pale buff color on the exterior. In years past, this type has been given the name *C. biflorus* var. *estriatus*. The most obvious characteristic of this species is the long foliage at flowering time and leaves that have no prominent ribs in the grooves on the underside—an important point in identification.

C. biflorus ssp. **weldenii.** Found in northeastern Italy, western Yugoslavia, and northern Albania. The subspecies is easy to distinguish by its white flowers, which are without any stripes and no yellow throat. A selection of this, 'Fairy,' has pure white flowers with a grayish tinge to the outer segments. It would appear that *C. pallidus* and *C. weldenii* f. *lutescens* are but variants of this subspecies. Flowers in Feb.–March.

C. biflorus ssp. **alexandri.** Native to southern Yugoslavia and southwestern Bulgaria. A robust plant that is quite like *C. biflorus* ssp. *biflorus* in the color of the flowers but without color in the throat, and thus, in this respect, like *C. b.* ssp. *weldenii*. The leaves are wider than in either of the other subspecies and the foliage is not well developed at the time of flowering, which is Feb.–March.

C. chrysanthus (syn. *C. annulatus* var. *chrysanthus*, *C. croceus*, *C. skorpilii*). Native to Greece and Asia Minor; introduced prior to 1847. The corm tunic is membranous to shell-like and annulate at the base. The leaves, up to 10 in. long, are produced with the flowers. The flowers are bright orange feathered with bronze with orange anthers. Mid-Feb.-flowering. This species is the parent of many fine garden hybrids. The heights vary a little but are under 6 in., most in the 3–4 in. range. Among them the following are offered by various nurseries:

'Advance'. Yellow inside, violet outside; very free-flowering.

'Blue Bird'. White inside with white margin to the petals, dark blue on the exterior.

'Blue Pearl'. Soft blue with bronzy base and golden throat, darker violet-blue on outside of petals.

'Blue Peter'. Soft blue with golden throat, purple exterior.

'Cream Beauty'. Soft, creamy yellow; free-flowering.

'E. A. Bowles'. Named after the legendary horticulturist who devoted so much of his time to the *Crocus*. Deep butter-yellow with bronze feathering, especially at the base; an outstanding variety.

'E. P. Bowles'. A little shorter than 'E. A. Bowles' but markings are more pronounced; bronze with a hint of purple.

'Fusco-tinctus'. Rich, golden yellow, thought to be a form of the species.

'Gipsy Girl'. Golden yellow outer perianth segments feathered with purplish-brown.

'Goldilocks'. Similar to 'Gipsy Girl'; long-lasting.

'Jeannine'. Recent introduction; light yellow with light crimson feathering.

'Ladykiller'. Rich purple edged with white, white interior.

'Mariette'. Large flowers; soft yellow inside, purplish outside.

'Princess Beatrix'. Clear, light blue exterior with darker feathers toward the golden yellow base.

'Saturnus'. Dark yellow inside, dark purple outside; earlier flowering than many.

'Skyline'. Soft blue, violet veining.

'Snow Bunting'. White inside, outer petals creamy with dark lilac feathering.

'Sunkist'. Golden yellow feathered bronze on the outside, inside pure golden yellow, yellow anthers.

'Violet Queen'. Very dwarf, only 1–2 inches high; violet-blue flowers.

'White Beauty'. White, outside purple striped.

'Zwanenburg Bronze'. Unusual color, a good bronze on the outside, interior yellow; a good variety and an eye-catcher.

C. corsicus (syn. *C. insularis*). Native to Corsica; introduced 1843. The small corm is silky and finely reticulated. The leaves are up to 8 in. long, height of flowers 2–4 in. The flowers have unusual markings with the outer perianth segments a pale lilac with a heavy feathering of deeper purple, and the inner segments mauve with a white throat tinged with yellow on the outside. This species is late-flowering, March–early April.

C. etruscus. Native to Italy; introduced 1877. The tunic of the corm is coarsely reticulated. The 8–10 in. long leaves have a white band. The large flowers are lilac or lavender, some striped more heavily than others with deeper lilac; the throat is yellow. March-flowering. The form 'Zwanenburg' is an excellent plant being vigorous with good textured flowers.

C. flavus. Native to much of Europe; it has been in cultivation for at least 400 years. The most popular and most commonly grown yellow *Crocus*, which is quite variable in the wild. For many years selections from this species have been sold by nurseries under such names as 'Dutch Yellow' and 'Yellow Giant.' Brian Mathew feels that this is a hybrid between *C. flavus* and *C. angustifolius* (syn. *C. susianus*). This species has been divided into two subspecies: *C. flavus* ssp. *flavus* (syn. *C. aureus*, *C. lacteus*, *C. lagenaeflorus*, *C. luteus*, *C. maesiacus*) has a style obscurely divided into 3 short branches with perianth segments ¾–1¼ in. long, while *C. f.* ssp. *dissectus* (syn. *C. mouradii*) has the style distinctly divided into 6 or more slender branches, and the perianth segments a little shorter.

The flowers of both of the subspecies are pale to deep orange-yellow, sometimes with brownish stripes or a suffusion of this color on the perianth tube and base of the perianth segments. The filaments and anther are yellow. The foliage is about the same length as the height of the flowers at flowering time, 3–4 in., but extend greatly later, often being as much as 12 in. in length. The flowering time is March–April.

It is not surprising that, because of the great variation that occurs in this species, over the years many of the variants had been given species rank. These include: *C. aureus*; *C. lagenaeflorus*, a larger-flowered form that was unstriped; *C. sulphureus* var. *striatus*, a variant with pale yellow flowers with 3 brown stripes on the outer perianth segments; *C. sulphureus* var. *striatellus*, which is similar to var. *striatus* but the stripes are visible only at the top of the perianth tube; *C. sulphureus* var. *concolor*, which is of a paler color with the tips of the petals almost white and without stripes; *C. luteus*, with large flowers and olive-green stripes on the outside of the outer perianth segments; *C. lacteus* var. *concolor*, with flowers of pale cream with a yellow throat; and *C. lacteus* var. *penicillatus*, a pale cream flowered form with a hint of pale blue striping on the perianth segments. In addition to these, others have made divisions but under the species *C. lagenaeflorus* and included *C. graveolens* and *C. olivieri* as varieties of this species.

Thus we have in the literture such names as *C. lagenaeflorus* var. *aureus* for the gold colored form, *C. l.* var. *aureus trilineatus* for those flowers that are gold with the exterior of the segments having bluish lines, *C. l.* var. *aureus sulphurascens* for the form with pale sulfur flowers, and *C. l.* var. *aureus albus* for the white-flowered form.

Also under *C. lagenaeflorus* var. *lacteus*, the name given to the form with creamy-white flowers, are *C. l.* var. *lacteus concolor* for cream colored flowers with a yellow throat; *C. l.* var. *lacteus penicillatus* for the cream colored form with bluish lines on the exterior; and *C. l.* var. *lacteus lutescens* for the pale yellow form which has a deep yellow throat, this color being confined to a deep stripe in each perianth segment with the color fading as it gets closer to the edges of the segments, and with pale bluish lines on the perianth tube.

Under *C. lagenaeflorus* var. *sulphureus* are flowers with a sulfur-yellow color with a deeper yellow throat: *C. l.* var. *sulphureus pallidus* for those that have a much paler flower, almost to white, and *C. l.* var. *sulphureus striatus* for those with sulfur-yellow flowers but striped on the exterior of the perianth segments.

Other authorities regarded certain of the above as species in their own right. It is therefore a distinct advantage to have the great work done by Mathew on record, as not only is the number of species reduced but his classification and logical approach to this entire genus makes eminent and practical sense.

C. fleischeri. Native to western and southern Turkey; introduced prior to 1827. The yellowish corms have a finely reticulated tunic and produce bulbils or cormlets at their bases. The leaves are 12 in. long. The flowers are white, striped with purple at the base; the anthers are orange-red. One of the first to flower in the early spring, late Jan. or early Feb. Height 2–3 in.

C. minimus (syn. *C. insularis*). Native to Corsica, Sardinia; intro-

Crinum scabrum

Plate 81

Crinum yuccaeflorum (?)

Plate 82

Crinum × *powellii album.* This is one of the best forms of this interspecific hybrid, and often is listed in catalogs. WARD

Crocosmia masonorum. This lovely plant, with its brilliant flowers, gives a grand display in late summer. In colder climates, it should be lifted in the fall and replanted in the spring. Despite this work, it is well worth a place in any garden. GRAF

Crocus oliveri ssp. *balansae* 'Zwanenburg'. One of the loveliest species, with superb color, it will remain in flower for a long period of time. 'Zwanenburg' is a good selection that adds much color to the early spring parade of flowers. WARD

Crocus boryi. A charming little species that is not commonly cultivated. Deserving of greater recognition as a garden plant. WARD

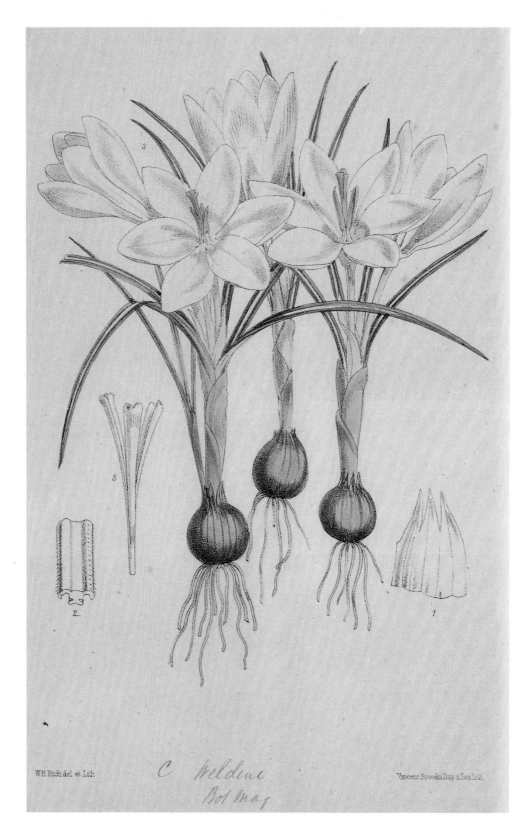

Crocus biflorus

Plate 84

Crocus cancellatus. The flowering time will vary by as much as two or three months. This species is always a welcome sight when in flower. WARD

Crocus cancellatus. This species, with the rather pointed perianth segments, is most likely *Crocus cancellatus* ssp *mazziaricus*. WARD

Crocus caspius. A rare species from the Caspian sea area, introduced at the turn of the century, but mostly found in bulb collections, not the average garden. WARD

Crocus chrysanthus 'Cream Beauty'. There are many cultivars listed in catalogs. All of them have merit, but some more than others. 'Cream Beauty' is free flowering, long lasting, but perhaps not showy if snow is on the ground. DE HERTOGH

Plate 85

Crocus chrysanthus **'Mariette'**. Another good plant that will never disappoint you. WARD

Crocus biflorus **ssp. *crewei***. Comparatively rare in cultivation, and not a likely candidate for general cultivation. WARD

Crocus hadriaticus **var. *chrysobelonicus***. The difference between the species and the var. *chrysobelonicus,* is that the species has markings on the outside of the perianth segments, while the variety does not have a yellow but pure white throat. WARD

Crocus kotschyanus. A great plant for any garden as it increases rapidly, producing lots of cormels. The color is a delight. This species will often come into flower in late summer. WARD

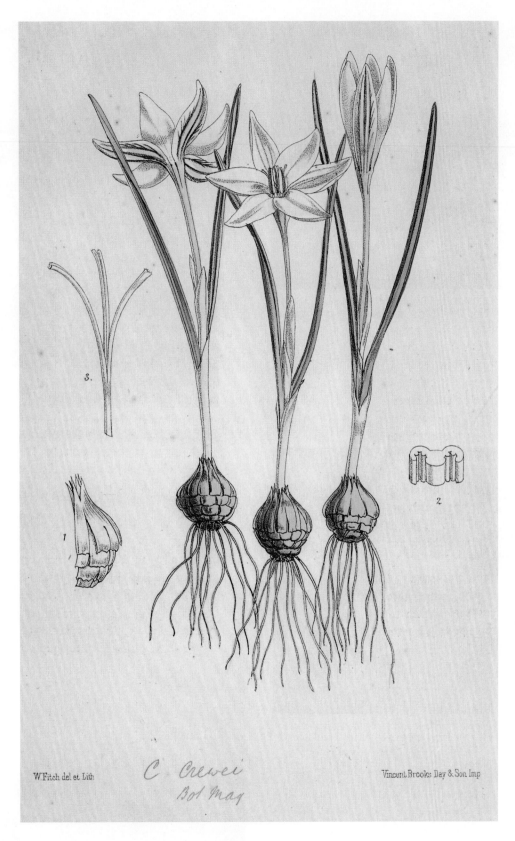

Crocus crewei

Plate 87

***Crocus laevigatus* var. *fontenayi*.** Even frosty temperatures will not stop this plant flowering. The variety is a little larger than the species type, and flowers profusely. WARD

Crocus medius. While only a few inches in height, the color of this species, from the southern part of France, is quite striking. WARD

Crocus minimus. Somewhat variable, as the markings vary in intensity, but no matter what the variation, this is a lovely species. WARD

Crocus pulchellus. One of the species of Autumn Flowering Crocus that has been long in cultivation, and deservedly so; a native of Greece. WARD

Crocus etruscus

Plate 89

***Crocus pulchellus* 'Zephyr'.** Having been long in cultivation, it is not surprising that great selections have been made from this species; 'Zephyr' is one of the best. WARD

***Crocus sieberi* 'Firefly'.** A good plant that should be planted and grown in the garden, as it is of quite easy culture. Many bulb fanciers like to enjoy the flowers indoors, and certainly these plants are happy in the container. AUTHOR

***Crocus sieberi* 'Violet Queen'.** A good garden plant that will stay around for a good number of years. The size of the planting will increase naturally if conditions are good. WARD

Crocus speciosus. Fortunately, this lovely plant is easy to grow, and, perhaps, should be the first Autumn Crocus to be tried in a garden. WARD

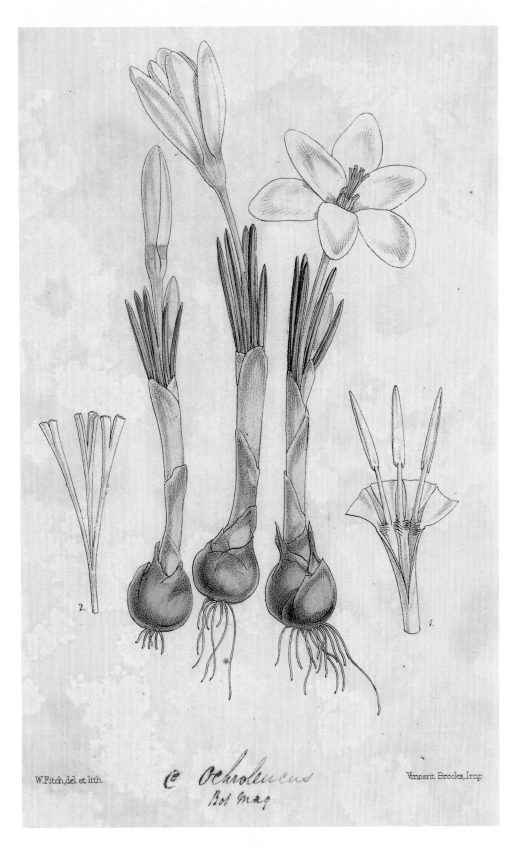

Crocus ochroleucus

Plate 91

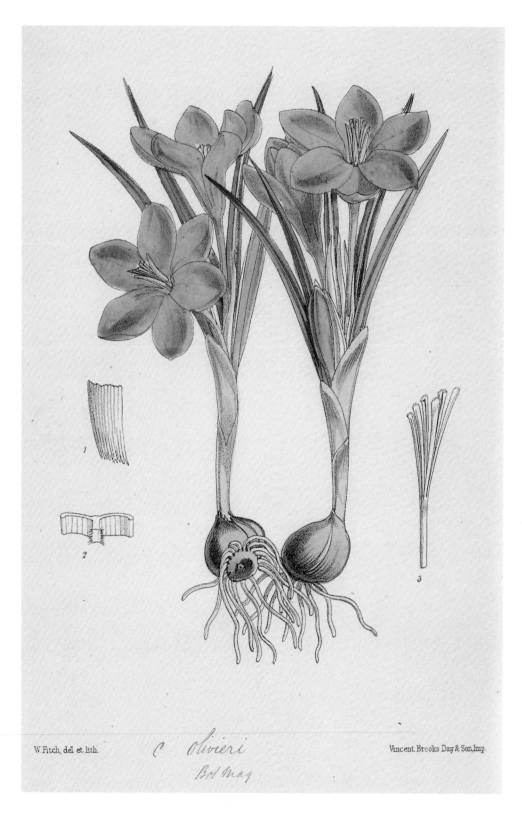

Crocus oliveri

Plate 92

Crocus salzmanni

Plate 93

Crocus tommasinianus. Another species that has been long grown in gardens. There are often several selections available for growers, all of them worthy garden plants. WARD

Crocus tournefortii. Not as widely grown as many of the Autumn flowering crocus, and like so many of the rarer species, best grown in containers. Such bulbs are prize possessions and worthy of extra attention and protection. WARD

***Crocus vernus* 'Flower Record'.** There are many selections of *Crocus vernus.* This is one of the darkest purples available, that does exceptionally well in containers as well as in the open ground. DE HERTOGH

***Crocus.* Mixed planting.** Almost any combination of species and cultivars or selections make a great early spring display. Harbinger of so many spring flowering bulbs yet to make their appearance. AUTHOR

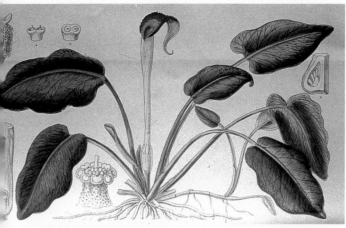

Cryptocoryne griffithii. These tropical plants need high temperatures in order to perform well. Rare in cultivation in temperate climates, and subjects for tropical greenhouses, I doubt if any are in cultivation in North America. J.H., by M. BIRDSAY

Curtonus paniculatus. There are few plants that will produce more flowers per stem. Couple this with their being quite long lasting, and you have plants worthy of being grown in many gardens. LOVELL

Cyanella alba. Despite the name 'alba', this species also has a yellow form, seen here in the wild. Even the white form has a hint of yellow in it. AUTHOR

Cyclamen africanum. While not a hardy species, and thus only to be grown outdoors in mild climates, this is perhaps the loveliest species, and should be included in any floral display, in greenhouses and conservatories. WARD

Cybistetes longifolia. This photograph was taken in the wild. The buds and young flowers are pink, changing to white with age. The dried flowerheads are frequently used in arrangements, and it is not uncommon to see heads of this species suspended from the ceilings of homes and used as mobiles. When dried, the heads are strong and rigid. DUNCAN

Cyclamen abchasicum. Native of the Caucasus, this is one of the species about which discussion rages. WARD

Cyclamen cilicium. Perhaps this graceful flower understates the beauty of many species which should be more popular in our gardens. WARD

Cyclamen graceum. While it is reported that, in the wild, the corms are found at considerable depth, this species grows well in containers, and the foliage is attractive. Needs a warm resting period in late summer, which it receives in its native habitat, Turkey. WARD

duced before 1804. The pear-shaped corm has a tunic with parallel fibers. The leaves are 8–9 in. long. This plant is quite dwarf, only 2–3 in. in height. The flowers are light violet on the outside, while the inside is lighter. Frequently they will have deep purple markings on the outside, covering almost the entire segments. The stigmata are reddish-orange. March–April-flowering.

C. olivieri. Native to Turkey, Greece, the Aegean Islands, Yugoslavia, Romania, and Bulgaria. This lovely and popular *Crocus* is a variable species. The variations are not only in the color of the flowers but also in the number of leaves and their degree of hairiness. Spring-flowering, the color ranges from pale lemon-yellow to deep orange, often with a dark brown or purplish-brown stain on the perianth tube sometimes extending into the lower parts of the perianth segments.

It is understandable that over the years a great number of the variants have been given species rank, but, now, after the work done by Mathew, many of these are placed into one of the three subspecies into which he divides this species.

The three subspecies are: *C. o.* ssp. *istanbulensis,* which has a tunic to the corm that is wholly and coarsely fibrous, that is weakly reticulated at the apex of the corm; *C. o.* ssp. *olivieri* (syn. *C. aucheri, C. suterianus*), which has the style divided into 6 branches and the exterior of the perianth segments usually unmarked; and *C. o.* ssp. *balansae* (syn. *C. balansae*), which has a style divided into 12–15 branches, and the exterior of the perianth segments usually are marked with brownish-purple stripes or a suffusion of this color.

C. o. ssp. *olivieri* has deep, shiny green foliage, which is about the same length as the height of the flowers at the time of flowering, 2–3 in., but lengthen to 8–10 in. From 1–4 flowers are produced, the most common color being orange-yellow, but sometimes paler and sometimes striped or suffused with brown on the perianth tube and on the outside of the outer perianth segments. The throat is yellow, as are the filaments and anthers. The style is divided into 6 slender, yellow or orange branches. Flowering in the wild in Feb.–April.

C. o. ssp. *balansae* is much the same as the above but the style is divided into 12–15 slender branches and the flowers usually are striped or suffused with purplish-brown or mahogany on the exterior of the outer perianth segments. It is a little earlier in flowering than the previous subspecies, coming into flower in Jan.–March.

C. o. ssp. *istanbulensis* differs from the previous subspecies by having upright-growing foliage, the flowers are a little larger, and the overall appearance is of a more robust plant. The period of flowering in the wild is shorter than the other species, it being in the month of March.

C. sieberi. Native to Crete and its subspecies to Greece, with one subspecies, *C. sieberi* ssp. *sublimis,* native to southern Yugoslavia, southern Bulgaria, and southern Albania. It is quite closely related to *C. veluchensis,* the habitats of the two overlapping, but is distinguished by having a yellow or golden throat. The form of both species is distinctive, being a tapering, rounded perianth tube which is most graceful. Four subspecies are recognized by Mathew.

C. sieberi ssp. *sieberi* (syn. *C. sieberi* var. *versicolor, C. sieberi* var. *heterochromus*). The corm has a fibrous tunic, produces from 4–7 leaves at the same time as the flowers are open and elongate to 8–10 in. Flowers are 2–3 in. in height, fragrant, with the interior white, usually stained purple on the exterior of the outer segments; the throat is deep yellow or orange. Filaments yellow, style divided into 3 yellow to orange-red branches, each expanded and either frilled or lobed. Spring-flowering, March or April in cultivation but much later in the wild, often as late as June in higher elevations.

A good garden plant, certain selections are offered in catalogs: 'Violet Queen,' a richly colored plant, and 'Bowles White,' a good white. A variant known previously as *C. sieberi* forma *tricolor* is a startling plant with a deep yellow throat then a band of white separating this color from the rich lilac-blue perianth segments, hence giving the impression of having three colors.

C. s. ssp. *atticus.* Native to Greece. This species has a corm that has a coarsely fibrous, reticulated tunic. The flowers are pale to deep lilac-blue or violet, with a yellow throat.

C. s. ssp. *sublimis* (syn. *C. maudii, C. athous*). This plant has from 2–8 leaves, lilac-blue flowers, often with darker tips to the segments, sometimes a white zone between this color and the pale to deep yellow throat.

C. s. ssp. nivalis. Differs from the above in having a tunic to the corm that is finely or sometimes coarsely reticulated. Produces from 3–6 leaves, with the flowers being lilac-blue and having a yellow throat. Flowering time in the wild from Feb.–June.

C. stellaris. This species has been long in cultivation and is unusual in that it has not been found in the wild. It is a sterile species and is thought to be a hybrid between *C. flavus* (syn. *C. aureus*) and *C. angustifolius* (syn. *C. susianus*). The tunic of the corm is coarsely reticulated. The leaves often are more than 10 in. long. It is a striking, low-growing plant, only 2–3 in. in height, with good orange flowers that are marked on the outside with dark mahogany lines that appear almost black. Feb.–March-flowering. Now regarded as a form of *C. flavus* (which see).

C. susianus (Cloth of Gold; syn. *C. angustifolius*). Native to Crimea, Caucasus; known before 1597. The corm is coarsely reticulated. The leaves reach a length of 10 in. This dwarf plant, only a couple of inches high, is a deep golden yellow tinged with mahogany on the outside. The stigma is orange-scarlet. An unusual feature is that the anthers are twice as long as the orange filaments. Early flowering in March. *C. s.* ssp. 'Minor'. A selection form the species, selected because the leaves are more upright and not quite so long. The flowers are deeper in color, as well, and the plant comes into flower later, height a little shorter.

C. tommasinianus. Native to Dalmatia; introduced before 1847. The nearly spherical corm has a finely reticulated tunic. The leaves are 8–9 in. long. The flowers reach a height of 6 in. and have a long, white throat and the petals are pale lavender on the inside, with the outside more silvery gray. March-flowering. This is one of the species that seeds freely and many forms and variations are known, all of which are early flowering. Among the best are:

'Barr's Purple'. Rich purple-lilac; large flowers.
'Ruby Giant'. Deep violet-purple; very free-flowering.
'Whitewall Purple'. Reddish-purple flowers.

C. vernus. This is the Wild Crocus of the Alps and the Pyrenees; introduced by 1765. The corm has a reticulated tunic. It can be found in large drifts with colors that vary from pure white to deep purple. Such is the variation that there are many cultivars offered that are of this parentage. There are some that surpass in beauty the species and they certainly have the vigor that ensures the gardener a good display the first year after planting. Should they appreciate their location, they will increase in quantity. Height ±3 in. Feb.–April-flowering.

The following are by no means all of the cultivars offered but they are among the finest:

White
'Joan of Arc'. One of the older cultivars and excellent.
'Peter Pan'. Flowers not so large but numerous.
'Snowstorm'. Ivory-white with a purple base.

Yellow

'Dutch Yellow' or 'Yellow Mammoth' or similar name, such as 'Large Yellow.'

Blue and Purple

'Early Perfection'. Violet-purple-blue, darker edges.
'Enchantress'. Light purple with silvery overhue.
'Flower Record'. Dark purple; good for containers.
'Grand Maitre'. An old but excellent lavender-violet.
'Purpureus Grandiflorus'. Very large, cup-shaped flowers; good for both pot culture and in the garden.
'Queen of the Blues'. Good-textured, long-lasting, large flowers.
'Remembrance'. Silvery purple; one of the earliest to flower.
'Vanguard'. Light grayish-blue on the outside.

Striped

'King of the Striped'. Violet with lighter stripes.
'Pickwick'. White and lilac; shorter flowering; exceptional plants.
'Striped Beauty'. Violet and white; good, rounded flower; lasts well.

CROCUS—Autumn Flowering

C. *aphyllus* syn. of C. *nudiflorus*.
C. *asturicus* syn. of C. *serotinus* ssp. *salzmannii*.
C. *asumaniae* Native to S. Turkey. Introduced 1979. White, occasionally with dark veins. Oct.–Nov. flowering.
C. *autranii* Native to U.S.S.R., Caucauss. Introduced 1893. Mid to deep violet with faint veining, base of segments white, throat white. Oct. flowering.
C. *banaticus* (syns. C. *byzantinus*, C. *iridiflorus*) Native to Romania, N. & W. Russia & Yugoslavia. Introduced 1831. Tube varies from white to deep purple, flowers lilac to purple. Anthers are yellow, stigma has lilac threadlike branches. 3–4 in. in height. Sept.–Oct.. flowering.
C. *biflorus* ssp. *melantherus* (syn. C. *melantherus*) Native to S. Greece. White ground, striped or speckeled with purple or gray on exterior, anthers blackish. Oct.–Nov. flowering.
C. *boryanus* = C. *boryi*.
C. *boryanus* var. *caspius* = C. *caspius*.
C. *boryanus* var. *cephalonensis* = C. *boryi*.
C. *boryanus* var. *graius* = C. *boryi*.
C. *boryi* (syn. C. *cretensis*, C. *ionicus*, C. *marathonisius*) Native to Western & Southern Greece & Crete. Introduced 1831. Flowers creamy white sometimes veined purple, throat yellow, anthers white, style orange-yellow & much dissected. 4 in. in height. Sept.–Dec. flowering.
C. *b.* var. *laevigatus* = C. *laevigatus*.
C. *b.* var. *marathoniseus* = C. *boryi*.
C. *b.* var. *orphanidis* = C. *tournefortii*.
C. *b.* var. *tournefortii* = C. *tournefortii*.
C. *byzantinus* syn. of C. *banaticus*.
C. *cambessedesanus* = C. *cambessedesii*.
C. *cambessedesii* Native to Balearic Islands. Introduced in 1831. White or creamy to deep lilac, striped purple, throat white, yellow anthers. 3 in. in height. Sept.–Dec. flowering.
C. *campestris* syn. of C. *pallasii* ssp. *pallasii*.
C. *cancellatus* var. *cilicicus* = C. *cancellatus* ssp. *cancellatus*.
C. *c. f. damascenus* = C. *cancellatus* ssp. *damascenus*.
C. *c.* ssp. *damascenus* (syn. C. *edulis*) Native from Turkey to Iran & Israel. Introduced 1845. Corm tunic very coarsely fibrous. Oct.–Dec. flowering.
C. *c.* var. *hermoneus* = C. *hermoneus*.
C. *c.* var. *kotschianus* syn. of C. *cancellatus* ssp. *cancellatus*.
C. *c.* ssp. *lycius* Native to S. Turkey. Segments white or very pale lilac, style equals or shorter than anthers. November flowering.
C. *c.* var. *margaritaceus* = C. *cancellatus* ssp. *mazziaricus*.
C. *c.* var. *mazziaricus* = C. *cancellatus* ssp. *mazziaricus*.
C. *c.* ssp. *mazziaricus* (syn. C. *schimperi*, C. *spruneri*) Native to Yugoslavia, Greece, S.W. Turkey. Introduced 1845. Segments white or mid to deep lilac, style much longer than the anthers. Sept.–Nov. flowering.
C. *c.* var. *naupliensis* = C. *cancellatus* ssp. *mazziaricus*.
C. *c.* ssp. *pamphylicus* see C. *cancellatus* ssp. *cancellatus*.
C. *c.* var. *persicus* = C. *cancellatus* ssp. *damascenus*.
C. *cartwrightianus* (syn. C. *sativus* var. *cartwrightianus*) Native to Greece, Attica, Crete, Cyclades. Introduced 1843. Pale to deep lilac-purple or white, strongly veined darker, throat white or lilac, anthers yellow. Oct.–Dec. flowering.
C. *c.* var. *leucadensis* = C. *hadriaticus*.
C. *cashmirianus* = C. *sativus*.
C. *caspius* Native to Iran, USSR on shores of Caspian Sea. Introduced 1838. Flowers white to pale lilac, throat, anthers & style yellow. 3–4 in. in height. Sept.–Nov. flowering.
C. *cilicicus* syn. of C. *cancellatus* ssp. *cancellatus*.
C. *clusianus* = C. *serotinus* ssp. *clusii*.
C. *clusii* syn. of C. *serotinus* ssp. *clusii*.
C. *constantinopolitanus* = C. *pulchellus*.
C. *cretensis* syn. of C. *boryi*.
C. *damascenus* = C. *cancellatus* ssp. *damascenus*.
C. *dispathaceus* = C. *pallasii* ssp. *dispathaceus*.
C. *edulis* syn. of C. *cancellatus* ssp. *damascenus*.
C. *elwesii* syn. of C. *pallasii* ssp. *pallasii*.
C. *fimbriatus* syn. of C. *nudiflorus*.
C. *fontenayi* syn. of C. *laevigatus*.
C. *gilanicus* Native to W. Iran. Introduced 1975. White with purple veins and may be marked with pale pinkish-lavender at tips of segments. October flowering.
C. *graecus* = C. *cartwrightianus*.

C. granatensis syn. of *C. serotinus* ssp. *salzmanii*.
C. hadriaticus var. *chrysobelonicus* = *C. hadriaticus*.
C. h. var. *saundersianus* = *C. hadriaticus* A yellow-throated variant.
C. haussknechtii syn. of *C. pallasii* ssp. *haussknechtii*.
C. herbertianus = *C. banaticus*.
C. hermoneus Native to Mt. Herman, Central Israel, W. Jordan. Introduced in 1881. White to lilac-blue with darker veining at base of segments. Sept.–Oct. flowering.
C. h. ssp. *hermoneus* see *C. hermoneus*.
C. h. ssp. *palaestinus* see *C. hermoneus*.
C. hybernus syn. of *C. pallasii* ssp. *pallasii*.
C. ionicus syn. of *C. boryi*.
C. iridiflorus syn. of *C. banaticus*.
C. karduchorum Native to S.E. Turkey. Introduced 1881. Pale to mid lilac-blue, finely veined darker, fading in color to white at base of segments, throat white, anthers creamy white, style white. Sept.–Oct. flowering.
C. karsianus syn. of *C. kotschyanus* ssp. *suworowianus*.
C. kotschyanus ssp. *cappadocicus* Native to Turkey. Introduced 1980. Like ssp. *kotschyanus* but corm asymmetrical, lying on its side; throat of perianth glabrous. Sept. flowering.
C. k. ssp. *hakkariensis* Native to S.E. Turkey. Introduced 1980. Flowers lilac but with faint yellow blotches in throat. Sept.–Oct. flowering.
C. k. ssp. *suworowianus* (syn. *C. karsianus*, *C. suworowianus*, *C. vallicola* var. *lilacinus*, *C. vallicola* var. *suworowianus*, *C. zohrabii*) Native to U.S.S.R. Introduced 1980. Creamy white sometimes with violet veins, throat pale yellow. Sept.–Oct. flowering.
C. laevigatus var. *albiflorus* = White variant of *C. laevigatus*.
C. l. var. *fontenayi* = *C. laevigatus*.
C. l. var. *major* = variant of *C. laevigatus*.
C. lazicus syn. of *C. scharojanii*.
C. libanoticus syn. of *C. pallasii* ssp. *pallasii*.
C. longiflorus var. *melitensis* = *C. longiflorus*.
C. l. var. *wilhelmii* = pale variant of *C. longiflorus*.
C. macrobolbos syn. of *C. pallasii* ssp. *turcicus*.
C. marathonisius syn. of *C. boryi*.
C. mazziaricus = *C. cancellatus* ssp. *mazziaricus*.
C. melantherus syn. of *C. biflorus* ssp. *melantherus*.
C. moabiticus Native to N. Jordan. Introduced 1912. White with purplish veins, throat white with pruple, style reddish-orange, anthers yellow. Nov.–Dec. flowering.
C. odorus var. *longiflorus* = *C. longiflorus*.
C. o. var. *melitensis* = *C. longiflorus*.
C. officinalis = *C. sativus*.
C. o. var. *sativus* = *C. sativus*.
C. o. silvestris = *C. vernus*.
C. olbanus syn. of *C. pallasii* ssp. *pallasii*.
C. oreocreticus Native to Crete. Introduced 1949. Mid-lilac to purple with darker veining, exterior pale silvery to buff, throat lilac, style red, anthers yellow. Oct.–Dec. flowering.
C. orphanidis syn. of *C. tournefortii*.
C. orsinii = *C. sativus*.
C. palaestinus = *C. hermoneus*.
C. pallasii ssp. *pallasii* (syn. *C. campestris*, *C. elwesii*, *C. hybernus*, *C. libanoticus*, *C. olbanus*, *C. thiebautii*) Native to S. Yugoslavia, E. Romania, Bulgaria, Crimea, Lebanon, Israel & Turkey. Introduced 1817. Pale pinkish-lilac to deep lilac-blue, veined darker; throat white or lilac, anthers yellow, style red. Oct.–Nov. flowering.
C. p. ssp. *dispathaceus* Native to S. Turkey. Introduced 1924. Flowers deep reddish-purple or mauve-pink; style yellow or pale orange. Sept.–Nov. flowering.
C. p. ssp. *haussknechtii* (syn. *C. haussknechtii*) Native to W. Iran, N.E. Iraq, S. Jordan. Introduced 1977. Much like ssp. *pallasii*, style very deep red. Oct.–Nov. flowering.
C. p. ssp. *turcicus* (syn. *C. macrobolbos*) Native to S.E. Turkey, Syria, Lebanon. Introduced 1977. Very similar to ssp. *pallasii*. Oct.–Nov. flowering.
C. peloponnesiacus syn. of *C. hadriaticus*.
C. pendulus = *C. sativus*.
C. pholegandrius syn. of *C. tournefortii*.
C. pylarum = *C. cancellatus* ssp. *cancellatus*.
C. pyrenaeus syn. of *C. nudiflorus*.
C. robertianus Native to Greece. Introduced 1973. Pale to deep lilac-blue, throat pale yellow or whitish, style deep orange, anthers yellow. Oct.–Nov. flowering.
C. salzmannianus = *C. serotinus* ssp. *salzmannii*.
C. salzmannii syn. of *C. serotinus* ssp. *salzmannii*.
C. s. var. *coloratus* = *C. serotinus* ssp. *salzmannii*.
C. s. var. *erectophyllus* = *C. serotinus* ssp. *salzmanii*.
C. s. var. *pallidus* = *C. serotinus* ssp. *salzmanii*.
C. sativus var. *cartwrightianus* syn. of *C. cartwrightianus*.
C. s. var. *cashmirianus* = *C. sativus*.
C. s. var. *elwesii* = *C. pallasii* ssp. *pallasii*.
C. s. var. *haussknechtii* = *C. pallasii* ssp. *haussknechtii*.
C. s. var. *officinalis* = *C. sativus*.
C. s. var. *pallasii* = *C. pallasii* ssp. *pallasii*.
C. s. var. *thomasii* = *C. thomasii*.
C. scharojanii (syn. *C. lazicus*) Native to U.S.S.R., N.E. Turkey. Introduced 1868. Deep yellow, often shading to orange at tips, throat yellow, style orange, anthers creamy or yellow. 4–5 in. in height. July–Oct. flowering.

C. s. var. *flavus* = *C. scharojanii* × *C. vallicola*.
C. schimperi syn. of *C. cancellatus* ssp. *mazziaricus*.
C. serotinus var. *salzmannii* = *C. serotinus* ssp. *salzmannii*.
C. setifolius = *C. sativus*.
C. speciosus var. *aitchisonii* variant of *C. speciosus* ssp. *speciosus*.
C. s. var. *albus* white variant of *C. speciosus* ssp. *speciosus*.
C. s. var. *caucasicus* = *C. speciosus* ssp. *speciosus*.
C. s. var. *globosus* variant of *C. speciosus* ssp. *speciosus*.
C. s. ssp. *ilgazensis* Native to N. Turkey. Much like *C. speciosus* ssp. *speciosus* with smaller flowers. Sept.–Oct. flowering.
C. s. var. *laxior* = *C. speciosus* ssp. *speciosus*.
C. s. var. *transylvanicus* = *C. speciosus* ssp. *speciosus*.
C. s. ssp. *xantholaimos* Native to N. Turkey. Like *C. speciosus* ssp. *speciosus* but throat of perianth deep yellow. Sept.–Oct. flowering.
C. spruneri syn. of *C. cancellatus* ssp. *mazziaricus*.
C. suworowianus syn. of *C. kotschyanus* ssp. *suworowianus*.
C. tauricus = *C. speciosus*.
C. thiebautii syn. of *C. pallasii* ssp. *pallasii*.
C. thomasianus = *C. thomasii*.
C. thomasii (syn. *C. visianicus*) Native to S. Italy, W. Yugoslavia. Introduced 1826. Pale to deep lilac, sometimes veined or stained with violet, throat pale yellow, style bright red, anthers yellow. Oct.–Nov. flowering.
C. t. var. *princeps* = variant of *C. thomasii*.
C. t. var. *laevis* = variant of *C. thomasii*.
C. tingitanus = *C. serotinus* ssp. *salzmannii*.
C. tournefortianus = *C. tournefortii*.
C. tournefortii (syn. *C. orphanidis, C. pholegandrius*) Native to Greek Islands. Introduced 1831. Lilac sometimes veined darker, throat yellow, anthers white, style orange, much divided. 3–4 in. in height. Sept.–Dec. flowering.
C. vallicola Native to N.E. Turkey, S. Caucasus, & U.S.S.R. Introduced 1845. Creamy-white, often with faint purple veins, throat white & hairy, anthers white, style cream to deep yellow. 4–5 in. in height. Aug.–Oct. flowering.
C. v. var. *lilacinus* syn. of *C. kotschyanus* ssp. *suworowianus*.
C. v. var. *suworowianus* syn. of *C. kotschyanus* ssp. *suworowianus*.
C. v. var. *zohrabii* = *C. kotschyanus* ssp. *suworowianus*.
C. veneris Native to Cyprus. Introduced 1842. White, usually with violet stripe or feathering; throat, anthers and style yellow. Nov. (–Jan.) flowering.
C. visianicus syn. of *C. thomasii*.
C. vitellinus Native from S. Turkey to Lebanon. Introduced 1828. Bright orange, sometimes with bronze markings, anthers & throat yellow, style yellow to orange, much dissected, fragrant. 3 in. in height. Nov.–Jan. flowering.
C. wilhelmii = *C. longiflorus*.
C. zohrabii syn. of *C. kotschyanus* ssp. *suworowianus*.
C. zonatus syn. of *C. kotschyanus* ssp. *kotschyanus*.
C. zorhabii = *C. kotschyanus* ssp. *suworowianus*.

CROCUS—Winter–Spring Flowering

C. abantensis Native to N.W. Turkey. Introduced 1975. Mid to deep blue. April flowering.
C. acutiflorus syn. of *C. vernus* ssp. *albiflorus*.
C. adamii syn. of *C. biflorus* ssp. *adamii*.
C. adanensis Native to S. Turkey. Introduced 1975. Pale lilac-blue with white or creamy base of segments. March flowering.
C. aerius (syn. *C. biliottii*) Native to N. Turkey. Introduced 1847. Deep blue to lilac purple, veined even darker, throat yellow. 4 in. in height. April–May flowering.
C. a. var. *cyprius* syn. of *C. cyprius*.
C. a. var. *pestalozzae* syn. of *C. pestalozzae*.
C. a. var. *pulchricolor* syn. of *C. biflorus* ssp. *pulchricolor*.
C. a. var. *major* variant of *C. biflorus?* ssp. *pulchricolor*.
C. a. var. *stauricus* = *C. aerius*.
C. alatavicus var. *albus* White flowers except for a yellow throat.
C. a. var. *ochroleucus* Like type but 3 outer perianth segments dull yellowish-white on exterior, fading to white at edges.
C. a. var. *porphyreus* Like type but flowers stippled outside with bright claret-purple.
C. a. var. *typicus* Blackish-violet stippling on the exterior, ground white or yellow.
C. albiflorus syn. of *C. vernus* ssp. *albiflorus*.
C. a. ssp. *heuffelianus* variant of *C. vernus* ssp. *vernus*.
C. a. ssp. *neapolitanus* = *C. vernus* ssp. *vernus*.
C. albinfrons = *C. vernus* ssp. *albiflorus*.
C. aleppicus (syn. *C. gaillardotii*) Native to Syria, Lebanon, N.E. Israel, Jordan. Introduced 1873. White, usually with blue, violet, or purple stripes, veining or suffusion on exterior. Dec.–Feb. flowering.
C. alexandri syn. of *C. biflorus* ssp. *alexandri*.
C. algeriensis syn. of *C. nevadensis*.
C. almehensis Native to N.E. Iran. Introduced 1973. Orange-yellow often marked with bronze on exterior. April–May flowering.
C. annulatus = *C. biflorus?* ssp. *adamii*.
C. a. var. *adamicus* = *C. biflorus* ssp. *adamii*.
C. a. var. *albus* = *C. biflorus* ssp. *weldenii*.
C. a. var. *biflorus* see *C. biflorus* ssp. *biflorus*.
C. a. var. *caerulescens* = *C. biflorus* ssp. *biflorus*.
C. a. var. *chrysanthus* syn. of *C. chrysanthus*.
C. a. var. *estriatus* = *C. biflorus* ssp. *biflorus*.
C. a. var. *graecus* = *C. biflorus* ssp. *melantherus*.

C. a. var. *nibigena* = ?*C. biflorus* ssp. *nubigena*.
C. a. var. *purpurascens* = *C. biflorus* ssp. *weldenii*.
C. a. var. *pusillus* = *C. biflorus* ssp. *biflorus*.
C. a. var. *tauricus* = *C. biflorus* ssp. *adamii*.
C. antalyensis Native to S. Turkey near Antalya. Introduced 1969. Flowers deep lilac with a yellow throat inside, biscuit-colored with purple veining outside. 4–5 in. in height. Spring flowering.
C. appendiculatus = variant of *C. vernus* ssp. *albiflorus*.
C. argenteus see *C. biflorus* ssp. *biflorus*.
C. artvinensis = *C. biflorus* ssp. *artvinensis*.
C. athous see *C. sieberi* ssp. *sublimis*.
C. atlanticus syn. of *C. nevadensis*.
C. atticus syn. of *C. sieberi* ssp. *atticus*.
C. aucheri syn. of *C. olivieri* ssp. *olivieri*.
C. aureus see *C. flavus* ssp. *flavus*.
C. a. var. *lacteus* = *C. lagenaeflorus* ssp. *flavus*.
C. babiogorensis = *C. vernus* ssp. *vernus*.
C. balansae syn. of *C. olivieri* ssp. *balansae*.
C. balcanensis syn. of *C. veluchensis*.
C. baytopiorum Native to S.W. Turkey. Introduced 1974. Pale blue with delicate veining. Feb.–April flowering.
C. biflorus var. *adamicus* = *C. biflorus* ssp. *adamii*.
C. b. ssp. *adamii* (syn. *C. adamii, C. biflorus* var. *violaceus, C. biflorus* var. *taurica, C. geghartii, C. tauricus*) Native to S. Yugoslavia, Bulgaria, N.W. Turkey, Crimea, Caucasus, N. Iran. Introduced 1831. Flowers lilac, sometimes white, exterior with purple stripes; leaves are prominently ribbed in grooves on underside. Feb.–April flowering.
C. b. var. *adamii* = *C. biflorus* ssp. *adamii*.
C. b. var. *alexandri* = *C. biflorus* ssp. *alexandri*.
C. b. var. *argenteus* = *C. biflorus* ssp. *biflorus*.
C. b. var. *barrii* = variant of *C. biflorus* ssp. *biflorus*.
C. b. ssp. *crewei* (syn. *C. crewei, C. tauri* var. *melanthorus*) Native to W. Turkey. Introduced 1875. Leaves gray-green with prominent ribs on underside. March–April flowering.
C. b. var. *estriatus* see *C. biflorus* ssp. *biflorus*.
C. b. ssp. *isauricus* (syn. *C. isauricus*) Native to S. Turkey. Introduced 1924. Lilac or white, striped or speckled on the exterior with purple or gray-purple. March–April flowering.
C. b. var. *leichtlinii* syn. of *C. leichtlinii*.
C. b. var. *nubigenus* = *C. biflorus* ssp. *nubigena*.
C. b. ssp. *nubigena* (syn. *C. nubigena*) Native to W. & S.W. Turkey, Aegean Islands. Introduced 1843. Bluish-lilac, tube with purple lines on outside. Perianth segments marked with 3 feathered purple lines. Orange yellow at base of inner segments. Anthers blackish-maroon. Feb.–March flowering.
C. b. var. *parkinsonii* = *C. biflorus* ssp. *biflorus*.
C. b. var. *pestalozzae* = *C. pestallozzae*.
C. b. var. *praecox* = variant of *C. biflorus* ssp. *biflorus*.
C. b. ssp. *pseudonubigena* Native to S.E. Turkey. Flowers scented of cloves, white to lilac with 3 dark stripes on exterior sometimes merging to form one broad stripe. Feb.–April flowering.
C. b. ssp. *pulchricolor* (syn. *C. aerius* var. *pulchricolor*) Native to N.W. Turkey. Introduced 1845. Rich blue-violet often darker at the base of segments. Unstriped on exterior and with a large deep yellow zone in the center. April–June flowering.
C. b. ssp. *punctatus* Native to S. Turkey. Small lilac-blue flower, flecked on exterior with violet so that the outer segments appear slightly darker. Feb.–March flowering.
C. b. var. *pusillus* = variant of *C. biflorus* ssp. *biflorus*.
C. b. ssp. *stridii* (syn. *C. stridii*) Native to N.E. Greece to Thessalonika. Introduced 1980. Like *C. b.* ssp. *nubigena* but leaves are not veined, anthers blackish-purple. Jan.–Feb. flowering.
C. b. ssp. *tauri* (syn. *C. roopiae*) Native to Asia Minor. Introduced 1881. Pale to mid lilac without dark stripes but may be finely veined or feathered darker. Throat pale yellow. April–May flowering.
C. b. var. *taurica* syn. of *C. biflorus* ssp. *adamii*.
C. b. subvar. *tenorianus* = variant of *C. biflorus* ssp. *biflorus*.
C. b. var. *violaceus* syn. of *C. biflorus* ssp. *adamii*.
C. b. var. *visianicus* = *C. biflorus* ssp. *weldenii*.
C. b. var. *weldenii* = *C. biflorus* ssp. *weldenii*.
C. biliottii syn. of *C. aerius*.
C. boulosii Native to Libya. Introduced 1968. White with a grayish-blue stain at base on back of outer perianth segments. Throat yellow or white. January flowering.
C. caeruleus syn. of *C. vernus* ssp. *albiflorus*.
C. candidus (syn. *C. kirkii*) Native to Northwestern Turkey. Introduced 1812. White, outer segments speckled or veined, sometimes feathered purple, anthers & styles yellow-orange, deep yellow throat. 3–4 in. in height. Feb.–March flowering.
C. c. var. *landerianus* = *C. flavus* ssp. *flavus*.
C. c. mountainii = *C. olivieri* ssp. *olivieri*.
C. c. var. *subflavus* = *C. olivieri* ssp. *olivieri*.
C. capensis = *Romulea rosea*.
C. carpetanus (syn. *C. lusitanicus*) Native to Spain, Portugal. Introduced 1842. Lilac to white shaded grayish, bluish or pinkish on outside. Anthers yellow, throat white or pale yellow. 4 in. in height. March–May flowering.
C. chrysanthus var. *albidus* = albino of *C. biflorus* ssp. *pulchricolor*.
C. c. var. *bicolor* = *C. olivieri* ssp. *olivieri*.
C. c. f. *brunellus* = variant of *C. chrysanthus*.
C. c. var. *caerulescens* = *C. biflorus* ssp. *pulchricolor*.
C. c. var. *citrinus* = pallid variant of *C. chrysanthus*.
C. c. var. *fuscolineatus* = variant of *C. chrysanthus*.

C. c. var. *fuscotinctus* = variant of *C. chrysanthus*.
C. c. f. lilacinus = variant of *C. chrysanthus*.
C. circumscissus see *C. biflorus* ssp. *biflorus*.
C. crestensis = *C. versicolor*.
C. crewei = *C. biflorus* ssp. *crewei*.
'*C. crewei*' of gardens = *C. biflorus* ssp. *melantherus*.
C. cristensis = *C. versicolor*.
C. croceus syn. of *C. chrysanthus*.
C. × *cultorum* = garden variants of *C. vernus*.
C. cvijicii Native to S. Yugoslavia, N. Greece, E. Albania. Introduced 1926. Pale to deep yellow, cream or white, throat yellow or white & hairy, anthers orange-yellow. May–June flowering.
C. cyprius Native to Cyprus. Introduced 1865. Soft lavender with deep purple blotches, throat orange-yellow, filaments scarlet, anthers yellow. 2–3 in. in height. Jan.–April flowering.
C. dalmaticus Native to S.W.Yugoslavia, N. Albania. Introduced 1842. Lilac or buff exterior or uniform grayish-lavender, usually veined or feathered purple at base, throat & anthers yellow, style orange. 3 in. in height. Feb.–April flowering.
C. d. f. albiflorus = variant of *C. dalmaticus*.
C. d. var. *niveus* = *C. biflorus* ssp. *weldenii*.
C. danfordiae Native to Central & Southern Turkey. Introduced 1881. Varies from sulphur yellow, rarely pale blue or whitish, sometimes brown or gray stippling on the outside, anthers yellow, style orange. 3 in. in height. Feb.–March flowering.
C. discolor = *C. vernus* ssp. *vernus*.
C. exiguus = *C. vernus* ssp. *vernus*.
C. fleischerianus = *C. fleischeri*.
C. floribundus = *C. flavus* ssp. *flavus*.
C. fragrans = *C. versicolor*.
C. × *fritschii* = *C. vernus* ssp. *vernus* × *C. vernus* ssp. *albiflorus*. Native to Austria.
C. fulvus = variant of *C. angustifolius*.
C. fussianus = *C. biflorus* ssp. *biflorus*.
C. gaillardotii syn. of *C. aleppicus*.
C. gargaricus (syn. *C. thirkeanus*) Native to Western Turkey, Mt. Gargarus & on Ulu Dag. Introduced 1841. Brilliant yellow or orange, anthers yellow, orange style. 3–4 in. in height. April–May flowering.
C. g. var. *citrinus* Lemon-yellow or pale forms of *C. gargaricus*.
C. g. var. *panchrysus* Deep yellow or orange colored forms of *C. gargaricus*.
C. geghartii syn. of *C. biflorus* ssp. *adamii*.
C. grandiflorus = *C. vernus* ssp. *vernus*.
C. graveolens (closely related to *C. vitellinus*) Native to S. Turkey, Syria, Lebanon, & N. Israel. Introduced 1882. Yellow, often striped or feathered brown. 2–3 in. in height. Feb.–April flowering.
C. hartmannianus Native to Cyprus. Introduced 1914. Soft lavender with purple markings, throat orange-yellow. 2–3 in. in height. Jan.–Feb. flowering.
C. heuffelianus syn. of *C. vernus* ssp. *vernus*.
C. h. ssp. *tommasinianus* = *C. tommasinianus*.
C. heuffelii = *C. vernus* ssp. *vernus*.
C. hittiticus syn. of *C. reticulatus* ssp. *hittiticus*.
C. hyemalis (syn. *C. hyemalis* var. *foxii*) Native from Israel to S. Syria. Introduced 1859. White, feathered purple, throat deep orange-yellow. 3 in. in height. Nov.–Dec. flowering.
C. h. var. *foxii* syn. of *C. hyemalis*.
C. imperati ssp. *imperati* (syn. *C. incurvus*, *C. neapolitanus*) Native to W. Italy. Introduced 1826. Color varies—usually bright lilac-purple inside with a yellowish, silvery or buff on outside, striped with purple lines & feathering, throat & anthers yellow, style orange. 3–4 in. in height. Jan.–March flowering.
C. i. var. *albiflos* Pure white with yellowish flush on outside.
C. i. var. *albus* White variant of *C. imperati* ssp. *imperati*.
C. i. var. *montanus* White variant of *C. imperati* ssp. *imperati*.
C. i. var. *nucerensis* White variant of *C. imperati* ssp. *imperati*.
C. i. var. *reidii* Variant of *C. imperati* ssp. *imperati* with flowers rose-pink with white spots inside at base of segments & buff with 3 brown lines outside.
C. i. ssp. *suaveolens* (syn. *C. suaveolens*) Native to W. Italy. Introduced 1977. Sweetly scented, somewhat paler in color than *C. imperati* ssp. *imperati*. Jan.–Feb. flowering.
C. i. var. *sardoa* = *C. minimus*.
C. imperationius = *C. imperati* ssp. *imperati*.
C. incurvus syn. of *C. imperati* ssp. *imperati*.
C. indivisus = *C. vernus*.
C. insularis syn. of *C. corsicus* & *C. minimus*.
C. i. var. *major* = *C. corsicus*.
C. i. var. *minimus* = *C. minimus*.
C. isauricus syn. of *C. biflorus* ssp. *isauricus*.
C. italicus = *C. biflorus* ssp. *biflorus*.
C. kirkii syn. of *C. candidus*.
C. korolkowii Native to Afghanistan, Pakistan, U.S.S.R. Introduced 1880. Deep yellow, outer segments marked or feathered brown or dark bronze, anthers yellow. 4 in. in height. Feb.–March flowering.
C. k. var. *vinosus* Richly purple marked variant of *C. korolkowii*.
C. kosaninii Native to Yugoslavia. Introduced 1976. Lilac-blue to mid violet, often with external dark stain or 3 dark stripes on lower ½ of outer segments, throat deep yellow, filaments & anthers yellow. March flowering.
C. lacteus syn. of *C. flavus* ssp. *flavus*.
C. l. var. *concolor* see *C. flavus*.
C. l. var. *penicillatus* see *C. flavus*.

C. lagenaeflorus see *C. flavus*.
C. l. var. *albidus* = variant of *C. flavus* ssp. *flavus*.
C. l. var. *aureus* see *C. flavus*.
C. l. var. *candidus* = *C. candidus*.
C. l. var. *haemicus* = *C. flavus* ssp. *flavus*.
C. l. var. *isabellinus* = variant of *C. flavus* ssp. *flavus*.
C. l. var. *lacteus* see *C. flavus*.
C. l. var. *landerianus* = *C. flavus* ssp. *flavus*.
C. l. var. *luteus* = *C. flavus* ssp. *flavus*.
C. l. var. *olivierianus* = *C. olivieri* ssp. *olivieri*.
C. l. var. *stellaris* see *C. stellaris*.
C. l. var. *striatellus* = variant of *C. flavus* ssp. *flavus*.
C. l. var. *sulphureus* see *C. flavus*.
C. l. var. *syriacus* = *C. graveolens*.
C. latifolius = *C. vernus* ssp. *vernus*.
C. leichtlinii Native to S.E. Turkey. Introduced 1924. Pale blue, lilac-blue, grayish slatey-blue, throat deep yellow, style yellow to orange. March–April flowering.
C. lineatus = *C. biflorus* ssp. *biflorus*.
C. lusitanicus syn. of *C. carpetanus*.
C. luteus see *C. flavus* ssp. *flavus*.
C. maesiacus see *C. flavus* ssp. *flavus*.
C. magontanus = *C. cambessedesii*.
C. malyi Native to W. Yugoslavia. Introduced 1871. Segments white, throat & anthers yellow, style orange. 4 in. in height. March–April flowering.
C. marcetii syn. of *C. nevadensis*.
C. maudii see *C. sieberi* ssp. *sublimis*.
C. meridionalis = *C. versicolor*.
C. mesiacus = *C. flavus* ssp. *flavus*.
C. michelsonii Native to Iran, USSR, S. Central Asia. Introduced 1932. White, very blue suffusion on external segments, style whitish or pale yellow, anthers yellow, rare. 4 in. in height. March-April flowering.
C. micranthus syn. of *C. reticulatus* ssp. *reticulatus*.
C. minimus f. *cambessedesii* = *C. cambessedesii*.
C. m. ssp. *cambessedesii* = *C. cambessedesii*.
C. m. var. *corsicus* = *C. corsicus*.
C. m. var. *italicus* = *C. imperati* ssp. *imperati* & *C. i.* ssp. *suaveolens*.
C. m. var. *major* = *C. fleischeri*.
C. m. var. *sardous* = variant of *C. minimus*.
C. moesiacus = *C. flavus* ssp. *flavus*.
C. montenegrinus variant of *C. vernus* ssp. *albiflorus*.
C. mouradii see *C. flavus* ssp. *dissectus*.
C. multifidus syn. of *C. nudiflorus*.
C. napolitanus syn. of *C. vernus* ssp. *vernus*.
C. nervifolius syn. of *C. vernus* ssp. *vernus*.
C. nevadensis (syn. *C. algeriensis, C. atlanticus, C. marcetii*) Native to Spain, Algeria, Morocco. Introduced 1871. Cream, white, or grayish lilac with darker veining, anthers yellow, throat whitish or pale yellow-green. 3–4 in. in height. Feb.–April flowering.
C. nubigena syn. of *C. biflorus* ssp. *nubigena*.
C. obovatus probably a cultivar of *C. vernus* ssp. *vernus*.
C. olivieri f. *balcanicus* = *C. olivieri* ssp. *olivieri*.
C. orbelicus syn. of *C. veluchensis*.
C. pallidus see *C. biflorus* ssp. *weldenii*.
C. parkinsonii = *C. biflorus* ssp. *biflorus*.
C. pelistericus Native to Macedonia, in S. Yugoslavia & N. Greece. Introduced 1976. Deep violet, finely veined darker, throat white, anthers yellow, style whitish or orange-yellow. May–June flowering.
C. penicillatus = *C. lacteus* var. *penicillatus* (see *C. flavus*)
C. pestalozzae Native to N.W. Turkey. Introduced 1853. White or blue-lilac, throat & anthers yellow, filaments with black spots. 1–2 in. in height. Jan.–March flowering.
C. p. var. *caeruleus* Blue flowered form of *C. pestalozzae*.
C. pictus variant of *C. vernus*.
C. ponticus = *C. kotschyanus* ssp. *kotschyanus*.
C. praecox see *C. biflorus* ssp. *biflorus*.
C. pulchricolor = *C. biflorus* ssp. *pulchricolor*.
C. purpureus syn. of *C. vernus* ssp. *vernus*.
C. pusillus see *C. biflorus* ssp. *biflorus*.
C. pygmaeus variant of *C. vernus* ssp. *albiflorus*.
C. recurvus = *C. imperati* ssp. *imperati*.
C. reflexus = *C. angustifolius*.
C. reineggeri = *C. vernus*.
C. reinwardtii = *C. versicolor*.
C. reticulatus f. *adsperus* = *C. reticulatus* ssp. *reticulatus*.
C. r. var. *albicans* = *C. reticulatus* ssp. *reticulatus*.
C. r. var. *ancyrensis* = *C. ancyrensis*.
C. r. var. *aureus* = *C. angustifolius*.
C. r. var. *auritextus* = *C. angustifolius*.
C. r. var. *dalmaticus* = *C. dalmaticus*.

C. r. etruscus = *C. etruscus*.
C. r. 'Dobrogensis' = *C. reticulatus* ssp. *reticulatus*.
C. r. ssp. *hittiticus* (syn. *C. hittiticus*) Native to S. Turkey. Introduced 1975. White or lilac & striped on the exterior, frequently speckled outside, anthers blackish or dark purple. Feb.–April flowering.
C. r. f. leptanthus = *C. reticulatus* ssp. *reticulatus*.
C. r. var. *micranthus* = *C. reticulatus* ssp. *reticulatus*.
C. r. f. pluriflorus = *C. reticulatus* ssp. *reticulatus*.
C. r. var. *rectilimbus* = *C. angustifolius*.
C. r. var. *reflexus* = *C. angustifolius*.
C. r. ssp. *reticulatus* (syn. *C. micranthus, C. variegatus, C. vittatus*) Native from N.E. Italy, Yugoslavia, Hungary, Bulgaria, Turkey & S.W. U.S.S.R. Introduced 1805. White or lilac, strongly striped purple on the outside, throat yellow or white, style in 3 branches yellow to deep orange, yellow anthers & pale yellow or white filaments. 3–4 in. in height. March–May flowering.
C. r. var. *variegatus* = *C. reticulatus* ssp. *reticulatus*.
C. revolutus = *C. angustifolius*.
C. roegnerianus = *C. angustifolius*.
C. roopiae syn. of *C. biflorus* ssp. *tauri*.
C. sativus var. *vernus* = *C. vernus* ssp. *vernus*.
C. scardicus Native to S. Yugoslavia, E. Albania. Introduced 1926. Yellow-orange, usually purple inside & out at base of segments, yellow anthers & filaments, style orange, throat yellow or white. 2–3 in. in height. May–June flowering.
C. scepusiensis syn. of *C. vernus* ssp. *vernus*.
C. serbicus = *C. tommasinianus*.
C. sibthorpianus var. *latifolius* = *C. sieberi* ssp. *sieberi*.
C. s. var. *pulchricolor* = *C. biflorus* ssp. *pulchricolor*.
C. s. var. *stauricus* = *C. aerius*.
C. siculus variant of *C. vernus* ssp. *albiflorus*.
C. sieberi var. *atticus* = *C. sieberi* ssp. *atticus*.
C. s. var. *atticus f. nivalis* = *C. sieberi* ssp. *nivalis*.
C. s. var. *heterochromus* syn. of *C. sieberi* ssp. *sieberi*.
C. s. var. *orbelicus* = *C. veluchensis*.
C. s. var. *sieberi* = *C. sieberi* ssp. *sieberi*.
C. s. var. *tenuifolius* = *C. veluchensis*.
C. s. var. *veluchensis* = *C. veluchensis*.
C. s. var. *versicolor* syn. of *C. sieberi* ssp. *sieberi*.
C. sieberianus = *C. sieberi* ssp. *sieberi*.
C. sieheanus Native to Turkey. Introduced 1939. Tube orange or purple, segments orange, throat yellow. 3–4 in. in height. April–May flowering.
C. skorpilii syn. of *C. chrysanthus*.
C. smyrnensis = *C. fleischeri*.
C. stridii syn. of *C. biflorus* ssp. *stridii*.
C. suaveolens syn. of *C. imperati* ssp. *suaveolens*.
C. sublimis = *C. sieberi* ssp. *sublimis*.
C. sulphureus see *C. flavus* ssp. *flavus*.
C. s. var. *concolor* see *C. flavus* ssp. *flavus*.
C. s. var. *striatellus* see *C. flavus* ssp. *flavus*.
C. s. var. *striatus* see *C. flavus* ssp. *flavus*.
C. susianus see *C. angustifolius*.
C. s. var. *ancyrensis* = *C. ancyrensis*.
C. s. var. *fulvus* = variant of *C. angustifolius*.
C. s. var. *immaculatus* = *C. ancyrensis*.
C. suterianus syn. of *C. olivieri* ssp. *olivieri*.
C. tauri = *C. biflorus* ssp. *tauri*.
C. t. var. *melanthorus* syn. of *C. biflorus* ssp. *crewei*.
C. tauricus syn. of *C. biflorus* ssp. *adamii*.
C. thessalus = *C. sieberi* ssp. *sublimis*.
C. thirkeanus syn. of *C. gargaricus*.
C. tommasinianus var. *pictus* variant of *C. tommasinianus*.
C. t. var. *roseus* variant of *C. tommasinianus*.
C. tomoricus syn. of *C. veluchensis*.
C. uniflorus = *C. vernus* ssp. *vernus*.
C. variegatus syn. of *C. reticulatus* ssp. *reticulatus*.
C. v. var. *micranthus* = *C. reticulatus* ssp. *reticulatus*.
C. veluchensis (syn. *C. balcanensis, C. orbelicus, C. tomoricus*) Native to Yugoslavia, Bulgaria, Greece & Albania. Introduced 1845. Color varies from pale silvery-lilac to a deep purple; white, hairy throat; anthers yellow; style orange. 3–4 in. in height. April–June flowering.
C. v. orbelicus = *C. veluchensis*.
C. vernus var. *albiflorus* = *C. vernus* ssp. *albiflorus*.
C. v. ssp. *albiflorus* (syn. *C. albiflorus, C. caeruleus, C. montenegrinus, C. siculus, C. vilmae*) Native to Albania, Austria, Czechoslovakia, Sicily, Italy, Yugoslavia, Germany, Switzerland, France, Spain. Introduced 1906. Often white but may be purple or striped. March–July flowering.
C. v. var. *communis* = *C. vernus* ssp. *albiflorus*.
C. v. var. *heuffelianus* variant of *C. vernus* ssp. *vernus*.
C. v. var. *leucorhyncus* variant of *C. vernus* ssp. *vernus*.
C. v. var. *leucostigma* variant of *C. vernus* ssp. *vernus*.
C. v. var. *neapolitanus* = *C. vernus* ssp. *vernus*.
C. v. var. *nivigena* = *C. vernus* ssp. *vernus*.
C. v. pictus variant of *C. vernus* ssp. *vernus*.

C. v. var. *siculus* = *C. vernus* ssp. *vernus*.
C. v. var. *tommasinianus* = *C. tommasinianus*.
C. v. var. *typicus* = *C. vernus* ssp. *vernus*.
C. versicolor Native to French Maritime Alps, S.E. France to N.W. Italy, Monaco. Introduced 1808. White to lilac, with strong purple veining, throat pale yellow to white, yellow or white filaments, anthers yellow, style orange or yellow. 5–6 in. in height. Feb.–April flowering.
C. v. var. *dalmaticus* = *C. thomasii*.
C. v. ssp. *marcetii* var. *marcetii* = variant of *C. nevadensis*.
C. v. ssp. *marcetii* var. *aragonensis* variant of *C. nevadensis*.
C. v. var. *picturatus* variant of *C. versicolor*.
C. v. var. *purpureus* variant of *C. versicolor*.
C. vilmae syn. of *C. vernus* ssp. *albiflorus*.
C. violaceus = *C. biflorus* ssp. *adamii*.
C. vitellinus Native from S. Turkey to Lebanon. Introduced 1828. Bright orange, sometimes with bronze markings, anthers & throat yellow, style yellow to orange, much dissected, fragrant. 3 in. in height. March.–April. flowering in cold climates.
C. vitellinus var. *balansae* = *C. olivieri* ssp. *balansae*.
C. weldenii see *C. biflorus* ssp. *weldenii*.

CRYPTOCORYNE — ARACEAE

The name is derived from the Greek *kryptos* ("hidden"), *koryne* ("club"). The spadix is completely hidden inside the spathe. While reportedly a genus with some 40 species of rhizomatous plants, they are all from the Tropics; many of them being water plants often with the foliage underwater, or bog plants requiring heat and humidity in order to grow. They are at home in tropical Asia and Malaya and it is doubtful that any are in cultivation, except in botanic gardens, and there only rarely.

There is great variation in leaf shape, from cordate to elliptic, from lanceolate to linear. The spadices are very slender, mostly brownish-purple, and often twisted and folded. They are not very attractive. *C. wendtii* perhaps has value as a tropical aquarium plant as it grows both underwater and with foliage resting on the water surface.

CULTURE Require high temperatures, in the 65°–70°F at night. Also need bog conditions. Rhizomes should be anchored to the bottom of pools, spaced some 18–24 in. apart, and just nestled into the soil. Bright to moderate light, as one would find in tropical conditions, is needed.

PROPAGATION Division of the rootstocks is the best.

PESTS & DISEASES No special problems.

USES In pools in tropical houses and possibly aquariums or bog areas in greenhouses where jungle appearance is required.

The flowering time of these tropical plants is sporadic, often in flower the year round with heaviest flowering in the summer.

SPECIES

C. affinis (Water Trumpet). Native to the Malay Peninsula. Strong rhizomatous roots; lanceolate leaf blades, dark green leaves with paler veins. The spathe a twisting, corkscrew shape, pale green outside with purple lines and inside very dark purple. In flower in native habitat much of the year. Height to 30 in., often less.

C. ciliata. Native to Bengal and India in monsoon area; introduced 1823. This is a tropical aquatic plant in flower in summer with sporadic flowers in other months reaching almost 20 in. in height, often less. The leaves have unequal sides, are stalked and lance-shaped, usually underwater; they are inclined to be yellowish-green, with the depressed midrib a darker green. The spathe is long, narrow, and an unpleasant green in color, although sometimes it is reddish-brown with lighter colored interior, often yellow. Reportedly fragrant.

C. griffithii (syn. *C. purpurea*). Native to Malaya. The small leaves are 2½ in. long and almost as wide. The lower part of the spathe tube is cylindrical, ½ in. in length; the upper about 1 in. The flower is rose colored and a good, deep red within. In flower much of the year.

C. petchii. Native to Sri Lanka. A tropical, robust plant forming a rosette of leaves 6–10 in. long and 1 in. or more wide. The spathe tube is tubular, up to 6 in. long, then opening, with the spathe beyond the opening tapering and pointed and extending some 4–6 in. Olive-green to brown in color with the inside a good, rich, reddish or purplish brown. Height of plant is ±20 in.

C. purpurea. See *C. griffithii*.

C. retrospiralis. Native to Malaya and India in tropical regions. Leaves are up to 12 in. long, narrow, rarely over ½ in. wide. The spathe tube is twisted 8 in. long and dark green-brownish, with the lower part narrow and the upper part tapering to a point, almost hairlike. Height 18–24 in.

C. spiralis (syn. *Arum spirale*). Native to India; introduced 1816. The leaves are about 6 in. long and up to ½ in. wide. The spathe tube is about an inch in length, purple, more reddish within, and much twisted in the lower part, with a straight tail above the opening.

C. wendtii. Native to Thailand. Sturdy, branched leaves are spear-shaped, 3–5 in. long, and just over an inch wide. The spathe is 4 in. long, brown-purplish-green with a twisted mouth and tail above it. Height of plant 12–18 in.

CURTONUS — IRIDACEAE

From the Greek *kurtos* ("bent") and *onos* ("axis"), an apt name for this monotypic genus as the flower spikes are bent in quite a curious manner. The plant is native to southern Africa, being found in Natal and the Transvaal. *C. paniculatus* is the only species in the genus. The corms are sub-globose. In appearance it is not unlike *Crocosmia*. The leaves, sword-shaped with distinct veining, clothe the lower part of the flower stalk. The basal leaves are held in a fan-shape, iris-like in appearance, and up to 3 in. wide.

The orange-red flowers are carried on an open branch stem that grows to a height of 48 in. The unusual feature is that the branches of the flower spike all turn toward one side, twisting around the stem as if blown by the wind. There are up to 100 individual flowers on each tube, with short bracts and petals about 2 in. long, the upper petals being much longer than the lower, but all are lobed. Flowers begin to open in August and are carried for a long time; the ends of the panicles being in blossom while the first ones to open will have long passed. The seeds will have formed before the last flowers are open.

CULTURE These plants are not fully hardy and need protection in colder climates but are at home in a sunny, well-drained border. They do need moisture in the early part of the year but can withstand drier conditions in the latter part of the year.

PROPAGATION The corms increase quite rapidly and, while it is not necessary to lift them each year, this is the best way to propagate the plants. They can be lifted after the foliage begins to ripen. The young corm is then separated from the parent and planted back. They also can be lifted and divided in early spring, prior to the plants starting into growth.

Seed should be harvested in the fall and held for sowing in the spring. The seeds germinate well in a light, sandy soil mix, and, by the end of the summer, the new corms should be grown on in a nursery row for another season before planting out.

PESTS & DISEASES No special problems.

USES Good in a narrow border against a wall where they can be warm and fairly dry in late summer/early fall, providing late-summer color to the open, sunny border. Due to its not being well known, it is more of a curiosity, but it is deserving of a place in the garden. It also can be grown easily in containers.

SPECIES

C. paniculatus. Described above.

CYANASTRUM — TECOPHILAEACEAE

Derived from Greek *kyanos* ("blue") and *astron* ("star"). This small genus is found in the tropical parts of Africa, from Nigeria to Mozambique. For the most part the plants inhabit damp areas and thus for successful culture require the warmth and humidity of a greenhouse. The rootstocks are corms, the flowers blue or white, the foliage thin, and the individual leaves are wide. On certain species only few leaves are produced and in some cases they are solitary, which are oblong-lanceolate to heart-shaped.

The information about this genus is rather scant, some authorities place it in Cyanastraceae others in Tecophilaeaceae. The ovaries are inferior, the flowers carried in a raceme or are in panicles, the inner and outer perianth segments are similar, there are 6 stamens, 6 petals.

CULTURE Require warmth and humidity and shade. Night temperatures should not drop below 55°F. The need for humidity requires that these plants are suitable for outdoor culture only in tropical to subtropical regions.

The corms should be planted 1–2 in. deep and spaced some 6–8 in. apart. The soil should be high in organic matter and should be kept moist with only a slightly less amount of moisture provided during the resting stage. Little or no feeding is required but if given should be as soon as growth commences. When being grown in containers feedings with an organic fertilizer are beneficial. This should start when growth commences and continue for 4–5 months. In the wild the plants are summer-flowering; their performance in the Northern Hemisphere would most likely be the same, but information in this regard is lacking.

PROPAGATION Cormels are produced and can be separated from the parent corms. Seed should be sown in warm temperatures in spring, the night temperatures in the 60°–65°F range. The soil mixture should contain a lot of humus and the seeds should be barely covered. After germination they should be transplanted into individual pots and as soon as of size planted out into their permanent positions.

PESTS & DISEASES No special problems.

USES For greenhouse culture only except in subtropical to tropical regions where they should be planted in the shade.

SPECIES

C. cordifolium. Native to Nigeria, Cameroun, and Gabon. Reaches a height of some 10–12 in. The flower spike held in a raceme which is up to 6 in. in length. Only one rather wide, glossy leaf is produced at the base and this can reach 6 in. in width, cordate with the length varying between 3–8 in. Leaf stalk of a few to up to 8 in. in length. A few smaller leaves reportedly are produced on the flowering stem. The flowers are just less than an inch in diameter, blue or purplish, fragrant, and are subtended by a large green to purplish-green bract. The corms are flattened and heart-shaped. Flowering time Feb.–May.

C. hostifolium. Native to Tanzania and Mozambique. Requires shade and moisture. The foliage is not fully developed when the white flowers are produced. The flowers are smaller than the previous species, but the height of the spike is a little more, around 6 in. Several leaves are produced from the corm, they are tapered at the base and 3–5 in. in width. Flowering Nov.–Jan.

C. johnstonii. Native to Tanzania and Zambia. Foliage similar to *C. cordifolium*. Flowers bright blue, in a raceme of 4–6 in., each flower about an inch in diameter. Flowering Oct.–Dec. in the wild.

CYANELLA — TECOPHILAEACEAE

The name is derived from the diminutive of Greek *kyanos* ("blue"), as some of the flowers are blue to violet-blue. A genus of some 7 species native to South Africa, mostly from the Cape Province, where they flower from September to November, this being springtime. They are quite common, often being found in rocky areas, with the corms quite deep in the ground. The foliage varies with the species; in some plants the basal leaves form a rosette on the ground with the flower spike being leafless, in others the foliage is quite grasslike, basal but erect with the leaves being almost cylindrical and tufted.

The flowers are attractive in most species, open to give a flat flower often as much as 1 in. in diameter. Many in this genus produce as many as 20 flowers per spike but others produce fewer, and in the case of *C. alba* the flowers are solitary. The perianth segments are free to their base, the lower 3 pointing downward, the other upward. In most species the segments are of the same size; in *C. alba* the inner are more rounded, the outer pointed. The stamens are unusual, 5 of them are short and of more or less equal size and length while the sixth is larger and longer and opposite the lower petal. In some cases the flowers are produced as the foliage starts to wither.

The corm is used by the South African natives to make a paste for the treatment of boils, carbuncles, and abscesses and is also reportedly used for the treatment of leg wounds. The Xhosa people use *C. lutea* to treat infertility. It is used as a bedding on which the patient sleeps and after, it is used in a ritual that takes place at the river. The common name given to some species is Five Fingers.

CULTURE The plants are not hardy, able to withstand only very light frosts. They like the sun and well-drained soil. As they are not tall-growing they should be grown in the front of the border or in the rock garden. The corms should be planted 3–4 in. deep, spaced some 6–8 in. apart, and given moisture throughout the winter and early spring, with less water being given in late summer so the corms can enjoy a resting period.

In climates where there is more than just an occasional or light frost, the plants should be grown under protection, either in a cool greenhouse or in a frame where frost can be excluded. In such climates they can be grown in containers and taken outside when all danger of frost has passed. Some feeding can be given in early spring when growth commences. As these plants are natives of rather poor soils, overfeeding should be avoided.

PROPAGATION Lifting the corms and separating the cormels, produced after the foliage has died down, is a good way to increase the stock, with the cormels planted in shallow drills and grown on to flowering size. Seed should be sown in the spring in a sandy soil mix, the seed being barely covered. They should be kept barely moist and allowed to become dry when the foliage dies. The small plants in the second season can be planted in individual containers and should be of size to plant out at the end of the second season of growth.

PESTS & DISEASES No special problems.

USES Rock garden or in the front of sunny borders, container plants in cooler climates or in cool greenhouse borders.

SPECIES

C. alba. Native to the Clanwilliam area of Cape Province on the Atlantic Ocean side of South Africa. The numerous leaves are grasslike, up to 5 in. in length, and grow in a basal tuft. The flowering stem reaches a height of some 10 in. and the flowers, which are solitary, are produced in early spring. The flowers are over an inch in diameter, the outer petals are pointed and above them the inner petals, which are rounded at their tips and a little cup-shaped, remain a little more erect, the outer petals being almost flat. The color is white with a hint of light yellow. The stamens are grouped in the center of the flower and the orange pollen is quite visible. The sixth stamen lies above the lowest petal and is whiter in color than the petals. This is a lovely plant and well worth growing, but is not in cultivation except in rare instances.

C. capensis. See *C. hyacinthoides*.

C. hyacinthoides (Lady's Hand; syn. *C. capensis*). Native to the Cape Peninsula of South Africa; introduced 1768. It reaches a height of 12 in., often a little more, the flowering being much branched and the flowers carried up and away from the stem on strong pedicels, the lower part sheathed by a small, narrow, pointed leaf. The slightly fragrant flowers, often as many as 8–10 per branch, are flat when open, the stamens being shown to advantage in July and August (Oct.–April in the wild). The color is slightly variable, pale lilac or bluish-violet with a small blotch of carmine often found at the base of the petals.

The foliage is lanceolate with wavy margins; the number of leaves varies between 5–8 and the length can be as much as 8–12 in. mostly basal, but sometimes 1 or 2 of smaller size found on the flowering stem. Some authorities regard *C. hyacinthoides* as a separate species, but it would appear that *C. capensis* and *C. hyacinthoides* are the same. *C. hyacinthoides* is the correct name due to it having historical preference. *C. pentheri* is now regarded as being in this species, (see below), the differences at best are slight.

C. lutea (Five Fingers). Found in the Outeniqua and Tsitsikamma Mountains in the southwestern part of Cape Province, South Africa, and along the coast in grassy areas and forest margins; introduced 1788. Flowers from September to December in the wild. The height of the flowering stem, which is sometimes branched, is 12 in. in height. The basal leaves are narrow with wavy edges and there are a few leaves on the stem which are narrower and more sharply pointed. The flowers are yellow, about an inch in diameter; on the outside of the outer tepals are noticeable brown markings caused by the veins. The smaller stamens are bunched in the center of the flower and the sixth and longer being quite noticeable. The outward-facing flowers are carried away from the stem by strong pedicels that arch up and then away. Some authorities regard this as a species that is synonymous with *C. odoratissima*, as this species is very similar. The color of this species is described as deep rose fading to pale blue then yellow. It would seem that in other respects the plants are similar and perhaps *C. odoratissima* is a form of *C. lutea*.

C. odoratissima. Regarded by some as synonymous with *C. lutea*. See above.

C. orchidiformis. Native to the Clanwilliam and Namaqualand area of the Cape Province of South Africa. Found growing in rocky areas. The corms produce a rosette of leaves that lie flat on the ground or nearly so, with usually 5 leaves with a length of some 5 in. and about 1 in. in width, green with a slightly grayish tinge. Inside this rosette 1 or 2 smaller, erect leaves sometimes are produced. The flowering stem is leafless reaching a height of 12–16 in., the lower part green and the upper purplish in color. The flowers, as many as 16 per stem, are carried on strong pedicels that arch away from the stem. Flowers are very light mauve with a darker ring of carmine-mauve in the center; quite flat when fully opened, with the petals quite widely spaced.

C. pentheri. Found in the same areas as *C. orchidiformis*. It is not as an attractive species, the tepals being narrow and somewhat reflexed, bluish-mauve in color, and carried on a spike that is sometimes, but rarely, branched. The height is some 10–12 in. The foliage is narrow, sometimes crisped, erect, and the leaves arise from a basal tuft. The flowers are about 1 in. in diameter before the tepals recurve and the stamens are quite prominent in the center of the flower with the longer stamen being placed down and away on top of the lower tepal. It is possibly the least attractive species of the genus. Now regarded as *C. hyacinthoides*.

C. uniflora. In the literature, this name is used for a shorter-growing form of *C. alba*. *C. uniflora* is an invalid name.

COMMENTS A genus that deserves more attention and might well prove to be a useful one for the warmer climates.

CYBISTETES — AMARYLLIDACEAE

The name is from the Greek *kubistetes* ("a tumbler"), an allusion to the umbel of flower, which becomes detached and is blown around by the wind. This is a monotypic genus, the only species is *Cybistetes longifolia* (Malagas Lily, St. Joseph's Lily; syn. *Amaryllis longifolia, A. falcata, Ammocharis falcata, Crinum falcatum*). Native to the southern part of South Africa, around the Cape and the wine-growing area of Stellenbosch and Paarl; introduced 1774. Note that in the genus *Amaryllis* the flowers are regular while in *Cybistetes* they are zygomorphic, therefore different from *Ammocharis*. Also, stamens declinate and pedicels lengthen.

The bulbs are large, up to 8 in. in diameter. The strap-shaped leaves, which vary in length from a few inches to more than a foot, lie prostrate on the ground. The are produced after the flowers have passed in the fall and persist throughout the winter and into spring, after which they die back.

The fragrant, sweet-scented flowers appear in mid- to late summer, June/July. They are carried on fleshy stalks, 9–12 in. high. The rounded heads may have 20 or more blossoms, each consisting of 6 petals, up to 3 in. in length and of equal width. In bud, the flowers have a pink flush, which is confined to the central rib when open. Once open, the petals curl back at the tips and open wide, almost back to where the flower joins the stalk. The inside of the petals and the reverse, except for the rib, are white. The stigma is longer than the stamens, which are held loosely in the opened flower. The pedicels lengthen as the flower fades and the fruit is carried in a pendant position. The seeds are fleshy.

CULTURE The bulbs are best planted at the beginning of summer, after the leaves have died down. Set them so the top is level with the soil. They are not fussy as to the type of soil, as long as it drains well, and they like to be planted in the sun. Moisture is required as soon as the leaves appear and throughout the fall and spring. A light feeding should be given after the flowers begin to fade, with occasional feedings during the winter and spring, using a liquid fertilizer if the soil is poor, otherwise no feeding necessary.

The bulbs become dormant during early summer and must be protected from frost during this resting period. The shelter of a greenhouse is needed as light is necessary to nurture the leaves. In warmer climates, with little or no frost, the bulbs can be left in the ground for many years.

PROPAGATION Seed can be sown as soon as ripe or saved until the following spring and sown in a sandy mixture. The seedlings should be left in the container for one whole year and then grown on for a season before planting out into flowering locations.

Offsets are not produced freely but, if they are present, they can be separated from the parent bulbs when the foliage has died down. Care must be taken not to break the rather fleshy roots. The offsets will flower in the second season if of fair size when planted.

PESTS & DISEASES Not subject to any particular pests or diseases, but, when the flowers appear, aphids may attack and should be treated if found.

USES A good plant to have in a sunny shrub border where additional color is needed during the summer. They should be planted toward the front of the border due to their low height, as well as in a location where the fragrance can be enjoyed. It is a good container plant, especially where winters are cold but it must be located where it can receive light throughout the winter.

SPECIES

C. longifolia. Described above.

CYCLAMEN — PRIMULACEAE

From the Greek *kyklos* ("circular"), so named because of the peculiar habit in some species of the spiral twisting of the flower stalk after the flowers have faded. *Cyclamen* are native to the lands around the Mediterranean, yet some species are very hardy and will grow in even very cold climates. For a comparatively small genus, about 20 species, they occupy an important position in horticulture. Florist varieties provide excellent house plants, while others are valuable in the woodland, rock garden, and those shady corners of the garden where a low-growing plant with decorative leaves is wanted.

There are many different strains, even among the same species, that are produced from the corms. One example is *C. persicum*, from which the florist hybrids are raised. Fragrance, or lack of same, is another peculiarity to be noted. While identical in all respects, even to the placement of the genes on the chromosomes, some species are fragrant, others are not. Efforts have been made to find out why, but, so far, they have been unsuccessful.

Botanists are continually naming new species on the basis of these slight differences. In my opinion they are merely variants of the principal species and should be listed as such.

The leaves of all species are radicle and broad, with a stalk. The flowers are always solitary, one per stalk. They vary in size but each has 5 petals, always reflexed, sometimes curled, and the lower part of each forms into a short tube. Flowering time varies with the different species. With careful selection it is possible to have them in flower throughout the year. Planted in a good location, they can establish themselves and spread over a large area.

The fruit is a capsule, often drawn down to soil level by the aforementioned twisting of the stalk.

CULTURE Florist *Cyclamen* require far different growing techniques than those grown in the garden. The portion dealing with florist varieties follows the general cultural guide below.

Plant in early fall as soon as corms are available. Loose soil, high in decayed or partially decayed organic matter, such as found in woodland areas, is ideal. Do not plant too deeply. Cover corms with soil or loose organic matter, such as leaves, to depth of ½ in., except for *C. persicum*, which should have corm breaking the surface. For most part, *Cyclamen* prefer shade, with the exception of *C. cilicium*, which enjoys sun, and *C. graecum*, which not only enjoys sun but also needs it during summer to perform well. All species appreciate moisture during autumn, winter, and into spring. While not fussy about drainage, it is best never to plant in an area where soil is not free-draining.

C. graecum should be allowed to become dry for a period of 6–8 weeks in midsummer. Other species like drier resting periods but not being baked by the sun. *C. coum* and its forms prefer to be kept slightly moist during summer.

Cyclamen grow very well in preferred locations but excessive debris over top of planting should be removed. This can swamp them. If in an area where some increase in organic matter does not occur each year, top dress with a little sifted leaf mold or peat moss. A little bonemeal added to the soil and used as a top dressing each spring will keep *Cyclamen* happy.

Once established, leave the plants undisturbed. Do not lift. Corms will increase in size but will not produce offsets so overcrowding is not a problem. If seedlings are numerous (*Cyclamen* set seed freely) some of them may have to be removed.

°Florist Varieties Most important aspects are to keep them growing at all times and give adequate light to avoid their becoming leggy. Temperatures during growing period must be moderate. Test of well-grown plant is to turn pot upside down; leaves should be able to support weight of plant and pot.

Seed is sown in August or September, in flats. Soil mix, which is 3 parts good topsoil, 1 part peat moss or leaf mold, and 1 part coarse sand, should be free-draining but kept moist. Night temperatures should be in the 55°F. range, perhaps a little higher during the day but not excessively so. Light is not needed for germination but must be given 6–8 hrs. per day as soon as germination begins, which should take about a month. Fresh seeds germinate well; older seeds take longer and are erratic. About 120 days after sowing, seedlings will be ready for transplanting.

Transplant into richer soil mixture. Space plants 2–3 in. apart. Top of little corm should be just breaking soil. Objective is to keep them growing slowly, so maintain 50°F at night but drop to 45°F as soon as plants are growing away. When of fairly good size and leaves are touching each other, move to 3–4 in. pots. This will be in about 60 days.

Some 60–90 days later, plants are ready to be moved to 6–8 in. flowering pots. As it is now late spring/early summer, temperatures will be warm. Move plants to north-facing frame where light intensity is high but without direct sun. Do not place where plants will tend to grow toward light as will occur if frames are deep. Temperatures should be kept as low as possible. Advantageous to syringe daily with cold water in morning and late afternoon.

Feed with liquid fertilizer toward end of summer. Bring into warmth of greenhouse in September. Night temperatures around 60°F; higher temperatures during day will not cause problem, but give shade and do not expose to bright sunlight and high temperatures. Ample ventilation must be given.

Flowering will take place about 15–16 months after sowing of seeds. After flowering, plants must still be kept moist. When leaves begin to die down, reduce water further. Allow plants to rest. Repot in July and start into growth again. Place in cool area so foliage does not become too leggy.

Where little or no winter frost is present, plants can be set out in shady spot. Do not plant too deeply; corm should be just breaking surface. Flowering will occur following spring and for a number of years thereafter.

Some growers prefer keeping plants in flats until final potting. This practice has disadvantages as roots dislike being disturbed and the larger the plants, the more disturbance. This is offset by considerable savings in labor costs.

PESTS & DISEASES Several fungal diseases attack *Cyclamen*. *Cercosporella* causes leaf spot. *Erwinia carotovora* causes soft rot of the corm. *Botrytis cinerea* is a gray mold that can be serious if good air circulation is not provided; if attack occurs, improve air circulation, keep water off foliage, and remove all diseased material. All can be controlled by use of fungicide, following directions on label. Another problem, most probably a virus, causes dwarfing and distortion. Rogue out and discard any plants so affected.

Mites also attack leaves but can be controlled by use of an insecticide/miticide. Larvae of vine weevil feed on roots and corms; control by sterilizing soil.

USES Excellent for woodland areas and shady spots. Must be allowed to stay in place and increase, which species will do by self-sowing. Plant in bold drifts for best effect. Lovely plants for small containers to be brought indoors for color, especially in colder areas where plants will perform well in frames protected from severe cold and snow. Home gardener can help production of seed by hand-pollinating flowers.

SPECIES & CULTIVARS

C. **africanum.** Native to North Africa. Will not stand frost so grow only where winters are not severe and where overhead vegetation will give protection from possible frost. Lives among pine trees and scrub in native habitats and is useful in such areas where few other plants will grow. Not unlike *C. neapolitanum* in appearance. Roots produced from all parts of large corm. Bright green, large leaves, 6 in. or more wide, with undulated margin; appear after flowers and last through winter into spring. Flower stems up to 6 in. in height; twist as seeds ripen. Rose-colored flowers, varying from light to dark, with deep carmine patch around base. Yellow anthers with a few violet lines at back. Flowering early fall, Sept. or Oct.

C. **alpinum.** Frequently listed in catalogs. Similar to *C. cilicium* (which see) but flowers are bright carmine or white with blotch of violet-purple at base of corolla lobes. Allied to *C. coum. C. trochopteranthum* would appear to be a geographic variation.

C. × **atkinsii.** Regarded by some authorities as an interspecific hybrid between *C. coum* × *C. vernum* as it carries characteristics of both; however, appears to be only a form of *C. coum*. There seems to be good reason to believe that this is but a particularly good form of *C. coum* as the seed is fertile. Seldom more than a few inches high. Rounded leaves, deep green with silvery markings on upper surface; produced before flowers. Flowers vary in color, mostly in pale rose shades but often listed in catalogs as crimson, which color can be found in lines or blotches on lower part of petals. Flowering early spring. Forms include *album*, white with red base; and *roseum*, good pink.

C. **balearicum.** Native to Balearic Islands. Needs warm climate where frost unlikely; will not be harmed by light frost if given protection of high shade. Unusual in that the leaf and flower stems creep underground before emerging. Small plant. Heart-shaped leaves very heavily silvered on upper side. Flowers are white with a rose colored throat, quite wide in the mouth, often more than ½ in.; petals seldom more than an inch long; 2–4 in. high; slightly fragrant. Flowering spring, Feb./March.

C. **cilicium.** Native to southern Asia Minor; introduced 1872. Pretty species, sometimes with slight fragrance. Leaves generally heart-shaped but with lobes close to stalk and sometimes overlapping, slightly toothed along edge. Flowers on 4–5-in. high stems; light pink with prominent deeper rose at base of petals; tube quite short; petals curl back gracefully. One form, smaller in overall size, sometimes listed as *C. c. alpinum* or *C. alpinum*. Flowering in autumn, winter, or early spring.

C. **coum.** Much diversity of opinion regarding this species. Many variants, including: *C. orbiculatum, C. ibericum, C. pseudibericum, C. hiemale, C. × atkinsii, C. parviflorum,* and *C. vernum;* and subspecies *C.* ssp. *caucasicum,* comprised of *C. elegans, C. adsharicum, C. caucasicum, C. abchasicum,* and *C. circassicum*. Not surprising there is much confusion, but my opinion is that these are merely geographic variants. Although most are frequently listed in catalogs, I feel that *C. abchasicum,* from the Caucusus; *C. adsharicum,* from along the U.S.S.R.-Iran border; *C. caucasicum,* from the Caspian region of Talysch; *C. circassicum,* from the western Caucusus; and *C. hiemale,* from Europe and parts of Asia, are the principal forms.

Thick, globose or flattish tuber, 1–1½ in. thick. Dark green leaves with dull crimson on undersides, occasionally few markings on upper surface, but not pronounced. Flowers variable in color, from pink to crimson, with deeper-colored spot at base; yellow anthers; on stems that reach a height of 4–5 in. Both leaves and flower stems creep for short distance underground before emerging. Flowers anytime between December and March. Very hardy; down to 0°F, provided snow cover present. Make excellent garden plants; will spread if left in favorable location. Forms listed as *album* (white) and *roseum* (pink) in catalogs.

C. **creticum.** Native to Crete. Regarded by some as synonymous with *C. repandum*, except that *C. creticum* has 2 extra chromosomes. Leaves ivy-shaped but sometimes more oval, toothed margin, good green with few markings on upper surface, dull red below. Flower colors variable—white, rose, and carmine forms have been recorded; petals form narrow tube and then extend backward to length 4 times that of tube. Flower stalks curl as seeds ripen. Spring-flowering, in March, but later where winters are very cold. Reaches 4–6 in. in height.

C. europaeum (syn. *C. littorale*, *C. purpurascens*). Found wild in northern Italy, in the Alps, and eastward into Yugoslavia. Has been long in cultivation. Seldom more than 4 in. in height. Rounded leaves, which may or may not have markings on surface. Flowers in shades of red, from light to dark, with more intense color at base of petals; white form also has been recorded. Owes much of its popularity to fragrance of flowers and to length of flowering—from late summer into fall. Best grown in some shade; should have moisture throughout year, with less in fall.

C. graecum. Native to Greece and western Turkey; introduced circa 1930. Corms often found at considerable depth in wild. Fleshy roots credited with being able to pull plant deeper into ground. Distinctive leaves are large and margin is rough to the touch; persist into winter; underneath surface red, then turns to green. Carmine flowers with added color at base; distinct swellings around mouth; not fragrant. Height to 3 in. Needs resting period with warmth in late summer. September-flowering. Many forms have been given rank such as *cypro-graecum*, *pseudo-graecum*. The distinction between these is minor at best.

C. hederifolium. See *C. neapolitanum*.

C. ibericum. See *C. coum*.

C. libanoticum. Native to Lebanon. One of the loveliest of species but not very hardy; grows well outdoors only where winters are mild and there is little frost. Leaves develop in earlier months; dark green with continuous white zone on upper surface and blotches of yellow-green, deep purple below; usually heart-shaped but may vary. Large, fragrant flowers, pure white to good rose with darker markings around mouth. Late winter/early spring-flowering, in Feb./March, sometimes earlier. Height 4 in.

C. neapolitanum (syn. *C. hederifolium*). Native to Europe, mainly in southern part from France to Greece, also naturalized in other parts. Distinctive tubers often more than 3 in. in diameter; produce roots from upper surface only; lower portion free of roots and somewhat rounded. When of good size will produce large number of flowers, more than 30 not uncommon. Leaves show great variation in shape and color, some round, others ivy- or lance-shaped, margins also vary from smooth to crinkled; persists from August throughout winter to May of following year; seem to last until seed pods ripen and about to uncurl from twisting flower stem. One of most free-flowering of species. Flowers vary from white to carmine, with auricles around mouth. White form often listed as *C. n. album*. Hardy even in colder areas, coming into flower in late summer (August), appearing before leaves, extending into fall, reaching peak in September with full complement of leaves. Spreads quite rapidly. Height 3–6 in. One of the finest of species and should be first grown by anyone considering starting collection of *Cyclamen*.

C. persicum. Despite name, native not to Iran (Persia) but to Cyprus, various Greek islands, Lebanon, and Tunisia; introduced 1731. Species from which popular house plants have been raised. Not hardy so should be considered for those areas where there is little or no frost. Prefer some shade and adequate amounts of moisture while in leaf and in flower. Rounded tubers root from the base. Plant tuber so top half is aboveground. One of tallest of species, with flowers held above foliage on stems 6 in. or more in height. Leaves varied in size and color, generally dark green zoned with lighter green markings, almost always heart-shaped. Fragrant flowers in wide color range—from white to all shades of rose to good dark colors. Petals are twisted. Flowering in March–April.

As noted, *C. persicum* is the parent of many florist varieties of hybrids. Among them:

'Afterglow'. Deep cherry-pink.
'Bonfire'. Scarlet-red.
'Grandia'. Snow-white; deeply frilled; 4 in. across; early flowering.
'Matador'. Intense scarlet; large; early flowering.
'Mauve Queen'. Purplish-mauve.
'Pink Pearl'. Deep salmon-pink.
'Rose Queen'. Deep salmon-pink.
'Salmon King'. Pale salmon-pink.
'Victoria'. White with rose-red eye and edging on petals; large-flowered.

Miniatures (5–6 in.)
'Fortissimo'. Bright scarlet-red.
'Largo'. Salmon.
'Piano'. White with cherry-red eye.

C. pseudibericum. Native to southern Turkey; introduced 1897. Hardiness of species questionable so plant outdoors only in areas where there is little or no frost. Set corms in protected locations. Leaves have serrated edge; often have yellowish-green mottling on surface with crimson undersurface. Large, fragrant flowers possibly darkest of all *Cyclamen*—dark purplish-carmine with even more intense color at base of petals which are about 1 in. long and ½ in. wide. Flowers carried above foliage. Feb.–March flowering. Possibly a variant of *C. coum*, which see.

C. repandum (Ivy-leaved Cyclamen). Native to Europe, from southern France eastward to Italy and the Greek islands. Leaves variable, mottled dark green, sometimes more, sometimes less but always reddish on underside; not very large. Flowers also variable, from rosy white to deep carmine with darker zone of color around mouth. Style longer than tube and quite prominent. Height 4–6 in. Flowering time March through early May, making it one of the last of the *Cyclamen* to bloom in spring/early summer. Needs protection from frost so plant in sheltered area. Variable species. Var. *rhodense* has darker color at mouth.

C. vernum. Very similar to *C. coum* and regarded as variant of it. Leaves, however, are more heart-shaped and appear before flowers. Flowers are carmine, but also variable, with yellow anthers. Grows wild in Bulgaria to eastern Turkey, in wooded areas, which it needs to perform well. Height 4–5 in.

CYCLAMEN

C. abchasicum see *C. coum*.
C. adsharicum see *C. coum*.
C. alphinum syn. of *C. trochopteranthum*.
C. caucasicum see *C. coum*.
C. circassicum see *C. coum*.
C. cyprium Native to Cyprus. Large white flowers with a red blotch at the base of petals. 2–3 in. in height. Autumn flowering.
C. elegans (see *C. coum*) Native to Caspian coast of Iran. Large pale pink flowers. Dec.–March flowering.
C. hiemale see *C. coum*.

C. littorale see *C. coum*.
C. mirabile Native to Western Asia Minor. Introduced 1906. Pink with red dots at base of petal. 2–3 in. in height. Sept.–Nov. flowering.
C. orbiculatum see *C. coum*.
C. parviflorum (see *C. coum*) Native to N. Turkey. Introduced 1946. Pale or deep rose-pink with purplish blotches. 2–3 in. in height. April–June flowering.
C. purpurascens see *C. coum*.
C. rohlfsianum Native to Libya, N. Africa. Pink or crimson. 4–6 in. in height. Sept.–Oct. flowering.
C. trochopteranthum (syn. *C. alpinum*) Native to S.W. Turkey. Introduced circa 1872. Carmine. 3–4 in. in height. Feb.–April flowering.

CYCLOBOTHRA — LILIACEAE

The name is derived from *cyclo* ("rounded") and *bothra* ("a pit" or "trench"). This refers to the center of the flowers which are sunken, much like a pit. This genus was held as separate from *Calochortus* for many years. Today, after the work of M. Ownbey of the Missouri Botanic Garden, one division of *Calochortus* is named *Cyclobothra*.

Certain authorities do not adhere to the division of this genus and its inclusion into *Calochortus*. In the literature 3 species sometimes are found; they are:

Cyclobothra amoena. See *Calochortus amoenus*
C. lutea. See *Calochortus barbatus*
C. barbata. See *Calochortus barbatus*
For culture and other information see *Calochortus*.

CYPELLA — IRIDACEAE

From the Greek, *kypellon*, for the form of the flowers, which form a goblet or cup. Although 8 species are listed in the genus, only 6 are described in the literature. All are native to South America. They are not commonly grown but deserve to be, as they can be treated much like *Gladiolus*—planted in the spring, lifted in the fall, and stored overwinter. In warmer climates they can be left in the ground.

The bulb is tunicated, and sometimes bulbils are produced in the axils of the leaves. The leaves are few, with several on a stem, and the linear length depends on the species. The flowers are brightly colored but not long-lived. They are terminal solitary or in corymbs; distinct from other genera in that they have 3 large, spreading petals that are well separated from one another. The 3 inner petals, erect and recurved at the tips, curl distinctively, forming a cup.

CULTURE Well-drained, sandy soil, preferably with good organic matter, in full sun suits these bulbs best. They should be planted 4 in. deep and 4 in. apart, in the fall in mild areas and in the spring after frosts are almost past in colder areas. They need moisture while establishing but, once the flowering period is over, the amount of water given can be reduced. It is best to remove spent flowers to conserve strength for the bulbs. They can be cultivated in containers as well, 5 bulbs to a 6-in. pot in a good, sandy soil mix.

PROPAGATION This is best done from offsets taken when the bulbs are lifted in the fall, stored overwinter, and planted in the spring. Those species that produce bulbils can be propagated in this way, the bulbils being taken when ripe and planted in a sandy soil mix. The bulbs should be allowed to develop good roots and size in a cooler temperature before being brought out into a sunnier location to grow and flower. They frequently will flower in the second season after being started.

Seed should be sown in spring in a sandy soil mix and then grown on in the same manner as the bulbils.

PESTS & DISEASES Not subject to any special pests or diseases.

USES These plants are mainly for the gardener who wishes to grow unusual plants. They may have a use in open grassland, especially *C. peruviana*, but would need some moisture in early summer when the plants make their growth. They also make unusual container plants but are little more than conversation pieces as the flowers do not last very long.

SPECIES

C. herbertii. Native to Uruguay and neighboring Argentina; introduced 1823. This is the most commonly cultivated species of the genus. Flower spike reaches a height of 12–14 in., with several flowers per stem, which often is multibranched. Flowers are 2 in. in diameter; deep yellow with purple toward the base confined to a line in the middle of the petal; inner petals upright and then curled inward, same color with purplish markings. Late-summer-flowering. Foliage is lanceolate, tapering to a fine point up to 20 in. in length.

C. herrerae. Native to Peru. Not commonly found in cultivation. Flower spike varies in height from 10–18 in. Flowers are blue; inner segments curled with a yellow crest, center is whitish with red spots. Basal leaves are narrow, stem leaves short, bulbils sometimes produced in axils of leaves. Fall-flowering, often around Christmastime. Needs greenhouse conditions except in warmest climates.

C. peruviana (syn. *Hesperoxiphion peruvianum*). Native to the Andes in Peru. Found growing on grassy slopes; could well be considered for such plantings in climates like those of California. Two or three flowers per unbranched stem; deep yellow spotted red-brown at base; inner segments have a distinct beard. Late-summer-flowering. Height 8–16 in. Foliage grasslike.

C. plumbea. Found wild in Brazil, Uruguay, and Argentina; introduced 1838. A strong grower, sometimes reaching 36 in. in height but generally less. Not too many sword-shaped leaves are produced. Flowers are blue-gray, red-brown colors at the base of the larger outer petals, inner ones yellow toward their bases; open wide up to 3 in. in diameter; 2 or 3 produced from the unbranched stem but do not last long. Late-summer-flowering.

CYPELLA

C. drummondii syn. of *Herbertia drummondii*.
C. gracilis Native to Paraguay. May be a dwarf form of *C. herbertii*, which see. 6 in. in height.
C. herbertii var. *brevicristata* Bright lemon yellow.
C. rosei Native to South America, Escuinapa, Mexico. Light lilac, rosy with orange markings. 6–12 in. in height. Late Summer flowering.

CYPHIA — CAMPANULACEAE

The derivation of the name of this genus is not known. It is a rather obscure genus containing some 24–50 species native to the southern part of Africa. They are slender and small plants with mostly pink and/or white flowers, the color being variable. Certain of the species are twining plants. Only 2 species are known to be tuberous.

The flowers are not unattractive; certain species, notably *C. volubilis*, perhaps deserving of cultivation. The foliage is linear or three-lobed, often varying on the same plant. The flowers are carried singly and well spaced along the stems, they open to give a flat-faced flower; the lower part of the petals forms into a tube. In the native habitat it is found on dry hillsides in areas where there is winter rain; flowers are produced in very early spring, climbing up the stems of small shrubs.

The non-climbing species will often finish flowering by midsummer and then go into dormancy until the following spring. The height of the climbers is 30–36 in., that of the perennials often less.

CULTURE The plants are not hardy and need protection from frost. The tubers can be lifted when dormant and planted in the spring after danger of frost is past. The tubers need a portion of the stem in order to produce shoots and thus is not unlike a dahlia and can be treated in the same way. They need well-drained, warm soils, adequate moisture in the spring, and a dry period in late summer when the plants will become dormant. They need full sun, but tolerate some shade.

PROPAGATION The tubers can be lifted and divided provided that each division has an "eye". Seed should be sown in the spring and placed in a sandy soil mix in bright light with a nighttime temperature of some 55°F. Individual plants should be transplanted to individual containers as soon as they are large enough to be handled and grown on in frost-free and warm conditions.

PESTS & DISEASES No special problems.

USES For the bulb fancier only. Needs protection in all but the frost-free areas or grow in a cool greenhouse.

SPECIES

C. bulbosa. Native to the Cape of Good Hope area of South Africa. A perennial with underground tuber. Produces pinkish-white flowers that form a perianth tube for over ½ the length of the perianth segments. The flowers are carried in small spikes from the main stems with as many as 16–18 flowers per spike. Height 6–20 in.

C. volubilis. Native to the Cape Peninsula and other parts of the southern tip of Africa. A climbing plant with tuberous rootstock with lovely pink flowers yet the color is variable, sometimes being much lighter or white. Flowers spread their petals wide and diameter of the flowers often over 1 in. The foliage is not very evident on the 30–36 in. stalks which twine tightly around their supports.

COMMENTS It must be noted that in the literature, which to say the least is scant about this genus, no record is given that all species have tuberous or bulbous rootstocks. Information is definite about certain species, notably those listed above. However, unless an examination is made of the plants in the wild, it will not be possible to say definitely that all of the following species have a bulbous rootstock.

Many of the species are rare. Herbarium specimens do not have a sample of or even a reference to the rootstock. It would seem most likely that these plants do have a bulbous rootstock, as it is recorded that they are most prolific and show to advantage immediately after fire. If the regeneration were from seed, more time would be needed for the plants to reach flowering size. In addition, since some species are known to have bulbous rootstocks, this would indicate that most also do. However, there are many genera, for example, *Pelargonium*, where only a few plants have rootstocks that are bulbous. The same applies to *Begonia* and *Iris* as well as others. However, the literature regarding such genera has received more attention than this rather rare, yet sometimes beautiful, genus.

CYPHIA

C. angustifolia Native to Southern Karoo. Pink. Twining. Flowering Feb.–April; Aug.–Oct (Northern Hemisphere).

C. campestris (syn. *C. linarioides*) Native to Eastern Cape, Orange Free State. Lilac or white. 6–12 in. in height. Flowering Aug.–Oct.; Feb.–April (Northern Hemisphere).

C. crenata Native to Namaqualand to Cape Peninsula. Mauve. Twining. Flowering Aug.–Oct.; Feb.–April (Northern Hemisphere).

C. dentariifolia Native to Swartberg Mts., Mossel Bay, South Africa. Mauve. Twining. Flowering July–Aug.; Dec.–Feb. (Northern Hemisphere).

C. dentata Native to Clanwilliam. Violet. Twining. September; March (Northern Hemisphere).

C. digitata Native to Karoo, Namaqualand. Lilac or white. Twining. Flowering June–Nov.; Dec.–May (Northern Hemisphere).

C. georgica Native to George, South Africa. Erect or twining. Flowering March; Sept. (Northern Hemisphere).

C. heterophylla Native to Eastern Cape Province. White or lilac. Twining. Flowering April–Oct.; Oct.–April (Northern Hemisphere).

C. incisa Native from Western Cape Province to Cape Peninsula. Whitish. 2–20 in. in height. Flowering Aug.–Oct.; Feb.–April (Northern Hemisphere).

C. linarioides syn of *C. campestris*.

C. oligotricha Native to S. Namaqualand. Whitish. 5–6 in. in height. Flowering July–Sept.; Dec.–March (Northern Hemisphere).

C. phyteuma Native to Cape Peninsula. Pink and white. 6–12 in. in height. Flowering Aug–Nov; Feb.–May (Northern Hemisphere).

C. ranunculifolia Native to Pakhuis Mts., South Africa. Reddish-violet. Twining to 30 in. in height. Flowering Sept.; March (Northern Hemisphere).

C. schlechteri Native to Namaqualand. Lilac. Twining. Flowering July–Sept.; Dec.–March (Northern Hemisphere).

C. subtubulata Native to Western Cape Province. White or Mauve. Twining to 24 in. in height. Flowering Aug.–Nov.; Feb.–May (Northern Hemisphere).

C. sylvatica Native to Eastern Cape Province. Lilac. Twining. Flowering April–July; Oct.–Dec. (Northern Hemisphere).

C. tortilis Native to Eastern Cape. Lilac. Twining. Flowering May; Nov. (Northern Hemisphere).

C. triphylla Native to E. Cape, Karoo, Lesotho. Pink to lilac. Erect or twining. Flowering April–May; Oct.–Nov. (Northern Hemisphere).

C. undulata Native to Cape Province. Lilac. Semi-twining. Flowering March–Sept.; Sept.–March (Northern Hemisphere).

C. zeyheriana Native to Western Cape Province. Yellowish. Twining. Flowering July–Oct.; Dec.–April (Northern Hemisphere).

Cybistetes longifolia

Cyphia phyteuma

Plate 98

***Cyclamen hederifolium* 'Album'.** This plant will often be found in catalogs listed as *C. neapolitanum album*. It is a hardy plant that should be among the first to be tried by gardners wishing to grow cyclamen outdoors. WARD

***Cyclamen hederifolium* 'Apollo'.** Sometimes considered a separate species, sometimes regarded as *C. neapolitanum,* this selection is a good plant. WARD

Cyclamen libanoticum. This lovely species should be grown in every garden where there is little or no frost. Unfortunately it is not a hardy plant, and has to be grown in containers for protection. WARD

Cyclamen persicum. The lovely florist's plants we all enjoy are descendants of this species which, despite its name, is native of Cypress and various Greek Islands, Lebanon and Tunisia. One of the taller growing species. WARD

Cyclamen pseudibericum. A very lovely little species which carries its flowers well above the foliage. The fragrant flowers are of the true 'cyclamen' color. WARD

Cyclamen repandum. It is easy to see why this species is called the 'Ivy-leaved Cyclamen'. Both the mottling of the leaves and the flower color are variable. WARD

Cyclamen trochopteranthum. Rather rare, and doubtful if it will ever become well known, this species from Turkey is prized by collectors. WARD

Cypella plumbea. It is unfortunate that these unusual flowers do not last a long time. Their flowering season, being late in summer, is at a time when bulbous plants are of particular interest. They can add interest to shrub borders, which have finished flowering. WARD

Cypella peruviana. It is a pity that species of this genus are not more commonly grown. They add much to the interest of the borders in late summer. It would be well worth growing in grassy, well drained areas in milder climate zones. BOUSSARD

Cypella rosei. While only reaching 12 in. in height, this species is a showy one and of easy culture. Despite these good points it is rare in cultivation. J.H., by R. CRUDEN

Cyrtanthus contractus. A lovely plant that deserves more attention both as a pot plant and for growing in grassland areas in warmer climates. ORNDUFF

Cyrtanthus contractus. A close up picture of the flowers. B.R.I.

Cyrtanthus mackenii. This species is at home on the banks of streams in South Africa and yet will grow just as well in dryer locations. B.R.I.

Cyrtanthus mackenii* var. *cooperi. The yellow form of the species which is perhaps not as attractive; there are a number of variants in the species. AUTHOR

Cyrtanthus stenanthus* var. *major. Free flowering, a little variable as regards color, this plant is a little more orange than the ususal var. *major,* but despite the beauty, this plant is rare and will no doubt remain a plant for the collector. AUTHOR

Plate 102

DAHLIAS

Single. A good clear yellow. AUTHOR

Collarette. Many people appreciate the simple form of these flowers. AUTHOR

Collarette. The classification of some cultivars is not easy, as they seem to cross the given lines. AUTHOR

Single. A good deep red color is always a pretty sight. AUTHOR

Pompom. These are very popular and very sturdy plants. Certain cultivars 'break' and such flowers are either appreciated or hated by gardeners. AUTHOR

Plate 103

DAHLIAS

Show or Ball. Dahlia flowers have a will of their own, they face in many directions. AUTHOR

Show or Ball. Naming of certain flowers is not easy, giant pompom, ball or show? In either case, these are pretty flowers. AUTHOR

Water-lily type. The name aptly describes the form of the flower, but is not a classification that is accepted by all. BARNHOORN

Cactus, dwarf bedding. 'Park Princess' is a great cultivar for mass planting. DE HERTOGH

Plate 104

DAHLIAS

Semi-cactus. 'Purple Gem' is a popular cultivar, but some dislike the open center. DE HERTOGH

Cactus. Yellow cultivars are not always of as clear a color as this. AUTHOR

'Mirimar'. This flower seems to be a lot different from the previous one 'Glory of Heemstede', but it's in the same class. DE HERTOGH

Small flowered decorative. 'Glory of Heemstede' is placed in the Small flowered decorative class, but here again the lines are not easy to draw. DE HERTOGH

Plate 105

DAHLIAS

Decorative. Some cultivars have rather small flowers & are often known as Small Decorative. The smaller decorative flowers are not as untidy as the larger. A wide range of colors is to be found in all classes. Dahlias make great cut flowers, and colors for every type of decor can be found. AUTHOR

Some of the cultivars available are quite compact growers. AUTHOR

Flower size can be increased by dis-budding, but the flowers do look lonely. AUTHOR

Almost a perfect plant—good color, healthy foliage, and a good flower count. AUTHOR

Dahlias in the garden. Dahlias can supply different forms and colors, and are always attractive, but they do want lots of sunshine and adequate moisture to grow well. AUTHOR

Plate 106

Daubenya aurea. Unusual and quite striking in its way, but not likely to become a best seller. Certainly of great interest to gardeners wishing unusual, but at the same time, lovely bulbs to grow. ORNDUFF

Dicentra canadensis. A lovely species that is native to the eastern United States. The foliage remains attractive after the flowers have passed. KITTY KOHOUT

Dicentra spectabilis. At home in containers or in the open ground, this plant, while it does produce fleshy tuberous roots, is more frequently listed as a perennial than a bulbous plant. None of the genus appreciate sunlight and heat, & should be planted in areas where there is shade and humidity. AUTHOR

Dierama medium. Native to the Transvaal, this is one of the shorter species, making it well suited for the rock garden. BOUSSARD

Dierama pendulum. This species is not as tall as the more commonly grown *D. pulcherrimum,* but it is a graceful plant and ideal for planting at the top of a wall. AUTHOR

Dierama pendulum. A closer look at the flowers, but it is the overall height of the plant that makes it easy to distinguish from *D. pulcherrimum.* AUTHOR

Dierama pulcherrimum. While there are a number of selections of these lovely plants listed in catalogs, the true beauty of the species, here photographed in the wild, lies not so much in the color, but in the form of the plants. AUTHOR

Dierama pulcherrimum. A closer view of the lovely shape of the flowers which sway in even a slight breeze, adding yet another dimension to their beauty. AUTHOR

Dietes bicolor. While the flowers may pass quickly, there are always a great number produced; as one fades others come into flower, so that seldom is the plant without color in the spring. AUTHOR

Dietes grandiflora. A lovely plant, both in and out of flower. It is a pity that this genus is not hardy, as it has flowers of great beauty. It should be grown in gardens in warm climates, such as California. AUTHOR

Dioscorea villosa. So often we think only of the flowers of a plant, but often the seed pods are of interest. JOHN KOHOUT

Dipcadi longifolium

Plate 110

Dipcadi serotinum

Dipidax triquetra. A lovely species from Cape Province. In flower for a long period in spring time (August-September) in the wild. K.N.B.G.

Dipidax triquetra. The individual flowers are attractive. While in the young stage the flowerhead is compact, it expands to allow each flower to be enjoyed. ORNDUFF

Dracunculus vulgaris. This species, while not having a pleasant smell, is widely grown. The bright red berries in the fall are an added attraction. Great plant for the edge of a woodland. WARD

Eleutherine plicata. Rare in cultivation, but not uncommon in tropical countries, especially South America. Not of great beauty, but the bulbs produce a great number of flowers in a year, the greatest number in spring. LOVELL

CYRTANTHUS — AMARYLLIDACEAE

From the Greek *kyrtos* ("curved") and *anthos* ("flower")—a description of the flower head, as the flowers all bend downward from the top of the stalk to which they are attached. This is a genus of some 40 species native to South Africa and to eastern Africa.

These are attractive plants that never have caught on to any degree, possibly due to their being a little on the tender side in temperate climates and thus not grown by many people. A number are listed by various Dutch nurseries, and they would be good additions to gardens where the climate is mild, with little or no frost in the winter. Some species do stand a few degrees of frost but, for the most part, should be regarded in the same light as dahlias and either lifted or protected during the winter. In some cases this would not be possible as they are late-fall-flowering; however, many of them should be grown, especially the selections offered by the nurseries. Although they are subject to differing climatic conditions in their native habitats, they do adapt to a wide range of growing conditions, e.g., *C. mackenii* is at home along stream beds in Natal and yet will grow well in drier locations.

The bulbs are tunicated. The leaves are narrow, sometimes grasslike, and persistent. They appear at different times, sometimes at the same time as the flowers and sometimes after the flowers are produced, depending on the species. The flowers are produced in an umbel. They are tubular, with the flower opening a little wider than the tube. The ends of the perianth segments separate from each other for only a short distance, the tube being as much as 10 times longer. The 3 outer segments curve in slightly at their tips. The stamens are inserted on the perianth tube. Certain species are fragrant, with the colors varying from red through pink to white. The fruit is an oblong capsule with many flattened, black seeds.

CULTURE The bulbs grow well in good garden soil but the richer the soil the better they perform, increasing in height and flower count. They respond to being fed while growing, therefore, light feedings of organic liquid fertilizer should be given.

The late-summer- and fall-flowering species should be planted in the spring and the spring species should be planted in the fall. Plant 2–3 in. deep, spacing them 5–6 in. apart, in a sunny area. Where the climate is marginal, i.e., frost expected at night but rarely lasting through the morning, they should be given the protection of a wall or other favored location.

If being grown in containers, plant in a rich but well-drained soil mix. Moisture should be adequate at all times that they are growing but more is needed while they are making their leaf growth. After the growth has finished, keep moist, with the exception of *C. obliquus*, which appreciates more water at all times. Temperatures should be in the 40°F range during the winter if being grown indoors. On the anniversary of their potting date they should be repotted, the offsets being removed for increasing the stock.

PROPAGATION The bulbs produce numerous offsets, which can be separated and grown on when the bulbs are lifted. They will flower the fourth season after germination. Seed of all species can be sown in spring, (with the spring-flowering species sown as soon as ripe) in a sandy mixture and kept moist and warm with temperatures in the 60°F range. Transplant when of good size and line out in nursery rows where they can be grown until of sufficient size for planting out.

PESTS & DISEASES Not subject to any particular pests or diseases.

USES Not so widely grown as they deserve. They are good pot plants, especially *C. obliquus*, which has the potential of being a good plant for open grassland areas where there is adequate moisture and the winters are not severe.

SPECIES & CULTIVARS

C. angustifolius. Native to the southwestern part of Cape Province, South Africa; introduced 1773. The leaves are most commonly produced after the flowers; they are narrow and up to 18 in. long. The flowers are carried on a hollow stem, which is reddish on the lower part and reaches a height of 12 in. Flowers are a little more than 2 in. long, flare at the mouth, and as many as 12 are produced on a stem. The reddish-orange blooms droop around the stem and appear in late summer/early autumn. This species is not suited for colder areas.

C. contractus. One of the most widely distributed species in South Africa. Found in eastern Cape Province and along the coast of Natal and into Transvaal; first described in 1898. The plant reaches a height of some 12 in. The leaves are grasslike and long, often more than 12–15 in., and appear soon after the flowers. Usually flowers in the spring and early summer, but may flower at any time of the year. The flowers are rose-colored, inclined to be a little redder on occasion, and, like the other species, are tubular, drooping from the stem, and up to 2 in. in length.

C. falcatus. Native to Natal; first described in 1939. The leaves, which emerge at about the same time as the flowers but persist longer into the summer, are broader than in many of the other species. The species has some of the longest flowers of the genus, each more than 2 in. Flower stems are spotted with purple and are up to 18 in. high. The pendant flowers are intermediate in color between red and pink, with a hint of yellow on the outside of the tube, and appear in early spring.

C. galpinii. Native to the eastern sections of the Transvaal and into Zululand; introduced 1892. This is one of the more dwarf species, 8 in. in height. Frequently only 1 leaf is produced, usually after the flower, but it can be up to 10 in. long. Also unlike the other species, only 1, seldom 2, flowers are produced per stem. The flower is scarlet to rich pink, and appears in August.

C. huttonii. Native to the eastern Cape Province and eastern Transvaal; introduced 1864. Leaves can be an inch wide and 20–24 in. long. They are carried with the flowers, up to but seldom more than an inch at the base, tapering quickly. The numerous flowers, often more than 20 in the head, vary in color from a good orange-red to deep red but yellow on the inside. A tube is formed but the segments open quickly into a wider flower than many of the other species. Flowering in late April–early May. The strong stem is hollow, often passing a foot in height.

C. mackenii. Native to Natal and eastern Cape Province where it is found along the coast; introduced 1868. The 8–12 in. long leaves are produced with the flowers in early summer–late spring, usually April–May. The fragrant flowers are in shades of light cream and white and are a little more than an inch long. Each 12 in. stem will produce 4 or more flowers, which are not so pendant as many of the other species but seldom are erect, more often being horizontal or a little upward-facing. Flowers remain tubular. Although this species generally is found along the banks of streams, it will flower well in the garden with appropriate moisture. Var. *mackenii* has white flowers; var. *cooperi* has yellow or cream flowers and has been successful in cultivation producing a clear pink form and one with orange perianth tubes and yellow segments. All are the same height—more or less as the species.

C. obliquus (Giant Cyrtanthus). Found growing in the veldt in Natal and down toward the Cape; introduced 1774. This attractive species has been grown as a pot plant in Europe for many years. The leaves are up to 1 in. wide and 18 in. long. The flowers are drooping and tubular, up to 3 in. long, yellow at the base, shading into red, with the tips of the perianth segments greenish-yellow where the segments divide. Height 4–12 in. Flowering in early summer.

C. o'brienii. Native of the eastern Cape and the Drakensberg Mountains. of Natal; introduced 1894. Described as having scarlet flowers 8 in. in height in a nodding umbel of 7–8 flowers; bright green leaves, 12 in. long and less than ½ in. wide, often produced before the flowers, which are produced in early Jan.–Feb. Possibly a form of *C. macowanii.*

C. sanguineus. Native to the eastern Cape and Natal to tropical eastern Africa, where it is now becoming rare; introduced 1860. The plant reaches a height of 12–14 in. The leaves, which are produced before the flowers, are up to 18 in. long. The brilliant scarlet flowers, seldom more than 1 or 2 in the flower head, are 4 in. long and open to more than 1 in. in diameter and appear in late summer–early fall. There are some color variations of the species noted, including one with glaucous leaves which is sometimes described as *C. s. glaucophyllus.* The plant should have adequate moisture when in leaf and until the flowers have passed. While it likes full sun in hot areas, it should be given high shade during the hottest part of the day.

C. tuckii (Brandlelie, Fire Lily). Native to southeastern Transvaal, Natal, and eastern Cape Province; introduced 1884. Possibly the hardiest of the species, growing in the wild where quite severe frost occurs. The spirally twisted leaves, up to 18 in. in length, are produced with the flowers in late spring. The blood-red flowers, with a hint of yellow at the base, are tubular, up to 2 in. in length, and reach a height of 12–18 in. They are numerous, often more than 10 in the head. There are two varieties noted: var. *transvaalensis,* which has blood-red flowers with no other color present, and var. *viridilobus,* which has green lobes. The common name refers to the plants seeming to profit from fire scorching their growing areas, coming back stronger and producing larger flowers. The following selections are offered by nurseries:

'Atlanta': Coral-pink petals.
'Euterpe': Small, ivory white flowers.
'Themis': Large flowers; outside pink, inside red.
'Yellow Star': Large, golden yellow flowers.

CYRTANTHUS

C. attenuatus Native to Orange Free State & Lesotho. Introduced 1939. Yellow to greenish-yellow. Peduncle 16–20 in. in height. Flowering Sept.–Dec.; March–June (Northern Hemisphere).
C. bicolor (syn. *C. parviflorus*) Native to E. Transvaal, Swaziland, N. Natal. Introduced 1896. Red or yellow or variable. 2–10 in. in height. Flowering most commonly Oct., but also throughout year. April (Northern Hemisphere).
C. brachyscyphus (syn. *C. parviflorus*) Native to S. Natal & E. Cape Province. Introduced 1888. Bright red. 12 in. in height. Flowering Sept.–Nov. sometimes to April; March–May (Northern Hemisphere).
C. breviflorus (syn. *C. luteus, Anoiganthus breviflorus*) Native to E. Cape Province, Natal, Transvaal & Tropical Africa. Bright yellow segments. 2–10 in. in height. Early Spring flowering.
C. carneus Native to S.W. Cape Province. Introduced 1831. Pink, paler at base of perianth. 12 in. in height. Flowering Dec.–Feb in wild; July–Aug. (Northern Hemisphere).
C. clavatus (syn. *C. uniflorus*) Native to Cape Province. Introduced 1816. White with six red, red-brown or green stripes. 3–8 in. in height. Flowering May; Aug. (Northern Hemisphere).
C. collinus Native to S.E. Cape Province. Introduced 1816. Bright red with white lines. 6–10 in. in height. Flowering Dec.; June (Northern Hemisphere).
C. dendrophilus syn. of *C. epiphyticus.*
C. epiphyticus (syn. *C. dendrophilus*) Native to Natal. Introduced 1913. Scarlet. 10–16 in. in height. September flowering.
C. erubescens Native to Natal. Introduced 1960. Pink. 8–20 in. in height. Flowering Oct.–Dec.; April–June (Northern Hemisphere).
C. eucallus Native to Transvaal. Red, lighter in the throat. 12 in. in height.
C. fergusoniae Native to Southern Cape. Introduced 1931. Bright red. 8–18 in. in height. Flowering Dec.–Jan.; June–July (Northern Hemisphere).
C. flanaganii Native to South Africa. Introduced 1938. Yellow. 8–12 in. in height. June flowering.
C. flavus Native to E. Cape Province. Introduced 1934. Canary yellow. 8–10 in. in height. Spring flowering.
C. guthrieae Native to S.W. Cape Province. Introduced 1921. Bright red with golden highlights. 4 in. in height. Flowering April; Oct. (Northern Hemisphere).
C. helictus Native to S.E. Cape Province. Introduced 1839. White with brown or green stripes. 3–5 in. in height. Flowering Sept.–Jan.; March–July (Northern Hemisphere).
C. herrei Native to N.W. Cape Province. Introduced 1932. Tube red, inner lobes greenish. 12–16 in. in height. Feb.–April; Aug.–Oct. (Northern Hemisphere).
C. inaequalis Native to S.E. Cape Province. Introduced 1939. Coral red. 12 in. in height. Flowering Jan. (Northern Hemisphere).
C. junodii Native to N.E. Transvaal. Introduced 1907. Tube red, lobes yellow. 10–20 in. in height. Spring flowering.
C. labiatus Native to S.E. Cape Province. Introduced 1980. Light coral red. 12 in. in height. Flowering Dec.–Jan.; June–July (Northern Hemisphere).
C. leucanthus Native to S.W. Cape Province. Introduced 1898. White, often tinged yellow. 6–10 in. in height. Flowering Late Nov.–Feb; Late May–Aug. (Northern Hemisphere).
C. loddigesianus Native to S.E. Cape Province. Introduced 1839. White with cream tinge in middle of segments, light green at base. 6–7 in. in height. Flowering Nov.–Feb.; June–Aug. (Northern Hemisphere).
C. lutescens syn. of *C. ochroleucus.*
C. luteus syn. of *C. breviflorus.*
C. macowanii Native to Cape Province. Introduced 1875. Bright scarlet. 12 in. in height. June–July flowering.
C. montanus Native to E. Cape Province. Introduced 1977. Red. 4 in. in height. September flowering.
C. nutans Native to Swaziland & N. Natal. Pale yellow. 6–8 in. in height. Flowering Late Aug.–Oct.; Late Feb.–April (Northern Hemisphere).
C. ochroleucus (syn. *C. lutescens*) Native to Cape Province. Introduced 1836. Pale yellow or yellowish white. 6–12 in. in height. Feb. flowering.
C. odorus Native to South Africa. Introduced 1818. Bright red. 6 in. in height. July–Aug. flowering.
C. pallidus Native to South Africa. Introduced 1824. Pale red. 6 in. in height. Jan. flowering (Northern Hemisphere).
C. parviflorus syn. of *C. bicolor* & *C. brachyscyphus.*
C. purpureus see *Vallota speciosa* & *Vallota purpurea.*
C. rectiflorus Native to E. Cape Province. Introduced 1896. Red. 12–18 in. in height. Spring flowering.
C. rhododactylus Native to South Africa. Introduced 1909. Rose scarlet on lobes, paler inside. 5–6 in. in height. Feb.–Aug. flowering (Northern Hemisphere).
C. rotundilobus Native to Transkei, (Africa). Introduced 1921. Scarlet red. 5 in. in height. Dec.–Jan.; June–July (Northern Hemisphere).

C. smithiae Native to South Africa. Introduced 1876. White or pale pink striped red or red-brown. 10–12 in. in height. Flowering April–June (Northern Hemisphere).
C. speciosus Native to S.E. Cape Province. Introduced 1942. Creamy white, striped with red or pink, green & pink at base. 6–7 in. in height. Flowering Dec.–Jan.; June–July (Northern Hemisphere).
C. spiralis Native to South Africa. Introduced 1815. Red. 10 in. in height. June–Aug. (Northern Hemisphere).
C. staadensis Native to S.E. Cape Province. Introduced 1914. 1½ in. in height. Flowering Mid Feb.–March; Mid Aug.–Sept. (Northern Hemisphere).
C. stenanthus var. *stenanthus* Native to S.E. Transvaal, Natal, Lesotho. Introduced 1897. Red or greenish red. 8–12 in. in height. Flowering Oct.–Dec.; April–June (Northern Hemisphere).
C. stenanthus var. *major* Native to E. Transvaal, E. Orange Free State, Swaziland, Natal. Introduced 1939. Yellow. 8–12 in. in height. Flowering Oct.–March; April–Sept. (Northern Hemisphere).
C. suaveolens Native to Lesotho & E. Cape Province. Introduced 1914. Red; scented of cloves. 6–8 in. in height. Flowering Nov.–Jan.; May–July (Northern Hemisphere).
C. thorncroftii Native to Transvaal. Introduced 1909. Pale salmon pink. 10 in. in height. Flowering Nov.–July; May–Dec. (Northern Hemisphere).
C. uniflorus syn. of *C. clavatus*.
C. ventricosus Native to Cape Peninsula. Introduced 1799. Red with pink. 4–8 in. in height. Flowering June–Nov. (Northern Hemisphere).

DAHLIA — COMPOSITAE

Named for Dr. Andreas Dahl, an 18th century. Swedish botanist and pupil of Linnaeus. The *Dahlia* was brought to Spain from its native Mexico in the late 1780s. Its first introduction to Britain was in 1798, which I have already described (see HISTORY). Three specimens reached Paris in 1802.

While there are perhaps a dozen species of *Dahlia* found in the wild in Mexico, Central America, and Colombia, the great majority of plants grown today are variations of the true species and hybrids raised from variants that have arisen in cultivation. Aztec gardeners already had selected variants or hybrids long before its introduction in Europe, making the actual parentage even more obscure.

The first cultivation, however, was to exploit the whole plant's potential as a vegetable, but it was judged to be inedible. The flower petals, though, can be eaten and give color to a salad. The roots, too, were used as food by the natives of Mexico. They contain a valuable and healthy starchy substance called inulin, which is used today in the manufacture of certain chemicals.

The fleshy roots of the *Dahlia* are tuberous, capable of holding considerable reserves of moisture and food for lengthy periods of time. In its native habitat, moisture is absorbed during the growing period when rain is plentiful and conserved during the dry season, which occurs in late summer and fall. The stems branch at the base and often are multibranched. They are produced only from the crown, which is the lower part of the flowering stem. This remains turgid during the resting period and must be protected if new shoots are to develop.

The flowers are carried on long stalks. Although most species in the wild produce only single flowers, some, such as *D. variabilis*, are found producing many semidoubles. In singles, those known as perfect flowers have a yellow disk in the center, which is surrounded by ray florets. These are male flowers, where the pistils are petaloid. There is great variation in color. In doubles, the number of male flowers that are petaloid is greatly increased. Sometimes only the disk florets are fertile, the female parts of the ray florets being rudimentary. Only rarely are all disk florets absent, and most flowers will produce at least some seeds, even in the double forms. The protective sheath that envelops the developing ovary(ies) is formed by bracteoles, which separate the individual florets. The *Dahlia* are not necessarily self-sterile, but the male pollen is ripe prior to the stigma being receptive, resulting in the flowers seldom being self-pollinated.

When first introduced, only 3 forms of *Dahlia* were identified: *D. coccinea*, with single red flowers; *D. rosea*, with single rose flowers; and *D. pinnata*, with double purple flowers. With the proliferation of hybrids, variations, forms, etc., a more extensive grouping of the garden forms was adopted by the National Dahlia Society and the Royal Horticultural Society.

Class I. SINGLE DAHLIAS
(a). *Show Singles.* Flower heads not more than 3 in. across, 8 (only) rays, smooth, somewhat recurved at the tips, overlapping to form perfectly round flower.
(b). *Singles.* Flower heads with rays not so completely overlapping as in (a), tips separated.
(c). *Mignon Dahlias.* Flower heads as in (b), however, plants not more than 18 in. high.

Class II. ANEMONE-FLOWERED DAHLIAS
Flower heads with one or more series of flat ray-florets surrounding dense group of tubular florets which are longer than disk-florets in Class I and usually different color than rays.

Class III. COLLARETTE DAHLIAS
Flower heads with one or more series of flat ray-florets as in Class I; ring of florets (the collarette) only half the length of rays and usually of different color above each series.
(a). *Collarette Singles.* Flower heads with single series of rays and one collarette with yellow disk.
(b). *Collarette Peony-flowered.* Flower heads with 2 or 3 series of rays and collars and yellow disk.
(c). *Collarette Decorative.* Similar to (b) but fully double.

Class IV. PEONY-FLOWERED DAHLIAS.
Flower heads with 2 or 3 series of ray-florets and central disk.
(a). *Large Peony-flowered.* Flower heads more than 7 in. across.
(b). *Medium Peony-flowered.* Flower heads from 5–7 in. across.
(c). *Small Peony-flowered.* Flower heads less than 5 in. across.
(d). *Dwarf Peony-flowered.* Plants not more than 30 in. high.

Class V. FORMAL DECORATIVE DAHLIAS.
Flower heads fully double, showing no disk; all florets regularly arranged with margins usually slightly incurved, flattened toward tips, which may be broadly rounded or pointed. Four subclasses, (a), (b), (c), and (d), correspond with those under Class IV.

Class VI. INFORMAL DECORATIVE DAHLIAS
Flower heads fully double, showing no disk; rays not regularly arranged, broad, more or less flat or slightly twisted, more or less acutely pointed. Four subclasses, (a), (b), (c), and (d), correspond with those under Class IV.

Class VII. SHOW DAHLIAS
Flower heads fully double, more than 3 in. across, almost globular, with central florets similar to outer but smaller florets with margins incurved, tubular or cup-shaped, short and blunt at mouth. Fancy dahlias form subclass here, with white-tipped or striped florets.

Class VIII. POMPON DAHLIAS
Flower heads similar to those in Class VII but smaller; for show purposes flower heads must not exceed 2 in. across.

Class IX. CACTUS DAHLIAS
Flower heads fully double, margins of flowers revolute for not less than three-quarters of their length, central florets form filbert-shaped group.
 (a). *Large-flowered Cactus.* Flower heads more than 4½ in. across.
 (b). *Small-flowered Cactus.* Flower heads not more than 4½ in. across.
 (c). *Dwarf Cactus.* Plant not more than 30 in. high.

Class X. SEMI-CACTUS DAHLIAS
Flower heads fully double, florets broad at base, margins revolute toward tip, slightly twisted for about half their lengths. Three subclasses, (a), (b), and (c), correspond with those under Class IX.

Class XI. STAR DAHLIAS
Flower heads small, with 2 or 3 series of slightly pointed rays, barely overlapping or not overlapping at all at their more-or-less recurving margins, forming cup-shaped flower head with disk.

CULTURE Dahlias are easy to grow. Tolerant of a wide range of soils, they prefer one that is cultivated to a fair depth and with a high organic content. They require sun in order to grow and flower well. While the rootstocks will overwinter in the ground providing the soil temperature stays above 34°F, it is preferable to lift them in the fall, after the foliage has started to die back, and store them in a frost-free location with good air circulation.

Planting is best in early spring, some 10 days or so before the last expected frost. The tubers should be set 3–4 in. deep. This will allow for a period of 14–20 days before the young shoots emerge. Thus they will not be damaged by frost, to which the young shoots are susceptible. The distances between the plants will vary according to height of the full-grown plants. Dwarf or low-growing types should be spaced 12–18 in. apart. The tallest-growing, those over 36 or 40 in. in height, space some 36 in. apart. The taller-growing cultivars will require support, especially if grown in a windy location.

Dahlias require regular watering in the summer months. To facilitate watering, plant the tubers in the center of a shallow depression or saucer of soil. This will allow easier watering as the water will stay where required, at the base of the plants above the tubers which will produce a large number of roots, but they stay quite close to the original tuber. The actual size of the depression is determined by the height of the plant; smaller plants that are set quite closely together can be placed in a shallow trench so they can be flooded.

Weak but regular feedings of fertilizer should be given up through midsummer. A formula in the N.15 P.15 K.15 range is suitable applied as a liquid or in granular form. As soon as the flower buds develop it is best to reduce the amount of fertilizer, especially nitrogen, but phosphates and potash, which play a part in the general health of the plant and in flower production, should be applied. This does not mean abandoning nitrogen altogether but the amount should be approximately halved. This will allow for continued but not excessive growth; the desired state as the plant, by the time it starts to produce flower buds, will be close to optimum size.

While some natural branching of the young plants may occur (to some degree this depends on the cultivars being grown), it is an advantage to pinch out the growth bud of young plants when they are some 4–6 in. in height. Such pinching of the growing tip encourages branching from the stem, and a bushier plant is the result. Not necessary with lower-growing cultivars that branch naturally themselves, but taller growing forms can become gangly, and pinching improves not only the appearance but increases flower production. A four- or five-branched plant is ideal.

Should the production of large flowers be desired, disbudding should be undertaken as soon as the buds are of fair size, about that of a small pea. The middle bud is left, those appearing in a cluster around this middle bud are removed by pinching, using the thumb and index finger. The bud that remains will produce a significantly larger flower than if all buds are allowed to develop. If a display of color is desired, no disbudding is needed.

Dahlias will continue to grow well into the fall. As the days shorten, the growth rate slows. In colder areas the first frost will kill the top growth but not harm the tuber in the soil. In warmer climates, where little or no frost is experienced, growth will come to a halt sometime in Oct.–Nov. When the top growth is harmed by frost or when growth stops, the stalks should be cut, leaving some 4 or 5 in. above the ground. The actual time of harvesting also will depend on the climate of an area. Working in rain is not conducive to good production, therefore, the harvesting of tubers should be prior to the onset of fall rains. The tubers are lifted with care so they are not damaged. The short length of stem left when cutting the plants will be useful when handling the plants. As soon as the soil around the tubers has dried they are cleaned by removing dried soil with a brush and stored in a well-ventilated, frost-free area. The tubers should be labeled so that when replanting, size, color, type, etc., are known.

During the storage period, the tubers should be examined on a regular basis. Any that begin to spoil should be cleaned by removing diseased tissue and dusting with a fungicide. Any that are severely attacked should be discarded. As all new shoots for the next season's growth will arise from "eyes" at the base of the stem, particular attention should be given to this area. Without such eyes the tuber is worthless as the tubers themselves have no buds; eyes are found only on the portion of the stem adjoining the tubers.

PROPAGATION Dahlias can be propagated in numerous ways. Many firms offer seed, generally of the low-growing, bedding types, seldom more than 18–20 in. in height. There are no single colors offered but a good selection within restricted color ranges. Large tubers can be divided, provided each tuber has an eye; this method is well suited to the home gardener. Cuttings provide an excellent method of propagation. Recently, tissue culture, which enables a rapid increase in stock, has been practiced commercially.

Seed
Seed sown in the early spring will produce flowers and tubers during the following summer. Seed should be sown in February or early March. Use a standard potting soil mix and space the seeds about 1 in. apart. Cover with ⅛ in. of soil and water the seed flat or container after sowing. A very high percentage of seed will germinate. With a nighttime temperature of 55°F all seed should be germinated in 21 days.

As soon as the seedlings are large enough to handle, transplant into individual containers using a standard potting soil mix. They should be grown on, protecting them from temperatures lower than 45°F at night. While lower temperatures, as long as not below 34°F, will not harm the young plants, it is best to maintain higher temperatures to ensure adequate and good growth. The young plants should be hardened off and can be planted out when all danger of frost is past. In areas where no frost is experienced, daylight hours are the governing factor. The young plants should not be set out until the night temperatures are constantly in the 45°–50°F range. Early April in the Northern Hemisphere is the approximate time when sufficient daylight length will ensure good and steady growth.

Division
Larger tubers provide the opportunity to propagate the tubers

by division. This should be done at the correct planting time; i.e., after all danger of frost is past. It is essential that each portion of the divided plant have an "eye," the name given to the dormant bud. With a sharp knife, cut the portion of the stem with an eye together with an attached tuber. In some instances more than one tuber will be joined with one or more eyes. For good growth, small tubers, even with an eye, will not produce as good a plant as larger tubers. After division the cut surfaces should be dusted with a fungicide and planted in well-prepared ground after danger of frost is past.

Cuttings

For the home gardener, growing dahlias from cuttings is a simple procedure. The method used follows the same treatment as for the commercial grower, however, the timing will differ.

In late February or early March, the tubers are placed in a container with the their collars exposed. A light sandy/peaty soil mix or pure peat can be used. Temperatures should be warm, in the 55°F range at night. The soil mix is kept moist, and adequate light must be given. After a week, young shoots will emerge from the collar, on the portion of the old stem immediately above the tubers themselves. These are allowed to develop until some 3–5 in. in length. They then are removed, together with a portion of the older tissue, which will form the "heel" of the cutting.

These cuttings then are inserted into a sandy soil mix, either in a flat or around the rim of a pot. It is essential that the humidity of the area in which the cuttings are placed be high. Bottom heat will speed rooting. Use of a root hormone also will speed root production. For small quantities, the placing of a plastic bag over the cuttings will speed rooting. When such a tent is made, care must be taken to ensure a good air space between the cutting and the plastic; a wire frame will keep the plastic away from the cuttings. Rooting will take some 10–18 days, temperature making the difference between slow and quick rooting. As soon as the cuttings regain turgidity, the plastic cover can be removed for periods of time to reduce the likelihood of fungal infection. After a week the plastic can be removed entirely.

As soon as growth is seen, the cuttings should be examined for root growth. As soon as rooted they should be transplanted into individual containers, hardened off, and planted outdoors when all danger of frost is past, or the daylight hours and temperature are suitable, as with plants raised from seed.

Commercial Propagation by Cuttings

The number of cuttings produced by tubers will vary according to variety. Such information can be obtained only by experience and the commercial grower is advised to allow for an adequate number of mother stock until such experience is gained. Full details of commercial production are covered in Chapter 4, page 00.

PESTS & DISEASES Virus diseases can be troublesome. The viruses are spread by sucking insects. In the garden it is advisable to discard all plants that show typical virus damage, such as mottled foliage and stunted and deformed growth. Contol of aphids and spider mites should be undertaken as soon as noticed; spraying with insecticidal soap is an effective control.

Commercial production demands somewhat more severe control. Regular spraying of the crop with an insecticide will keep sucking insects under control and reduce the spread of virus. The fields should be constantly inspected and all sections of infested plants removed at once. Particular attention should be given to the selection of tubers selected for propagation. Only the finest plants, free from all imperfections, should be considered.

Insects

The reader is referred to the chapter on Pests and Diseases for typical symptoms of insect attack. It should be remembered that good culture and the use of healthy stock, to a large degree, can negate problems caused by insects. The most likely pests are:

Aphids, controlled with insecticides or insecticidal soap.

Japanese beetles, controlled by trapping and with insecticides.

Thrips, causing the very visible streaks of gray or brown in the foliage, again controlled by insecticides. However, in commercial production, problems can be avoided by dusting or dipping in a control product the tubers prior to planting.

Wireworms, can attack but rarely; stems are eaten; should these pests be found, treating the soil, tubers, and aboveground stems is required.

Earwigs, are more unsightly than an actual problem. However, if flower production is required for shows or for the market as cut flowers, control will be necessary as these pests can wreck petals by eating them. Nocturnal in habit, earwigs can be trapped by the placing of a clay pot filled with straw on top of a cane in areas being attacked, which will attract the pests. The straw containing the earwigs can be removed and disposed of along with the insects, which is best done each morning. **Red spider mites,** seldom a problem.

Nematodes, can cause lesions to form on the leaves and often, in severe infestations, distort regular growth (stunting, etc.). Must be controlled by soil fumigation, best done by a professional. In the commercial production fields, such fumigation should be considered if any hint of nematodes is found.

Diseases

Fungus diseases such as mildew or rust are not frequently a problem. Good air circulation will keep down the likelihood of attacks. Removal of affected foliage is a reasonable control in the garden, and keeping water off the foliage in areas where mildew is a problem with neighboring plants is advised.

In the commercial production fields, a controlled program of fungicide spraying should be considered, especially if the fields are not in a location with good air circulation. Such attacks by fungus diseases are more likely to take place in summer when humidity is high. Good cultural practices can reduce the problems and, if experience shows that such attacks are likely, consideration should be given to leaving greater distances between the plants in the rows and between the rows themselves.

USES Popular show flower. Many varieties in all sizes and forms used to highlight perennial borders, in beds by themselves, or low-growing forms as border for flower beds. Also excellent container plants. Start in spring in good soil mix, give frequent feedings of weak liquid organic fertilizer. Need sun to flower well.

SPECIES

D. coccinea (syn. *D. cervantesii, D. crocea*). Native to Mexico; introduced 1798. Leaves pinnate or bipinnate. Erect stems carry flowers to height of 36 in. Flowers scarlet, sometimes with orange or yellow centers. Summer-flowering.

D. excelsa. Leaves often more than 24 in. long. Erect stems become quite woody at end of season; carry flowers some 20 ft. in air. Flowers produced in late summer are rather dull, pale purple. Grow only in frost-free areas so it will flower or else stem will not reach flowering height.

D. imperialis. Native to Mexico, Guatemala, San Salvador, Colombia, and Costa Rica. Extremely variable height—from 6–18 ft. Stems branched, usually only at base, with swollen nodes. Flowers 6 in. in diameter, white with red markings at base. Flowers in late summer, leaves have 2–3 leaflets, total length 3 ft.

D. juarezii. From Mexico; introduced 1864. Cactus type, with florets of irregular length, overlapping. Reaches height of 3–5 ft. Mid- to late-summer-flowering. Bright scarlet.

D. pinnata. From Mexico; introduced 1798. Low-growing, seldom more than 36 in. in height. Large flowers, red with bluish tinge. Leaves have rounded, toothed leaflets to 10 in. long, including petiole.

D. rosea (syn. *D. superflua*). Native to Mexico; introduced 1794. Regarded as parent of present garden varieties. 3–5 ft. in height. Broad, toothed leaves. Flowers in summer. Variable in color.

D. variabilis (syn. of *D. rosea*). Name sometimes given to semidouble forms of *D. rosea* found in wild. Name of *D. variabilis* given in 1829 by Desfontaines, but called *D. rosea* by Cavanille in 1794 and takes precedence.

CULTIVARS

The number of cultivars and forms offered today by nurseries is extensive. In making selections, it is best to determine first the form, color, and height desired. Select stock from first-class nurseries and seed firms. While bargain rates sometimes are advertised, the purchase of quality stock from established companies will give better assurance that stock is virus-free and true to name.

The following is a list of *Dahlia* cultivars commonly found in the catalog of major suppliers. All are summer-flowering.

Single Dahlias
'Bambino'. White. 10 in. in height.
'Chessy'. Bronze-yellow. 10 in. in height.
'G. F. Hemerik'. Soft orange. 14–20 in. in height.
'Irene van der Zwet'. Soft yellow. 14–20 in. in height.
'Murillo'. Lilac-pink with darker center. 14–20 in. in height.
'Reddy'. Red. 10 in. in height.

Anemone-flowered Dahlias
'Bridesmaid'. Ivory white. 10–18 in. in height.
'Guinea'. Yellow. 10–18 in. in height.
'Honey'. Apricot-pink. 10–18 in. in height.
'Siemen Doorenbosch'. Light magenta. 10–18 in. in height.

Collarette Dahlias
'Clair de Lune'. Pale sulfur-yellow with cream collar. 30–40 in. in height.
'La Cierva'. Purple with white collar. 30–40 in. in height.
'La Giaconda'. Scarlet with golden yellow collar. 30–40 in. in height.

Decorative Dahlias
Large-flowered
'Jocondo'. Reddish-purple. 36–60 in. in height.
'Lavender Perfection'. Lavender-pink. 36–60 in. in height.
Medium-flowered
'Deuil du Roi Albert'. Violet-purple tipped white. 48 in. in height.
'Edinburgh'. Maroon and white. 36 in. in height.
'Majuba'. Blood-red. 48 in. in height.
'Peter'. Purple-pink. 48 in. in height.
'Peter's Glory'. Lilac-rose with white center. 48 in. in height.
'Red and White'. Clear red and white. 36 in. in height.
'Rosella'. Lilac-pink. 42–48 in. in height.
'Snow Country'. Pure white. 36 in. in height.
Small-flowered
'Arabian Night'. Deep maroon, almost black. 42 in. in height.
'Chinese Lantern'. Orange-red on yellow background. 48 in. in height.
'Gerrie Hoek'. Deep pink. 42 in. in height.
'House of Orange'. Soft amber-orange. 48 in. in height.
Miniature-flowered
'David Howard'. Deep orange-yellow. 24 in. in height.
'Lilianne Ballego'. Bronze. 36–42 in. in height.

Pompon Dahlias
'Albino'. Pure white. 36–42 in. in height.
'Deepest Yellow'. Deep yellow. 36–48 in. in height.
'Lydia'. Showy red. 36–42 in. in height.
'Nero'. Dark red with purple sheen. 36–48 in. in height.
'Potgieter'. Primrose yellow. 36–42 in. in height.
'Stolze von Berlin'. Pink. 36–48 in. in height.
'Zonnegoud'. Canary yellow. 36–48 in. in height.

Cactus Dahlias
'Apple Blossom'. Rose with lighter center. 36–60 in. in height.
'Border Princess'. Salmon orange. 24 in. in height.
'Doris Day'. Cardinal red. 42 in. in height.
'Good Earth'. Pure lilac-pink. 48 in. in height.
'Park Princess'. Pink and rose. 24 in. in height.

Semi-Cactus Dahlias
'Apache'. Bright red, fimbriated. 42 in. in height.
'Belle Dame'. Salmon pink. 48 in. in height.
'Firebird' (syn. 'Vuurvogel'). Primrose yellow shading to orange-red. 42–48 in. in height.
'Hazard'. Mandarin red and apricot. 48 in. in height.
'Highness'. Pure white. 48 in. in height.
'Hit Parade. Signal red. 48 in. in height.
'Hoek's Yellow'. Creamy primrose. 48 in. in height.
'Moonglow'. Pure white. 48 in. in height.
'My Love'. Creamy white. 36 in. in height.
'Popular Guest'. Purple-rose with lighter center, fimbriated. 48 in. in height.
'Rotterdam'. Dark-velvet red. 48 in. in height.
Cactus-flowered. Mixed colors. 4 ft. in height.
Double pompon. Mixed. 3 ft. in height.
Large-flowered. Double, mixed. 5 ft. in height.

Dahlias From Seed
'Redskin'. Mixed colors, double and semidouble. 12–14 in. in height.
'Rigoletto'. Mixed colors, double flowered (yellow, red, orange, pink, and white). 12 in. in height.

DAHLIA

D. gracilis Native to Mexico. Brilliant orange-scarlet. 48 in. in height. Summer flowering.
D. merckii Native to Mexico. Introduced before 1839. Lilac. 24–36 in. in height. Summer–Autumn flowering.
D. scapigera Native to Mexico. Introduced 1838. Various colors. 24 in. in height. Summer flowering.
D. superflua see *D. rosea*.
D. zimapanii syn. of *Cosmos diversifolius*.

DAUBENYA — LILIACEAE

Named in honor of Dr. Charles Daubeny (1795–1867), professor of botany at Oxford University. A genus of a single species, however, in the literature, other species appear, but today only one, *D. aurea,* is recognized. This is a most attractive plant, looking much like a *Massonia* or *Whiteheadia.* It has 2, dark green leaves, rounded at the tips, resting on the ground, produced opposite each other and touching at their base. Between these leaves, the flowers are produced. *D. aurea* is found in the Cape Province of South Africa.

The leaves lie on the ground but arch a little, are some 4 in. in length, 2–3 in. in width. Flowers are a bright golden yellow with a hint of orange, almost sessile, and the tips of the petals reach only 4 in. in height. The perianth segments and stamens unite into a ring about 2 in. in diameter. Flowers in June. Height 3–4 in.

CULTURE A sandy, well-drained soil in sun, with the bulbs being set 2–3 in. deep and spaced some 6 in. apart. Not hardy, only for frost-free areas and elsewhere in the cool greenhouse where they can be protected from frost and temperatures below 40°F. Moisture required during the winter months and in the spring, but, after the foliage begins to die back, water should be held and the bulbs allowed to become dry while dormant. Grows well in containers in a sandy soil mix. Little or no feeding required.

PROPAGATION Offsets produced by the parent bulbs is the best way to propagate these plants. Seed can be sown in the spring, only very lightly covered in a sandy soil mix and given temperatures in the 50°F range at night and bright light. After germination and as soon as the plants are easy to handle, transplant to individual containers and grow them on until they reach flowering stage and they can then be set outside in warmer climates.

PESTS & DISEASES No special problems but watch out for slugs and snails as the foliage is close to the ground.

USES For the bulb fancier only and in rock gardens in warm climates.

SPECIES

D. aurea (syn. *D. coccinea*). Native to South Africa; introduced 1832. Described above.

D. coccinea. See *D. aurea.*

D. fulva. is described as being a little taller and having more reddish flowers, but this is most likely only a variation of the type. As this was introduced in 1836, it would appear that *D. aurea* is the name of standing.

COMMENTS A great little plant for the collector of bulbs because it is rare, but will never become a best seller or much cultivated.

DICENTRA — FUMARIACEAE*

The name is derived from Greek *dis* ("twice") and *kentron* ("spur"), the flowers being two-spurred. A genus of some 18–19 species, some with tuberous rootstocks, some with a rootstock that is a rhizome which is used by the plant to overwinter. It should be noted that in some of the literature the spelling of the genus is *Dielytra*, a mistake in the spelling. The genus is quite well distributed from North America to Japan to the Himalayas.

The flowers are most attractive, the 2 outer petals being reflexed or spreading, while the inner are held together over the style and anthers. This configuration gives the shape that is known as Dutchman's Breeches.

The leaves are stalked and much divided. Flowers are held in terminal racemes and are pink or yellow. These plants have a number of common names, Lady's Locket, Bleeding Heart, Lyre Flower, these being given to various species; Turkey Corn and Staggerweed refer to *D. exima.* Spring/early summer-flowering.

CULTURE All of the species prefer a cooler place in the garden with shade and a rich, deep humus with good moisture. The rootstocks should be planted some 2–3 in. deep and the plants spaced some 8–12 in. apart. They should not be exposed to full sun. Except where noted, the species are hardy. They make good container plants. Fertilizing is not required in good soil but some liquid organic fertilizer can be applied as soon as growth is noticed in the spring. As these plants are rather fleshy, care should be taken not to overfeed or the plants can become a little rank in growth.

PROPAGATION Lifting and dividing the tuberous and rhizomatous rootstocks in the early spring is an easy way to propagate the species. They also can be propagated from seed; best sown as soon as possible after harvest and grown under cool conditions with good moisture and a soil mix rich in organic matter.

*See Appendix for discussion of the family.

PESTS & DISEASES Being rather fleshy plants they provide an attractive target for slugs and snails and precautions should be taken against attack from these pests.

USES Great plants for the cool, shady parts of the garden, in woodlands, and on the north side of buildings.

SPECIES

D. canadensis. Native to the eastern part of North America, from Nova Scotia south to North Carolina; introduced 1822. Rootstock small, rounded, yellow tuber. The plants grow to 6–12 in. in height. Usually only one leaf is produced and this is grayish in color and finely cut, always basal. Flowers are in a raceme with the spurs rounded, inner petals notedly crested, greenish-white with hint of purple, fragrant. Flowering April–May.

D. cucullaria. Native to Nova Scotia south to North Carolina and westward into Kansas; introduced 1731. Rootstock is a cluster of small tubers. Quite similar to *D. canadensis,* but not quite so tall-growing and has more foliage. The spurs are held at right angles to each other and the flowers are some 1 in. in length, white tinged with yellow. Flowering in April–May.

D. eximia. Native to the eastern part of the United States from New York to Georgia; introduced 1812. Develops from a short, fleshy rhizome and reaches up to 18 in. in height. A constant supply of pinkish flowers are produced in panicles, the tips of the outer petals are spreading; flowering early summer. A soft pink selection has been made; this also has foliage that is glaucous and has been given the name 'Spring Morning'.

D. formosa. Native to the western area of the United States and north to lower British Columbia; introduced 1796. Rootstock is a fleshy rhizome. The flowers are produced in a raceme on a leafless flowering stem 18 in. in height. Foliage is basal and forms a hummock above which the drooping flowers are displayed. The flowers are not large, vary in color from pink to white, and resemble tiny lockets, a little over ½ in. in length. Flowering May–June. A number of selections have been made: 'Alba', a pure white; 'Adrian Bloom', grayer foliage and flowers of rich red/carmine; 'Paramount', dark carmine flowers. The selection 'Bountiful' is well named as it spreads quite rapidly by seed.

D. peregrina (syn. *D. p.* var. *pusilla, D. pusilla*). Native to eastern Siberia to Japan. Unlike many of these species, is at home on sandy, well-drained slopes and is later in flower, coming into bloom in midsummer. The rootstock is a short, erect rhizome. This is a dwarf species barely reaching over 3 in. in height, with few grayish-blue leaves in a tuft. Flowers are pink and quite large for the overall size of the plant, being up to 1 in. in length, but only 2–3 are produced. A lovely little plant for the cool greenhouse; the protection being needed as it is such a small plant and easily upset by too much moisture, and being low-growing the beauty is missed in a woodland border.

D. spectabilis (Bleeding Heart, Seal Flower). Native to Siberia and Japan; introduced 1816. A plant that produces fleshy tubers on its roots which, when being divided and having a piece of root with an "eye" or bud above the tuber, will produce a new plant. 18–30 in. in height and flowering in spring and early summer. The foliage is much cut, the segments ovate and wedge-shaped. The flowers are rosy crimson and held in a graceful arching raceme, the individual flowers being an inch in length. White forms are known and many selections have been made from this species including 'Pantaloons', a pure white, and ideal for shady and moist locations.

DICENTRA

D. formosa ssp. *oregana* Native to N.W. California, S.W. Oregon. Introduced 1927?. Cream, inner petals tipped pink. 10 in. in height. April–May flowering.
D. nevadensis (syn. *D. formosa* ssp. *nevadensis*) Native to Sierra Mts. of Central California. Outer petals cream to pink; inner white, pale yellow at tips. To 18 in. in height. July flowering.
D. oregana syn. of *D. formosa* ssp. *oregana*.
D. pauciflora Native to California. Pale pink-purple, tipped darker. 2–4 in. in height. May–Aug. flowering.
D. pusilla see *D. peregrina*.
D. uniflora Native to Western America from British Columbia to California. White to pink or lilac. 2–3 in. in height. May–July flowering.

DICHELOSTEMMA — AMARYLLIDACEAE

The name is derived from the Greek *dichelos* ("bifid"), *stemma* ("crown"). This refers to the bifid (divided halfway down into two parts) staminodes of one of the species, *D. congestum*.

This genus formerly was included in *Brodiaea* and is so described in this book. Great discussion has occurred over the family of the genus. Some botanists place it in Amaryllidaceae due to the umbellate inflorescence; others in Liliaceae due to the flowers having superior ovaries; and still others in Alliaceae because the characteristics of both families are combined. Thus confusion reigns.

The 5 species are native to the Pacific coastline of the United States. They are (synonyms in parentheses): *Dichelostemma congesta* (*Brodiaea pulchella*), *D. ida-maia* (*Brodiaea coccinea, Brodiaea ida-maia*), *D. multiflorum* (*Brodiaea multiflora*), *D. pulchellum* (*Brodiaea capitata*), and *D. volubile* (*Brodiaea volubilis*).

DIERAMA — IRIDACEAE

From the Greek *dierama* ("like a bell," "funnel"), because of the shape of the perianth. There is a great deal of discussion about the number of species that belong to this genus. Some authorities are of the opinion that there are only a few; others believe there are as many as 30. There is no doubt that, over the years, the numbers will keep changing until a final resolution is made. All of the species are native to southern Africa.

The corm is large. The leaves remain on the plants all year so that the corm is never really in a dormant state. The leaves are long, more than 20 in. in most species, narrow, rigid, and quite grasslike. The outstanding feature of *Dierama* is the arching flower spike, which rises above the foliage, bending with the pendant flowers. The slender flower stems move constantly in even the slightest breeze, calling attention to them. The flowers open successively so that the spikes are of interest over a long period of time. The flowers are carried in a panicle, with nearly equal segments. The shape is like that of a bell (hence the name) with colors generally in the purple range, varying from very pale to very dark.

CULTURE Needs good, well-drained garden soil; adequate moisture during spring and early summer; sun, but in hottest climates should receive some shade during height of day.

Plant spring-flowering species in late summer/fall; fall-flowering in spring. For best display, set corms in groups of 5–7, 3–5 in. deep, and 12–18 in. apart. As plants will remain in ground for several years, advantageous to incorporate bonemeal into soil at planting time and continue to feed every spring as new growth is made.

Not completely hardy but can withstand some frost, but protect where cold weather is protracted. Arching stems are best seen over water so in such location place just above waterline, but not where soil remains constantly wet. Resent being transplanted; lift only after growing in same location for number of years.

PROPAGATION Easiest method is by lifting and separating offsets on parent and replanting. Best done after flowering, for both spring- and fall-flowering species.

Seed can be sown late summer, or as soon as ripe, for spring-flowering, or in spring for fall-flowering types. Use sandy soil mix, barely covering seeds. Preferable to transplant to individual containers when large enough to handle. Plant out into final location when of good size.

PESTS & DISEASES No special problems, but sometimes prone to aphid attack.

USES Effective in herbaceous perennial border or as isolated species alongside pond or stream. Should be planted in groups so the graceful, long-arching stems are seen to advantage. As foliage remains during entire year, also effective in isolated beds.

Can be grown in containers; large enough so plants will have ample room to expand because they will remain therein for several seasons.

SPECIES

D. grandiflorum. Native to area around Cape Town. Foliage up to 24 in. long. Thin flower spike reaches more than 36 in. in height. Flowers about 2 in. long, light mauve with darker markings. Late-spring/early summer-flowering. Once established can withstand drought. Rootstock a corm.

D. pendulum. Native to eastern coastal region of South Africa. Corm covered with fibers. Stiff leaves, attached to base of stalk, which reach about 3 ft. in height. Flowers, which appear above foliage, up to an inch long, vary in color from whitish to pink to purple. Late-spring/early summer-flowering. Selections have been made and are listed as var. *album* (white flower) and var. *roseum* (rosy purple).

D. pulcherrimum. Native to eastern Cape Province of South Africa. Most commonly grown species from which many hybrids and selections have been made and now are offered by nurseries.

Whitish corms, older ones have thick cover of dry, parallel fibers. Foliage not so high as in other species and a little wider. Flower spikes reach a height of 6 ft. Quite large flowers, bright purple to good carmine-red; distinctive bracts, white with some browning at base which show well against colorful perianth segments. Late-summer/early fall-flowering (Sept.–Oct.). In wild found growing in damp locations so adequate moisture must be given during summer when in most active growth. Among varieties offered, all with characteristics that resemble *D. pulcherrimum* and flower at same time, are:

'Heron'. Wine-red.
'Kingfisher'. Pale pinkish-purple.
'Port Wine'. Deep wine-red.
'Skylark'. Purple-violet.
'Windhover'. Bright rose-pink.

Hybrids recently have been introduced that are between *D. pulcherrimum* and *D. pumilum* (see below). Intermediate in height being 4–5 ft. between two parents. Better suited for smaller gardens than selections from *D. pulcherrimum*.

'Ceres'. Pale cobalt-violet.
'Oberon'. Carmine-purple.
'Puck'. Light rose-pink.
'Titania'. Light pink.

DIERAMA

D. igneum Native to Eastern Cape, Natal, Lesotho, Transvaal. Pinky-mauve turning deep mauve [corms whitish, with many coats of dry, parallel fibres.]. 36 in. in height. Mid to late Spring flowering.
D. medium Native to South Africa. Rose. 36 in. in height. Spring to Summer flowering.
D. pumilum Native to Natal. White or yellowish. 24–32 in. in height. Late Summer flowering.

DIETES — IRIDACEAE

The name is derived from the Greek *dis* ("twice") and *etes* ("an associate"), describing the petals. It is native to South Africa, Lord Howe Island, and Australia. Many authorities include this genus in *Moraea* but others separate it because the rootstock is a rhizome, not a corm as in *Moraea*. This seems to me a reasonable division and so the genus is listed here separately. The foliage of these plants is evergreen whereas in *Moraea* it is deciduous.

The foliage is sword-shaped, quite broad, tapering to a fine point. It is often quite long and leathery in texture. The numerous flowers are very iris-like in appearance but, in most species, last for only a day. The 3 outer perianth segments are broad, the 3 inner narrower but often of the same length as the outer; the syle is in 3 branches and petaloid, giving a very full flower appearance.

CULTURE Hardy only in warm climates. The rhizomes should be planted just below the surface of the soil, spacing the smaller species some 4–6 in. apart, the taller-growing 18 in. apart, in full sun. They thrive in almost any soil type but, to grow well, must have moisture, but can survive periods of drought. Planting time is almost any time but late summer is best. Fertilizer should be given during the winter and spring months and the amount of water and feeding reduced during the summer months. In climates where frost is experienced the plants can be grown in containers and taken indoors or given the protection of a cool greenhouse during the winter and taken outdoors again in the spring.

PROPAGATION Lifting and dividing the rhizomes in the late summer is the best way to propagate these plants. Seed is also a possibility and easy to raise: sow in a soil mix high in organic matter, with gentle heat in the spring. This will give good results and plants suitable for planting out will often be obtained in one season or two at most.

PESTS & DISEASES No special problems.

USES Great plants for the border in warm climates and for containers in cooler climates if they can be given winter protection. Ideal for planting alongside the banks of streams and around pools.

SPECIES

D. bicolor (Peacock Flower; syn. *Moraea bicolor*). Native to the eastern Cape Province; introduced circa 1886. It reaches a height of some 24 in. or more. Flowers are a light cream with brown blotches at the base of the broader petals. The flowering stem is much branched and many flowers are produced. The leaves are 30 in. in length and some 2 in. in width. The main flowering time in the wild is Sept.–Oct., in early spring, but a number of flowers are produced throughout the summer months.

D. catenulata. See *D. iridioides*.

D. grandiflora. Native to the eastern Cape Province to Natal and eastern Africa. Has perhaps the loveliest flowers of the genus. They are pure white with an orange blotch extending up into the middle of the blade on the outer perianth segments and surrounded on the edges of the tepals by white. The inner parts have just a touch of orange-brown at the base and the petaloid styles have a pinkish hue to them. Flowering midspring–summer. The foliage is almost 36 in. in length and some 4 in. or more in width and makes this plant attractive even when not in flower. Flowering stems reach to 48 in. in height.

D. iridioides (Wild Iris; syn. *D. catenulata, D. vegeta, M. catenulata, M. iridioides, M. vegeta*). Native to the eastern Cape Province and northward into subtropical regions. It has a creeping rhizome. Growing in the wild it forms a great ground cover along shady forest streams, yet also is found in drier regions. Slender stems carry light white-bluish flowers with yellow-orange markings on the lower perianth segments. The petaloid styles are purplish in color and forked. Flowering in late winter; in the wild Aug.–March, and some flowers are produced throughout the summer months. Reaches 18–24 in. in height. The appearance of these plants is more like an *Iris* than any other species in this genus. Foliage is basal, fanlike, 18 in. in length, and sword-shaped. *D. vegeta* also native to the eastern Cape Province. The foliage is dark green, leathery, and arranged like a fan. The flower stalks are slightly branched and reach up to 18 in. in height. The plants given this name are *D. iridioides,* which is now the valid name but in the literature often listed separately. Two forms are known: one flowering Aug.–Nov. in the wild, the other Nov.–Feb. The later flowering type has much broader leaves but the flowers are identical. The Bantu use the thicker foliage for mats.

D. robinsoniana (syn. *Iris robinsoniana, Moraea robinsoniana*). Native to Lord Howe Island; introduced 1877. A tall-growing species reaching up to 72 in. in height. Stems much branched and carry pure white flowers often 4 in. in diameter. The foliage is wide and some 72 in. in length. It becomes a large plant and should be used at the back of a border. Summer-flowering.

D. vegeta. See *D. iridioides.*

COMMENTS These are great plants for warmer climates and no doubt interesting hybrids could be obtained by breeding.

DIOSCOREA — DIOSCOREACEAE

The genus is named in honor of the Greek Pedanios Dioscorides, a native of Anazauba, Cilicia, and the author of a book on medicinal herbs which provided the foundation of all botanic knowledge up to modern times. The genus contains several hundred species, many of them of great economic importance as they produce large, edible tubers that take the place of the potato in warmer climates. Perhaps the Chinese yam, *Dioscorea batatas,* is the best known; the tubers of this plant are often over 36 in. in length. The name *yam* ("to eat") is derived from a dialect of Guinea.

Certain species, such as *D. discolor* and *D. multicolor,* have been grown for their decorative leaves. In certain catalogs many of these have been given Latin names but many of these are not valid. The name *D. sativa* has been used indiscriminately to cover species with edible tubers, and this name also would seem to be without validity.

The plants are twining, with flowers that are either monoecious or dioecious, the male flowers carrying 3 fertile and 3 infertile stamens. The seeds produced are winged. It is interesting to note that certain species have stems that twist counterclockwise. Many of the plants contain chemicals that are used in medicine.

CULTURE Only one species can be claimed to be hardy to any reasonable degree, this being *D. batatas,* but even this species should receive protection in the coldest climates; the top growth will be cut by frost but new growth will emerge each spring. The other species require winter protection to a more or lesser degree. They need deep soil rich in humus and abundant moisture throughout the growing season. It is possible to lift the plants in the late summer-early fall and store them in a frost-proof area in dry sand. The tubers should be planted with some 3–4 in. of soil over them.

PROPAGATION Division of the tubers is the best and easiest way to propagate these plants. This is most effectively done in the spring but must be accomplished while plants are dormant. Cuttings can be taken and rooted in sharp sand in moderate to warm temperatures. Seed can be sown in the spring and should be covered with ¼ in. of sandy loam and kept moist and given temperatures in the 55°F range at night. The plants should be transplanted as soon as large enough to handle but room must be given for the roots to develop, thus, deep containers must be used.

PESTS & DISEASES No special problems.

USES Many species are of great economic importance. Certain species have ornamental foliage and are interesting vining plants.

SPECIES

D. alata (White Yam). Native to India and South Sea Islands; long in cultivation. Tremendous tuberous roots, many feet in length and often weighing over 100 lbs. Stem is angled, and the flowers are monoecious and widely separated, males flower on branched stems, female flowers are simple. Often tubers are produced on the stems. The tuber is much used in the Tropics as food. Leaves oblong or ovate, heart-shaped at base, 7 veins.

D. batatas (Chinese Yam). Native to the Philippines. Tubers 2–3 ft. in length and produced deep in the ground. Thus protected from cold, this plant can be grown in colder climates withstanding temperatures down to 0°F but preferring warmer climates. The smooth stem reaches a height of 6–10 ft. The leaves are deep green, often with tubers in the axils, and heart-shaped at base. Flowers are white and very small, fragrant, in racemes. Var. *decaisneana* has shorter tubers. All flower in summer months.

D. bulbifera (Air Potato). Native to the East Indies. Small tubers sometimes form on the roots and aerial tubers are produced which often reach several pounds in weight. The leaves are alternate, on long stalks, and are heart-shaped. The flowers are greenish, in racemes, produced from the stem in summer. Not hardy, requiring warm conditions.

D. cotinifolia (syn. *D. malifolia*). Native to South Africa. Found growing in wooded areas, in deep shade, with its stems twining for many feet through the undergrowth. The rootstock consists of several fleshy tubers a little less than an inch in diameter and about 2 in. in length. The flowers are monoecious and insignificant, produced in Dec.–Jan. in the wild. According to native custom the foliage is gathered to make beds on which women lie to increase their fertility. Foliage is shiny, heart-shaped, and pale green.

D. elephantipes (Hottentot Bread, Elephants Foot). Introduced 1747. It produces a very large tuber which is baked and eaten by the Hottentots, a tribe native to the Cape Province in the southern part of Africa. Stems are produced annually reaching up to 10 ft high in one season. The flowers are greenish-yellow, sometimes spotted, produced in racemes thrown out from the axils of the heart-shaped leaves produced in summer. There are both male and female flowers. Needs protection from the cold; suitable for greenhouse conditions or warm climates.

COMMENTS A genus containing species of great economic importance and likely to remain so. Interest lies in the size of the tuber produced.

The following species are listed in the literature. The information available is sparse and is given here only to acquaint the reader with species listed. Many of them are of doubtful standing.

DIOSCOREA

D. aculeata (Birch-rind Yam, Goa Potato) Native to Tropical Asia. According to Dr. Seeman, at Viti, it never flowers or fruits. Root sweetish, edible.

D. argyraea (Possibly a form of *D. discolor*) Native to Colombia. Leaves green, heart-shaped.

D. atropurpurea (Malacca Yam, Rangoon Yam) Native to Siam, Burma. Edible.

D. balcanica (Balkan Yam) Native to N. Albania, S.W. Yugoslavia. Insignificant, green flowers, tuberous, rare. Climbers to 24 in. in height. June–July flowering.

D. caucasica Native to Caucasus. Introduced 1894. Rhizome thick, horizontal. Flower greenish. May be used as a climber in shady areas.

D. cayenensis (Yellow Yam) Native to Tropical S. America. Probably of hybrid origin. Tubers superficial. Edible.

D. crinita Native to Natal. Introduced 1884. White, racemes many, drooping. Climber. Sept. flowering.

D. daemona Native to the East Indies. Bitter root eaten only when food is scarce.

D. decaisneana variety of *D. batatas,* which see.

D. deltoidea Wild & cultivated in the East Indies. Edible.

D. discolor Native to Ecuador, South America. Root tuberous, flowers greenish.

D. divaricata (Chinese Potato, Cinnamon Vine) Native to China, Philippines. Introduced to U.S.A. in 1855. Used as an oriental climber more so than a yam.

D. fargesii Native to W. China. Has both aerial & root tubers, edible.

D. fasciculata Native to Tropical E. Asia. These edible tubers resemble potatoes more than any other yam.

D. glauca syn. of *D. quaternata.*

D. globosa Native to East Indies. White rooted yam, edible.

D. hastifolia Native to Australia. Tubers cultivated by aborigines.

D. hirticaulis ("Wild Yam") Native from New Jersey to Georgia. Stem twines counterclockwise.

D. illustrata (Probably a form of *D. discolor*) Native to Brazil. Introduced 1873.

D. japonica Native to Japan. The root is sliced, boiled & eaten.

D. malifolia see *D. cotinifolia.*

D. multicolor Native to Brazil. Introduced 1868. The leaves vary; variegated, blotched, veined.

D. pyrenaica Native to N. Spain, S. France, Pyrenees. Green flowers; tiny rare yam. Up to 6 in. in height. June–Sept. flowering.

D. oppositifolia Edible yam from East Indies.

D. pentaphylla Native to Tropical Asia. Tubers may be eaten, but said to be bitter.

D. piperifolia Edible roots from South America.

D. purpurea Native to East Indies. Tubers crimson-red outside & white inside; cultivated as sweet potatoes.

D. quaternata (syn. *D. glauca*) Wild yam is found in woods from Pennsylvania to Florida, West to Missouri & Oklahoma.

D. quinqueloba Edible yam from Japan.

D. racemosa Native to Central America. Introduced 1850. Flower yellow & purple, tuberous. To 96 in. in height.

D. retusa Native to South Africa. Introduced 1870. Dull yellowish. Twining.

D. rubella Native to East Indies. Large tubers, said to be excellent eating.

D. spicata Native to East Indies. Edible.

D. tomentosa Doyala Yam of India.

D. trifida Native to South America, West Indies. Small tubers cultivated.

D. triloba Native to Jamaica. Small yam—9 in. × 3 in., edible.

D. villosa Native to Eastern U.S.A. Wild yam.

D. vittata Leaves heart-shaped, variegated red & white.

DIPCADI — LILIACEAE

Genus name is derived from an old Oriental name for a species of *Muscari* and is comprised of some 25–30 species found in South Africa, southern Europe and in the East Indies. The plants are found in dry, often rocky, ground, and the plants are generally coastal. In the literature, this genus is sometimes listed under *Uropetalum.*

The rootstock is a bulb, generally rounded, almost spherical. The flowers are not unlike those of *Hyacinthus* but are not nearly so attractive since the coloring of the flowers is rather dull, mostly green-brownish or yellowish-brown. The flowers are tubular at the base, the tepals then reflexed, the flowers seemingly pinched at the point where the tepals reflex. The outer segments much reflexed, the inner held more erect. The inner and outer tepals are held in a tube for the lower third of length. The flowers are drooping, held in a loose raceme, produced in the late spring or early summer. Foliage is basal, 3–5 leaves, linear, and shorter in length than the flower spikes.

CULTURE Bulbs should be planted some 1–3 in. deep in well-drained soil in sun. The bulbs are not hardy; the European species, *D. serotinum,* is perhaps the hardiest but even this species must be given winter protection and grown outside only where there is little or no frost. Moisture is needed in winter and spring with a definite drying period after the foliage dies down in the summer.

PROPAGATION Offsets provide the best means of propagation, these being removed from the parent bulbs in the spring. They are then grown on until they reach flowering size. Seed can be sown in the spring in sandy soil with temperatures in the 50°F range at night. The seed should be barely covered, preferably with a little fine sand. Moisture sufficient to keep barely moist and bright light given as soon as the seed germinates. The seedlings should be potted into individual containers the second spring just before growth commences. The plants can be set out at the end of the second or third year from seed.

PESTS & DISEASES No special problems.

USES For the collector only.

SPECIES

D. balfourii. Native to Socotra; introduced 1880. Reaches some 20 in. in height. Flowers greenish-yellow in a loose raceme of 8–12 flowers. The flowers are well spaced, occupying a good third of the flowering spike. 3–4 basal leaves, quite wide, often 1 in. or a little more, erect, and well below the height of the flowering spike. A tropical species flowering in late summer.

D. brevifolium. (Brown Bells, Curly-curly). Native to the coastal areas of South Africa. Bulb is almost spherical. Leaves basal, 2–3 held almost erect, linear, and 10–15 in. in length. The flowers are carried to a height of some 20 in. with 6–8 flowers per flowering stem. Flowers are greenish with a hint of bronze on the outer segments; perianth tube is much pinched just below the point where the tepals reflex. Flowering in the spring to early summer, Sept.–Nov. in the wild.

D. serotinum. (syn. *Uropetalum serotina*). Native to southwestern Europe and north Africa. Reaches some 18 in. in height. Flower color is variable, being in the yellowish-brown, greenish-brown range, with a hint of red sometimes apparent. Flowers are drooping, about 1 in. in length, the outer tepals recurving while inner tepals are held more erect. Up to 5 leaves, basal, linear, and shorter than the flowering stem. Flowering late spring–early summer, May–July.

D. viride. (syn. *D. umbonatum*). Native to the eastern coast of South Africa, from Port Elizabeth to Natal; introduced 1865. In its habitat it is a common plant, flowering in November and reaching a height of some 15 in. It is unusual in flower in that the inner segments are held together and form an urn shape, the outer tepals much reflexed and bending to touch the pedicel; while interesting they are not significant. Foliage clasps the flowering stem at the base, just exceeds the height of the flowering stem; folds to form a tubular leaf for much of its length. The bulbs are used by the natives of South Africa for their babies, possibly to cure colic.

COMMENTS These will never be widely grown, except by collectors who will appreciate the rather interesting flower shapes.

DIPCADI

D. ciliare Native to Karoo, Transvaal, Natal. Brown, green or yellowish. To 16 in. in height. Summer flowering.
D. crispum Native to Namaqualand, Namibia, W. Karoo. Brown to orange. To 12 in. in height. Winter flowering, May–July in wild.
D. glaucum (syn. *Uropetalum glaucum*) Native to South Africa. Introduced 1814. Brownish-green. 24–36 in. in height. August flowering.
D. longifolium (syn. *Uropetalum longifolium*) Widespread in tropical Africa. Green. 12–40 in. in height. Summer flowering.
D. tacazzeanum (syn. *Uropetalum tacazzeanum*) Native to Northeast Africa, drier parts of Tropical Africa. Introduced 1892. Green. 6–9 in. in height. Summer flowering.
D. umbonatum see *D. viride*.
D. welwitschii (syn. *Uropetalum welwitschii*) Native to Angola. Introduced 1867. Green. 12 in. in height. Summer flowering.

DIPIDAX — LILIACEAE

The name is derived from Greek *dis* ("twice") and *pidax* ("spring"). It is a genus of but 2 species, both native to South Africa, and related to *Wurmbea*. They like to be near or submerged occasionally in water. The rootstock is a corm. In one species, *D. triquetra*, the foliage is rush-like and the flower spike is enveloped in the sheath formed by one of the leaves. In the other species the leaves clasp the stem but do not cover it entirely. Now authorities place this genus in *Onixotis*.

The flowers are in terminal spikes, either white or white with pink, and often over an inch in diameter; 15 or more flowers in a spike open over a long period of time; in *D. punctata* the number of flowers is not so high. The flowers form an open cup-shape, the petals forming a shallow cup at their base; there is no perianth tube, the petals being separate from one another.

CULTURE The plants are not hardy in temperatures below the 35°F range and thus are suitable only for those areas where very little or no frost is experienced. They require moisture and have no objection to very wet conditions during the winter months. They should be planted along the sides of pools or in moist areas. The corms should be set some 2–3 in. deep, spacing them some 4–6 in. apart, the distance depending on the ultimate height of the species grown. The soil preferred is high in organic matter and if sandy, fertilizer can be given as soon as growth is seen in the spring.

PROPAGATION Cormels produced by the parent corms can be removed in the spring and grown on in soil that is kept moist. Seed can be sown in the spring, barely covering the seed and using a soil that holds moisture. Temperature for germination should be in the 55°F range at night. The plants should be transplanted again as soon as large enough to handle and then transplanted into individual containers or spaced 2–3 in. apart in deeper containers. The plants should reach flowering size in 2–3 seasons of growth.

PESTS & DISEASES No special problems.

USES Around the edges of ponds or in damp places either in full sun or light shade. The flowers last quite a long time and these plants are quite attractive.

SPECIES

D. punctata (syn. *D. ciliata, Onixotis punctata*). Native to the Cape Peninsula. Flowers in August in the wild, in very eary spring. It reaches a height of some 6–12 in. Foliage is carried around the stem, clasping the stem at the base of the leaves and then growing outward. The leaves are broad at the base and taper to a point, a light grayish-green in color. The flowers are white; petals spread to give a flattish flower that is slightly upward and outward facing. The flowers are less than an inch in diameter and 10–15 are in the flower head. The yellow stamens are quite prominent.

D. triquetra (Waterflower; syn. *Onixotis triquetra*). Native to the Cape Province. Widely distributed along the Indian Ocean seaboard of South Africa. Always found close to water or growing in water in its native habitat. The dark green foliage is reed-like, and few leaves are produced, often only 1, sometimes 2. The leaf will clasp the flower spike and from the curled leaf the flowering spike emerges subtended by a leaflike bract. The foliage will rise above the flowers and reaches a height of some 18 in. The flower spike will exceed the height of the foliage when fully developed and the top flowers are open, but then only just. The flowers are each over an inch in diameter, as many as 18–25 per 4–6 in. spike. The individual flowers are an inch or a little more in diameter, the petals opening to give a shallow, bowl-shaped flower, the perianth segments not joined. The color is white, with a pinkish hue sometimes suffused through the petals, and with a definite carmine color band at the base. The superior ovary is also carmine in color. Petals frequently will remain as the ovary swells and give a prominent carmine center to the flower. The flowers are spaced close together on the stem and are sessile or nearly so. Some variation in the color is known, some of the flowers being almost a light purple.

COMMENTS *D. triquetra* is a plant that deserves to be grown in moist areas in warmer climates; it is presently rare in cultivation.

DRACUNCULUS — ARACEAE

The name was used by Pliny for a plant with a curved rhizome. The genus was formerly included in *Arum,* and is commonly called the Dragon Arum. In general form the *Dracunculus* look like calla lilies, except for the leaves, which are handsome and deeply divided. There are only 2 species, 1 from the Mediterranean, which is the most commonly grown, and the other from the Canary Islands.

The male and female flowers on the spadix are next to one another and are not separated by sterile or rudimentary flowers. Both species flower in the summer, and both are noted for their unpleasant smell, which attracts flies, which, in turn, pollinate the plants.

CULTURE Rich soil with plenty of organic matter and abundant moisture, good drainage, and full sun suit plants best. Set rhizomes 5–6 in. deep and 18–24 in. apart. Appreciate top dressing of organic matter every spring. Lift and divide every 3–4 years or will become overcrowded and flowering will be reduced.

PROPAGATION Rhizomes are divided, cut into sections each with a bud, and planted back. Best done in early spring. Seed can be sown in spring in fairly rich soil mix. Keep moist. Transplant when seedlings large enough to handle. Grow on in nursery rows or individually grown on until planting-out size.

USES Striking plant when in good location surrounded by plants that complement the bold foliage and lovely flowers. Set out where bothersome flies will not be attracted and where odor will not cause problem. Also bold plant on edge of woodland areas.

SPECIES

D. canariensis. Native to the Canary Islands. Pale green leaves, on long stalks, covered with purplish spots where they wrap around stem of flowering spike; divided into several segments with middle ones up to 6 in. long, often 4–5 segments found on end of a foot-long leaf stalk. Narrow spathe, tube at base couple of inches long, with blade up to 12 in. long; yellow spadix ends in long tail, more than 6 in. long; flower stalk heavily spotted; 24–36 in. high. Flowering mid- to late summer. Total height of plant 36–48 in.

D. vulgaris (syn. *Arum dracunculus*). Native to Mediterranean region; introduced 1910. Most commonly grown of the 2 species. Tuber large and rounded. Height about 36 in. Leaves divided into several lobes, usually 5–7; base wraps around stem of flower spike; leaf stalk pale green marked with darker green. Striking plant with crimson-red spathe, very dark red, almost black spadix; spadix stalk good red, surrounded by tube of lower part of spathe; tube 2 in. long, striped purple at mouth; spathe often 12 in. long and more than 6 in. wide, greenish-purple exterior. Flowering early to midsummer. Flowers followed by a crop of scarlet berries in late summer/early fall.

DRIMIA — LILIACEAE

The name is given due to the sap from the roots being acrid and causing inflammation of the skin, *drimys* ("acrid"). The genus is related to *Scilla* but has compressed seeds and also the base of the perianth segments are united. There are some 8 or more species, native to South Africa and to tropical Africa, few if any in cultivation.

The rootstock is a scaly bulb with roots that are produced from the bottom and from among the lower scales. The size of the bulb varies with the species, those of *Drimia hyacinthoides* being 3 in. in diameter, while those of *D. minor* are barely ½ in. Generally, 3 basal leaves are produced and these are soft, smooth, and linear. A long raceme is produced in summer, often with 20 or more flowers, but they are not long-lasting. Color varies from pinkish-purple to yellow. Petals are united at the base; flowers are not wide opening but in certain species, notably *D. media*, the tepals are much reflexed. Few if any of these bulbs are in cultivation and the plants are not of great horticultural merit.

CULTURE Free-draining soil, full sun, and moisture during the spring and summer are needed by these plants. They are not hardy and will not take any frost. Bulbs should be planted 2–3 in. deep, the smaller species being planted just below soil level. Plants require a dry period at the end of summer after the foliage has died down.

PROPAGATION Offsets are produced and these should be removed during the dormant season. Seed can be sown on sandy soil, barely covering the seed. They should be given 55°F at night, kept barely moist, and then transplanted to individual containers as soon as growth starts in the second season.

PESTS & DISEASES No special problems.

USES For the collector only.

SPECIES

D. anomala. Native to South Africa; introduced 1862. One or two leaves, 18–20 in. in length, quite thick and fleshy. Up to 30 yellow flowers produced in a loose raceme on a stem that reaches to a height of some 18 in. Summer-flowering.

D. coleae. Native to South Africa. Very small plant, only some 8–10 in. in height. Flowers greenish and anthers purple. Summer-flowering.

D. elata. Native to sandy areas quite close to the coast in the Cape Peninsula. Up to 20 flowers in a 4 in. long raceme, tepals much reflexed, white, and the stamens thrust well forward and held tightly together. Up to 18–35 in. in total height, sometimes more. Flowering in late summer with the leaves appearing after the flowers. Foliage basal, 2–3 leaves, linear, and up to 14–18 in. in length.

D. hyacinthoides. Native to the eastern Cape Province of South Africa. Found growing in full sun and reaching a height of some 18–20 in. Flowers purple, hardly opening; outer tepals margined white at the tips, the inner with a hint of white. Number of flowers often more than 30, drooping as they open, and produced in midsummer. Bulb large, often 3 in. in diameter. Foliage erect and basal, 3 leaves produced, soft, green, and smooth.

D. media. Native to the eastern Cape Province on the Indian Ocean side. Found growing in the coastal scrub along the seashore. One of the taller species, often over 20 in. in height. Flowers purplish, tepals much reflexed, and stamens prominent. Stamen filaments are blue-gray in color, anthers are black. Foliage narrow and linear, produced at the base of the stems and 20–24 in. in length. Summer-flowering; Jan.–March, in the wild.

D. minor (syn. *Urginea pygmaea*). Native to the Cape Peninsula. Found growing along the coast on the eastern side. Dwarf plant barely 3 in. in height. Flowers less than ¼ in. in diameter; tepals open to produce tiny flat flower, 2–3 per plant. Foliage also very dwarf, about 1 in. in length with bulb ½ in. in diameter or less. Flowering in late summer; March in the native habitat.

COMMENTS Not of great interest and doubtful if worthy of much attention.

DRIMIA

D. ciliaris Native to S. Namaqualand. Brown-purple. To 6 in. in height. Summer flowering.
D. ensifolia syn. of *Ledebouria undulata* which see.
D. forsteri Native to S.E. part of Cape Province. Cream & green. 36 + in. in height. Late Summer flowering.
D. haworthioides Native to E. Cape & Karoo. Greenish brown. 8–16 in. in height. Summer flowering.
D. villosa Native to Cape of Good Hope. Introduced circa 1824. Green filaments, pinkish flowers. To 10–12 in. in height. Flowering early summer, before leaves.

ELEUTHERINE — IRIDACEAE

The name is derived from Greek *eleutheros* ("free"), a reference to free filaments. A tropical South American genus comprised of only a few, perhaps 2, species. Not commonly grown in cultivation if at all, but quite widespread in the Tropics and reportedly in many parts of the world. The rootstock is a bulb composed of red scales loosely held. The white flowers are produced in stalked clusters, each lasting but a short time but many are produced. Being a tropical plant, the flowering time is throughout much of the year. Flowers are some ½ in. in diameter and each flowering branch is subtended by a bract. The flowering stalk can be some 10 in. in height but is often less. The bulbs produce 1 or 2 leaves, up to 18 in. in length and folded.

CULTURE Needs heat and moist air, suitable only for greenhouse conditions where the temperature falls no lower than 55°F with good humidity. Requires a soil mix that is quite free draining. Bulbs should be set 1–3 in. deep and never allowed to dry out. Only in very warm climates can these plants be grown outdoors.

PROPAGATION By seed, sown in free-draining soil mix in warm temperatures (60°–65°F at night) in the spring; plants transplanted as soon as large enough to handle and when of size set out. Established plants can be lifted and the smaller bulbs separated from parent bulbs.

PESTS & DISEASES No special problems.

USES Need tropical conditions, but then more for the fancier of bulbous plants as there are many tropical plants with more merit.

SPECIES

E. anomala. Listed in the literature but is regarded as similar to *E. plicata*.

E. plicata. (syn. *E. bulbosa, Keitia natalensis, Sisyrinchium palmifolium*). Native to tropical South America. Red-scaled bulb produces 1 or 2 linear leaves, 1 in. wide and 12–18 in. long. Many white flowers, ½ in. in diameter, grow in clusters; last only a short while, but are produced for much of the year, spring being the most common time. Plant reaches 12–18 in. in height.

COMMENTS This genus, while apparently quite common in the Tropics, is not generally referred to in the literature; a not unusual occurrence with plants from South America.

EMINIUM — ARACEAE

The name is an ancient one and was used by Dioscorides. A genus of perhaps some 3–4 species, but information is rather scant. They are native to central Asia and are very rare in cultivation. The rootstock is a tuber that produces 2–3 leaves which have the habit of twisting around the midrib. The spathe is tubular at the base, which is often at or even below ground level, flattens out and is held away from the spadix. The erect spadix is slender and shorter than the spathe. The sterile flowers are awl-shaped and thus differ from *Sauromatum* in which they are club-shaped. The plants have an unpleasant smell, and are quite hardy, provided their need to be kept dry during the summer months is taken into account.

CULTURE Sun and well-drained soil are essential. The tubers should be set an inch or so deep into the soil. Water is needed in the early part of the year but the plants require a dry period in the summer. In most soils no feeding is required.

PROPAGATION The best means is by offsets produced by the parent tubers and these are best taken in the spring prior to growth beginning. Seed is produced and this should be sown in the fall and given the protection of a frame overwinter so that the plants do not become too moist.

PESTS & DISEASES No special problem.

USES With their rather rank smell, they are plants for the collector and should be considered as a rare and not-too-attractive plant. However, the spathe of *E. albertii* is quite attractive.

SPECIES

E. albertii. Native to central Asia; introduced 1884. Found growing on slopes that are both earthy and rocky. The tuber is flattish. The spathe is deep carmine and the spadix is also dark. The spathe carried at ground level is some 8–10 in. in length. The foliage is usually much twisted around the midribs and is a lighter green in color with the blades being some 3–5 in. in length. The spadix is about half the length of the spathe. The rudimentary sterile flowers between the male and female on the spadix are awl-shaped. Flowering in early summer.

E. intortum. Native to southern Turkey and Syria. Found growing in poor land in full sun, where it flowers from March–May. The foliage is grayish-green and lies flat on the ground. The spathe forms a hood over the spadix. The spathe carried at ground level is some 6–8 in. in length, crisped along the edges, with a bluish-green hue on the outside with purple veining and a deep crimson, almost black, color on the interior. The tuber of this species is unusual in that it is covered with a white powder. The shape is spherical.

ENDYMION — LILIACEAE

The English Bluebell and the Spanish Bluebell were at one time in this genus. These plants were moved from *Scilla* to *Endymion*, and now they are placed in the genus *Hyacinthoides*, hopefully to have found a permanent place there. See *Hyacinthoides*.

E. campanulatus. See *Hyacinthoides hispanica*.

E. hispanicus. See *Hyacinthoides hispanica*.

E. non-scriptus. Ssee *Hyacinthoides non-scripta*.

E. nutans. See *Hyacinthoides non-scripta*.

ERANTHIS — RANUNCULACEAE

From the Greek *er* ("Spring") and *anthos* ("flower"), name given due to their very early flowering. The species name *Eranthis hyemalis* translates as "flower of spring" and is commonly known as winter aconite. *E. hyemalis* is native to Europe and there are several other species in this genus that are native to Japan, the eastern Mediterranean, and one species from Siberia.

Despite there being so few species, *Eranthis* has been hybridized. *E.* × *tubergenii*, a hybrid between *E. cilicica* and *E. hyemalis*, has the largest flowers of this genus. This, however, is not so much of an advantage as *E. hyemalis* spreads so quickly by seed and natural increase that the size of the individual flowers is of secondary importance.

The *Eranthis* have tuberous rootstocks. The tubers are small, about the size of a pea, irregularly shaped, and whitish, with a hint of brown in older plants. The leaves form a rosette, lying on the soil, and act almost as a calyx for the flowers. Flowers grow on short stalks, and have petal-like sepals and numerous stamens.

The lovely, carpeting winter aconite is often in flower even before the Snowdrops and adds color to the garden early in the year. It is fully hardy and is the most popular. *E. hyemalis* generally is the only species listed in catalogs.

The plants are best grown in areas where they will not be disturbed, such as established shrub borders. Light is essential, however, so deciduous shrubs are preferable. If planted in conjunction with evergreen shrubs, they should be placed where they will receive some sun. The high canopy of deciduous trees suits them well.

CULTURE Not too fond of acid soils with pH below 5.5. Plant tubers in late summer or early fall. Cover with 1 in. of soil, which should be loose and friable, such as that found on floor of mixed forests. Like ample moisture in winter and early spring. Place tubers close together, some 3 in. apart. Best to plant in clumps rather than formal pattern. Once established, leave undisturbed.

Can be cultivated in containers, crowding tubers together in soil mix that is high in humus; equal parts of leaf mold, good topsoil, and sharp sand is good mix. Cover tubers with just a little soil, allow to become established in cool location, and bring indoors when buds are developed and only one or two are in flower. After flowering, should be taken outside and planted in garden. Tubers so treated should not be grown for indoor culture again for several seasons.

PROPAGATION Can be lifted and divided as soon as finished flowering, but best done only if plantings are large. Seed can be harvested and sown directly outdoors. Just scatter over loose soil so that seeds can filter down a little. If to be grown in containers, prepare a flat or pot with loose mixture, place sifted layer of leaf mold on top, firm lightly, and sprinkle seeds over surface. Barely cover seed by sifting small amount of leaf mold over them. Keep cool and moist. Transplant seedlings directly to location where they are to be grown or where they will have sufficient room to develop into small plants. Plant about 2–3 in. apart.

PESTS & DISEASES No special problems.

USES While in flower, look nice when planted with other early flowering spring bulbs, such as *Crocus* and *Cyclamen coum*. Interesting planting can be made in garden using these plants together, especially when combined with early flowering shrubs such as *Hamamelis* (witch hazel). Good plants for woodland areas where they get good light and adequate moisture early in the year. Make interesting container plants.

SPECIES

E. cilicica. Native to Greece and Asia Minor; introduced 1892. Similar to *E. hyemalis* but more robust. Foliage bronzy green, finely cut. Deep yellow flowers carried on short stems, only reach a little more than 2–3 in. in height. Flowering late Jan./Feb.

E. hyemalis. Native to much of western Europe; long in cultivation. Some regard *E. cilicica* as synonymous. Grows to height of 4 in. Leaves form rosette on ground and are stalked if plants not of flowering size. Bright yellow flowers vary in size, depending on location; those with ample early spring moisture are larger. Flowering in Northern Hemisphere Feb./March, Southern Hemisphere, June/July.

E. pinnatifida. Native to Japan. Found growing in woodlands. Smaller and not so strong-growing as other species. White flowers only 1 in. in diameter. Leaves have bluish tinge on undersides. Spring-flowering. Reaches 4–6 in. in height.

E. × **tubergenii** (hybrid between *E. cilicica* and *E. hyemalis*). Introduced 1923 by Van Tubergen. Flowers are light yellow, larger and the appearance of the plants is intermediate between the parents. A selection sometimes offered under name of 'Guinea Gold.' Although excellent plants, do not naturalize as readily as species. Reaches up to 8 in. in height. Leaves intermediate between those of the parents.

ERANTHIS

E. longistiptitatus Native to Central Asia. Introduced 1979. Yellow. 2–4 in. in height. Feb.–May flowering.

D. sibirica Habit of *E. hyemalis*. Native to E. Siberia.

EREMURUS — LILIACEAE

From the Greek *eremos* ("solitary") and *oura* ("tail"), in reference to the flower spike, which towers above the foliage. Common names include Foxtail Lilies and Desert Candles. *Eremurus* are among the most spectacular of the early summer-flowering plants that have a tuberous rootstock. They are native to western and central Asia—Afghanistan, northern India, Turkestan, and Siberia—with 1 species from China. The genus includes approximately 40 species, many not bulbous or tuberous, and many hybrids, but only about 5 or 6 species and several hybrids are generally available.

The rootstock is thick, often looking like a starfish, and must be handled with great care as they are very brittle. The bud is situated where the tubers meet and is rather large, much like half of an egg. The tall, unbranched flower spike arises from a cluster of leaves, which are linear and strap-shaped and vary in length according to the species. White, yellow, or pink flowers are densely packed on the spike. The perianth segments of the flowers often are joined for a short distance at the base, and all open wide.

The plants are quite hardy but, if they come into growth while there is still some danger of frost, they should be given protection, such as a covering of straw. It also is best not to grow them in areas where the early morning sun will strike them while frost still persists. Due to their 6–10 ft. height, they also should be protected from strong winds.

CULTURE Rootstock best planted in September in rich, well-drained, sandy soil in sunny location. Space about 36 in. apart; cover with at least 6 in. of soil. Like to have ample moisture early in the year, while spike is growing and flowering. After flowering, reduce water slightly and allow plants to harden off.

Once planted, allow to remain in the same location for a number of years. Each fall, just prior to the onset of winter, top-dress with good organic mulch. In spring, clear away mulch from flower spike and emerging leaves to avoid rotting, especially if there has been a lot of rain. As mentioned, the individual flower spikes should be supported in windy areas; not necessary in other locations. Needs cold winters to grow well; not suited for frost-free climates.

PROPAGATION Best way is to lift and divide rootstocks. After 3 years of good growth, each plant generally will provide 2 or 3 additional plants. Lifting should be done as the foliage dies down; replant young plants as soon as possible.

Seed can be sown as soon as ripe in fall. Use well-drained soil mix. Place in individual pots when seedlings of sufficient size to handle. After second or third season of growth, plant out, but will need another season or two to flower. Because roots are so fragile, many growers plant seedlings in protected frame, allowing them to stay there until sufficient size for ultimate bedding spot.

PESTS & DISEASES No special problems.

USES Great accent plant, especially if placed with a background of dark foliage. Height of flower spikes must be kept in mind as they can dominate a border, so plant where they will not be out of scale with companion plants. Sunny, well-drained border among earlier flowering shrubs or in middle of perennial border are suitable locations. *E. stenophyllus* makes excellent cut flower.

SPECIES & CULTIVARS

E. elwesii. Introduced 1884. One of tallest of species, reaching more than 6 ft., often as much as 10 ft. when full grown. Flower spikes tower over slightly fleshy leaves, which will reach 36 in. in length. Flower stalks quite long, so, in consequence, spike carries large number of pink blossoms. May flowering. Some authorities consider this species as hybrid between *E. himalaicus* and *E. robustus*, but experimental crosses of the two have not produced *E. elwesii*. Possibly variety of *E. robustus* or identical with *E. aitchisonii*. Pure white form is var. *albus*.

E. himalaicus. Native to Himalayas; introduced 1881. Strap-shaped leaves, little more than foot in length. Pure white flowers crowded on flower spike, which reaches height of 36 in. or more. May flowering.

E. olgae (syn. *E. angustifolius*). Rare species from Iran, to Tadzhikistan in USSR; introduced 1881. Height 36 in. or more. Pale whitish-pink flowers. Very late flowering—July or August. Leaves linear, 12 in. in length, with rough margins.

E. robustus. Native to the Tien Shan and Pamir-Alai regions; introduced 1874. Another tall-growing species, more than 6 ft. Has widest leaves of species, as much as 4 in. wide and 48 in. long; bright green. Flower spike very large, crowded with many deep pink flowers; lowest flowers have long flower spikes while those at top a little shorter. Flowers often cover more than 4 ft. of spike. June-flowering.

E. × shelford (syn. *E. × isabellinus*). Named after the garden at Great Shelford in Cambridge, Eng., owned by Sir Michael Foster. Cross between *E. stenophyllus* and *E. olgae*. Type has orange-buff flowers, but many colors found among Shelford-Ruiter hybrids, all of which are of medium height and all early summer-flowering. Among these hybrids offered are:

> 'Cleopatra'. Orange flowers with darker red midrib on the exterior of the petals; orange anthers.
> 'Image'. Yellow flowers; outside of petals green.
> 'Obelisk'. White flowers with green rib on exterior.
> 'Parade'. Light pink flowers.

E. stenophyllus (*E. bungei*). Native to central Asia, especially Kopet-Dag; introduced 1885. One of shorter species, 24–36 in. in height. Leaves about 12–15 in. long, narrow, many produced. Bright yellow flowers, each with flower stalk several inches long, carried over three-fourths of length of flower spike; open slowly and last quite a while, allowing spikes to remain attractive for extensive period. Early June-flowering. Several selections offered, among them:

> 'Highdown Gold'. Darker yellow flowers than species.
> 'Magnificus'. Brighter yellow flowers than species.
> 'Sulphureus'. Clear, sulfur-yellow flowers.

These are some subspecies, basically slight variations due to geographic location. *E. s.* ssp. *stenophyllus* of Iran and *E. s.* ssp. *aurantiacus* from Afghanistan, W. Pakistan, and the Pamir Alai.

EREMURUS

E. afghanicus Native to E. Afghanistan. White. 40–80 in. in height. April–May flowering.
E. aitchisonii Native to E. Afghanistan, Tadjikistan. Dense spikes of pale reddish. 30–80 in. in height. May–June flowering.
E. albertii Native to N. Afghanistan. Pink. 16–40 in. in height. May–June flowering.
E. angustifolius see *E. olgae*.
E. aurantiacus subspecies of *E. stenophyllus*, which see.
E. bungei see *E. stenophyllus*.
E. cristatus Native to Western & Central Tien Shan. Deep magenta. 18–24 in. in height. May flowering.
E. furseorum Native to N.E. Afghanistan. Introduced 1964. White. 36 in. in height. June flowering.
E. giselae Native to N. Persia. White, tinted red. 12 in. in height. May–June flowering.

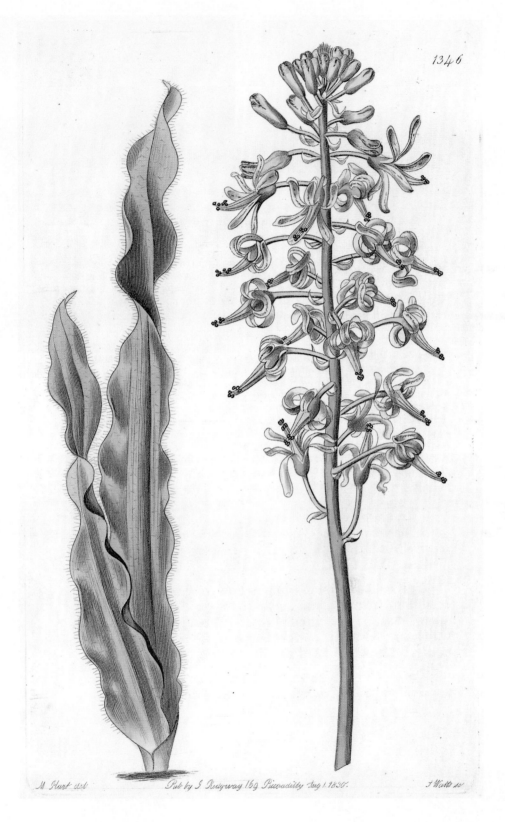

Drimia villosa

Plate 113

Eminium spiculatum **ssp.** *albovirens.* This is a rare species. I am indebted to Chris Lovell for bringing it to my attention. Native of the eastern Mediterranean region, it exhibits the typical habit of having twisting foliage, and of the spathe forming a hood over the spadix. The nomenclature of such rare genera is, to say the least, suspect. LOVELL

Endymion, **now** *Hyacinthioides hispanica.* For many years these plants were known as *Endymion hispanicus,* on their way from *Scilla* to their present home in *Hyacinthioides.* I.B.

Eranthis cilicica. One of the great attributes of this plant is its early flowering, it can start in January, and last for a number of weeks. I.B.

Eranthis cilicica. The bright yellow flowers are a welcome sight early in the year, fully hardy and easy to grow, a plant deserving of attention. DE HERTOGH

Eremurus stenophyllus (E. bungei). This plant is worth a place in any garden, even though the foliage is a little untidy. There are many selections listed in catalogs, but for pure lemon colored flowers, the species itself is hard to beat. Plants do need a cold period during the winter, makin them difficult to grow in mild climates. DE HERTOGH

Eriospermum abyssinicum. This is an extremely rare species found in southern Africa and western Cameroon. It flowers in midsummer, December to January, in its native habitat. The color is a good, clean yellow; height is 12–18 in. BOUSSARD

Erythronium dens-canis. This European species has been long in cultivation, and rightly so. Today there are a number of selections offered in catalogs; all of them are good garden plants. WARD

Erythronium grandiflorum. A lovely species found in California, Oregon and Washington. Flower in March/April. J.H., by L. R. HECKARD

Plate 115

Eriospermum pubescens

Plate 116

Erythronium montanum. A lovely alpine species found at higher elevations on the mountains in the Pacific northwest. This picture was taken on Mt. Rainer in Washington, where it grows in abundance. J.H., by CHARLES S. WEBBER

Erythronium multiscapoideum. The stem will branch before it leaves the protection of the foliage. This lovely species is native to the western regions of the United States. WARD

Erythronium revolutum. The color of the flowers can vary from the purest of whites to a good pink. Several selections are offered in catalogs. AUTHOR

Erythronium tuolumnense* 'Pagoda'** (foreground) ***Erythronium tuolumnense (background). The difference between the species, and the selection known as 'Pagoda' can easily be seen; the flowers are larger and of a purer yellow in the selected form. AUTHOR

Plate 117

Erythronium dens-canis

Plate 118

Erythronium grandiflorum

Plate 119

***Eucomis* hybrids.** The potential of this genus is such that, nurseries are crossing various species. Here are some results, photographed in South Africa at a nursery near Johannesburg. BARNHOORN

Eucharis grandiflora. The most popular species of the genus. While for many years a favorite of florists, it is not quite as popular today. It is still listed in the majority of bulb catalogs. DE HERTOGH

Ferraria crispa* ssp. *crispa. This lovely plant, together with other species of the genus, deserve more consideration for warmer climates. Some slight differences in color of the flowers occur, but all are attractive and, hopefully, will one day be better known. ORNDUFF

Plate 120

Freesia fergusoniae. BOUSSARD   *Freesia occidentalis.* BOUSSARD

There are only minor differences between these 2 species. For the specialist and the botanist, such slight botanical differences merit separate and different species names. Are such always merited? You be the judge.

Freesia. Modern cultivars of this genus are among the most popular of bulbs, being used as cut flowers as well as container plants. Here are beds used for commercial production. AUTHOR

Freesia **cultivars.** The wide range of colors available today speak well of the work of various hybridizers over the years. This, plus being able to have them in flower at any month of the year, has made the *Freesia* one of the most popular of bulbs, not to mention the fragrance! AUTHOR

Plate 121

Fritillaria acmopetala. A pale form of this species, an excellent plant for the woodland. WARD

Fritillaria assyriaca. A lovely species, not too commonly grown, that deserves more attention. WARD

Fritillaria aurea. A Turkish species, of good color, first introduced in 1854. Despite being known for many years it is found in collections, rather than gardens. WARD

Plate 122

Fritillaria bucharica. A rare species, but a delightful plant. The white flowers will often show tinges of green. WARD

Fritillaria camschatcensis. One of the species with the widest distribution, from the Pacific Northwest to Japan, and one of the darkest colors in the genus. WARD

***Fritillaria graeca* var. *thessalica*.** There are a number of forms of the species, perhaps attributable to geographic distribution more than anything else. WARD

***Fritillaria imperialis* 'Aurora'.** These striking plants are very popular, and are at their best when planted in groups of at least 5 or 7 with other spring flowering bulbs. AUTHOR

Fritillaria imperialis **'Lutea'**. Not only striking, but long lasting in flower. Remember the fragrance of these flowers is not very pleasant, so do not plant them close to the home. AUTHOR

Fritillaria involucrata. If traveling in the south of France, in April or early May, keep your eyes open for this plant growing in woodland areas. WARD

Fritillaria meleagris. Probably the most popular species. There are a great number of selections listed in bulb catalogs, all deserving consideration. DE HERTOGH

Fritillaria michailovskyi. Despite its most attractive flowers, this species from N.E. Turkey is rarely grown, but is a favorite of those that know the genus. WARD

Fritillaria nigra

Plate 125

Fritillaria persica

Plate 126

Fritillaria persica. An excellent garden plant, and one that is frequently seen in display gardens of spring flowering bulbs. A little tricky to grow, but if the plants are happy, they will put on a great show each year. DE HERTOGH

Fritillaria persica (**pale form**). While not quite as vigorous, and fewer flowers per spike, this plant catches the eye in a spring garden. Planted correctly I still feel *F. persica* type is a better garden plant. WARD

Fritillaria pinetorum. One of the few species that does not mind being quite dry later in the year, rather a distinctive habit. J.H., by G. T. ROBBINS

Fritillaria pyrenaica. Mostly found in collections of Botanical Gardens, this photo was taken in the botanic garden in Munich. Apparently they have labels to spare! AUTHOR

Gagea minima

Plate 128

E. × *himrob* = *E. himalaicus* × *E. robustus* (syn. *E. robustus superbus*). Garden origin. Light bluish-pink. May flowering.
E. inderiensis Native to Turkestan, U.S.S.R., Iran, Afghanistan. Yellow or white. 36–48 in. in height. Spring flowering.
E. × *isabellinus* see *E.* × *shelford*.
E. kaufmannii Native to Afghanistan, Tadjikistan. White or yellow. 36–48 in. in height. Spring flowering.
E. korshinskii Native to Afghanistan, Tadjikistan. Brownish-orange. 48 in. in height.
E. persicus Native to Iran, Afghanistan. White, tinted red. 12–28 in. in height.
E. regelii Native to Central Asia, Tien Shan. Brownish magenta. Up to 72 in. in height. April–May flowering.
E. × *tubergenii* (*E. stenophyllus* × *E. himalaicus*) Pale yellow. Spikes 18 in. long. May flowering.
E. × *warei* (*E. stenophyllus* × *E. olgae*) Native to Turkestan. Introduced 1900. Orange, a natural hybrid. 96 in. in height. June flowering.

ERIOSPERMUM — LILIACEAE

The name is derived from the Latin *erion* ("wool") and *sperma* ("seeds") referring to the densely woolly seeds of these plants. While there are reportedly some 60 or more species in this genus, few have been described in the literature and it is doubtful if any are of horticultural merit. These plants are found in coastal regions mostly in the Cape Province, South Africa, and tropical Africa. Flowers are produced in midsummer, from December to January, in its native habitat.

The flowers are whitish or yellowish, in a raceme of up to 12–18 flowers; perianth segments free from one another but the densely woolly seeds are distinct. The white, fluffy seeds, carried on the plant after flowering, are the most noticeable feature. The height varies a little with the species but most flower spikes reach some 18–20 in. The flowering spike is often produced after the foliage has died down. The rootstocks are fleshy tubers, often large. The leaves are often solitary, rounded and basal.

CULTURE The plants are not hardy and must be protected against low temperatures in the 30°F range. They need well-drained soil and full sun. Tubers should be planted 1–3 in. deep. Moisture is required in spring and early summer but after the flowers have started to fade water should be withheld. Need greenhouse conditions in colder climates. No fertilizer required if in moderately sandy soils.

PROPAGATION Division of the tubers in spring provides the easiest means of propagation, or seed can be sown in the spring, in moderate heat, in sandy soil mix with the seed barely covered. Temperature at nights 55°F range, and kept moist but never wet. When growth starts in the second spring, transplant young plants into larger containers with the soil being a sandy, well-draining mix. Plants should be spaced 1–2 in. apart and grown on until tubers are 1 in. or longer before being planted out.

PESTS & DISEASES No special problems.

USES For the collector only.

SPECIES

E. brevipes. Native to the eastern coastal regions of South Africa; introduced circa 1880. Rootstock is a large, fleshy tuber. Leaf is some 4 in. in length and nearly as wide. Flowers are small and greenish, produced after the single, rounded leaf has started to die back. This is the most interesting species as the seeds are quite attractive.

Other species, all of them flowering in the summer months, include:

E. bellendenii. Native to South Africa; introduced 1800. 12 in. in height; flowers light blue.

E. mackenii. Native to Natal; introduced 1871. 8–10 in. in height; flowers bright yellow.

E. proliferum. Native to South Africa; introduced 1821. 6–9 in. in height; foliage sessile; flowers green/white.

E. pubescens. Native to South Africa, Cape of Good Hope; introduced 1820. The plant was first introduced into culture by the firm of Colvill of Chelsea, London. Large tuber, irregular in shape, brown in color with the inside being reddish. The leaves are upright, with a long leaf stalk, the venation is distinct and the leaf is covered with many fine hairs. The flowering spike reaches 9–12 in. in height and carries 15 or more flowers. These have long stalks that curve up and away from the stem giving upright-facing flowers. The perianth segments are free to the base, white with a distinct green keel. The flowers are some ½ in. in diameter, with the stamens being a little shorter and held inside the flower. The stigma is longer than the stamens.

ERIOSPERMUM

E. alcicorne Native to Karoo. White with dark keels. 6–8 in. in height. Summer flowering.
E. bifidum Native to E. Cape Province. Flowers green with orange filaments. 8–10 in. in height. Summer flowering.
E. bowieanum Native to South Africa. White. 4–6 in. in height. Summer flowering.
E. capense Native to Namaqualand, E. Cape Province, Karoo. Flowers whitish. 7–8 in. in height. Summer flowering.
E. cordiforme Native to South Africa. White. 4–5 in. in height. Summer flowering.
E. distiflorum Native to South Africa. White. 5–7 in. in height. Summer flowering.
E. dregi Native to South Africa. White. 6–8 in. in height. Summer flowering.
E. dyeri Native to South Africa. White. 5–7 in. in height. Summer flowering.
E. graminifolium Native to South Africa. White. 6–7 in. in height. Summer flowering.
E. lancifolium Native to Cape Province. Flowers white with reddish keels. 6 in. in height. Summer flowering.
E. nanum Native to Cape Peninsula. Flowers yellowish with reddish keels. 6–8 in. in height. Summer flowering.
E. paradoxum Native to S.W. Cape Province. White. 6–8 in. in height. Late Summer flowering.
E. parvifolium Native to South Africa. White. 6–7 in. in height. Summer flowering.
E. patentiflorum Native to South Africa. White. 7–8 in. in height. Summer flowering.
E. porphyrium Native to E. Cape Province. Flowers cream to green. 6–8 in. in height. Summer flowering.
E. pumilum Native to South Africa. White. 3–4 in. in height. Summer flowering.
E. schlechteri Native to Cape Province. Yellow. 6–8 in. in height. Summer flowering.
E. spirale Native to South Africa. White. 6 in. in height. Summer flowering.
E. zeyheri Native to South Africa. White. 4–6 in. in height. Summer flowering.

ERYTHRONIUM — LILIACEAE

Genus name is derived from the Greek *erythos* ("red"). There are about 20 species, the majority native to the west coast of the United States. It is to be hoped that these lovely spring-flowering bulbs will one day be more widely grown in cultivation than at present. Few spring flowers have the beauty and grace found in *Erythronium*, and this is true of all these species and selections. They are not difficult to grow and enhance the garden, in both sun and shade, with nodding, lily-like flowers.

The European species, which also is found in Asia, *E. dens-canis*, is the one to which the common name, and the one now used for the entire genus, Dog's Tooth Violet, was given originally. It was so called, not because of the shape of the flower but rather that of the corm, which is not unlike a dog's tooth, rather pointed, with the size varying from species to species. Other common names are due to their habitats and foliage— Avalanche Lily, because it appears on the mountains as the snows melt; Trout Lily and Fawn Lily, because of the mottled markings on the leaves; and Adder's Tongue and Lamb's Tongue, because of the shape of the bright green leaves.

As mentioned, the corm is rather pointed and is fleshy. The leaves are radical, ovate, or ovate-lanceolate, except in *E. montanum*, in which they are cordate. Generally, two attractive, mottled leaves are produced at ground level, above which the pendulous flowers are carried on leafless stalks. Flowers can be single on a stem or as many as 8, but seldom more. Rarely do the flowers reach a height of more than 18 in.; the average is about 12 in. The flowers have 6 segments, and the petals are reflexed (prominent stamens and stigma). Most of the species flower in March and April.

Erythronium are found in all types of soil—some very poor and rocky, some quite rich, with much organic matter. In their native habitats, many of the western American species are subject to hot, dry conditions during the summer months, covered with snow during the winter, frost during the spring. These have little effect on the plants, which indicates they are very hardy and tolerant of quite severe weather conditions. They like ample moisture in the spring, while they are making their growth, and this must be provided if they are successfully naturalized.

Plants set seed freely. In the wild they are most commonly found in groupings protected by scrub against the heat of the sun, as well as excessive moisture, which is absorbed by the shrubs.

CULTURE Best time for planting is early fall, as soon as plants become available in nurseries. Set 3 in. deep in good, well-drained soil where partial shade is provided during hottest part of day. As noted above, some species more tolerant of moist conditions than others, but must receive adequate moisture in early spring when leaves are making growth. Appreciate additional dressings of fallen leaves when planted in woodland gardens. Best placed where plants can be left undisturbed.

PROPAGATION Lift and divide established plantings when foliage is almost, but not quite, withered. Replant at once. Seed should be sown as soon as it is ripe. Use soil mix with adequate organic matter and once germinated keep plants growing by avoiding any check caused by lack of moisture. Place container in shady location. Seed sometimes is slow (6–8 weeks) to germinate. Seedlings should be transplanted to individual pots and planted out in their second full season. Not quick to flower; may take 3 or more seasons. But perseverance will pay off.

Bulbs in native habitats of certain species become quite dry during late summer; however, when naturalizing, best to store in barely moist material so they do not dry out completely, if held for any time before being planted.

PESTS & DISEASES No special problems.

USES For most effective presentation, plant in groups where they can be left undisturbed. Look well in front of mixed shrub borders that are lightly shaded or in woodland settings. In areas where the need of warmth in summer (of certain species) can be coupled with good winter and spring moisture, should be tried on grass-covered slopes. Will stand full sun, but light shade is preferable.

SPECIES & CULTIVARS

E. albidum (Adder's Tongue). Native to the eastern part of North America from Ontario, Canada, south to Texas; introduced 1824. Corm produces stolons, thus spreading easily and quickly if location is right. Narrow leaves, 4–6 in. long, give rise to its common name. Foliage sometimes mottled, sometimes not; mottled-leaf varieties generally offered by nurseries. White flowers, some with bluish tinge on outside; 1 flower per stalk; height of plants 6–9 in., sometimes more. March-flowering. Many catalogs list white form under species name; *E. mesochoreum* is name given to light lavender form with unmottled leaves.

E. americanum. Native from Ontario, Canada, down the eastern Atlantic seacoast to Florida. Varied habitats, sometimes near the banks of streams, sometimes in rocky ground. Needs moist conditions throughout year. Corms produce stolons, enabling plant to spread. Narrow leaves, mottled with purple and white. Solitary flowers, on stems 4–8 in. high, bright yellow. April-flowering. *E. a.* var. *castaneum* has more brownish-orange flowers.

E. californicum. Native to northwestern California; introduced 1904. One of the taller species, reaching up to 18 in. in height. Richly mottled foliage; longer leaf stalks than many others in genus. Flowers cream colored with large orange-yellow markings at base; usually 1–3 flowers per stalk, although some records show as many as 12, which is very rare. April-flowering. This is often grown as *E. giganteum*.

E. citrinum. Native to Oregon and northern California. Similar to *E. californicum*; height a little less, seldom more than 12 in. in total height, often less. White or creamy white flowers; base of petals deep yellow, often with slight swelling. Stigma is entire. Most commonly 3 flowers per stalk; sometimes more when well established. In catalogs often confused with *E. tuolumnense*.

E. dens-canis. Native to Europe; cultivated since 1596. Most commonly grown *Erythronium* in Europe. Leaves attractively marked, brown to bluish-green, heavily marbled. One flower per stem, not unlike a large cyclamen in shape, with well-reflexed petals, bluish or purplish anthers. Colors range from white through pink to deep purple, but all flowers have ring of red-purple markings around base. Easily grown, loving woodland settings, where it flowers in March or April. Form from Siberia, var. *sibiricum*, has yellow anthers; that from Japan, var. *japonicum*, has deep violet flowers with an even darker center. Many selections and varieties, such as *E. d.* var. *album* (white), *E. d.* var. *longiflorum*, similar to *E. d.* var. *majus*, which both have larger flowers than type, have been named and are offered by nurseries. Included are:

'Frans Hals'. Outside petals imperial purple with greenish-bronze basal spot, inside petals purple with greenish-yellow circle in center; attractively marked leaves.
'Lilac Wonder'. Purple with chocolate-brown basal spot.
'Pink Perfection'. Clear, bright pink.
'Rose Queen'. Pink.
'Snowflake'. Pure white.

E. giganteum. Name used formerly for both *E. oregonum* and *E. grandiflorum* but now regarded as invalid. Often confused with *E. californicum*, probably because of its large size.

E. grandiflorum. Native to the Cascade Range of northwestern California and Oregon and north into Washington. No mottling on leaves, which are 6–8 in. long. One to 5 flowers carried on stems that reach 12 in. in height. Large, golden yellow flowers; red anthers. March–April flowering. Resembles *E. tuolumnense,* except that *E. grandiflorum* has lobed stigmas; the yellow anthers of the former also is a distinguishing feature. Anther color is also a distinguishing point between *E. grandiflorum* varieties; e.g., var. *chrysandrum* has golden yellow anthers; var. *candidum* and var. *pallidum* have white anthers. Two varieties with red anthers, var. *grandiflorum* and var. *nuttallianum,* often confused with species.

E. helenae. Native to California, where it is found growing on wooded slopes on the coast. Much like *E. californicum* but color zones (of white and yellow) are definite, not merging into one another; this has led to species sometimes being called *E. californicum bicolor.* Flowers white with yellow center, giving appearance of cream color; 10–15 in. high; 1–3 per stem. Spring-flowering. No mottling of foliage and thus similar to *E. californicum.*

E. hendersonii. Native to southwestern Oregon and adjacent areas in California; introduced 1887. Enjoys damp spring and dry summer in native habitat, conditions it likes when cultivated in garden. Dark green leaves, mottled with brown-purple and lines of lighter green. Several flowers per stem, 10 or more not uncommon; stems up to 10 in. high; lavender or mauve-lilac with darker purple at base of petals. March–April-flowering.

E. howellii. Native to Oregon. Flowers almost same as *E. citrinum,* but perianth segments without appendages or swellings at base. Height 6–8 in. Mottled leaves. One to 4 flowers per stem; white, with hint of yellow caused by striations of pale yellow, but turn pink as plant matures; white anthers. Flowering April–May.

E. idahoense. Native to border areas of Washington and Idaho. Needs dry summer as in native habitat. Leaves not mottled. Rarely more than 1 flower per stem which reaches 8–10 in. in height. White anthers show well against pale green center of ivory white petals. March–April-flowering.

E. klamathense. Native to Klamath area of Oregon and to northern California. Bright yellow-green leaves, no markings. Solitary flower, rarely more; creamy white or very pale yellow, with darker center; petals have inflated appendages at base; yellow anthers. Height 3–7 in. June–July-flowering.

E. montanum (Avalanche Lily). Native to mountains of Pacific Northwest in the United States and north into British Columbia, Canada. One of the true alpine species, found at high mountain elevations and growing profusely on Mt. Rainier. Obviously needs cold during winter, so performance in gardens is open to question. Unmottled, broad leaves narrow abruptly at base.

White flowers on stems 8–12 in. in height with definite orange band form a circle in center; broad petals; 1–5 flowers per stem. Flowers open very wide so almost flat; when bending over often face outward and sometimes a little upright, looking at the sun. Flowering late in summer when high snows melt, so can be found June–Sept., depending on altitude.

E. multiscapoideum (syn. *Fritillaria multiscapoideum*). Native to Sierra Nevada of California; introduced 1898. Unique among species growing in western areas because bulb produces stolons. Leaves green and purplish, heavily mottled. Creamy flowers with deeper yellow markings. Distinguishing habit is that flower stem branches before it emerges from cover of leaves, giving impression of solitary flower on stalk when actually there are at least 3, sometimes more. February–March-flowering. Formerly known as *E. hartwegii.*

E. oregonum. Native to the Pacific Northwest from Oregon to British Columbia; introduced 1868. Considered one of the finest species, easy to grow and increases well in cultivation. Height more than 12 in. Foliage mottled light brown and white. Creamy yellow flowers with yellow base; as many as 6 per stem; white anthers. 'B. C.' form offered by nurseries has distinctive reddish base to petals. April-flowering. Ssp. *leucandrum* has white anthers, not yellow as in type.

E. revolutum. Native to Pacific coastal areas; introduced circa 1895. Found growing where there is summer moisture provided by fog along Pacific Coast or summer rainfall of most northerly portion of coast. Height 10–12 in. Mottled leaves. Flower color ranges from pure white to rose-pink, some with yellow, almost orange, centers; others with brownish-red basal markings. Number of flowers produced varies, but usually 4–5. Late-April-flowering. Flowers can become quite large and some of largest of species have been selected and offered by nurseries, among them:
  'Johnsonii'. Deep rose with orange centers; 10–15 in. in height.
  'Rose Beauty'. Good rose; 12–15 in. in height.
  'White Beauty'. Pure white with colored basal markings; dwarf, 4–6 in. in height.

E. tuolumnense. Native to eastern California. Found growing in woods at base of Sierra Nevada. Yellow-green leaves not mottled. Deep golden yellow flowers open wide to present an almost flat flower with pale greenish-yellow base; yellow anthers; 1–4 per stem which reaches to 12 in. in height. Easy to cultivate; increases rapidly. Flowering late March–early April. Selection 'Pagoda' is taller, with larger flowers than species; color described as sulfur-yellow; listed in catalogs as hybrid between *E. tuolumnense* and *E. revolutum.*

ERYTHRONIUM

E. hartwegii see *E. multiscapoideum.*
E. japonicum Native to Japan—Islands of: Hokkaido, Honshu & Shikoku. Introduced 1897. Pink. 5 in. in height. April flowering.
E. johnsonii see *E. revolutum* 'Johnsonii'.
E. mesochoreum Native to Kansas, Iowa. Light lavender. 5–6 in. in height. March flowering.
E. nudopetalum Native to Idaho. Bright yellow. 3–5 in. in height. March–April flowering.
E. nuttallianum see *E. grandiflorum* var. *nuttallianum.*
E. parviflorum Native to Central Rocky Mts. Bright yellow. 4–10 in. in height. April flowering.
E. propullans Native to Minnesota, Ontario. Rose-purple, yellow at base. 2–3 in. in height. May flowering.
E. purdyi Native to California. Cream colored with light lemon yellow center. 6–8 in. in height. April–May flowering.
E. purpurascens Native to Sierra Nevada, California. Introduced 1881. Light yellow, tinged purple, orange base. 6–8 in. in height. May flowering.
E. sibiricum see *E. dens-canis* var. *sibiricum.*

EUCHARIS — AMARYLLIDACEAE

Eucharis is the Greek word for "very graceful." There are only a few species grown in cultivation of the some 8 species known. The large, long-necked bulbs need very warm temperatures in order to grow and flower as they are native to the warm regions of South America. The minimum night temperature is 60°–65°F.

The leaves are broad and dark green. The beautiful flowers are white, very fragrant, and unusual in appearance. The perianth tube ends with 6 petals opening to form a flower that often is 5 in. across, depending on the species. In the center of each flower base the stamens form a cup from which the anthers are poised on slender points, rising up from the cup. In most species, 8–10 flowers are borne above the foliage on strong stems 18–24 in. long.

CULTURE If grown outside, must receive warm temperatures as mentioned. In a greenhouse the minimum should be 65°F at night, with high humidity. Plant bulbs in spring. If grown in containers use a rich soil mix, with plenty of organic matter. Large container is best to allow ample root space; 6 bulbs to a 12-in. pot is about right. Cover bulb only with enough soil to anchor it and allow it to remain upright. Give day temperature of ±70°F. Water in well. Cover bulbs with rich soil mix; when growth has commenced and is above rim of container daytime temperatures can be raised to 80°F. Liquid feedings of organic fertilizer can be given once root growth is active. Plants will start to flower in early summer; 2 crops of flowers can be produced. After flowering reduce water given and lower the temperature. Keep humidity high during summer. Do not allow bulbs to become dry at any time. In spring remove some of the older surface soil and top-dress with fresh soil. Bulbs do not like to be disturbed so leave in same pot for several years.

PROPAGATION Bulbs will produce offsets. Separate in spring, before growth begins. Grow offsets on to full size before letting them flower. Seed is produced sometimes, but, because of requirements of high temperatures, high humidity, and lots of care, gardener is better advised to purchase bulbs or increase stock by production of offsets.

PESTS & DISEASES Greenhouse insects, such as mealy bugs and red spider mites, likely to attack plants. Control such infestation immediately.

USES Great plant for growing in warm greenhouse. Combine with certain species of orchids as both need high humidity; stagger planting so *Eucharis* is in flower while orchids are not of such great interest. Excellent cut flower, very fragrant, especially *E. grandiflora*, which is best species to grow.

SPECIES

E. bakeriana. Native to Colombia; introduced 1890. 4–5 dark green leaves, 18 in. long, held close to ground. White flowers flushed with green in center, semi-pendant; 4–5 per stem, which reaches 18 in. in height. Starts to flower early summer from spring planting; established bulbs bloom earlier in year.

E. candida. Native to Colombia; introduced 1851. Flower stem reaches height of 24 in., flowers held well above foliage. Solitary broad leaf remains flat. Flowers white, pendant, 3–4 in. in diameter, 6 or more produced on spike. Early summer-flowering and will give 2 flowering periods. Plentiful water needed while in growth but keep plants barely moist after flowering season; do not allow to dry out completely. *E. c. grandiflora*: see *E. grandiflora*.

E. grandiflora (syn. *E. amazonica, E. candida grandiflora*). Native to Colombia; introduced 1854. Species most popularly grown for its flowers. Very popular florists' item in years gone by, but not so widely grown today. Several leaves produced that remain close to ground; together with leaf stalk measures 20 in., stalk being a little longer than leaf blade. Flowers white, drooping slightly, 5 in. in diameter, very fragrant; 5–6 flowers on stalks reaching 24 in. in height. Flowers produced throughout winter so temperatures must be maintained in 65°–70°F range. Several selections have been made, including: var. *fragrans*, even more fragrant than species; and var. *moorei*, with smaller leaves and flowers.

E. sanderi. Native to Colombia; introduced 1885. White flowers without distinct cup in center; not so fragrant as others; at least 7–8 produced on 18 in. tall spike. Late-fall-flowering. Foliage same as *E. grandiflora*.

EUCHARIS

E. amazonica see *E. grandiflora*.
E. × *burfordiensis* (*E. mastersii* × *E.* × *stevensii*) Introduced 1899. White. 24 in. in height. Dec.–Feb. flowering.
E. × *elemetana* (*E. grandiflora* × *E. sanderi*) Introduced 1899. White. 24 in. in height. Winter flowering.
E. lehmannii related to *E. candida*. Native to Popayan. Introduced 1889. White. 24 in. in height. Winter flowering.
E. × *lowii* (probably *E. grandiflora* × *E. sanderi*) Native to Peru. Introduced 1912. White. 24 in. in height. Winter flowering.
E. mastersii Native to Colombia. Introduced 1885. White. 12 in. in height. Feb. flowering.
E. × *stevensii* (*E. candida* × *E. sanderi*) Introduced 1883. White. 18 in. in height. Winter flowering.
E. subedentata Native to Colombia. Introduced 1876. White. 18 in. in height. Winter flowering.

EUCOMIS — LILIACEAE

From the Greek *eukomes*, which means "beautiful headed," a reference to the tuft of leaves at the crown of the flower spike. It is commonly called the Pineapple Flower. These unusual-looking plants come from tropical and South Africa. There are about 10 species. The bulbs are large and tunicated, ovoid or globose in shape.

Leaves are green, basal, and arching, curving back to the ground, resting there a little untidily. The predominantly green flowers form a cylinder of color on stout, strong stems with a tuft of hairs above the flowers. In most species the flowers are arranged all around the pale lime-green flower spike. The stamens are quite stubby and spaced so that they are quite prominent in the center of the flowers; the perianth segments are almost equal.

The plants remain of interest for a long period of time. They come into flower in early summer, and, even after the flowers have faded, the seed pods are attractive—the swollen green fruits of *E. autumnalis* are almost triangular.

CULTURE Grow well in any rich, well-drained soil. Not completely hardy, so require protection in area where winters are cold and temperatures remain below freezing during the daytime.

Plant bulbs 5 in. deep, 12–24 in. apart, except for *E. pole-evansii*, which needs more (24 in.) room. Needs sun and moisture during spring and early summer. Reduce amount of water toward end of summer. Allow bulbs to remain in ground for several seasons, lifting only when overcrowded.

Good container plants. Best to plant 3–5 bulbs in large container rather than individually, especially if to be used on patio or deck.

PROPAGATION Best method is by offsets. Do this operation in autumn, after leaves have started to die back. Also can be propagated from seed. Sow in fall or spring. After plants are a couple of inches high, transplant to individual pots and grow on for planting out following spring. Will flower second season, i.e., 4 years from time seeds are first planted.

PESTS & DISEASES No special problems.

USES Look well when set by themselves with background of large rocks, which brings out colors of flowers and also provides extra protection. Excellent accent plant for front part of perennial border. Very fine container plant. As cut flowers, will last for several weeks.

SPECIES

E. autumnalis ssp. **autumnalis** (syn. *E. clavata; E. undulata*). Native to the eastern Cape Province, to Transvaal, Natal, and tropical South Africa; introduced 1760. Crown consists of 15–20 leaves, which are 2–3 in. wide and 24 in. long; extend a little way from stem then fall back and lie along ground. Flower spike reaches 18 in. in height; top 12 in. covered with flowers, each of which is ½ in. in diameter, green, fading to lighter yellow-green as it matures. Comes into flower March–April; continues to be of interest for many weeks. Some authorities list *E. clavata* and *E. undulata* as subspecies of *E. autumnalis*.

E. bicolor. Native to Natal; introduced 1878. Broad leaves, dark green with base clasping flower spike. Flowering stems spotted dark brown. Robust spikes of flowers, 12 in. high, but not quite so densely packed as others of species; distinctive flowers edged with purple. Needs moisture throughout summer but can be allowed to dry out some in fall. Flowering mid- to late summer.

E. comosa var. **comosa** (syn. *E. pallidiflora, E. punctata*). Native to the eastern Cape and Natal; introduced 1783. One of most distinctive of genus and one most commonly grown. Leaves spotted purple at base, 18–20 in. long. Flower stems 24 in. in height, half covered with blooms spotted slightly with purple. Flowers have violet-purple ovaries, light green petals that curve back slightly, throwing the stamens well out in front and showing off the color of the ovaries to advantage. Flower color sometimes varies, being tinged with pink; some plants in New Zealand have been selected that show this shade to advantage. In *E. p. striata*, purple spotting on leaves develops into stripes.

E. pole-evansii. Native to the Transvaal, Swaziland. Tallest growing of genus, commonly grown in gardens. Leaves 6 in. wide, 24 in. long. Flower spikes reach up to 6 ft., the top 24 in. of which covered with wide-opening green flowers. Thickness of cylinder formed by flowers can be as much as 8–10 in. Midsummer-flowering.

E. regia (syn. *E. nana*). Native to western Karroo, in the Cape Province of South Africa; introduced into Europe 1702. Dwarf, only 12 in. in height, half covered with green flowers. Leaves 24 in. long, 3–4 in. wide. Summer-flowering.

EUCOMIS

E. autumnalis ssp. *amaryllidifolia* Native to South Africa. Introduced 1878. Green. 12 in. in height. August flowering.
E. clavata see *E. autumnalis* ssp. *autumnalis*.
E. nana syn. of *E. regia*.
E. pallidiflora see *E. comosa* var. *comosa*.
E. punctata see *E. comosa* var. *comosa*.
E. robusta see *E. autumnalis* ssp. *autumnalis*.
E. undulata see *E. autumnalis* ssp. *autumnalis*.
E. zambesiaca much like *E. comosa* var. *comosa*. Native to E. Tropical Africa. Introduced 1886. Green. Flushes of flowers throughout Summer.

EUCROSIA — AMARYLLIDACEAE

The name is derived from Greek *eu* ("good") and *krossos* ("fringe") in reference to the fringe formed by the staminal cup. A monotypic genus formerly containing 4 species, the others now being grouped in *Callipsyche* (which see). The 1 species is *E. bicolor* (syn. *Callipsyche bicolor*), kept in this genus as it produces many leaves, not generally few, and for its stamens being joined and forming a corona. Introduced in 1817 from Peru. The flowers are a bright orange with green veining, carried in umbels, nodding, with stamens extending beyond the flowers by up to 4 in., they are joined at the base into an imperfect corona. The height of the flowering stem reaches 12 in. in height and is produced in April. It does not often produce more than 1 flowering stem per year.

CULTURE Not hardy, needing temperatures above 50°F and thus suitable for warm climates or greenhouses. Bulbs small, about 1 in. in diameter, which should be set just at soil level and best planted in the fall. Well-drained soil mix with good organic content required. Moisture should be given during the growing period but bulbs given a resting period late in the year. Needs sunlight but not a dry atmosphere.

PROPAGATION Small bulbs produced at the base of the older bulbs should be removed during the resting period. Seed can be sown in spring in temperatures of 65°F at night in a sandy soil mix and barely covered. After one season of growth, the small bulblets should be grown as offsets, giving them individual containers.

PESTS & DISEASES No special problems.

USES In containers or in borders in a greenhouse. In warm climates a fine border plant for sunny borders where good humidity can be maintained.

SPECIES

E. bicolor. Described above.

COMMENTS Rare and unusual, not frequently found in cultivation.

EURYCLES — AMARYLLIDACEAE

Name derived from Greek *eurys* ("broad"), referring to the foliage, and *kleio* ("close up"). This apparently refers to close examination of the cup revealing the base of the filaments of the flowers being more or less united into a cup-like form. *Eurycles* is rare in cultivation; there are but 2 species. Plants have tunicated bulbs, and, though bulbous, have only a brief resting period and are in fact evergreen. Flowers are produced in June-August on the Northern Hemisphere species, *E. sylvestris* (syn. *E. amboinensis*). The Southern Hemisphere species, *E. cunninghamii*, is native to Queensland and flowers in March.

Many white flowers are produced in umbels, tubular at base then perianth segments separate but not reflexed. The filaments are inserted at the throat of the perianth tube and are shorter than the segments; the filaments are margined on their lower half and form a more or less distinct cup.

The evergreen foliage is broad and long-stalked, stalks longer than the blades of the leaves, of good dark green with quite noticeable veining equally distributed on both sides of the midrib.

CULTURE Being evergreen there is not a definite resting period, except for a brief period after flowering when they should be lifted, divided, and replanted. If the bulbs are obtained without foliage they should be set at soil level or barely covered; if with foliage, plant so bulb is at soil level. They require a soil mix rich in organic matter with good drainage. They are not hardy and must receive nighttime temperatures in the 55°F–60°F range. They are thus only suitable for very warm climates or greenhouse growing. *E. cunninghamii* will grow in a little lower temperatures than *E. sylvestris*.

Adequate moisture must be provided at all times, except for a period of a few weeks after flowering when moisture can be reduced to allow a resting period. Weak feedings of liquid organic fertilizer should be given from spring throughout the flowering time.

PROPAGATION Lifting and dividing the bulbs provides the best way to increase stocks. This normally is every 3–5 years, but can be more frequent if increase of stock is needed. Seed can be sown in the spring in sandy soil mix; require high light levels and adequate temperature of 65°F at night and good humidity. The seedlings should be grown on in individual containers as soon as large enough to handle.

PESTS & DISEASES No special problems.

USES Good container plants in greenhouse conditions and for warm climates outside where humidity and good light conditions can be given. *E. cunninghamii* can be grown outdoors in summer but care should be given not to expose the plants to temperatures below 55°F at night.

SPECIES

E. amboinensis. See *E. sylvestris*

E. cunninghamii (Brisbane Lily). Native to Queensland. The foliage is evergreen, oblong, and some 10 in. in length. Flowers are carried in umbels of up to 18 white flowers, which are each about 2–3 in. in length; half of the length is formed into a tube, above which the cup, formed by the bases of the filaments, protrudes while the segments flair. Height of flowering stalk 12 in. Summer-flowering.

E. sylvestris (syn. *E. amboinensis*). Native to the Malay Peninsula and the Philippines; introduced 1759. The bulb is tunicated. Foliage broad, up to 12 in. in length, petiole longer than the blade. Some 12–15 veins are found on either side of the midrib. Perianth segments up to 3 in. long, half of their length being a tube, at the mouth of which are the united bases of the filaments forming a distinct 12-toothed cup. As many as 30 flowers in an umbel, color white. Plants reach 12–24 in. in height. Flowers in March. In the book *A Flora of Manila*, by E. D. Merrill, it is noted that this species is commonly in cultivation and has spread to northern Australia.

COMMENTS Suitable for greenhouse conditions and warm climates. Pretty plants well worth growing if such conditions as their culture requires can be given. Quite attractive plants even when not in flower.

EUSTEPHIA — AMARYLLIDACEAE

The name is derived from Greek *eu* ("good") and *stephos* ("crown") in reference to the crown-like appearance of the circle of stamens. A genus that is native to Peru. Similar in many ways to *Phaedranassa* but differing by having sessile leaves and winged filaments, which features also distinguish it from *Eucrosia*. It is distinguished from *Urceolina* by its shorter perianth tube.

The flowers are red and green, perianth tube narrow, and flowers funnel-shaped carried in one-sided umbels, tube short, segments quite long. Foliage linear, bulbs small and tunicated, flowering in the spring.

CULTURE Bulbs should be planted in late fall to early spring, the former being the better time. Plant the small bulbs 1–2 in. deep in a soil mix having good drainage and good organic matter. Water to settle bulbs and then only to keep barely moist until growth starts. Temperatures in the 55°–60°F range at night are required. They should be given bright, indirect light and weak feedings of liquid organic fertilizer as soon as growth commences. After growth has finished, gradually reduce the amount of moisture given to allow the bulbs to dry and enter a resting perod. If growing in containers plant 6–8 bulbs to a 12 in. pot.

PROPAGATION Offsets provide the best means of propagation; these should be removed from the parent bulbs during the dormant season and then grown on. Seed can be sown in the spring in a sandy soil mix with night temperatures in the 65°F range. Keep moist and transplant as soon as the seedlings are large enough to handle.

PESTS & DISEASES No special problems.

USES For the warm greenhouse or sunny border in warm areas.

SPECIES

E. coccinea. Native to Peru. Small, ovoid bulb 1 in. in diameter. Leaves narrow and linear, ¼ in. in width and 12–18 in. in length, bright green. Flowers a little over an inch in length, carried on stems to 16–18 in. in height. Flowers appear before the leaves, up to 8–10 per umbel with pedicels some ¾ in. in length, drooping, with green keel on the upper part and segments bright red except for shadings of green at tips and base. Flowering time is late spring-early summer.

E. pamiana. Native to Argentina; introduced 1926. A long, ovoid bulb some 2½ in. in diameter and 3 in. in length. Some 6–8 leaves are produced which are 14–18 in. in length, linear, and narrow, covered with a sheath at the base which is red-streaked. Flowers a little over an inch in length, carried in an umbel which is on a rounded stem to a height of some 18 in. The flowers are pendant, the tips of the petals are red, the lower parts purplish-red, with the central part greenish fading into the red and purplish colors. The inner segments have red coloring at the tips which is darker than that of the outer petals, making a most attractive combination. Flowering in spring, April–May.

COMMENTS   Very unusual flowers but unfortunately rare in cultivation.

EUSTYLIS — IRIDACEAE

The name is derived from the Greek words *eu* ("good") and *stylis* ("little pillar") in reference to the flower style. The genus is entirely American and contains perhaps some 4–5 species; only 1 species, *E. purpurea*, is cultivated to any degree and that rarely. For many years it was known as *Nemastylis purpurea* but was transfered to this genus because the stamens are not united and are pressed against the style with the anthers being sessile.

E. purpurea is native to the open woodlands of Texas and Louisiana. Its brown, ovoid bulbs are less than an inch in diameter. The leaves are sheathed at the base, pleated, about 24 in. in length, and a little less than an inch in width. The flowering spike is thin, zigzags, and often branches, reaching 16–18 in. in height. The flowers, which remain open for a few hours in mornings and then fade, are blue-purple, with the inner perianth segments smaller than the outer, cupped and crimped along their edges, with brown mottling on the lower parts, which mottling is also found on the outer segments at their base. The open flowers are 2 in. in diameter. There is some variation in the color, some plants having paler, almost white, flowers, with the mottling on a more yellow or orange base or reddish-brown. The common name is Pinewoods Lily. Flowering May–June.

CULTURE   The bulbs should be planted in the fall, in a sandy soil mix with good drainage. Moisture should be given as soon as growth is noticed in the spring. Set the bulbs 2 in. deep and space some 12 in. apart. They are not hardy and suitable only for areas where there is little or no frost and the ground never freezes. In cold areas they can be grown in deep containers in a sandy soil mix. They can be set outside when danger of frost has passed and brought indoors in the fall and overwintered in a frost-free area.

PROPAGATION   The bulbs are small and not many offsets are produced, thus, seed sown in the spring in a sandy soil mix with night temperatures in the 45°F range is preferred. The seed should be barely covered. Sown thinly the seedlings can remain in the container for one season and then the small bulblets should be transplanted to individual containers the following spring. Grown on until they are ½ in. in diameter and then planted out or placed in containers.

PESTS & DISEASES   No special problems.

SPECIES

E. **purpurea.** The only *Eustylis* species in cultivation. Described above.

COMMENTS   This genus is rare in cultivation and never likely to be widely grown. It should be considered for those creating an unusual garden of rare plants.

FERRARIA — IRIDACEAE

This genus was named in honor of Giovanni Battista Ferrari (1584–1655), an Italian botanist. A small genus of bulbous plants with a rootstock that is a corm, often much misshapen and without a tunic. They are native to South Africa and the southern tropical regions of Africa.

The majority of these plants are quite dwarf. The stems are branched; at the base of the plants are several large leaves that clasp the stem and are arranged as if to protect it. The leaves at the base are stiff and quite hard, most commonly grayish-green. Those leaves located on the stems are reduced in size, the smallest being at the top of the stem. The impression of the plant is of being rugged, stiff, and strong.

Flowers are quite large, often as much as 2 in. in diameter, and a number are produced in early to mid-spring. The flowers are mostly in the brown, green, and purple colors and carried on sturdy pedicels. Frequently there are attractive mottlings on the flowers. The petals are crisped, and the flowers are held upright except in a few species, but even then, when first open, they are held erect. The flowers have an unpleasant smell and are pollinated by flies attracted to the smell.

I have seen these plants close to the Cape of Good Hope. While not striking to the point of being much noticed, they are quite attractive and fascinating and once noticed the eye seems attracted to them, as soon as one spots one then others seem to jump out at you. They grow mostly by the sea in stony or sandy ground.

CULTURE   The plants need full sun and well-drained soil. The corms are found deep in the soil in their native habitat and should be planted 3–4 in. deep, spacing them some 6–8 in. apart. They are not fully hardy and should only be grown outdoors in areas where there are not severe frosts or frost is experienced early in the year before the new growth in the spring is aboveground. They require water in the early part of the year but during the summer and early fall they should be allowed to become dry and warm. In other areas they will need the protection of a cool greenhouse where they will perform better in borders than in containers as the corms prefer to be deep in the soil.

Unless the soil is poor they will need little fertilizer, which if given should be applied early in the year as soon as the plants appear above the soil. In warmer climates they should be planted near rocks so they enjoy the reflected heat. As their native habitat is close to the sea, they are good plants for mild maritime climates being unaffected by salt-laden breezes. In bold clumps in the rock garden they can be most effective, and great conversation topics.

PROPAGATION   The easiest means of propagation is the removal of the small cormels from the parent corms. This should be done in the very early spring with the small cormels being grown on in shallow drills. After a season or two they can be planted out into flowering positions. Seed is quite freely produced. This should be sown in the spring in a sandy soil mix with the seeds only just being covered, kept warm and with bright light they will germinate quite quickly. By the beginning of the second season plants can be individually potted. They will begin to flower in the third season.

PESTS & DISEASES No special problems.

USES In bold clumps in the rock garden where they will enjoy the reflected warmth of the rocks, in sunny borders in mild climates, and in the cool greenhouse, preferably planted in the ground. While they are reportedly smelly plants, you do have to get quite close to them to notice the fetid odor.

SPECIES

F. antherosa (syn. *F. ferrariola, F. viridiflora*). Native to South Africa and quite widespread from the Cape inland; introduced 1800. A dwarf species inhabiting dry, open ground and flowering in the wild from August–October. Stem is unbranched and reaches a height of only 4–5 in. Its 5–6 rigid leaves are each only 3–4 in. in length, about 1½ in. in width and protect the stem. Flowers are dull greenish along the margins of the tepals with a dull blue-purple stripe in the center of the tepals. The tepals are much curved and crisped at the edges. Not the most attractive of the species.

F. crispa ssp. **crispa;** often called *F. crispa* (syn. *F. punctata, F. undulata, Tigridia undulata*). Widely distributed in South Africa; introduced 1755. Plants are found growing in dunes and sandy places, in the Little Karroo in the veldt around Oudtshoorn, and on grassy mountain slopes such as the Table Mountain. It is a strange but attractive plant and can reach a height of some 18 in. but is often much less. There are often 2 or 3 stems enclosed by strong-looking foliage which clasps the stem. Leaves overlap each other, being up to 12 in. long, shorter further up the stem, with the uppermost being bractlike and surrounding the flowers. The flowers are 2 in. or more in diameter, have an appearance of velvet, brownish-purple in color with a greenish-white, V-shaped marking in the middle; the tips curl back and the edges are very crisped. The brown stigma is much dissected at the tip with the lower part surrounded by a tube formed by the anthers. In the wild flowering Oct.–early Dec.

F. divaricata. A South African species native to the western coast and Kalahari Desert area; introduced 1825. It reaches a height of some 14–18 in. The stems are usually branched, the leaves linear to oblong, sheathed at the base and not as much stem is hidden as in some species. The basal leaves are inclined to spread outward and are not erect; they are bluish-green in color. The stigma is much fringed, looking like a miniature shaving brush, and purplish brown; tepals are yellowish-green with triangular brown markings, much crisped along the edges. The lower part of the tepals are erect and form a small cup, then the tepals recurve to an almost flat or horizontal position with the tips curling back into an upright position. The diameter of the upright-facing, open flowers is 2 in. or a little more. This plant is most attractive and deserves to be more widely cultivated. Flowering in the wild Sept.–Oct.

F. uncinata (syn. *F. crispulata; F. framesii*). Native to sandy areas in the Clanwilliam area of western South Africa; introduced 1825. Unlike the other species described the leaves are taller than the flowers. The stems are branched and reach a height of some 8–10 in. The bluish-green leaves are linear with rough margins, often curved, and sheathed at the base. The tepals are much crisped along their edges, held horizontally with the edges curving down, but the tips curl upward and are much crenulated and yellowish-orange in color. The orange-yellow color is continued down the edges of the tepals and the remainder of the tepal is greenish with blue blotches or markings—an unusual color combination. The flowers will reach 2 in. in diameter when fully opened. Another unusual-looking flower that deserves to be in cultivation. This species seems to thrive on sandy soil and possibly should be considered for dune plantings in warmer climates. Flowers in the spring–early summer.

COMMENTS Not frequently found in cultivation but these plants deserve more consideration and wider distribution. While not brightly colored their distinct form and markings are most unusual and I find them attractive.

FERRARIA

F. atrata Native to South Africa. Reddish-purple fringed brownish-green. ± 6 in. in height. Spring flowering.
F. ferrariola see *F. antherosa*.
F. framesii see *F. uncinata*.
F. glutinosa see *F. welwitschii*.
F. obtusifolia Native to South Africa. Introduced 1825. Branched/ brown. ± 6 in. in height. May–July flowering.
F. punctata see *F. crispa* ssp. *crispa*.
F. undulata see *F. crispa* ssp. *crispa*.
F. viridiflora see *F. antherosa*.
F. welwitschii (now *F. glutinosa*) Native to South Africa. Introduced 1871. Bright yellow, dotted brown. ± 6 in. in height. July flowering.

FREESIA — IRIDACEAE

Named by plant collector C. E. Ecklon, an apothecary born in 1795 in Northern Schleswig and died in South Africa in 1868, in honor of a German doctor who was his student, Friedrich Heinrich Theodor Freese, who died in 1876. There are only about 11 species in the genus, all native to South Africa. While the white species were introduced early into cultivation, being sent to England in 1816, it was not until the introduction of *F. armstrongii* in 1898 that the development began of the plants we know today. This species has pink flowers and a delightful fragrance. Crossed with *F. refracta*, which had been introduced earlier, the hybrid known as *F.* × *kewensis* was developed. The work was done in England, France, Italy, and Holland. Presently, there are many named hybrids, all lovely plants, as well as many doubles with very large flowers, offered for sale by nurseries.

The *Freesia* are unable to withstand frost. They need warmth to grow well; therefore, they can be grown only for winter and spring gardens where winters are mild. In other areas they can be planted in late spring for summer flowering; however, they are considered more of a florist flower at that time of year.

The round or ovoid corms are loosely covered by a netted tunic. A number of leaves are produced, which are narrow, like a thick blade of grass, and are shorter than the flower spikes. The foliage continues to grow long after the flowers have died off.

The zygomorphic flowers are carried on strong, branched stems, with one much larger than the other and bearing more flowers. The hybrids generally carry more flowers in the spike than the species, which seldom have more than 4–5 flowers, but are quite crowded, with just enough room for them to develop. The one-sided spike bends just below the point where the flowers grow, so that they are held in an upright positon. The tip of the spike also bends slightly so that the flowers are almost outward-facing.

CULTURE   In warm areas corms can be planted in September, 2 in. deep and 3 in. apart, in well-drained soil, in full sun. Feed liquid organic fertilizer as soon as growth appears. Ample moisture must be made available throughout the time leaves are green; feeding can be done once a month. Stop feeding and diminish water when foliage starts to die down. When foliage is completely dry, withhold water entirely. Begin again in late August so corms can commence flowering cycle once more. Not unusual for corms to die down in early summer, start into growth again in late summer, and give another display of flowers in late fall, which will not be so prolific, however, as that of spring. For this reason, feeding is necessary for corms to gather strength. In colder areas, lift corms after foliage dies, store overwinter, and replant in spring.

Corms that have been lifted and stored under exact temperature controls to retard development are available from The Netherlands. Can be planted outdoors after danger of frost is past. Will flower during summer months.

If grown indoors, corms are best planted in deep container to allow roots to expand—6 corms to 6-in. pot is about right. If a flat is used, should be at least 4–5 in. deep, setting corms 2–3 in. apart. Plant in September. Bring indoors when nighttime temperatures drop below 45°F. Keep at 50°F at night, in bright sunlight if possible during day. Feed once a week and keep moist. Continue growing in same container until foliage begins to die down. Reduce water and feeding. Plants can then be set outside if no frost is expected. Planted out in garden, another flowering may be had if corms were not grown in too high temperatures indoors.

Plants can be brought into flower from seed. Soak seeds in warm water for 24 hours before sowing. Sow indoors at anytime of year in rich soil mix—well-rotted compost, good topsoil, and sharp sand. Temperature 65°F at night, which can be lowered when germination starts. Keep as cool as possible during summer; high temperatures cause lankiness and so plants are less attractive. Give ample air circulation and good light at all times. Cannot be allowed to dry out or have growth interrupted by low temperatures; must be kept growing. In areas where there is some frost, can be grown in a cold frame. If required for outdoor planting, set out when foliage has died down. Do not move while growing as plants resent being disturbed.

PROPAGATION   Corms will produce number of offsets, which can be separated and planted back. Do this after foliage has died down. Newly formed corms should not be transplanted until dormant. Propagate from seed as described above.

PESTS & DISEASES   Leaves can be attacked by fungus diseases if not given adequate ventilation during growing season. Aphids, mites, and thrips can be a problem, so put proper control into effect as soon as noticed.

USES   Excellent house plants and cut flowers. A must for warm areas as they produce flowers during winter and early spring, when color is most welcome. Good in rock garden and in front of borders; however, may not be worth time and effort needed if cannot be left in ground.

FREESIA

F. alba Native to South Africa. Creamy yellow flushed with purple. 6–12 in. in height. March–May flowering.
F. andersoniae see *F. leichtlinii.*
F. brevis see *F. corymbosa.*
F. corymbosa now includes *F. armstrongii, F. brevis, F. metelerkampiae.*
F. elimensis Native to South Africa. 6–12 in. in height. See photo. Spring flowering.
F. fergusoniae Native to S.W. Cape Province. 6–12 in. in height. See photo. Spring flowering.
F. flava see *F. speciosa.*
F. hurlingii syn. *F. refracta.*

SPECIES & CULTIVARS

F. armstrongii. Now regarded as being a variant of *F. corymbosa*. Native to the coast of eastern South Africa; introduced 1898. Grows to height of 15–18 in. Grasslike foliage. Pale pink flowers but very variable. Spring-flowering. Parent of many hybrids.

F. corymbosa (syn. *F. brevis, F. metelerkampiae*). Native to eastern Cape Province, spring-flowering, foliage an erect, spiral fan, 10 in. long. One of the most fragrant, especially in evening. 12 in. in height. Flowers golden yellow, variable. Aside from fragrance, has little to recommend it.

F. leichtlinii (syn. *F. andersoniae, F. middlemostii*). Native to the southern coastal areas of South Africa. Dwarf species up to 12 in. in height, sometimes more but often much less; found growing in sandy and stony ground. Foliage quite tall. Flowers are prominent but do not rise above foliage as much as in other species. Flowers 2 in. long; somewhat variable in color, from cream to purple, most commonly off-white or creamy with darker yellow-cream on inside of petals with purple flushes on exterior. Very fragrant. April–May flowering but can start into flower in March in warmer areas. Good plant for rock garden.

F. speciosa (syn. *F. flava.*). Native to southeastern coast of South Africa. Very fragrant, creamy yellow flowers. Not often found in culture. Spring-flowering, 10–12 in. in height. Foliage held in fan shape.

The following are some of the named hybrids offered by nurseries:

White
'Ballerina'
'Diana'. Double-flowered.
'Marie'
'Snow Queen'
'White Swan'

Yellow
'Carmelita'. Golden yellow.
'Corona'. Double-flowered.
'Fantasy'. Creamy; double-flowered.
'Golden Melody'
'Rijnveld's'. Golden yellow.
'Royal Gold'. Golden yellow.

Orange
'Princess Marijke'. Flame-orange.

Red & Pink
'Pimpernel'. Flame-scarlet.
'Rose Marie'. Dark pink; double-flowered.
'Stockholm'. Chrysanthemum-red, yellow throat.
'Viking'. Pink.

Purple & Blue
'Blue Wimple'. Violet-blue.
'Romany'. Blue; double-flowered.
'Royal Blue'. Campanula blue, white throat striped violet.

F. metelerkampiae see *F. corymbosa*.

F. middlemostii see *F. leichtlinii*.

F. occidentalis Native to South Africa. 6–12 in. in height. See photo. Spring flowering.

F. refracta (syn. *F. hurlingii*) Native from Worcester to Swellendam, South Africa. Introduced 1816. Lime-yellow tinted with dull mauve. 12 in. in height. Late Winter—early Spring flowering.

FRITILLARIA — LILIACEAE

Name derived from the Latin word *fritillus,* which means a "dice box." This genus is widely distributed yet limited to the Northern Hemisphere. There are over 80 species; no doubt more will be created when the genus is examined in depth. Many are found in Europe, especially around the Mediterranean, eastward to China and Japan, while others are found in North America, particularly along the western coastal states of the United States. *Fritillaria* are related to the genera *Lilium, Notholirion,* and *Tulipa,* with their bell-shaped flowers, but the flowers are not so brilliantly colored. The plants are very graceful, however, and the checkering found on many of the flowers is most attractive.

The stems are unbranched and leafy. Depending on the species, the leaves are arranged sometimes in whorls and sometimes in pairs. Not infrequently basal leaves are produced, which are broader and longer than the stem leaves. These are seen most often in younger bulbs, prior to sending up the flower spike. The flowers are made up of 6 segments, which are equal or nearly so. Each petal has a nectary at the base, which often is very conspicuous. The 'Crown Imperial' is the only one of the genus with the flowers grouped on top of a strong stem, topped by a tuft of leaves. All others have flowers up the stem, in a variety of sizes. Some even have been mistaken for dwarf lilies. The fruit is a capsule with numerous seeds. The bulbs of all *Fritillaria* are very fragile and must be handled with care. The fleshy bulbs will dry out in a short space of time. It is, therefore, important that they be protected and, if shipped, wrapped so that little or no moisture is lost. The European and Asiatic species have somewhat larger bulbs than the American species, and are composed of fleshy scales, generally no more than 3 or 4. The American species have more numerous scales, and on the outside of the bulb can be found numerous fleshy, smaller scales which detach easily and can be grown on to flowering size.

CULTURE Varies with each species but hot, dry conditions are not ideal for any *Fritillaria.* Like abundant moisture in spring and early summer. Allow to dry out after flowering. Moisture must be present throughout life of leaves.

Plant bulbs some 3 in. deep and about 6–8 in. apart, with exception of *F. imperialis* and its cultivars which should be spaced 14–18 in. apart. Set out in fall as soon as bulbs available. Hardiness of some species questionable, especially in areas where spring frosts common. High-organic-content soil an advantage; give weak feeding of organic fertilizer when bulbs emerge if soil rather poor. Good drainage essential so avoid waterlogged soil. Planting in neighborhood of shrubs is good to protect from full sun; *F. imperialis* can stand full sun, except in very hot locations where high shade should be given. Leave bulbs in position for a number of years.

PROPAGATION Seed produced, but raising plants from seed to flowering is slow process, often taking 5 years or more. Seed can be sown when ripe in late summer or saved and sown in spring. Use sandy soil mix, barely covering seeds. Place in cool spot and keep moist. Germination may take at least 6 months, perhaps more. Leave in original container for one complete season before transplanting. In some species seedlings send out "sinker" roots and newly formed, little bulb will be found deep in container. Transplant to individual pots; grow on for at least one more season before planting out. Will generally flower in third or fourth season.

Bulbs that are fleshy will produce many offsets. Separate these from parent and line out in rows in garden or place in container at depth of ½ in. Do not plant in full, hot sun outdoors. Less moisture needed in late summer but do not allow to dry out completely. This method will save at least 2 seasons over planting of seedlings.

Scales also can be propagated, as in the lilies (see *Lilium*). Number of scales taken from each bulb must be limited in order not to weaken small parent bulb.

USES Excellent for both woodland and container. Can be grown in protection of cold frame, getting them into flower a little earlier and so brought indoors to be enjoyed longer.

SPECIES & CULTIVARS

F. acmopetala (syn. *F. lycia*). Native to Cyprus, Syria, and Lebanon; introduced 1874. Described in many books as reaching height of only 12–16 in. but nurseries, over the years, have selected strongest forms and developed strains that reach more than 30 in. Narrow leaves, up to 3 in. long, produced on stem. Flowers quite large; 3 to a stem; recurved, pointed tips; olive-green streaked with brown or purple on outside, shining olive-green on inside; inner segments inclined to be more brownish-green on outside. Flowering in April.

F. alburyana (syn. *F. erzurumica*). Recent introduction collected in Turkey by the Cheese, Albury, and Watson expedition in 1966. Rather distinct in that bright pink flowers are lightly checkered and open very wide. Spring-flowering. Height 6–10 in.

F. armena. Native to Asia Minor, introduced 1846. Given various names by botanists: some claim it is close to *F. caucasica,* although it is smaller in all parts; others say dark purple form of *F. caucasica* is actually *F. armena*—to be resolved only when definitive monograph published, as is the case with so many other genera and species. *F. armena* seldom more than 6–8 in. in height. Gray-green leaves scattered on stem; lower ones more clustered than upper ones, which are also somewhat narrower. Flowers pendant to somewhat outward-facing; seldom more than 1 per stem; dark blue with streaks of purple on exterior, interior reddish-brown or with hint of yellow. Flowering in April.

F. assyriaca (syn. *F. canaliculata*). Native to Iraq and western Iran; introduced 1874. There is quite a lot of confusion over this name. It is a valid one and belongs to a species that is 5–10 in. in height, with narrow lanceolate leaves which are gray-green and appear mostly on the upper part of the stem. The flowers are usually carried singly and are somewhat variable in color, being sometimes greenish or with two colors—green and violet, to dull violet with yellowish margins. The plants that are often in culture and given the name *F. assyriaca* should be named *F. uva-vulpis.* There are two subspecies of the true *F. assyriaca.*—*F. a.* ssp *melananthera,* which has greenish flowers with a black interior, and *F. a.* ssp *assyriaca,* which has a yellow interior to the flower and yellowish anthers.

F. biflora (Mission Bells; syn. *F. agrestis, F. grayana, F. roderickii, F. succulenta*). Native to southern California, northward into southern

Oregon. Its common name, Mission Bells, has been given to several other species as well. Height 12 in. Leaves at base of flower spike. Dark brownish, unmottled flowers, often with some shading of green. Early spring-flowering.

F. bithynica. (syn. *F. citrina, F. dasyphylla, F. pineticola,* and *F. schliemannii.*). Wide distribution—Greek Islands to Turkey; introduced 1874. One of the yellow-flowered species. Separation of species, one from the other, perhaps of greater concern to botanists than to gardeners and now all regarded as being *F. bithynica.* Height 6–8 in. Inside of flower has green markings. Spring-flowering.

F. camschatcensis (syn. *F. kamtschatcensis, Lilium camschatcense*). Probably has widest distribution of any species in this genus—from Alaska to Japan, Washington state, and Oregon. Some forms reach 12 in. in height, but most commonly less, with some not reaching more than 6 in. Leaves carried in whorls on strong stalks. Flowers among darkest of all *Fritillaria*, deep maroon-purple, almost black; 1–3 per stalk. Prefers shade and moist soil and no bright sunlight. Flowering from early spring through mid-spring to almost early summer in very cold climates.

F. cirrhosa. Native to the central and eastern Himalayas. Up to 18 in. in height. Rather unique characteristic in that top leaves, which are scattered on stem, sometimes in whorls, have tendril at tip—or cirrus, hence the name. Flower color either brown with green or green with brown markings. Flowering in March–May.

F. gracilis. Native to Yugoslavia. Height 12 in. Narrow leaves. Generally 1 flower, rarely 2–5; purplish with brown and yellow markings, inside markings lighter than those on outside. Spring-flowering.

F. imperialis (Crown Imperial). Native to western Himalayas; introduced before 1590. Grows in wide range of soils, will tolerate quite heavy soils. Should be placed in sunny location and left undisturbed. Large bulbs, often take a year to settle in after planting. Strong, 4 ft. high stems carry many pendant flowers that are topped off with tufts of leaves, making the species distinct. Leaves carried at base on stem, either in whorls or alternate, but upper part of stem is bare except for tuft at top. Nectaries white and very obvious. Bulbs and flowers have smell of rotten meat but still make very showy plants for gardens. Flowering in April–May and are long-lasting. Many selections offered by nurseries, among them:
- 'Aurora'. Very strong grower with orange-red flowers.
- 'Lutea'. Good yellow flowers.
- 'Lutea maxima'. Even stronger grower than 'Lutea'; forms very good heads.
- 'Rubra'. Orange flower with a hint of brown.

F. involucrata. Native to the French Riviera, growing in Maritime Alps. Little more than 12 in. high. Leaves produced toward middle of flower spike. Flowers pale green with purple checkering. Flowering in April–May.

F. lanceolata (Checker-Lily; syn. *F. mutica*). Native to much of western seaboard of North America, from California to British Columbia. Leaves 6–10 in. long, usually in 3 whorls. Generally up to 3 flowers produced on stems that reach 18 in. or more in height; dark purple mottled with yellow-green. April-flowering, sometimes later.

F. latifolia. Native to the Caucasus and Turkey; introduced 1799. Grows up to 12 in. in height. Flowers usually solitary, carried on strong stems; up to 2 in. in diameter; deep purple-maroon with strong green-yellow checkering. April–May-flowering. Species once so popular that many selections were made and offered by nurseries; not so widely grown today but still very fine for gardens. Var. *nobilis* has large flower that is carried just above ground; 2 yellow forms, *F. l.* var. *lutea* and *F. l.* var. *aurea*, are known; the former is the taller.

F. lusitanica (syn. *F. hispanica; F. pyrenaica*). Found in the Pyrenees of France and Spain. Most often listed as *F. lusitanica* but alternate names probably given due to location in which found, with only slight variations. Narrow leaves, 5–7 in. long, scattered toward upper part of flower stem, which reaches about 15–18 in. high. Flowers deep purplish-brown or deep maroon, green striped with yellow, checkered with purple on inside. April-flowering.

F. meleagris (Snake's Head Fritillary). Native to most of Europe, long been a favorite in European gardens. Flourishes in damp places, so plant only where moisture is abundant throughout year. Prefers high shade such as that provided by open woodland, and soil that is high in leaf mold or other organic matter. Leaves, generally 4–6, about 6 in. long, narrow, found in middle of flower stem. Flowers carried on strong stems up to 15 in. high; pale to deep pink with hint of purple, checkered with blackish dots, yellowish-green interior. April-flowering; earlier in warmer climates. Many forms have been selected, among them:
- 'Alba'. Pure white.
- 'Aphrodite'. Large, pure-white flower.
- 'Artemis'. Checkered purple with green.
- 'Charon'. Darkest form offered; light purple checkered with black.
- 'Orion'. Dull purple.
- 'Saturnus'. Brightest colored, violet with reddish hue.

F. messanensis. Found in wild in Crete woodlands; introduced 1939. Very similar to *F. gracilis*. Leaves give glaucous appearance. Stems reach up to 20 in. Flowers mostly solitary, brownish-purple with yellow-green markings. Flowers late March–April. *F. sphaciotica* is similar but more delicate.

F. multiscapoideum (syn. *Erythronium multiscapoideum,* which see).

F. pallidiflora. Native to Siberia. Very hardy. Strong, wiry stems, reach up to 10–15 in. high. Leaves in 2 whorls, generally 6 leaves per whorl. Pale yellow flowers spotted with reddish-crimson-brown on inside. Flowering early May.

F. persica (syn. *F. arabica, F. eggeri, F. libanotica*). Native to Cypress, southern Turkey, to Iran; introduced 1753. One of tallest of genus, sometimes reaching 36 in. in height. Gray-green leaves cover stem and sometimes twist a little. Many small flowers per spike, often more than 30 on a well-grown plant; dark plum, opening wide to expose hint of green on inside, some variation with some darker than others. Makes interesting plant set among other spring-flowering bulbs. Good form is *F. p.* 'Adiyaman,' named for town near which it is found; taller than most, reaching 48 in.; May-flowering; give light shade to prolong length of flowering time.

F. pontica. Native to southeastern Europe and Asia Minor; introduced 1826. Easy to grow if given light shade. Height 15–18 in. Leaves located on upper part of stem with top ones often in a whorl. Flowers unspectacular, in April–May, dull green with light brown tips to segments; 1–4 on stalk.

F. pudica (Johnny-Jump-Up). Native to the western United States and Canada, from California north to British Columbia. Tolerates wide range of climates, with or without summer moisture, but never in drought situations. Height about 6–8 in. Leaves found near base of stem, almost erect, commonly whorled. Flowers in April–June, yellow with purple tint, 1–2 per stem.

F. purdyi. Native to northern California. Lovely and unusual-looking species with leaves that form a rosette at base of flower stalk. Seldom more than 6 in. in height. May have only 1 or as many as 6–7 flowers, white with broad chocolate-crimson stripe and dots on interior of petals with a distinct green circle at base of petals, which open wide. Flowering March–June depending on location, earliest flowering at lower altitudes.

F. pyrenaica (see also *F. lusitanica*). Variable species, name given to form with plum-colored flowers that have golden bronze interior and petals whose edges recurve slightly. Up to 18 in. in height. April-flowering.

F. raddeana (*F. askabadensis micheli*). Form of *F. imperialis* from the Transcaspian region of the USSR. Height 24–30 in. Pale greenish-yellow flowers produced in April.

F. recurva. Native to Oregon and California; introduced 1870. One of finest of genus. Up to 24 in. in height. Leaves scattered in whorls around center of plant. Many scarlet-red flowers, orange and red-flecked interior, and recurved petals. A very striking plant. May-flowering. Form known as 'Mt. Hood' even brighter in color. Var. *coccinea* has mottled flowers, yellow and scarlet.

F. uva-vulpis. This is the correct name for the plant that often is cultivated under the name *F. assyriaca*. It is native to northern Iraq, western Iran, and southeastern Turkey; introduced 1974. A great number of offsets are produced. The plant reaches 12–14 in. in height; has glaucous, green foliage; pendulous maroon-colored flowers, the interior of which is golden bronze with many faint black lines. Flowering time is April-May, a little earlier in its native habitat. A vigorous species.

FRITILLARIA

F. adamantina see *F. atropurpurea*.
F. alpina syn. of *F. pinardii*.
F. amabilis Native to S. Japan. Bluish. 6–10 in. in height. March flowering.
F. arabica see *F. persica*.
F. ariana Native to Central Asia, N.E. Iran, N.W. Afghanistan. Cream. 4–10 in. in height. March–April flowering.
F. askabadensis micheli see *F. raddeana*.
F. atropurpurea (a robust form is known as *F. adamantina*) Native to N. America from Oregon & California east to the Dakotas & Nebraska & south to New Mexico. Purplish brown, spotted yellow & white. 6–20 in. in height. April–July flowering.
F. aurea Native to Turkey. Introduced 1854. Yellow with red-brown veining. 2–6 in. in height. Spring flowering.
F. brandegei (syn. *F. hutchinsonii*) Native to S. Sierra Nevada, California. Purplish pink, 12 in. in height. April–June flowering.
F. bucharica Native to N.E. Afghanistan & Central Asia. White, tinted green. 12 in. in height. March–May flowering.
F. canaliculata see *F. assyriaca*.
F. carduchorum syn. of *F. nigra*.
F. carica Native to Turkey & the eastern Aegean Islands. Introduced 1975. Yellow or greenish-yellow. 2–6 in. in height. April flowering.
F. caucasica see *F. armena*.
F. caussolensis syn. of *F. nigra*.
F. crassifolia Native to Lebanon & S. Turkey to Iran. Yellow-green checked purple. Up to 15 in. in height but usually 2–6 in. in height.
F. c. ssp. *crassifolia* Native to Turkey. Greener in color.
F. c. ssp. *hakkarensis* Native to Turkey & Iraq. Greener in color but shorter, less shiny.
F. c. ssp. *kurdica* syn. *F. kurdica* which see.
F. c. ssp. *poluninii* Native to N.E. Iraq. White-bluish. April–June flowering.
F. citrina see *F. bithynica*.
F. dasyphylla see *F. bithynica*.
F. davisii Native to Mani Peninsula of Greece. Introduced 1940. Deep chocolate brown, checkered. 6–10 in. in height. Feb.–March flowering.
F. delphinensis syn. of *F. tubiformis*.
F. discolor syn. of *F. sewerzowii*.
F. eduardii Native to S. U.S.S.R. Brick-red. Intermediate between *F. imperialis* & *F. raddenna*—form like *F. imperialis*. April flowering.
F. eggeri see *F. persica*.
F. ehrhartii Native to Greek Islands. Metallic purplish-brown & yellow-green inside. 2½–12 in. in height. Feb.–May flowering.
F. elwesii Native to S.W. Turkey. Introduced 1884. Purplish-blue & green stripes. 6–12 in. in height. March–May flowering.
F. epirotica Native to N.W. Greece. Deep brown-purple & checked inside. Dwarf. June flowering.
F. erzurumica see *F. alburyana*.
F. euboeica Native to S. Greece. Yellow with grayish-green leaves. 2–4 in. in height. May–June flowering.
F. falcata Native to California. Greenish without, mottled rusty-brown & yellow within. 2–4 in. in height. March–May flowering.
F. ferganensis syn. of *F. walujewii*.
F. fleischeri syn. of *F. pinardii*.
F. fleischeriana Native to Turkey. Introduced 1829. Purple brown with green stripe. 2–6 in. in height. April flowering.
F. forbesii Native to S.W. Turkey. Introduced 1874. Pale yellow-green to brownish purple. 6 in. in height. Feb.–March flowering.
F. gentneri Native to Oregon. Dark reddish-purplish tinge, spotted yellow. 12 in. in height. April–May flowering.
F. gibbosa Native to Iran, Afghanistan & Caucasus. Pale pink to brick colored, spotted. 5–7 in. in height. March–May flowering.
F. glauca Native to S. Oregon to N. California. Purplish or greenish marked with yellow. 3–6 in. in height. April–July flowering.
F. glaucoviridis Native to S. Turkey & N. Syria. Introduced 1930. Green without, yellowish-green within. 6–15 in. in height. April–May flowering.
F. graeca (robust form called *F. guicciardii*) Native to Crete & Greece. Deep purplish-brown, sometimes checkered. 3–10 in. in height. May–June flowering.
F. graeca ssp. *gussichiae* Native to N. Greece, S. Bulgaria & Yugoslavia. Pale green & light brown shading. 24 in. in height. April–May flowering.
F. guicciardii see *F. graeca*.
F. hookeri see *Notholirion macrophyllum*.
F. kamtschatcensis see *F. camschatcensis*.
F. korolkovia sewerzowii syn. of *F. sewerzowii*.
F. kurdica (syn. *F. crassifolia* ssp. *kurdica*, *F. grossheimiana*, *F. karadaghensis*). Native to E. Turkey & N. Iran. Introduced 1974. Chocolate-purple to green, tipped with yellow. 3–6 in. in height. April flowering.
F. latakiensis Native to S. Turkey, Syria & Lebanon. Introduced 1975. Brownish-purple. 18 in. in height. April flowering.

F. libanotica see *F. persica*.
F. liliacea Native to California. Cream with greenish lines & suffusions. 3–12 in. in height. Feb.–April flowering.
F. lycia see *F. acmopetala*.
F. macedonica Native to E. Albania, S.W. Yugoslavia. Purple checked on yellow ground. 2–5 in. in height. May–June flowering.
F. meleagroides Native to S.W. Bulgaria, Russia. Claret. 8–15 in. in height. April–June flowering.
F. michailovskyi Native to N.E. Turkey. Introduced 1905. Dark reddish-purple with a gray bloom, upper ⅓ of segment bright yellow. 4–8 in. in height. May–June flowering.
F. micrantha (syn. *F. parviflora*). Native to California's Sierra Nevada. Purplish or greenish-white faintly mottled. 16–36 in. in height. April–June flowering.
F. minima Native to S.E. Turkey. Introduced 1971. Yellow & green. 4 in. in height. June & July flowering.
F. minuta (syn. *F. carduchorum*). Native to E. Turkey & N.W. Iran. Introduced 1859. Brick red or yellow-brown tesselated. 2–5 in. in height. April–July flowering.
F. nigra (syn. *F. caussolensis*, *F. tenella*). Native to France, Italy, Balkans to Russia. Dull purple or greenish-yellow with a purple flush, or checked. 6–12 in. in height. April–June flowering.
F. obliqua Native to S. Greece. Dark mahogany with gray bloom. 12–15 in. in height. March–April flowering.
F. olivieri Native to W. Iran. Green with purple or chocolate stripes. 20 in. in height. May–June flowering.
F. parviflora syn. of *F. micrantha*.
F. phaeanthera Native to Northern California. Greenish-yellow to speckled red-purple. 12–16 in. in height. March–June flowering.
F. pinardii (syn. *F. alpina*, *F. syriaca* & *F. fleischeri*). Native to Turkey, Syria, Lebanon & S. Armenia. Introduced 1846. Yellow-green to brownish-red or green with a brown-purple suffusion. May–June flowering.
F. pineticola see *F. bithynica*.
F. pinetorum Native to Eastern California. Dull purple, mottled greenish-yellow. 4–12 in. in height. May–July flowering.
F. pluriflora Native to Northern California. Pinkish-purple. 6–12 in. in height. Feb.–April flowering.
F. rhodocanakis Native to Greece. Purplish-maroon with yellow tips. 6 in. in height. March–April flowering.
F. roylei (similar to *F. cirrhosa*) Native to W. Himalayas. Variable. 18 in. in height. May–July flowering.
F. ruthenica Native to E. Europe, Russia. Introduced 1826. Dark purple with obscure checkering. 12–24 in. in height. April–May flowering.
F. schliemannii see *F. bithynica*.
F. sewerzowii (syn. *F. korolkovia sewerzowii* & *F. discolor*). Native to Central Asia, U.S.S.R. Introduced 1874. Green or brownish with a gray bloom. 18 in. in height. March–July flowering.
F. sibthorpiana Native to S.W. Turkey, Greece. Introduced 1874. Bright yellow. 5–10 in. in height. April flowering.
F. sphaciotica see *F. messanensis*.
F. stenanthera Native to Central Asia. Creamy pink. ≅ 5 in. in height. March–April flowering.
F. straussii Native to S.E. Turkey, N.W. Iran. Introduced 1904. Dark purple, or green aging to purple. 6–12 in. in height. May–June flowering.
F. striata Native to Southern California. White to pink often with red stripes. 8–16 in. in height. March–April flowering.
F. stribrnyi Native to Turkey & S. Bulgaria. Introduced 1893. Green with purplish suffusion. 8–16 in. in height. March–April flowering.
F. syriaca syn. of *F. pinardii*.
F. thomsoniana see *Notholirion thomsonianum*.
F. tubiformis (syn. *F. delphinensis*). Native to S.E. France, N.E. Italy, Austria. Purplish, checked obscurely with yellow. 6–12 in. in height. May–June flowering.
F. t. ssp. *moggridgei* Flowers yellow.
F. t. var. *burneti* Smaller than type with a deeper coloring.
F. tuntasia Native to Greece. Black-purple. 12 in. in height. March–April flowering.
F. verticillata Native to C. Russia, east to China & Japan. White or yellowish with green base & some purple spotting. 12–24 in. in height. April flowering.
F. viridiflora Native to Turkey. Introduced 1895. Green. 4–10 in. in height. Spring flowering.
F. walujewii (syn. *F. ferganensis*). Native to Turkey, Central Asia. Introduced 1970. White, shaded gray-purple, purple-white spotted within. 10–16 in. in height. April–May flowering.
F. wanensis Native to S.E. Turkey. Introduced 1974. Externally maroon, internally red with some yellow at tips, slightly tessellated. 3–6 in. in height. April flowering. Regarded by some authorities as syn. with *F. kurdica* or form of *F. kurdica*.
F. whittallii Native to S.W. Turkey. Introduced 1893. Greenish, spotted & tipped blackish. 8–12 in. in height. April–June flowering.
F. zagrica Native to W. Iran, Turkey. Introduced 1881. Dark chocolate purple, tipped yellow. 4 in. in height. March–May flowering.

GAGEA — LILIACEAE

The genus is named in honor of Sir Thomas Gage (1781–1820) of Hengrave Hall in Suffolk, England, who botanized in Suffolk, Ireland, and Portugal. While often regarded as a small genus, some authorities now estimate there are over a 100 species. As the identification often is based upon the number of leaves produced by the small bulbs, it would appear likely that upon closer examination many of the species now regarded as separate will be grouped together.

The plants are native to Europe, found in alpine meadows and close to the sea. Most of the species' flowers are similar in that they are starry and yellow. There are only 2 species, *G. graeca* and *G. trinervia*, having white, bell-shaped flowers. Some of the species grow in quite moist areas, others on rocky, stony ground, and all of them would appear to be fully hardy. The distribution extends from the British Isles through most of Europe to Turkey. Several species produce bulbils in the axils of the leaves and sometimes plants produce bulbils instead of flowers (*G. fistulosa*, as an example). The flowers have 6 tepals, yellow except as noted, greenish-yellow on the outside, and carried in a terminal umbel which is accompanied by a bract. There can be as many as 15 flowers per plant, mostly upright-facing, segments separated, but often the tepals reflex to give a bowl shape to the flowers.

The difficulty in identifying the many species perhaps can be better understood when it is realized that the number of leaves produced is an important identification point and this characteristic is variable!

CULTURE The plants are hardy. The soil requirements will vary a little with the species but most if not all will grow well in sandy loams, with good moisture in the spring and summer, with good drainage and sun. The bulbs should be planted some 1–2 in. deep, spacing them 3–4 in. apart. Little or no feeding is required and if given should be during the growing season.

PROPAGATION Bulbils where produced provide an excellent means of propagation. These should be separated at the end of the season and grown on in nursery rows or in containers. Separating the bulblets produced by the parent bulbs and seed are the other methods. Seed should be sown in the spring in a sandy soil mix, barely covering the seed and given moderately warm night temperatures. The bulbs should be of flowering size in 2–3 seasons of growth.

PESTS & DISEASES No special problems.

USES Grown in very bold groups in a sunny location or in the rock garden they can be quite attractive. Not a group of plants that is commonly grown.

SPECIES

G. bohemica (syn. *G. saxatilis, G. smyrnaea, G. szovitsii*). Widely distributed from Great Britain to France and east to Turkey and Syria; introduced 1829. Found growing in rocky places often in small pockets of soil. A low-growing plant reaching only a little over 2 in. in height. Bright yellow flowers with greenish exterior; up to 4 flowers are produced but often less, each of them about 1 in. in diameter, with the tepals being blunt and curling to give a bowl-shaped flower. There are 2 basal leaves produced and these are thin, thread-like, and lie on the ground. Flowering in March. It seems to like to be very dry during the summer months.

G. fistulosa (syn. *G. anisanthos, G. liotardii*). Native to Europe, from the Pyrenees to the U.S.S.R.; introduced 1816. Found growing in mountain meadows. It is distinguished by its hollow but fleshy leaves; only 2 are produced, sometimes only 1. The yellow flowers are a little over an inch in diameter and have a greenish exterior. Though small, this plant is a species that looks quite sturdy. Flowering from May–June depending on altitude. Sometimes bulbils are produced instead of flowers. Reaches 3–7 in. in height.

G. graeca (syn. *Lloydia graeca*). Native to Greece, Crete, and Turkey; introduced 1905. Flowering in March and April and likes to be dry during the summer months. 2–4 basal leaves are produced and these are narrow and grasslike. The plant reaches a height of no more than 4 in. Flowers are white with purple stripes, the flowering stem usually having 1 or more small leaves, the shape of the flowers narrow, bell-shaped, or funnel-like. This plant is easily distinguished by its unusual color.

G. minima. Native to eastern parts of Europe, to Switzerland and north to Norway; long in cultivation. Rarely reaching almost 6 in. in height. Flowering in April–June in the northern habitats. The flowers are yellow with green on the outside of the tepals which are long and pointed, sometimes reflexed, and the easiest way to recognize this species. One basal leaf is produced.

G. pratensis (syn. *G. stenopetala*). Native to much of eastern Europe; introduced in 1827. Found growing in grassy places. A plant that perhaps will reach 6 in. in height and produces up to 5 flowers. The flowers are quite large, 1 in. or a little more in diameter. The basal leaf is solitary with fine hairs along its margin. Flower size and single leaf are the distinguishing points. Flowering April–May.

G. trinervia. Native to North Africa and Sicily. Similar to *G. graeca* in form and habit but seldom produces more than 1 flower but sometimes will; this habit and the longer style and pointed stamens being the distinguishing points. It also is unusual in that this plant is possibly the only species in this genus appearing on the African continent. *G. graeca* has 4–5 flowers.

COMMENTS These plants are not large but as they are widespread their potential for growing in poorer soils and lack of demanding culture might make them interesting for areas where few other species of plants now grow. Worthwhile but not overwhelming in the excitement they create.

GAGEA

G. amblyopetala syn. of *G. chrysantha*.
G. anisanthos see *G. fistulosa*.
G. arvensis syn. of *G. villosa*.
G. bulbifera Native to Turkey, Romania, Russia east to China. Introduced 1829. Yellow, with green stripe on outside of segments, 1–2 leaves, 1–3 flowers. 2–6 in. in height. April flowering.
G. bithynica very similar to *G. chrysantha*. Native to Turkey.
G. boissieri syn. of *G. villosa*.
G. chanae similar to *G. helenae* but with hairy stems & gray leaves.
G. chlorantha (syn. *G. damascena*). Native to Turkey, W. Asia, Russian central Asia. Introduced 1829. Yellow, green outside, 2 leaves, 2–5 flowers. 2–8 in. in height. Spring flowering.
G. chrysantha (syn. *G. amblyopetala*). Native to Turkey, Balkans, Italy & N. Africa. Introduced 1829. Yellow sometimes reddish-brown outside, 2 leaves, 1–7 flowers. 2–4 in. in height. Spring flowering.
G. commutata syn. of *G. fibrosa*.
G. confusa (syn. *G. minimoides*). Native to E. Turkey, N. Iraq, W. Iran, Kapet Dag & S. Transcaucasia. Introduced 1904. Yellow inside, brown or green outside, 1 leaf, 1–5 flowers. 2–6 in. in height. April–July flowering.
G. damascena syn.of *G. chlorantha*.
G. dubia syn. of *G. granatellii*.
G. fibrosa (syn. *G. rigida, G. commutata*). Native to S. Caucasus, Turkey, Aegean Islands, Greece, Russia, Iran, Israel, N. Africa. Introduced 1829. Yellow inside, green outside, usually 1 leaf, 1–5 flowers. To 6 in. in height. March–April flowering.
G. foliosa Native to Mediterranean Region. Introduced 1830. Yellow, 2 leaves, rarely one leaf, 1–5 flowers. 2–5 in. in height. April–July flowering.
G. gageoides Native to Turkey, Iran to Central Asia. Introduced 1932. Pale yellow, one or no leaf, 1 flower per stem. 2–6 in. in height. April–May flowering.
G. glacialis Native to Caucasus, Iran, N. Turkey. Yellow, 1 leaf, 1 or seldom 2 flowers. 2–6 in. in height. May–June flowering.
G. granatellii (syn. *G. pinardii, G. dubia*). Native to Mediterranean Region. Introduced 1845. Yellow, 2 leaves, 3–13 flowers. 3–5 in. in height. Spring flowering.
G. helenae Native to Caucasus, Turkey. Introduced 1924. Yellow, green outside, 1 leaf, 2–8 flowers. 2–5 in. in height. Spring flowering.
G. joannis syn. of *G. luteoides*.
G. juliae very similar to *G. peduncularis*. Native to Turkey & Cyprus.
G. linearifolia syn. of *G. luteoides*.
G. liotardii see *G. fistulosa*.
G. lutea (syn. *G. sylvatica*). Native throughout Europe. Yellow, green exterior, 1 leaf, to 10 flowers. 5–8 in. in height. April–June flowering.

G. *luteoides* (syn. *G. joannis, G. linearifolia, G. sintenissi* & *G. syriaca*). Native to Turkey, Caucasus & Iran. Introduced 1885. Pale yellow, 1 leaf, 2–6 flowers. 2–6 in. in height. Spring flowering.

G. *minimoides* syn. of *G. confusa*.

G. *nevadensis* (syn. *G. soleirolii*) like *G. peduncularis*. Native to Balearic Islands, Corsica, Spain, France, Portugal, Sardinia. June–Aug. flowering.

G. *peduncularis* Native to Bulgaria, Crete, N. Africa, Greece, Turkey, Yugoslavia. Introduced 1904. Yellow with green outside, 2 leaves, 1–3 flowers. 1–2 in. in height. March–April flowering.

G. *pinardii* syn. of *G. granatellii*.

G. *polymorpha* similar to *G. foliosa*. Native to Spain, E. Portugal.

G. *pusilla* Native to Central & Southern Europe. Yellow with green stripe on back of segments, 1 leaf, 2–3 flowers. 1–3 in. in height. Feb.–April flowering.

G. *reticulata* (syn. *G. tenuifolia*). Native to N. Africa, Turkey, Syria, Iran, USSR. Introduced 1829. Yellow greenish outside, 1 leaf, 2–4 flowers. 2–6 in. in height. April–May flowering.

G. *rigida* syn. of *G. fibrosa*.

G. *saxatilis* see *G. bohemica*.

G. *sintenisii* syn. of *G. luteoides*.

G. *smyrnaea* see *G. bohemica*.

G. *soleiroli* syn. of *G. nevadensis*.

G. *spathacea* Native to Europe. Yellowish, green outside, 2 leaves, 2 flowers. 4–6 in. in height. April–June flowering.

G. *stenopetala* see *G. pratensis*.

G. *sylvatica* syn. of *G. lutea*.

G. *syriaca* syn. of *G. luteoides*.

G. *szovitsii* see *G. bohemica*.

G. *taurica* very similar to *G. reticulata*. Native to Western & Central Europe.

G. *tenera* very similar to *G. chrysantha*. Native to Turkey, Iran, east to Russian central Asia.

G. *tenuissima* very similar to *G. chrysantha*. Native to Turkey.

G. *tenuifolia* syn. of *G. reticulata*.

G. *uliginosa* Native to Turkey, N. Iraq & Iran. Introduced 1904. Bright yellow, brownish-red outside, 1 leaf, 1 (rarely 2) flowers. 1–2 in. in height. Spring flowering.

G. *villosa* (syn. *G. arvensis, G. boissieri*). Native to Turkey, Europe, N. Africa, W. Asia. Introduced 1828. Yellow, 2 leaves, to 15 flowers. 1–6 in. in height. Spring flowering.

GALANTHUS — AMARYLLIDACEAE

From the Greek *gala* ("milk") and *anthos* ("flower"), the name given because of the milky white color of the flowers. They are commonly known as the Snowdrop. *Galanthus* is native to the eastern Mediterranean and into the USSR. All of the species are hardy, seemingly impervious to winter weather. Many of the 15 species and subspecies produce their flowers in late fall. They can grace the garden for many months if the selection of species is made with care.

The bulbs are round and tunicated. The various species are distinguished from one another by leaf characteristics. This is especially noticeable as they emerge from the ground. Leaves vary in length and in their position in relation to one another. This genus is related to the *Leucojum* (the Snowflakes); the difference lies in the perianth—the inner segments of *Galanthus* differ in length from the outer segments; those of *Leucojum* are of equal length.

The flowers are always white, with the outer segments longer than the inner. The inner petals are notched at the tips, with a green marking around this notch extending over the tip. In some species the green color continues down the inner petal. The flowers are usually solitary and drooping, produced from a spathe that holds the unopened bloom.

CULTURE   Plant 2–3 in. deep and 4–5 in. apart as soon as available in fall. Bulbs like humus, and incorporate bonemeal into soil when first planting. In rich soils they need no feeding but in poorer soils plants can be given weak feedings of organic fertilizer as they emerge. While tolerant of some shade they prefer conditions where they can enjoy sun for at least part of the day. In hot, very dry climates plant in shade and add organic matter to soil. Do not allow to become dry while growth is being made.

Once planted should be left in position for a number of years, lifting only if increase in stock is desired. Do not take kindly to being forced, but can be grown in containers and brought indoors just before buds open.

PROPAGATION   Lift and divide after flowering but while foliage is still green. In this they differ from nearly all other bulbs. This is best means of propagation. Plant back at once if desired.

Will naturalize by themselves since they produce good quantities of seed. If raised from seed for increase of stock, sow as soon as ripe in sandy/leafy soil mix. Will germinate that spring. Transplant into individual pots or into open ground. Grow on for another year before setting out in final locations. Seedlings will flower in 4–5 seasons.

PESTS & DISEASES   Shoots sometimes attacked by a fungus as they emerge from soil, covering them with gray mold. This causes rot to set in, possibly destroying bulb. Remove attacked plants and discard. Slugs and snails may be a problem.

USES   Excellent plants for any part of garden, as some bloom early and some late. Place on the edge of woodland areas or in front of borders in bold numbers. Fine container plants. Last long as cut flowers.

SPECIES

G. **byzantinus**. Native to western Turkey; introduced 1893. It prefers heavy soils. The gray-green leaves are broad and fold back along their edges. On emergence the leaves are flat and do not curl around each other. The outer segments of the flowers are an inch in length, sometimes a little longer, and the inner segments are marked with green at both the base and tip; flowers are carried to a height of some 6–8 in. Flowering Jan. or early Feb. *G. b.* var. *hiemalis* differs only in being earlier into flower.

G. **caucasicus**. Native to the Caucasus. Found growing in woodlands. Has many forms, one var. *hiemalis* flowering in late fall and winter, while the type flowers in the spring. Leaves gray, 1 in. in width and this species differs from almost all others in this genus in that sometimes 3 leaves are produced. Outer leaf/leaves clasp the base of the inner. The outer perianth segments are an inch in length, the inner are marked only at the apex with green. Height 6–9 in.

G. elwesii (Giant Snowdrop). Native to southeastern Europe and Turkey; introduced 1875. The most commonly grown snowdrop. Foliage is gray-green and fold around each other. The leaves are quite wide, over an inch. The outer perianth segments are up to an inch in length; the inner are marked with green at the base and apex, but this color is variable and often will appear suffused throughout the inner segments. 10 in. in height. Flowering in Feb.–March. *G. e.* var. *elwesii* has larger flowers. *G.* var. *maximus* has twisted leaves; ssp. *minor* (syn. *G. graecus*) has leaves that are twisted; yellow-green ovary.

G. ikariae (syn. *G. woronowii*). Native to Turkey, Iran, and the Caucasus. Often found growing in heavy clay soils and among scrub flowers Feb.–March. The leaves are wrapped around each other while young and are distinguished by being bright shiny green, thus differing from *G. elwesii*. Leaves are strap-like, 1 in. in width, and recurve at the tip. It also differs from *G. elwesii* in that the inner perianth segments are not green at their base, but the green at the apex of these inner segments can extend down the tepal. Outer segments are 1 in. in length. Height of flowering stem 8–10 in. and flowers produced in late Feb.–March. Ssp. *latifolius* often listed in catalogs.

G. nivalis ssp. **nivalis** (Common Snowdrop). Native to much of Europe. This widely distributed species has been in cultivation for a long time and is one of the finest species for the garden. The gray leaves are held pressed flat to each other at the base and are narrow, being only some ¼ in. in width. The flowers are 1 in. in length with the inner perianth segments having only a narrow green mark at the apex. The flowers reach a height of some 4–8 in. but are more commonly around 6 in. The flowers are produced in late Jan.–Feb. Ssp. *cilicicus*, which is at home in Lebanon, varies in having markings on the inner segments that are more arrow-shaped with pedicels shorter than the spathe. This is undoubtedly a geographic variant. Selections have been made and many are listed in catalogs. That some of the offerings are hybrids is a distinct possibility but the parents are unknown. All are great garden plants.

Among the forms or selections, variants that have been described over the years include: *viridans*, which has inner segments green with white margins; *virescens*, not unlike *viridans* but with outer segments also striped green; *lutescens*, with inner segments tipped yellow and a yellow ovary, *albus*, mostly pure white with only a tiny tip of green; *viridapicis*, with outer segments green tipped. In catalogs will be found in listings such as 'Sam Arnott', 'Neill Frazer', and 'Scharlokii,' which is somewhat later flowering; long bifid spathe and pale green markings on outer petals, confined to the tips; the overall color of the plant tends toward being yellow-green.

G. n. ssp. **reginae-olgae** is unusual in that it flowers in the fall before the leaves have fully emerged. Offerings in catalogs under such names as "*G. n. octobernsis*" are most likely this plant. There is also a slight difference in the foliage with the type, the leaves on this plant are linear with a glaucous central stripe. Whether this is a variant of *G. corcyrensis* or vice versa is a good question. *G. corcyrensis* flowers in the fall, is native to Corfu and Sicily, has glaucous foliage, and the inner perianth segments are marked toward the tips with green.

COMMENTS With this genus having been so long in cultivation it is quite possible that most if not all species, subspecies, and cultivars be grouped under *G. nivalis* ssp. *nivalis*. There is much work that needs to be done on this genus as variants are not uncommon and plants are difficult to distinguish with certainty.

With the right selection *Galanthus* can be in flower in the garden from early fall through to spring. They should be planted in bold groupings of at least 10–15, and look at their best when near woodland areas. Once planted they should be left undisturbed. Great container plants and good cut flowers.

GALANTHUS

G. allenii Native to Caucasus. Introduced 1883. Green mark at apex. 6 in. in height. March flowering.

G. corcyrensis—See *G. nivalis* ssp. *reginae-olgae*.

G. fosteri Native to Turkey, Syria & Lebanon. Introduced 1889. Leaves bright green, flowers have green spot on apex and at base. 3 in. in height. Jan.–March flowering.

G. graecus see *G. elwesii* ssp. *minor*.

G. plicatus Native to Crimea, Romania, W. USSR. Introduced 1818. Green marking at apex. 6 in. in height. Feb.–April flowering.

G. rizehensis Native to Turkey. Introduced 1933. Leaves—dull green, flower apex marked green. 3 in. in height. Feb.–March flowering.

The Following List Includes Introductions From USSR since 1967.

G. alpinus Native to USSR, Caucasus. Introduced since 1967. Inner segments have horseshoe shaped green patch. 3 in. in height. Spring flowering.

G. bortkewitschianus Native to USSR, N. Caucasus, Kabardian Republic. Introduced since 1967. Creamy white V-shape green mark, scented. 8 in. in height. Spring flowering.

G. krasnovii Native to USSR, Caucasus, Adjar Republic. Introduced since 1967. Leaves may have longitudinal crinkles; hoof-shaped green patch "extremely attractive." Spring flowering.

G. lagodechianus Native to USSR, E. Georgia, Province of Tiflis. Introduced since 1967. Leaves dark green; kidney shaped or oblong green marking on flower. Spring flowering.

G. transcaucasicus (syn. *G. caspius*). Native to USSR, Talysh, S. Transcaucasicus & near the Black Sea & Caspian Regions. Introduced since 1967. Flower has a semi-circular green patch; similar to *G. rizehensis*. Spring flowering.

G. woronowii see *G. ikariae*.

Galanthus caucasicus. Now regarded as a form of *G. nivalis*. It is one of the earliest in flower, and is often listed in catalogs. WARD

Galanthus elwesii. Perhaps the best known snowdrop, but many times does not do as well as expected, no doubt due to the bulbs requiring time to become acclimated. WARD

Galanthus nivalis. The common snowdrop, which is found throughout much of Europe. A great plant to naturalize in the garden. I.B.

***Galanthus* 'Atkinsii'.** Regarded by some authorities as one of the finest hybrids, while others consider it a form of *G. nivalis,* it does not matter, as this is a superb plant! WARD

Plate 129

***Galanthus nivalis* 'Lady Elphinstone'.** A double form with yellow markings, both on the inner and outer segments. Not a strong plant, and one which will frequently revert to the more common, green markings. WARD

***Galanthus nivalis* 'Lutescens'.** This form has yellow markings on the inner perianth segments, instead of the more common green. Unfortunately these plants are often weak. WARD

***Galanthus* 'S. Arnott'.** A selection of *G. nivalis,* a very robust plant. WARD

Galanthus cultivars and selections. The snowdrops are beloved garden plants that have enjoyed great popularity over the years. Catalogs list many under many different names. Shown here are some good examples of such offerings. WARD

'Brenda Troyle'. WARD

'Cordelia'. WARD

'Dionoysis'. WARD

'Hypolyta'. WARD

'John Gray'. WARD

Plate 131

'Lady Beatrix Stanley'. WARD

'Lavinia'. WARD

'Magnet'. WARD

'Mighty Atom'. WARD

'Maidwell'. WARD

Plate 132

Galaxia fugacissima. Rare in cultivation, but a charming little species. ORNDUFF

Galaxia graminea. More colorful than the previous species, and thus more noticeable in the wild and in the garden. A good species for the bulb lover who lives in warmer climates. ORNDUFF

Galtonia candicans. While not fully hardy, these bulbs are well worth growing in the flower border. In cold climates they will have to be lifted and stored overwinter in a frost free area and replanted in the spring. AUTHOR

Geissorhiza inflexa var. *erosa.* The common name for this plant is apt, Red Sequins. One of the many lovely spring flowering bulbs from South Africa that, unfortunately, is not hardy. ORNDUFF

Geissorhiza rochensis. Known as Winecups, these plants grow exceptionally well in containers. ORNDUFF

Geissorhiza splendidissima. This rare species is native to the Cape Province, and flowers in August–September in the wild. It reaches 6–8 in. in height. K.N.B.G.

Gelasine azurea. While there may be other species of this genus, this one is, to my knowledge, the only one in cultivation. Although introduced into cultivation in 1838, easy to cultivate and having a lovely color, this plant is rare. Perhaps in another 150 years it will become popular!! LOVELL

Gethyllis afra. I doubt if any of these plants are in cultivation, a pity, as their fragrance is so pleasant. ORNDUFF

Gethyllis cilaris? The fruit of the various species has a distinct fragrance. The size of the fruit is surprising, considering the small size of the plants. All of them need a dry period after flowering. ORNDUFF

Gladiolus angustus. While this species has been long in cultivation, it is still a rare species. It is seen here growing with *Iris bucharica.*
BOUSSARD

Gladiolus carmineus. A lovely species that will form good-sized clumps if planted, and left undisturbed. Unfortunately it is not a hardy species, and needs lifting and overwintering in a frost free location.
AUTHOR

Gladiolus citrinus. This lovely species is now unfortunately extinct, or nearly so, in the wild. It has color that is superior to *G.* × *citrinus*, one of the early hybrids. Also it has an orange tinge to the petals.
BOUSSARD

Gladiolus debilis var. *cochleatus.* The differences between the three varieties of *G. debilis* are subtle, but quite distinct. *G. d.* var. *debilis* has rose stripes with darker blotches, while the species illustrated here has distinct blotches with little striping. BOUSSARD

Gladiolus blandus carneus

Gladiolus carophyllaceus

Gladiolus carophyllaceus

Plate 138

Gladiolus debilis **var. *variegatus*.** This variety of *Gladiolus debilis* has the color arranged in spots, not in streaks, making it quite distinct. BOUSSARD

Gladiolus gracilis. While not producing many flowers, the color of this species is delightful. ORNDUFF

Gladiolus liliaceus. It is not only the modern hybrids that have flowers in many colors. Some of the species are themselves, most unusual. WARD

Gladiolus hollandii. While a great deal of fanfare was made when striped hybrids were introduced, we can see from this species that nature, as usual, was first! B.R.I.

Gladiolus italicus

Plate 140

G. italicus (syn. *G. segetum*). A well known *Gladiolus*, that grows wild in southern Europe. One of the advantages of growing this species, is that it faces in more than one direction. It still needs more attributes to tempt all but the ardent bulb collector. WARD

Gladiolus primulinus. Photographed in the wild at Victoria Falls, this lovely yellow species has been much used in the production of the modern hybrids. AUTHOR

Gladiolus punctulatus. The markings on the flowers makes this species well worth growing, and it does produce a good number of flowers. ORNDUFF

Gladiolus subcaeruleus. This autumn flowering species is rare, both in the wild and in cultivation. BOUSSARD

Plate 141

Gladiolus orchidiflorus

Gladiolus tristis

Plate 143

Gladiolus vaginatus. The delicate colors of this autumn flowering species vary a little, but are always attractive. BOUSSARD

***Gladiolus* 'Candy Stripe'.** There are so many cultivars listed today, that selection is often difficult. This cultivar shows the qualities looked for in new introductions: good and unusual color, great texture to the flowers, many flowers open while the lower ones are still attractive, vigor, and good placement of the flowers on the stem. AUTHOR

***Gladiolus* cultivar.** A good cultivar in a great color; even the topmost bud will give a good flower, the mark of a good introduction. AUTHOR

Plate 144

GALAXIA — IRIDACEAE

The name is derived from the Greek *galaktos* ("milk") in reference to the milky sap of the plants. This is a genus of South African bulbous plants, first introduced into culture in the late 18th century, with perhaps but 4 or 5 species, of which only 3 are described below. There may well be even fewer species but, as the plants are not widely grown even though they deserve more consideration in warmer climates, they have not been given much attention by botanists.

The small corms, rarely more than ½ in. in diameter, are covered with a coarse, fibrous tunic. The plants themselves also are small, not more than 3 in. in any direction. The leaves form a small rosette, lying close to the ground in most cases. Most species have yellow flowers, up to 1 in. in diameter, which appear in winter and into early spring—December to March. They enjoy a mild Mediterranean climate, with good moisture during the winter, warm springs, and dry, almost hot, summers.

CULTURE The small corms should be set just under the surface of the soil, within an inch or two of one another. They are best planted in early fall or late summer, in well-drained soil, preferably in full sun. They are unable to withstand temperatures below 35°F, except in the dormant state, but even then the temperature of the soil should not fall below 32°F. Summer heat will not bother them. Good moisture is needed during the winter months and in the early spring, with less moisture in the summer but not allowed to completely dry out until autumn. Even in soil that is low in fertility, these plants will do well. Only in the poorest soil should any fertilizer be used and that should be just weak dilutions given in the early stages of growth.

They should be left undisturbed, being lifted only to divide the corms; this is done in late summer. Being small, the plants must be kept weed free. Cultivation in containers should be practiced in all cold areas, the plants being kept under glass, except when temperatures are in the 50°F+ range.

PROPAGATION Lifting and separating out the young cormels is the best means of propagation. This should be done as soon as the foliage has begun to die down. These cormels can be sown in flats, barely covered with a sandy soil mix, and grown on for a season in the flats. They should be of sufficient size to be planted out at the end of their first complete season of growth. Some corms will be of sufficient size when lifted for immediate replanting.

PESTS & DISEASES Although there are no particular problems, especially from insects and diseases, care must be exercised to keep the plants open to good air circulation as they are easily smothered. In the wild, they are exposed to excellent air circulation and strong winds.

USES While never likely to take the world by storm, these rather pleasant plants do deserve a place in gardens where the climate is suitable if one has more than a passing interest in bulbs. They should be considered only for rock gardens, in the front of sunny borders, or in locations where they can be easily seen because they are so small. Their rather long flowering season makes them a possibly good bulb for warmer areas.

SPECIES

G. fugacissima. Native to Signal Hill at the foot of Table Mountain in the Cape Province. The foliage is of two kinds—the bottom broader leaves are about 1 in. long and lie flat, or nearly so, while those above are curled and extend above the flower to a height of some 2½ in. and are more iris-like in appearance. The white with yellowish center or yellow flowers are carried singly on a short stalk, nestling in the foliage. They are about an inch in diameter, the petals held loosely together in segments about ¼ in. wide, ½ in. long, with just a slight curve to give a flat look. Flowering time is Dec.–March; June–Sept. in its native habitat. Height of flower 1½–2 in. above the ground.

G. graminea. Native to the Cape. The foliage is quite grasslike and so the species is quite aptly named. A tufted form, the rather dainty and attractive flowers grow in rosettes surrounded by leaflike bracts. Several are produced at one time and they vary in color from shades of white tinged with pink with a yellow throat, or are entirely yellow. They are more funnel-shaped than the other species, rather more than an inch in diameter, with a perianth tube up to an inch long but opening to give an almost flat flower. Flowering time is Dec.–March; June–Sept. in native habitat. Height of plant only 1–3 in.

G. ovata. Native to the Cape. Very similar to *G. graminea*. The rosette of grasslike leaves is up to 2½ in. across and reaches only 1–2 in. in height. The flowers vary in color from purple to pink with a yellow throat and also sometimes a pure yellow, but darker than *G. graminea*. Although several flowers are produced, they usually are in flower only one at a time. Leaflike bracts form a small cup from which the flowers arise. The flowers are larger than one would expect from the size of the foliage, almost out of scale. They like to grow in well-drained, sandy or gritty soil. *G. ovata* var. *grandiflora* has been described as having larger flowers, but it is more than likely that this is just a fine specimen of the species. Flowering time is the same as the species above. By some authorities this variety has been raised to species rank.

GALTONIA — LILIACEAE

This lovely genus of summer- and late-summer-flowering bulbs is named in honor of Francis Galton (1822–1911), a British scientist who not only travelled widely in South Africa, the home of the *Galtonia*, but also was an advocate of the fingerprint method of identification. There are 3 species. They are related to the hyacinth and indeed look like giant hyacinths, except that the flowers are always white. Few leaves are produced and vary in length according to the species, inclined to become a little untidy after the plants have flowered. Flowers are held in a loose raceme and, depending on the species, can be well over 24 in. in length. Flowers pendulous, carried on long pedicels well away from the stem, tubular at the base but then widening with the lobes longer than the tube, perianth segments quite pointed with overall length being some 1½–2 in. Flowers open to give pleasant, bell-shaped flower.

CULTURE While not fully hardy, they can be grown outdoors in cold areas as long as the bulbs are lifted in the fall before severe frost arrives, stored overwinter, and planted again in the spring. In those areas where winters are not severe, i.e., temperatures seldom fall below 14°–20°F, they can be grown in a sunny border and given a protection of heavy mulch in winter. They require good drainage and full sun. The bulbs are large, often 5 in. in diameter, and should be planted with their tops just below the surface of the soil, spacing them 18–24 in. apart. They must receive good moisture during the growing season and be allowed to dry out after the foliage has died back. While reaching a considerable height, no staking is needed. Weak feedings of liquid fertilizer should be given as soon as growth is seen in the spring; these feedings can be discontinued as soon as the flower spikes emerge.

PROPAGATION Established bulbs will produce offsets, but these are not freely produced. Seed provides the best way to increase stock. Seed is best sown in late summer in containers with a soil mix that is porous but with enough organic matter for moisture retention. Seeds should be lightly covered with soil and placed in a temperature of some 55°F at night in bright light. Keep the seedlings growing until they start to die back, usually some 12–14 months after germination, the same time as mature plants die back. By then the bulbs should be of sufficient size to plant out. Transplanting the young bulbs to individual containers should be done with care as the bulbs do not like disturbance.

PESTS & DISEASES No special problems, but, if not receiving good drainage, rotting of the bulbs will occur.

USES Great plants for the summer border where the towering spikes will make quite a display and last a period of many weeks.

SPECIES

G. **candicans** (Cape Hyacinth, Berg Lily, Summer Hyacinth; syn. *Hyacinthus candicans*). Native to the Cape Province of South Africa; introduced circa 1860. The best species in this genus and the only one readily available. Can reach 50 in. in height and have as many as 30–40 flowers. The 2 in. long, pendant flowers start to open in August or late July and continue into early September. Flowers are pure white sometimes with green tips to the petals, which color is sometimes also seen at the base, on pedicels over 2 in. in length. Flowers have pleasant fragrance. Leaves are 2 in. in width, tapering gradually to fine point, some 30 in. in length. The leaves are sometimes erect but more frequently bend back toward the ground and can become untidy. A worthy garden plant.

G. **princeps**. Native to S. Africa—Natal and Cape Province. A smaller and less desirable species. Flowers are just over an inch in length, with a greenish tinge to the white or sometimes creamy white. Flowers in August; the flower spikes only reach up to 36 in. in height and are inclined to be less upright and not so strong-growing. Foliage is similar to *G. candicans* and some authorities regard this as a form of *G. candicans*

G. **viridiflora**. Native to South Africa in the same regions as other species. Rare in cultivation and perhaps rightly so as its only merit is the rather unusual greenish flowers, with leaves 3–5 inches wide that end abruptly in a blunt tip. Flowering in late summer; 24–36 in. in height.

COMMENTS *G. candicans* is the only species worth growing in the garden and this plant deserves space. Ideal for the perennial border and among shrubs as long as they do not crowd the towering flower spikes. Grows well in large containers such as half wine barrels.

GEISSORHIZA — IRIDACEAE

The name is derived from the Greek *geisson* ("tile") and *rhiza* ("root") because a tunic envelops the bulbs like tiles are laid on a roof. Native to the Cape Province area of South Africa, there are many species—some authorities mention that there may be as many as 60—but few are in cultivation and only a few more are to be found listed in books on the flora of the Cape at the southern tip of South Africa.

In their native habitat, these plants are commonly known as Wine Cups, Red Sequins, or Blue Sequins. Flowers are *Ixia*-like, but larger. *Hesperantha* is a closely related genus, but in *Geissorhiza* the style branches are much shorter. The *Ixia* have spathes which are thin and papery; in *Geissorhiza* they are green.

Unlike many bulbs from this part of the world, *Geissorhiza* frequently are found in very moist areas, especially during the growing season, standing in or close to ponds on marshy ground. They flower in early spring, at the time when there is much moisture.

The corms are small, rarely more than an inch in diameter. The foliage of these mostly dwarf plants—they vary in height from 4–5 in. to about 12—is reedlike, covering the stem. In certain species it rises to a sharp point above the flowers; in others it remains below the spike. The flower spikes emerge from the sheath formed by the foliage and produce buds protected by a green spathe. The buds open to flowers in various shapes—from a star form to a goblet form. The colors, too, are variable—sometimes of a single color, sometimes tricolored in shades of white to pale blue to purple to red-wine. In comparison to the height of the plants, the flowers are quite large, often more than ½ in. in diameter. Some species produce only 1 or 2 flowers at a time, others 4–5.

CULTURE Plant corms in late summer or early fall, 2 in. deep and 3–4 in. apart. They are native to areas where temperatures seldom drop below 32°F and thus are not hardy and should be grown in either warm areas or in a frost-free greenhouse. They like full sun but some shade in the hottest parts of the country during the midday period. During the growing season they must receive plenty of water and, while some will withstand dry periods after the foliage and flowers have passed, summer moisture will not harm the corms; but toward the end of summer a drier period should be given. Those that are found in damp, peaty soil are noted in the *Species* section. For the remainder, well-drained but moisture-holding soil is preferred. Little or no feeding is required.

They can be grown in areas where there are very light frosts. In such areas they should be planted in locations where the early morning sun will not strike them and quick warming is avoided. As they are small plants, large groupings are necessary, at least 25 or more in each group. They can be left in the ground in mild areas and, if grown in the soil in greenhouses, they can be left undisturbed. As these corms make their growth in the winter time, they should not be overwintered in storage. They grow well in containers in a sandy-peaty mix.

PROPAGATION Seed sown in late summer in a sandy-peaty mix germinates easily and will often produce flowering plants by the following spring and certainly by the second year. In warm areas they can be sown outdoors in rows or in containers with the seed being barely covered. Transplant into outdoor beds when large enough to handle or plant several into a container. When this is done, be sure to firm the soil ball well.

PESTS & DISEASES No special problems.

USES These are fine plants for the sunny border in warm climates and good container plants for areas where a frost-free greenhouse is available. They must be planted in bold groupings to be effective.

SPECIES

G. **geminata**. Native to the Cape Peninsula, South Africa. Found growing around seasonal pools and occasionally in shallow water. The tapering, pointed foliage forms a cylinder around the flowering stem. The flower spike emerges from this, carrying 3–5 star-shaped flowers with hints of pink in the petals. The green spathes that surround the flower buds are cup-shaped. The plant is about 9–12 in. in height and the flowers are a little more than ½ in. in diameter. Flowering time is August–Sept., March–April in Northern Hemisphere.

G. inflexa var. **erosa** (Red Sequins; syn. *G. hirta*). Native to South Africa. This species has pure carmine-colored flowers that are cup-shaped and grows to a height of some 10–14 in. The veins on the slender grasslike, foliage are prominent; leaves 10–12 in. long. Flowering in Sept. in native habitat; April in Northern Hemisphere.

G. juncea. Native to the Cape Peninsula, South Africa. Damp, peaty soil is home to this species. The flowers are carried some 10–12 in. above the ground, just a little higher than the reedlike foliage, which is pointed and clothes the stem. Shaped halfway between a bowl and a star, the small, white flowers, 4–6 per stem, have their petals arranged above each other on a slender stem. The spathes are not very prominent. Flowering time is Aug.–Sept. in native habitat, or March–April in Northern Hemisphere.

G. mathewsii var. **eurystigma** (Winecups). Native to Cape Peninsula of South Africa. Found in heavy clay soils, this slender plant generally has but 1 leaf, which sheathes the zigzag-shaped flowering spike. On occasion more than 1 spike is produced. If a second spike is produced from 1 spike, it will have fewer flowers than the main spike, which carries up to 9 flowers; produced in spring. Height 6–8 in. The cup-shaped flowers are purple-blue on the outside with deep crimson interiors; the top of the inside of the segments also is purple-blue. The common name for this species also is used for *G. rochensis*, which leads to the belief that this may be just another form of *G. rochensis*. In illustrations of this species, however, there is no band of color between the colors mentioned, thus differing from *G. rochensis* (see below).

G. ovata. Native to Cape Province of South Africa. This plant is only a few inches tall and is found growing in moist rock crevices up to 5,000 ft. The leaves are small, up to 2 in. long and ¼ in. wide, with a blunt tip. Unlike many of the other species, they do not sheathe the flower stem but are distinctly attached to it. Two to four flowers are carried above the foliage on a brownish-green stem that sometimes branches. The flowers, almost 1 in. in diameter, open wide to snow white flushed with pink and the exterior of the segments is pinkish. The sepals overlap at the lower third. Flowering Aug.–Dec. in the wild; March–May–June in Northern Hemisphere.

G. rochensis (Winecups; syn. *G. radians*. Native to Cape Province of South Africa; introduced 1790. In the wild it is frequently found in damp areas among grass. Its name accurately describes the flowers, which are goblet-shaped, purple-blue on the exterior and tip of the interior. On the inside of the segments a thin, white band separates this color from the red eye of the lower portion of the segments. While reaching only 4–8 in. in height, the individual flowers can be as much as just over an inch in diameter. Due to the overlapping perianth segments, the flowers appear quite substantial. Narrow, grasslike leaves with tapering points clasp the stem at the base. This species does well in containers. Flowering July–Aug. in the wild; Feb.–March, in Northern Hemisphere.

G. secunda (syn. *G. aspera*). Native to the Cape Peninsula, South Africa; introduced 1795. This plant is found growing in flat, sandy areas and on the slopes of mountains. While it will tolerate summer moisture, it grows well without it but must have good moisture during the fall and winter months. Generally only 2 leaves, which are like broad blades of grass, are carried on the flowering stem. The flower spike reaches to about 6–10 in. in height; emerges from the curling upper leaf to rise above the foliage. There are 3–6 flowers produced on the simple stem. These are star-shaped in a lovely violet-blue shade. The perianth segments are joined only at their bases. Flowering Aug.–Oct. in the wild; March–May, Northern Hemisphere.

COMMENTS The plants of this genus are highly regarded by those who grow them. It is doubtful that they ever will become as popular as their form and colors deserve. *G. rochensis* and *G. secunda* should be the first species tried as they are of easy culture and good plants.

GEISSORHIZA

G. aspera see *G. secunda*.
G. excisa Not a valid name, possibly *G. ovata*. Described as: Native to South Africa. Introduced 1789. White. 6 in. in height. April–May flowering.
G. grandis Not a valid name, possibly *G. geminata*. Described as: Native to South Africa. Introduced 1868. Straw with red midrib. 6–10 in. in height. April–May flowering.
G. hirta see *G. inflexa* var. *erosa*.
G. humilis Native to South Africa. Introduced 1809. Sulphur. 12 in. in height. June–July flowering.
G. inflexa var. *inflexa* (syn. *G. vaginata*) Native to South Africa. Bright yellow with dark purple spot at base of segments. 18 in. in height. May flowering.
G. radians see *G. rochensis*.
G. tenella Native to S.W. Cape Province. White or pink with a pink veined reverse. 4–12 in. in height. April–June flowering.
G. vaginata syn. of *G. inflexa* var. *inflexa*.

GELASINE — IRIDACEAE

The name is derived from *gelasinos* ("a smiling dimple"), but just what this has to do with the genus is difficult to determine. This is a small genus of South American bulbs, few plants being in cultivation and the literature is rather barren on this entire genus, a not altogether uncommon occurrence with South American genera. Only *G. azurea* is, to my knowledge, in cultivation and described in the literature. The rootstock is a bulb comprised of loose scales. The flowers are of a deep and intense blue and, while individually not long in flower, lasting barely a day, a goodly number of flowers are produced, so the plants remain in flower for a considerable perod of time, often as long as a month.

Flowers are carried on long branches from the central flowering spike and at the base of these branches are small, leaflike bracts. The flowers on the ends of these branches are enclosed in bracts with many individual flowers that are quite large, over an inch in diameter and sometimes almost 2 in. The stamens are inserted at the base of the short perianth tube and are united at the top into a cylindrical column. This surrounds the style, which is divided into 3 lobes at the apex. The color of the flowers also is present in the anthers and style, the only relief from the intense blue being the pollen and the white petal bases.

Foliage is green with a hint of blue, some 3–4 basal leaves are produced and these are up to 24 in. in length and folded along their length. The height of the flowering stem will reach 24–30 in. The records show that G. *azurea* was introduced in 1838. It is surprising that this plant has not become more widely cultivated as the color and apparent ease of culture would seem to merit wider acceptance.

CULTURE Not a plant for cold climates as it needs to be planted early in the spring, setting the bulbs some 3 in. deep in well-drained soil with good moisture retention, and preferably in a protected area such as near a wall. In climates where colder winters prevail, the bulbs are best lifted in the fall and stored overwinter and replanted in the following spring. In warmer climates the bulbs can be left in the ground. Moisture is required during the growing season and throughout the flowering time, which is May–July. Good light and full sun with some shade during the hottest part of the day in warm climates is needed. As this plant is rare in cultivation, I would tend to give the plant more shade until its performance was noted.

PESTS & DISEASES No special problems.

PROPAGATION Being a rare plant, it is suggested that if bulbs are obtained they be left undisturbed to become established, propagation being confined to seed. This is best sown in the spring, in moderate temperatures with good indirect light. Young seedlings can be grown on in flats until of size to merit individual containers.

USES A plant for the keen collector but, from all descriptions, well worth the effort in growing; under protection in colder climates, in a sunny border where spring comes early in the year, and in a sunny border with some shade in warmer climates.

SPECIES

G. **azurea**. Described above.

G. **coerulea**. Sometimes mentioned in the literature, having similar habit and of similar color but not quite so robust a plant as *C. azurea*.

GETHYLLIS — AMARYLLIDACEAE

The name is from an old Greek diminutive of the word *gethuon* ("leek"). The literature states that there are some 8–9 species of this genus, all native to South Africa, however, it is most likely that there are fewer. Differences are not so much with the flowers, fragrant and nearly all white or with a slight yellow hue to the petals, as in the foliage, some leaves having hairs. The plants are all dwarf, flower before the foliage is produced, and the flowers are produced in late spring/early summer. The bulbs produce a single flower, its long perianth tube uniform in size until the segments separate to give flowers some 3 in. in diameter about 3–6 in. above the ground. The petals recurve and are pointed, often with a stripe of carmine on the reverse along the midrib.

The fruit is produced some 7–9 weeks after the flowers and is carried above ground and is sweetly scented. The stamens lie close to the petals. The foliage is linear, mostly spirally twisted, sometimes hairy, sometimes without hairs. The leaves are sometimes sheathed at the base and usually are produced in pairs. The common name is Kukumakranka.

CULTURE The plants do not like to be in a soil that is too wet, preferring, as in the wild, sandy soil with good drainage. They like to have a dry period as the fruit is produced. They are not hardy and therefore are suitable for outdoor growing only in mild climates without frost. In other areas they should be grown in a greenhouse to give protection against frost. The bulbs should be planted 1–2 in. deep. Water should be given after the foliage starts to emerge; plants are native to an area with winter rains and the foliage appears at that time.

PROPAGATION Offsets from the parent bulbs can be taken after the foliage has died down. Seeds can be sown in moderate heat, 50°–55°F night temperatures, in a sandy soil mix in the spring with good light. The young seedlings should be transplanted into individual pots as soon as large enough to handle and then grown on for 2 seasons before being planted out.

PESTS & DISEASES No special problems.

USES For the sunny borders in warm climates in well-draining soil. They should be planted where the fragrance of the fruit and the flowers can be appreciated. Good pot plants for the cool greenhouse.

SPECIES

G. **afra**. Native to Cape Province; introduced 1820. Found growing in sandy soil. In the wild the flowers are produced in December, June in the Northern Hemisphere, and are some 4–6 in. above the ground, appearing before the foliage. The flowers are white, with a carmine stripe on the exterior of the petals above the perianth tube, where the petals reflex; the petals are pointed. The 12 stamens are held close to the petals. After the flowers are finished, the foliage appears. Some 12–20 leaves are narrow and linear, twisted, in pairs. The fruit appears in 7–9 weeks and is yellowish in color and club-shaped *G. britteniana* is very similar.

G. **cilaris**. Native to the Cape Province of South Africa; introduced 1788. Flowers similar to the above species, but the foliage is fringed with hairs, twisted; flowers in June–July. The fruit is bright reddish-orange.

G. **spiralis**. Native to Cape Province; introduced 1780. Flowers similar to above species but plant is reportedly taller, reaching some 9 in. in height. Flowers in June and July. Leaves twisted, linear, smooth.

G. **villosa**. Native to Cape Province; introduced 1787. Also about 9 in. in height, with very thin, threadlike leaves which are twisted and hairy. Flowers in June–July.

COMMENTS As will be seen from the above descriptions, the principal differences between the species is in the foliage. There also would appear to be some difference in the shape and color of the fruit. I would not be surprised if upon close examination of the genus 2 species remain, *G. afra* and *G. ciliaris*, thus accommodating the difference in hairy or non-hairy foliage and color variations of the fruit. Possibly *G. villosa*, with its threadlike foliage, also will be considered a separate and distinct species or perhaps a form of *G. ciliaris*.

GLADIOLUS — IRIDACEAE

From the Latin *gladius* ("sword"), the name used by Pliny, referring to the shape of the leaves. There are as many as 150 species and many more named hybrids of these lovely flowers. The species are found only in South Africa, tropical Africa, Europe, and around the Mediterranean. Few of the species are grown today, but the varicolored hybrids are known the world over.

The history of the hybrid *Gladiolus* is long and interesting. The first was *G.* × *colvilli*, named in 1823 for a nurseryman from Chelsea, in London, England. It was a cross between two South African species, *G. cardinalis* and *G. tristis*. The hybrids produced by this cross were the first of a series of early-flowering plants that were popular as cut flowers and were used for forcing. Two that have been grown for many years, still listed in catalogs and still in cultivation, are *G. ackermanii* and 'The Bride.' New varieties are offered each year, frequently listed as *G. nanus*.

The species used in the large-flowered hybrids are many. Van Houtte used *G. cardinalis* × *G. psittacinus*, an African species. Crosses within these were selected and offered under such names as × *citrinus*, with yellow flowers, and × *brenchleyensis*, with light red flowers. Further work in many countries produced hybrids from M. Souchet in France and Max Leichtlin in Germany. In 1878 M. Lemoine, of the famous nursery in Nancy, France, brought forth the robust × *lemoinei*, early flowering with large blossoms; some as large as 6 in. in diameter, the result of a cross between *G.* × *gandavensis* and *G. purpureo-auratus*.

Breeding continued apace, taken up by many horticulturists in Europe and America. One great input of new blood was provided by *G. primulinus*, sometimes called *G. nebulicola*, from Zimbabwe, where it grows in the vicinity of Victoria Falls. I have seen this graceful plant in its native habitat—pale lemon, ruffled flowers, hooded with a few leaves, but not too sturdy.

It can be easily seen why the classification of the many hybrids we enjoy today is difficult; however, the authorities have categorized them into the following six types for ease in identification:

1. Early flowering—*Nanus*; dwarf; slender; April–May flowering.
2. Summer- and autumn-flowering—*Primulinus*; mostly 36–56 in. high; spikes can be branched; 10–14 flowers, grouped into colors of white, yellow, etc.; upper segments horizontal or hooded.
3. Large-flowered—31 varieties listed; 36–54 in. high; strong stems; 14–20 flowers, up to 6 open at one time; upper segments often erect, sometimes hooded.
4. *Gladiolus tubergenii*—about 30 in. high; thin leaves; flowers not packed; mid June flowering.
5. Herald Gladioli—about 36 in. high; strong; erect; large flowers; early, often appearing even before *nanus*.
6. *Primulinus grandiflorus*—40–56 in. high; strong stems; usually 12–18 flowers, 3½–4½ inches in diameter; upper segments grouped into colors of yellow, cream, shaded pink, etc.

The following general description applies to all *Gladiolus*. The base of the stems are swollen, forming a corm. The leaves are always sword-shaped and vary in length according to species. Those of the hybrids often are coarse to the touch. The flower spike often is one-sided, with the individual flowers being zygomorphic. The tube formed by the perianth segments is curved. In many species the flowers open wide, while in the hybrids they are not so open and also are more crowded on the stem. The seed capsule is three-celled, producing a great number of seeds.

Only one species—*G. edulis*—has any value as an edible plant. According to Dr. E. Lewis Sturtevant, in his book "Edible Plants of the World," the corms taste like chestnuts when roasted. It seems they also are commonly eaten by baboons.

CULTURE "The root consists of two bulbes, one set upon the other; the uppermost in the spring is lesser... ; the lower greater...which shortly after perisheth." So is the "bulb" of *Gladiolus* described in *Gerard's Herbal* of 1597, offering a good idea of what happens to what actually are corms.

Easy to grow. Prefer sandy, rich soil with good drainage and sunny location. Can be planted anytime of year as long as frost is not likely, allowing several crops or seasons of flowers to be had by successive plantings. Flower in about 80–90 days.

Set corms at least 4 in. deep, about 6–10 in. apart. April/May normal planting time in most areas, however, in warmer regions, where little or no frost occurs, can be grown year-round. In colder areas corms best lifted after flowering. Dry and store in well-ventilated, warm place where temperature is in 50°F range. Some nurseries offer "Winter Hardy" *Gladiolus*—quite dwarf, only about 18–20 in. high; these can be planted in fall and will flower in spring.

PROPAGATION Easy to raise from seed, which are plentifully produced. Germinate readily and may flower within the year. Best sown in March in sandy soil mix. Sow thinly, barely covering. Give as much light as possible after germination, which is best in nighttime temperature of 45°F, or even a little warmer.

Seedlings can remain in containers in which germinated, but should be thinned out and given good air circulation. Harden off and stand outside in sunny location. Keep moist. As summer progresses hold back on water and allow small corms that have formed to dry out. They can then be harvested and stored as are the larger corms. Planting out can be done in spring in the garden or in rows in the nursery.

The many little corms produced around parents can be lined out in a row in the spring and grown on. Will reach good size by that fall, when they should be lifted. Plant again following spring and will flower that summer.

The amateur hybridizer can see satisfying results in short time because *Gladiolus* increase so easily and rapidly. Cost of individual corms is not high. Offered in various sizes and all will bloom. Unless gardener wishes extra-sturdy stalks or to subject them to forcing, smaller-sized corms are adequate.

PESTS & DISEASES
Dry Rot (*Stromatinia gladioli*): Also known as Stromatinia Neck Rot. Inner leaves yellow, starting at tips. Outer leaf parts underground and just above ground turn brown; browning continues to entire leaf. Sclerotia, light yellow pustules, will be seen, generally starting at base; roots also will be affected, with pustules turning black. Discard diseased bulbs. If attack not too severe, treat with good fungicide, such as Benlate, at recommended dilutions. The fungus can persist on corms left in the ground.

Botrytis Rot (*Botrytis gladiolorum*): Appears with moist and cool conditions. First sign is light colored blotch on leaves, which soon turns darker and spreads. All parts of leaves affected, especially base. Causes tissue to die. Good aeration helps. Destroy all infested leaves. Treat with fungicide; follow directions on label. This fungal problem has various names, such as Botryties (leaf and flower spot).

Fusarium Rot (*Fusarium oxysporum*): Plants will produce leaves but little root activity will occur. Outer leaves turn yellow and then brown, beginning at tip first; yellow streaking found on inner leaves. Treat corms with fungicide; follow directions on label. Due to yellowing of leaves often referred to as Fusarium Yellows.

Rust (*Uromyces transversalis*): Most common under warm, muggy conditions. Yellow spots develop on surface of leaves. May not affect flowering spikes, which can be harvested, but waste foliage must be discarded. Fungicides can be used but apply as soon as spots are seen.

Thrips (*Taeniothrips simplex*): Cause damage to both leaves and flowers. Most noticeable on foliage where damage occurs first. Identifiable as silver-gray spots that darken as insect spreads.

Tissue of leaves eaten away and plant just looks sick. Flower damage is seen as areas devoid of color, with dry look to flowers in general. In severe cases, flowers will not open. Insecticide must be applied. Treat corms with insecticidal powder if thought to be infected.

Corn Borer: The mature insect, the Pyralid Moth, lays eggs on leaves. Caterpillars that hatch eat into plant, feeding on leaves and stalks. Leaves turn yellow at tips; in severe cases flowers will die. Little, white grubs, about ¼ inch long, can be seen if infested plants are cut open. Do not plant corms where corn has been grown previously. If attack is noticed and cannot be controlled by removing infested plants, spray with an insecticide, following directions carefully.

Aphids: Not generally a serious problem.

With all the pests and diseases listed above, it might seem that *Gladiolus* is not worth raising. Nothing could be further from the truth. Purchase stock from reliable suppliers, rotate the location in which plants are grown, and keep off debris that might harbor the various pests and diseases. All will help minimize the control problem. Also keep a sharp eye out for the symptoms before they have a chance to develop.

USES Ever popular in gardens and as cut flowers. Striking accent plant among summer-flowering annuals and useful addition to perennial borders. Very good container plants but take care when selecting container as plants can become quite heavy when in full bloom. Cut flowers will last long time in floral arrangements; cut as soon as first buds show color with own foliage acting as complement to flowers.

SPECIES & CULTIVARS It would not be possible to list all the species and hybrids of *Gladiolus* in this volume, so I will attempt to include those I consider of greater importance.

G. alatus. Native to southwestern Cape Province, South Africa. Narrow, stiff leaves, 3–4 produced. Striking flower, unusual in coloring; upper 3 perianth segments salmon pink, sometimes a little lighter, strongly bent tube; lower segments much narrower, yellow with tips same color as upper segments; coloring can vary slightly, sometimes the yellow has green tinge with tips leaning toward orange. 8–10 flowers, carried on stem that zigzags up to height of some 12 in. Flowers are slightly hooded, slightly fragrant. Spring-flowering.

G. atroviolaceus (syn. *G. aleppicus*). Native to Turkey, Iran, and Iraq; introduced 1889. Flower spike one-sided, 2–3 ft. high, much longer than the 2–3 leaves which are rarely more than 12 in. long. Flowers, usually 6–10, very dark purple/violet shade. April/May flowering. Likes to be dry at end of summer. Hardiness questionable in extreme climates but can survive where winters not too severe. Best to lift and replant.

G. blandus. (Broad-leafed Painted Lady; see also *G. carneus* and note below). Native to the western Cape Province of South Africa; introduced 1774. The plants are found in different soil types, from rocky to very sandy, and are not difficult to grow. There are 4 leaves and as many as 10 flowers on a spike, which reaches 18 in. in height. Flower colors vary from white to pale pink to the purple hues. The lower perianth segments are marked with a darker blotch of color. This was one of the first *Gladiolus* to be introduced and now is known as *G. carneus*, which originally was applied to the pinkish form. Flowering in March–April. Several forms of *G. blandus* sometimes are offered for sale: *albidus*, white; *erubescens*, suffused with bright carmine-pink; and *excelsus*, a stouter plant.

Note: Great confusion exists about the names *G. carneus* and *G. blandus*. It would appear that the first plants introduced of this kind were known as *G. blandus*. In this variable species the red forms were called *G. b.* var. *carneus*. Certain authorities elevated this variety to specific rank and it became *G. carneus*, the valid name today. No doubt this is historically correct as the red form was known originally as *G. blandus carneus*. The red form of *G. cuspidatus* var. *ventricosus* was introduced and described as *G. blandus carneus* and illustrated under that name in the *Botanic Magazine* in 1801. Later, however, it was determined that this was indeed the red form of *G. cuspidatus*. In certain literature the name *G. carneus* can in fact refer to *G. blandus carneus* or to *G. cuspidatus* var. *ventricosus*. Is it any wonder that confusion reigns in the nomenclature, especially among species that vary greatly in the wild!

G. byzantinus. Native to Spain, Italy, North Africa, Corsica, and Malta; introduced 1629. I remember seeing it growing in the stony ground in Malta; the color is a striking burgundy red, with as many as 15 flowers; more can be expected in the garden. They are 3 in. long and, unlike many Gladiolus, the flower spike is not so one-sided, the flowers facing in two or three directions but not all around the spike. Leaves form a fan at the base of the stem and there are usually 3 or 4, sometimes 5. Height of the spike can be 24 in. but is usually less. There has been a white form of this plant recorded but it is very rare. Flowering in May or early June.

G. cardinalis. Native to South Africa; introduced 1789. This enjoys the name of *Nuwejaarsblom* (New Years Flower) in its native habitat, where it flowers in December. It also is called the Waterfall Gladiolus as it grows along the stream banks in the southwestern Cape. The flowers are cardinal-red, up to 24 in. in height, up to 3 in. in diameter, often with a blotch of white on the lower segments. One of the few Gladiolus that likes a little shade; much used in the production of our garden hybrids. It does require moisture throughout its growing season and is quite hardy. The 4–6 leaves are wide, sword-shaped, and up to 20 in. in length. Flowering July in the Northern Hemisphere.

G. carinatus (Mauve Afrikander). Native to the southwestern Cape. 12–18 in. in height. 2–9 flowers produced early in the spring, mauve with yellow markings on the lower segments. One of its best attributes is the fragrance, a trait that has been transmitted to some of the hybrids raised by using this species as a parent. Foliage—leaves 3, linear, the upper some 18 in. in length, others a little less.

G. carneus (syn. *G. blandus carneus, G. cuspidatus ventricosus, G. macowanianus, G. poppei, G. prismatosiphon*, however it is still listed in some catalogs as *G. carneus*). The species offered is generally the white form of *G. blandus* (which see), i.e., *G. b. albidus*, which has purple flakes on the lower segments. It responds to forcing. The many variations, varieties, and subspecies of *G. blandus* and *G. carneus* should rightly be under *G. carneus*.

G. carophyllaceus (Pink Afrikander; syn. *G. hirsutus*). Native to the western Cape Province; introduced 1795. Up to 20 in. in height with showy flowers that vary in color from pink, to rose, to cyclamen-purple, often with deeper shades in the center. It has a strong fragrance reminiscent of carnations, making it a popular variety with breeders. The flowers are 2–3 in. in diameter but seldom more than 6–8 are produced on a spike. Flowers are more upward-facing than those of many of the other species. Foliage—4–6 sword-shaped leaves—short and hairy.

G. × citrinus. One of the earliest hybrids produced and much used as a parent in breeding. Offered by several nurseries, sometimes listed as *G. symmetranthus*. This is a low-growing variety, up to 8–10 in. in height, with pale yellow flowers produced in June. This should not be confused with *G. citrinus*, which is native to the Cape Province and is now rare, if not extinct, in the wild.

G. × colvilli. Introduced 1823. The result of crossing *G. cardinalis* with *G. tristis*. The flowers are bright red. There are not many flowers per stem, which reaches 18 in. in height. Sometimes narrow leaves are found on the stem. Flowering in late June/early July.

The same cross has produced a number of hybrids that are often listed incorrectly under G. *nanus* or G. *colvilli*. Among the most commonly offered are the following:

G. *colvilli* 'The Bride'. Pure white; a very old hybrid and very good.

G. *c. albus*. Pure white with a yellow stripe on lower segments.

G. *c. roseus*. Soft pink.

G. *nanus*
- 'Amanda Mahy'. Bright salmon-red with violet blotches.
- 'Atom'. Red with a creamy border.
- 'Comet'. Bright red.
- 'Elvira'. Pale pink, red blotches, and pale yellow center.
- 'Fair Lady'. Double pink.
- 'Floriade'. Clear salmon-pink.
- 'Guernsey Glory'. Deep pink, pale purple with red edges and creamy blotches.
- 'Impressive'. Orange-pink, purple-rose blotch, edged with brown.
- 'Nymph'. White with crimson markings.
- 'Prins Claus'. Ivory white, edged with purple on lower segments, center of purple blotches white.
- 'Rose Marie'. Bright pink with scarlet blotches on lower segments.
- 'Sunmaid'. Outside pink, interior rose with white center, anthers violet; very attractive color combination.
- Mixed. A combination of the above is frequently offered.

In many catalogs the G. *nanus* types are listed as 'Baby Gladiolus' or some such similar name.

G. **cruentus** (Red Gladiolus). Native to South Africa; introduced 1868. Its large, scarlet flowers are often up to 5 in. in diameter, with white markings on the lower segments. Flowers produced mid- to late-summer; 36 in. high; it is dormant during the Winter months, is quite hardy, and can be left in the ground. The leaves are 12–18 in. in length.

G. **debilis** var. **debilis** (Painted Lady). Native to the western Cape Province; introduced 1820. This is a dwarf species, seldom more than 12 in. in height, pale pink or off-white, marked with rose stripes and blotches on the lower petals. Early spring-flowering, March–April. Usually 3 stiff leaves, 8–10 in. long.

G. **gracilis**. Native to the Cape Province of South Africa; introduced 1800. Grows to 24 in. in height and the leaves are more cylindrical than in other species. The stem is very slender but quite strong, and, while it supports the 30 flowers produced, the main attraction of this plant is its flower color—blue—although at times this is a little variable and sometimes appears washed out or pale. Perhaps one day this color will be obtained in the hybrids. April–May flowering. May need staking in exposed locations.

G. **illyricus**. Native to northern Africa and southern Europe. It is rare but sometimes is found in the wild in England. This hardy species is up to 18 in. in height, producing leaves that are very narrow and short. As many as 10 purple-red flowers with some white markings on the lower segments are produced. May-flowering.

G. **italicus** (syn. G. *dubius*, G. *segetum*). A well-known plant in southern Europe, spreading to Asia Minor, the USSR, and Afghanistan. The 4–5 leaves form a fan shape. Lilac flowers with a hint of red are found in a loose arrangement on the flower stalk; not particularly attractive as they are small and look out in more than one direction. June-flowering.

G. **liliaceus** (syn. G. *grandis*; regarded as a variation of G. *tristis* and sometimes placed in that species). Native to South Africa along the east coast from the Cape up to Port Elizabeth; introduced 1749. Grows to a height of 24 in. in sandy soil. Usually 6 scented flowers produced on a stem; individual flowers 3–4 in. in diameter. Flowers are variable in color, ranging from brown to pale yellow with light green at the throat, sometimes a dirty white with a brown/crimson stripe in the center of the petals or the brown/crimson color is in a blotch on the tips. The flowers open wide and turn up a little. Spring-flowering.

G. **natalensis** (now G. *daleni*; syn. G. *dracocephalus*, G. *psittacinus*, also, by some authorities, G. *quartinianus*). Widely distributed in Africa, from the Cape northward to Ethiopia. A strong grower with stems up to 48 in. high. Often more than 20 flowers are produced on a stem. Color varies from red to orange to yellow, sometimes streaked with maroon. The 3 upper petals, the center petal of which is much larger than the other 2, form a hood. The lower 3 petals are narrower and often have a patch of yellow covering the ends of the petal, though not so pronounced on a central lower segment. The flowers are carried on two sides of the stalk, well spaced from each other. The leaves are narrow, about 30 in. in length, and the number varies, but generally there are 4. Flowering in late summer. The species has been used by breeders. G. *n. cooperi* has lemon-yellow flowers.

G. **oppositiflorus** (Transkei Gladiolus). Found in the Cape region and up the coast to the Transkei; introduced 1892. Up to 24 in. in height. The leaves, 3–4, are up to 18 in. long. A great number, often more than 30, white or pale-pink, small flowers are produced. The lower segments are marked with maroon stripes. Flowering June–July.

G. **orchidiflorus** (syn. G. *dregei*, G. *viperatus*). Native to the western Cape. This is a short species, reaching only to 12 in. It has scented, pale green to amber flowers, a maroon stripe on the side petals, with maroon mixed with a lime-green on the lower segments. Flowering in early spring, March–April. Foliage—3–4 basal leaves, linear, up to 12 in. in length.

G. **papilio** (syn. G. *purpureo-auratus*). Native to South Africa; introduced 1866. Widely used in breeding and one of the parents of the hybrids known as Butterfly Hybrids. Unusual in that the corm produces stolons. Strong grower, reaching well over 36 in. 4–5 leaves are produced, up to 36 in. in length and not over an inch in width. Flower color is greenish-yellow, sometimes a little more green, other times a little more yellow. The margin of the flowers is purplish and this color may sometimes mark the entire flower. The flowers do not open very wide, being almost bell-shaped. This plant can produce many young corms and, if left in the ground, will soon produce quite a planting. Likes moisture throughout the growing season as it inhabits moist areas in its native habitat. Quite hardy except in the very coldest climates.

G. **primulinus**. Native to Zimbawbe; introduced 1879. A species found growing in the rain forest around Victoria Falls. Reaches a height of 30 in. on a slender stem. The pale yellow flowers are well spaced. Leaves are 18 in. long. This species has been much used in breeding, but is not hardy. Summer-flowering.

G. **saundersii**. Native to South Africa; introduced 1870. It is hardy and reaches 24 in. in height. Up to 12 flowers, loosely held in the spike, are a good vermilion, and the lower segments have white blotches spotted with red. Leaves are long and rigid. Flowering in late summer.

G. **sempervirens** (Cliff Gladiolus; syn. G. *splendens*). Native to the eastern Cape Province. An unusual evergreen species and produces stolons. The bright red flowers open wide, except for the center upper segment which curves over the stamens, there is a white stripe on the lower segments. Seldom more than 6–8 flowers are found per stem. The leaves persist throughout the winter and thus it is not hardy in the cooler climates but will stand a little frost as the accumulation of older leaves protects the corm; it will need protection where there is severe frost. Height up to 36 in. Flowering in late summer.

G. tenellus (syn. *G. trichonemifolius*). Native to the Cape Province. It has funnel-shaped flowers of a lovely sulfur-yellow. It attains 20 in. in height; is fragrant. Summer-flowering. Very rare. Foliage—leaves 3, cylindrical, up to 12–14 in. in length.

G. tristis. Native to Cape Province; introduced 1745. This is one of the sweetest-smelling species, especially at night. Spikes are up to 40 in. but more commonly around 24 in. The flowers are pale yellow, 3 in. in diameter, with as many as 15 flowers per spike, sometimes but rarely more, the 3–4 leaves are more or less cylindrical. This plant is not fully hardy and needs a sheltered spot in areas where frost in the wintertime is light and grown in milder areas in a sunny border. It appreciates moisture throughout the year and flowers in late summer. There are several varieties: var. *tristis* has flowers of green and brown; var. *concolor* has very pale yellow flowers.

G. × ramosus 'Robinetta'. A selection made between *G. cardinalis* × *G. oppositiflorus* seedlings. Introduced 1839. Flowers are currant-red, tall-growing, many on strong spikes 36 in. or more in height. Summer-flowering. Leaves 24 in. in length, sword-shaped.

G. undulatus (syn. *G. cuspidatus*). Native to the Cape Province; introduced 1790(?). The name refers to the petals being quite wavy. Tall, 36 in. high spikes carry as many as 8 slender, creamy flowers borne in a zigzag fashion. Flowers can be variable in color, from cream to greenish-white, but nurseries generally offer the cream-colored form. It makes a good cut flower. A little dubious as to hardiness and should be grown only in protected areas and given adequate moisture during the growing season. Summer-flowering.

HYBRIDS The number of hybrids offered today by the various nurseries covers the whole range of colors found in these fine plants. For convenience the following listed is by height and by color. The smallest varieties have already been listed under the species *nanus* found in the *G.* × *colvilli* section. These are but a few of the many hybrids offered by leading nurseries. **24–30 in.**

Red:
 'Small World'. Bright red.
 'Tom Thumb'. Deep red.
Orange-red:
 'Brightside'. Scarlet-orange-yellow throat.

'Petite Orange'. Lightly ruffled; deeper colored throat.
'Twinkle'. Apricot-orange.
Pink-rose:
 'China Doll'. Ruffled; pink-cream throat.
 'Gallery'. Rose-pink.
Yellow:
 'Golden Angel'. Deep yellow.
White:
 'Wood Violet'. White with purple throat.
Purple and violet:
 'Elfin Orchid'. Ruffled, lavender; white throat.
Green:
 'Jade'. Clear lime-green.

More than 36 in.

Red:
 'Black Velvet'. Silver edge to flowers.
 'Blaze'. Bright scarlet.
 'Cherry Flip'. Medium red.
 'Intrepid'. Clean red; great cut flower.
 'President de Gaulle'. Vermilion.
 'Red Freckles'. White striping in flowers.
Orange-red:
 'Halloween'. Bright orange-yellow center.
 'Orange Juice'. Good clean color.
 'Sunbeam'. Bright yellow blotch in center of flowers.
Pink-rose:
 'First Prize'. Ruffled; rose-pink.
 'Prima Donna'. Very large flowers; ruffled.
 'Royal Rose'. Light picotee edge to the petals.
Salmon:
 'Bon Voyage'. A lovely soft color.
 'Peter Pears'. Early flowering.
Yellow:
 'Golden Gate'. Rich golden yellow.
 'Lemonade'. Bright; ruffled.
 'Mardi-Gras'. Ruffled; cherry-red center.
 'Yellow Wonder'. Taller than most.
White:
 'Birthday Cake'. Ruffled; red throat.
 'King of Whites'. Large flowers and of good texture.
 'White Friendship'. Creamy.
 'White Surf'. Long lasting in flower.
Purple:
 'Blue Robe'. Violet-blue.
 'Misty Blue'. Medium blue.
Bicolors and striped:
 'Circus'. Salmon and red.
 'Jazz Age'. Rose and silver.
 'Peppermint Twist'. Red-rose and white.
 'Royal Tapestry'. Purple with silver.

GLADIOLUS

G. acuminatus Native to S.W. South Africa. Pale yellow. 12–22 in. in height. Sept.–Nov. flowering in the wild.
G. adlami Native to South Africa. Introduced 1889. Greenish yellow, red spotted at tip. 12–18 in. in height.
G. alatus var. *namaquensis* syn. of *G. equitans*.
G. a. var. *pulcherrimus* Native to South Africa. Salmon pink. 5–12 in. in height. Aug.–Oct. flowering in the wild.
G. a. var. *speciosus* (syn. *G. speciosus*). Native to South Africa. Deep salmon pink, bright yellow back of upper perianth outer lobes. 5–12 in. in height. Aug.–Oct. flowering in the wild.
G. albidus see *G. carneus* & *G. blandus*.
G. aleppicus see *G. atroviolaceus*.
G. anatolicus syn. of *G. illyricus* var. *anatolicus*.
G. angustus Native to South Africa. Introduced 1756. White with heart-shaped purple mark center of 3 lower perianth segments. 12–26 in. in height. Sept.–Dec. flowering in the wild.
G. antakiensis Native to S. Turkey, Lebanon. Introduced 1804. Pink, reddish pink. 12–30 in. in height. March–May flowering in the wild.
G. aurantiacus Native to South Africa. Introduced 1894. Orange yellow tinged red. 36 in. in height. Sept.–Nov. flowering in the wild.
G. a. var. *rubrotinctus* Native to South Africa. Orange yellow thickly dotted red. 36 in. in height. Sept.–Nov. flowering in the wild.
G. aureus (also see *Homoglossum aureum*) Native to Cape Peninsula, South Africa. Bright golden yellow. 12–20 in. in height. April flowering, Aug.–Sept. in the wild.
G. biflorus syn. of *G. quadrangulus*.
G. blandus var. *mortonius* (syn. *G. mortonius*) Native to South Africa. White, fine rose stripes. 12–18 in. in height. June flowering, Sept. in the wild.

G. b. niveus see *G. b.* var. *albidus*.
G. b. purpureo-albescens see *G. b.* var. *carneus*.
G. blommesteinii Native to South Africa. White, fine maroon stripes. 14–22 in. in height. Aug.–Oct. flowering in the wild.
G. bolusii syn. of *G. inflatus*.
G. brachyandrus Native to Tropical Africa. Introduced 1879. Bright pale scarlet. 24 in. in height. July flowering, Oct. in the wild.
G. brevifolius Native to Cape Peninsula, South Africa. Pale pink. 10–24 in. in height. Feb.–May flowering in the wild.
G. b. var. *obscurus* Native to South Africa. Pale pink. 10–24 in. in height. Feb.–May flowering in the wild.
G. brevitubus Native to South Africa. Pale scarlet. 10–14 in. in height. Aug. flowering, Nov. in the wild.
G. buckerveldii Native to Cedarberg Mts., South Africa. Apricot. 24–36 in. in height. Sept. flowering, Dec. in the wild.
G. bullatus (syn. *G. spathaceus*). Native to South Africa. Lilac, speckled purple. 12–24 in. in height. May–June flowering, Aug–Sept. in the wild.
G. callianthus (see also *Acidanthera*) Native from Ethiopia to Malawi. Introduced 1896. White, center purple blotch. To 44 in. in height. Sept.–Oct. flowering.
G. callistus (a form of *G. carneus* which see) Native to South Africa. Introduced 1930. Pale pink purple spot in throat. 18–48 in. in height. Sept.–Dec. flowering in the wild.
G. carmineus Native to Cape of Good Hope, South Africa. Deep pink, carmine; white marks on lower perianth segments. 10–14 in. in height. Sept. flowering, Feb.–May in the wild.
G. citrinus Native to Stellenbosch, South Africa. Pale yellow, maroon throat, long rounded segments. 5–12 in. in height. Aug.–Sept. flowering in the wild.
G. communis Native to S. Europe—Spain to Crimea. Introduced 1596. Rose pink, white or bright rose blotches, rust streaks on lower perianth segments. 12–24 in. in height. April–June flowering.
G. comptonii Native to S.W. South Africa. Introduced 1941. Bright yellow, rust streaks on lower perianth segments. 24–32 in. in height. July flowering in the wild.
G. confusus syn. of *G. hyalinus*.
G. cooperi (see *G. natalensis*) Native to Natal. Yellow, with red striped upper perianth segments. To 36 in. in height. Autumn flowering.
G. crispiflorus syn. of *G. imbricatus* var. *crispiflorus*.
G. cuspidatus (Also listed as syn. of *G. undulata*) Native to Cape Peninsula, South Africa. Introduced 1759. White with yellow spots, or lilac with deep purple spots on lower perianth segments. 10–36 in. in height. May–June flowering, Oct.–Nov. in the wild.
G. c. var. *ventricosus* (also see *G. blandus* & *G. carneus*) Purple rose, purple central spots.
G. daleni see *G. natalensis*.
G. dracocephalus see *G. natalensis*.
G. dregei see *G. orchidiflorus*.
G. dubius see *G. italicus*.
G. ecklonii Native from E. Cape to Natal. Introduced 1862. White closely speckled pink, purple, or deep maroon. 12–26 in. in height. Autumn flowering, Dec.–March in the wild.
G. edulis syn. of *G. permeabilis* ssp. *edulis*.
G. emiliae Native to S.W. South Africa. Deep brownish orange. 12–24 in. in height. Feb.–April flowering in the wild.
G. engysiphon Native to S.W. South Africa. White, deep maroon markings on lower perianth segments. 12–24 in. in height. March flowering in the wild.
G. equitans (syn. *G. alatus* var. *namaquensis*) Native to N.W. South Africa. Pale pinkish salmon, lemon markings lower perianth segments. 5–14 in. in height. Aug.–Oct. flowering in the wild.
G. exilis Native to South Africa. White, minute purple marks at lower throat. 18–36 in. in height. May flowering in the wild.
G. floribundus Native to S. South Africa. Introduced 1788. White, purple stripe each segment. 18 in. in height. May flowering; Sept.–Nov. in the wild.
G. f. ssp. *fasciatus* Native to S. coastal South Africa. Pink or mauve, dark red at throat & diamond shapes on lower perianth segments. 10–18 in. in height. Oct.–Jan. flowering in the wild.
G. f. ssp. *milleri* (syn. *G. milleri*). Native to S.W. South Africa. Creamy white. To 18 in. in height. Sept.–Nov. flowering in the wild.
G. f. ssp. *miniatus* (syn. *G. miniatus*). Native to S.W. coastal South Africa. Purplish red. To 18 in. in height. Sept.–Nov. flowering in the wild.
G. f. ssp. *rudis* (syn. *G. hastatus*, *G. rudis*, *G. vomerculus*). Native to South Africa. Introduced 1812. Creamy white, yellow splotch lower perianth segments. To 18 in. in height. Sept.–Nov. flowering in the wild.
G. fredericii (Possibly only a variety of *G. permeabilis*) Native to S. South Africa. Pale blue, deep stripe each segment. 18 in. in height. Oct.–Nov. flowering in the wild.
G. garnieri Native to Madagascar. Clear orange pink, yellow center. To 38 in. in height. Flowering in the wild varies with rainy season.
G. grandis see *G. liliaceus*.
G. guthriei Native to S.W. South Africa. Pale salmon. 12–36 in. in height. May–June flowering in the wild.
G. halophilus Native to S.E. Turkey, Iran, Iraq. Introduced 1854. Bright pink to red purple. 18–24 in. in height. April–May flowering.
G. hastatus syn. of *G. floribundus* ssp. *rudis*.
G. hirsutus see *G. caryophyllaceus*.
G. hirsutus roseus see *Homoglossum merianellum*.
G. hollandii Native to Transvaal, Swaziland, Mocambique. Pale pink streaked or speckled maroon or red. 24–36 in. in height. Feb.–April flowering in the wild.
G. hyalinus (syn. *G. confusus*) Native to W. Cape Peninsula, South Africa. Greenish yellow, pale yellow lower perianth segments, dark center streaks. 12–18 in. in height. May–July flowering in the wild.
G. illyricus var. *anatolicus* (syn. *G. anatolicus*). Same as *G. illyricus* but narrower segments. Native to W. Europe, Mediterranean.
G. i. var. *reuteri* Same as *G. illyricus* but more slender. Native to Spain & Portugal.
G. imbricatus Native to Central & Eastern Europe, Turkey. Introduced 1820. Pale carmine to reddish violet or purple. 12–32 in. in height. April–July flowering.
G. i. crispiflorus Same as *G. imbricatus* but smaller. Native to S. Russia.
G. i. galiciensis Same as *G. imbricatus* but smaller, more erect. Native to Galicia.
G. inflatus (syn. *G. bolusii*) Native to Western mts., South Africa. Rosy pink to lilac, white spear on segments, dark border. 10–18 in. in height. Sept.–Oct. flowering in the wild.
G. involutus Native to S.W. South Africa. White, red brown striped, center lower perianth segments marked yellow & red. 12–24 in. in height. Sept. flowering in the wild.
G. jonquilliodorus Native to Cape Peninsula, South Africa. Creamy white, streaked with maroon. 18–26 in. in height. Dec.–Jan. flowering in the wild.

G. kotschyanus Native to E. Turkey, Iran, Iraq. Introduced 1854. Deep violet paling at apex. 12–34 in. in height. April–June flowering.
G. laccatus Native to Southwest Africa. Bright red to lilac, pale spots at segment base. To 18 in. in height. April–May flowering.
G. lapeirousioides Native to W. South Africa. Introduced 1970. Pale lilac, maroon edged gold splotches on lower perianth segments. 5–7 in. in height. Sept. flowering in the wild.
G. leichtlinii (Not a valid name, possibly *G. liliaceus*). Native to South Africa. Introduced 1889. Bright red, shading to yellow at base of lower perianth segments. 24 in. in height.
G. lewisii Native to South Africa. Creamy shading to yellow base, upper perianth segments with mauve stripe. 12–24 in. in height. Sept.–Oct. flowering in the wild.
G. linearis syn. of *G. quadrangulus*.
G. longicollis Native from S. South Africa to Natal. Creamy yellow, dark central stripes. 12–22 in. in height. Sept.–Oct. flowering in the wild.
G. ludwigii syn. of *G. sericeo-villosus*.
G. ludwigii calvatus syn. of *G. sericeo-villosus* var. *calvatus*.
G. macowanianus see *G. carneus*.
G. macowanii (syn. *G. ochroleucus* var. *macowanii*). Native to S.E. South Africa. Deep salmon pink, darker stripes lower perianth segments. To 14 in. in height. Feb. flowering in the wild.
G. maculatus Native to S. South Africa. Yellow with brown or maroon spots or stripes. 12–28 in. in height. Mar.–July flowering in the wild.
G. m. ssp. *meridionalis* (syn. *G. meridionalis*) Native to S.W. South Africa. Pink or red. 12–28 in. in height. May–July flowering in the wild.
G. martleyi Native to S.W. South Africa. White, maroon bordered yellow blotches on lower perianth segments. 8–12 in. in height. Mar.–April flowering in the wild.
G. masoniorum Native to E. Cape Province, Transkei, South Africa. Introduced 1913. Cream, pale yellow in throat. 12 in. in height. Spring flowering in the wild.
G. melleri Native to E. & S. tropical Africa. Introduced 1913. Red, deep salmon, cream, yellow. 12 in. in height. October flowering in the wild.
G. meridionalis syn. of *G. maculatus* ssp. *meridionalis*.
G. milleri syn. of *G. floribundus* ssp. *milleri*.
G. miniatus syn. of *G. floribundus* ssp. *miniatus*.
G. monticola (syn. *G. tabularis*). Native to Cape Peninsula, South Africa. Very pale pink. 12–24 in. in height. Jan.–Mar. flowering in the wild.
G. mortonius syn. of *G. blandus* var. *mortonius*.
G. mutabilis Native to S. South Africa. Very pale lilac, lower perianth segments streaked maroon. 12–18 in. in height. July–Aug. flowering in the wild.
G. nerineoides Native to S.W. South Africa. Deep orange red. To 14 in. in height. Jan.–Mar. flowering in the wild.
G. ochroleucus var. *macowanii* syn. of *G. macowanii*.
G. ochroleucus var. *ochroleucus* syn. of *G. stanfordiae*.
G. odoratus Native to W. South Africa. Cream, heavily striped & speckled purple. 12–36 in. in height. May–June flowering in the wild.
G. oppositiflorus ssp. *salmoneus* syn. of *G. salmoneus*.
G. oreocharis Native to Mt. peaks, W. South Africa. Mauve pink. 8–12 in. in height. Dec.–Jan. flowering in the wild.
G. ornatus Native to Cape Peninsula, South Africa. Red purple, lower perianth segments white spears bordered deep purple. 10–24 in. in height. Aug.–Oct. flowering in the wild.
G. pallidus Native to Angola. Rose purple. 12–24 in. in height. Sept.–Nov. flowering in the wild.
G. palustris Native from W. Europe to Balkans. Reddish purple. 12–20 in. in height. April–July flowering.
G. pappei see *G. carneus*.
G. permeabilis ssp. *permeabilis* Native to S. coastal South Africa. Cream, pink, or mauve; dark central veins, lower perianth segments yellow bands. To 30 in. in height. Aug.–Nov. flowering in the wild.
G. p. ssp. *edulis* (syn. *G. edulis*) Native to South Africa. White, gray, mauve, yellow. 8–12 in. in height. Aug.–Nov. flowering in the wild.
G. persicus Native to Iran. Dark violet purple. Late Spring–Early Summer flowering.
G. pillansii Native to S.W. South Africa. Pale pink, lower perianth segments blotched yellow. 10–24 in. in height. Feb.–May flowering.
G. pilosus syn. of *G. punctulatus*.
G. platyphyllus (May be color form of *G. tysonii*). Native to S.E. South Africa. Bright orange, lower perianth segments green streaked on yellow. To 28 in. in height. Dec. flowering in the wild.
G. praecox see *Homoglossum watsonium*.
G. prismatosiphon see *G. carneus*.
G. pritzelii Native to W. mts., South Africa. Clear yellow, lower perianth segments streaked maroon. 12–14 in. in height. Aug.–Oct. flowering in the wild.
G. psittacinus see *G. natalensis*.
G. pulchellus syn. of *G. virescens*.
G. punctatus syn. of *G. recurvus*.
G. punctulatus (syn. *G. pilosus, G. villosus*). Native to South Africa. Mauve to purple; lower perianth segments white streaked purple. 12–24 in. in height. July–Nov. flowering in the wild.
G. purpureo-auratus see *G. papilio*.
G. quadrangulus (syn. *G. biflorus, G. linearis*). Native to Cape Peninsula, South Africa. White, streaked violet. 8–14 in. in height. Aug.–Oct. flowering in the wild.
G. quartinianus see *G. natalensis*.
G. recurvus (syn. *G. punctatus*). Native to S.W. South Africa. Introduced 1758. Yellow, thickly dotted blue, "eventually blue". 12–24 in. in height. June–July flowering in the wild.
G. rogersii Native from S.W. to S. South Africa. Mauve to purple, lower perianth segments yellow base, purple spots in throat. 12–24 in. in height. Aug.–Nov. flowering in the wild.
G. rudis syn. of *G. floribundus* ssp. *rudis*.
G. sabulosus Native to E. coastal South Africa. Pale lilac, maroon-bordered cream spears on segments. 8–20 in. in height. Oct. flowering in the wild.
G. salmoneus (syn. *G. oppositiflorus* ssp. *salmoneus*). Native to E. South Africa. Bright salmon, red streak center lower perianth segments. 18–26 in. in height. Feb.–Mar. flowering in the wild.
G. scullyi (syn. *G. venustus*) Native to S.W. South Africa. Yellow; tips dark maroon. 6–12 in. in height. Sept.–Oct. flowering in the wild.
G. segetum see *G. italicus*.
G. sericeo-villosus (syn. *G. ludwigii*). Native to Transvaal, South Africa. Introduced 1864. Pale mauve, lower perianth segments faint yellow-green marks. 36–72 in. in height. Late Summer flowering, April–May in the wild.

G. s. var. *calvatus* (syn. *G. ludwigii calvatus*). Same as *G. sericeo-villosus* but upper & lower perianth segments unequal.

G. spathaceus syn. of *G. bullatus*.

G. speciosus syn. of *G. alatus* var. *speciosus*.

G. splendens see of *G. sempervirens*.

G. stanfordiae (syn. *G. ochroleucus* var. *ochroleucus*). Native to E. Cape Province, South Africa. Introduced 1936. Pale salmon, white markings at throat. To 24 in. in height. March flowering in the wild.

G. stefaniae Native to South Africa. Bright pink, white streaks center lower perianth segments. 12–26 in. in height. Mar.–April flowering in the wild.

G. stellatus Native from S.W. to S. South Africa. White, pink streaks center segments. 12–24 in. in height. Aug.–Nov. flowering in the wild.

G. stokoei Native to S.W. South Africa. Scarlet. 10–12 in. in height. March flowering in the wild.

G. subcaeruleus Native to S.W. South Africa. White, yellow with brown specks base of lower perianth segments. 10–12 in. in height. April–May flowering in the wild.

G. symmetranthus see *G.* × *citrinus*.

G. tabularis syn. of *G. monticola*.

G. templemanii syn. of *G. virescens*.

G. tenuiflorus Native to W. Turkey. Introduced 1848.

G. trichonemifolius see *G. tenellus*.

G. triphyllus Native to S. Turkey, Cyprus. Introduced 1806. Pale to deep pink. 5–14 in. in height. April–May flowering.

G. tysonii (similar to *G. natalensis*) Native to E. South Africa. Salmon, green flushes base of lower perianth segments. To 36 in. in height. Dec.–Jan. flowering in the wild.

G. ukambanensis (syn. *Acidanthera laxiflora*) Native from Ethiopia to Tanzania. Cream or yellow, center blotched purple. To 36 in. in height. Summer flowering.

G. uysiae Native to South Africa. Upper perianth segments salmon streaked with white, lower perianth segments yellow with green blotch. 4–12 in. in height. Aug.–Sept. flowering in the wild.

G. vaginatus Native from S.W. to S. South Africa. White or pale mauve, lower perianth segments finely dotted in brown lines. 12–14 in. in height. Feb.–May flowering in the wild.

G. varius var. *varius* Native to South Africa. Clear pink, deep pink streaks lower perianth segments. Jan.–May flowering in the wild.

G. venustus syn. of *G. scullyi*.

G. vigilans Native to South Africa. Pink, white blotches edged in deep pink lower perianth segments. 12–24 in. in height. Nov. flowering in the wild.

G. villosus syn. of *G. punctulatus*.

G. vinulus syn. of *G. vittatus*.

G. violaceo-lineatus Native to South Africa. Very pale lilac, dark purple streaks of tiny dots. 18–28 in. in height. July flowering in the wild.

G. viperatus see *G. orchidiflorus*.

G. virescens (syn. *G. pulchellus*, *G. templemanii*). Native to W. South Africa. Yellow closely striped brown. 5–18 in. in height. Aug.–Sept. flowering in the wild.

G. viridiflorus Native to S.W. South Africa. Yellow green, yellow blotches bordered purple lower perianth segments. 5–10 in. in height. May–July flowering in the wild.

G. vittatus (syn. *G. vinulus*). Native to South West Africa. Introduced 1760. White to pink, red or lilac middle stripe. 12–15 in. in height. April–May flowering; Sept.–Nov. in the wild.

G. vomerculus syn. of *G. floribundus* ssp. *rudis*.

G. watermeyeri Native to W. South Africa. White, upper perianth segments red-veined, lower perianth segments deep yellow with white tips. 5–14 in. in height. Aug. flowering in the wild.

G. watsonioides see *Homoglossum watsonioides*.

G. watsonius see *Homoglossum watsonium*.

G. woodii Native to N.E. South Africa. Upper perianth segments gold brown, lower perianth segments yellow. 18 in. in height. Oct.–Nov. flowering in the wild.

GLORIOSA — LILIACEAE

Genus name derives from the Latin term *gloriosus* meaning "full of glory." This is a very good name for these plants, which produce the lovely cut flowers used so much by florists. They are grown in quantities in greenhouses and sold by the blossom in Holland and elsewhere.

There are many species, at one time as many as 30. Dr. D.V. Field, Royal Botanic Gardens at Kew, however, has made a very good case for there being only one species—*G. superba*—a very apt name. He explained the problems in classifying the species in the Royal Horticultural Society's publication *Lilies and other Liliaceae*, published in 1973. The colors and presence or absence of tendrils has been thought to be a characteristic of the various species, but not necessarily, as forms have been found with and without them. The likelihood that the variations are strictly geographic appears to be the most reasonable, however.

Gloriosa are native to Africa and India. I have seen them growing in very poor soil along the roads in South Africa and in Kruger National Park, climbing up shrubby plants, finding a way to always keep the flowers in the sun. These species were, of course, the ones that produce tendrils at the tips of the leaves. Others sprawl along the ground, not showing even a vestige of a tendril. Under cultivation, however, they may reach a height of 8 ft. or more.

Tubers are fleshy and multiply quickly. The stems are light green and look fragile, but are not. The flowers have 6 segments that are thin and reflex so the stamens are prominent. A peculiar characteristic is that the stigma bends at a right angle as it leaves the ovary. The petals are yellow or red, or a combination of these colors; they often are crinkled along the edges, and, even when fully reflexed, the tips will sometimes curl back. All parts of the plants are poisonous.

CULTURE Can be grown outdoors only where there is no danger of frost. Perform well in cool greenhouse. Started in pots and planted outdoors, will flower in summer but must be taken indoors in late summer/early fall.

Plant so tubers are an inch or two under surface, in well-drained, sunny location. Give moisture to get them growing, after which they can withstand drier conditions. Do appreciate moisture during summer, however. Best kept dry but never cold in pots overwinter.

G. superba performs well in containers but must be given porous soil with ample organic matter mixed with sand and good topsoil. Some liquid feeding with fertilizer will help when growth commences in spring. Some support needed so plants can scramble on it.

PROPAGATION Seed can be set and grown at temperature of 65°F in soil mix of sand and well-rotted organic matter. Best method of propagation, however, is to lift and divide rootstock.

PESTS & DISEASES No serious problems, but aphids can be a problem if grown under glass.

USES Excellent plants for the cool greenhouse and for cut flowers. Good container plants but must be grown in large containers to leave room for expansion. Adds touch of class to garden, more as a specimen plant than for mass of color.

SPECIES

G. abyssinica. Native to tropical Africa; introduced 1894. Leaves 4–6 in. long with tendrils at tips. Flower stalks 3–4 in. long. Deep yellow flowers; perianth segments reflexed and crinkled along edges. Summer-flowering, but first flowering period entirely dependent on when plant started into growth and temperature it receives. Can climb to a height of several feet.

G. carsonii. Native to central Africa; introduced 1904. Taller-growing than most *Gloriosa* species; often with no leaf tendrils. Flowers reddish-purple with yellow edge to petal. Summer-flowering; can reach 5 ft. or much more.

G. minor. Native to Kenya. Lower-growing, only 20–24 in., as name implies. Leaves closely packed on stem. Red flowers; not many produced. Summer-flowering.

G. rothschildiana. Native to South Africa; introduced circa 1900. Leaves have tendrils. Bright red flowers with crinkled edges, yellow margins; yellow at base of petals when first open, which disappears as it ages. Flower stalks quite long, up to 4 in., sometimes more, throwing flowers in late spring and summer well away from stem, which either crawls or sprawls over low-growing shrubs often for several feet. Var. *citrina* has yellow flowers splashed with crimson-purple; excellent for cool greenhouse; same habit and flowering time as species and the sooner started into growth the more flowers are produced.

G. superba. Considered the only *Gloriosa* species by some authorities. Introduced 1690. Excellent variety that will grow up to 6 ft. in one summer if started early. Leaves have tendrils. Late-spring- or early-summer-flowering. While flowers sometimes described as deep orange-red, I believe that the yellow found on edges of petals and at base give more an appearance of being orange. Var. *superba* has brilliant yellow flowers. All parts of this plant are poisonous and can be fatal if ingested.

G. virescens (Climbing Lily; syn. *G. simplex* but name not valid). Native to South Africa, Natal. Leaves have tendrils. Petals differ from some other species in that they are narrow at base, widen in middle, and then taper again. Flower tips less curled but still wavy along edges of petals; yellow at base but crimson at tips. Spring–summer-flowering.

GLORIOSA

All species and varieties are now regarded as forms of *G. superba*, which see.

G. lutea Native to South Africa. Yellow flowers. Medium height. Summer flowering.
G. leopoldii see *G. simplex* ssp. *leopoldii*.
G. plantii see *G. simplex* f. *plantii*.
G. simplex (see *G. virescens*) Native to Tropical South Africa. Introduced 1823. Orange & yellow. 48 in. in height. Summer flowering.
G. simplex f. *plantii* Native to Tropical Africa. Reddish yellow flowers. Summer flowering.
G. simplex ssp. *leopoldii* Native to Tropical Africa. Segments inrolled. Summer flowering.

GRIFFINIA — AMARYLLIDACEAE

The genus is named in honor of William Griffin (?–1827), who introduced these plants into cultivation. The plants are native to Brazil and there are some 6–7 species. The bulbs are quite large in some species, of moderate size in others, and are of a grayish-brown color in appearance, not unlike *Hippeastrum*.

Griffinia flower in spring and into summer. Flowers are all in the blue-lilac color range, often fading to white, loosely held in umbels containing 6–10 flowers; pedicels curve as though from the weight. The upper segments of the flowers are broader than the lower, the petals are free to their base or nearly so. Lower petals are arranged so that one petal points directly downward, the others at or near a right angle. The foliage mostly has lattice-like markings caused by the veins; leaves are quite wide, ovate-oblong in shape, with a distinct stalk.

CULTURE Coming from a warm climate these plants cannot stand cold weather and should be grown either in a warm greenhouse, with temperatures above 55°F, or in warm areas where temperatures do not fall below 45°F.

They require a well-draining, organically rich soil mix and bright but mostly indirect light. During their growing season they require goodly amounts of moisture but should be kept on the dry side during their resting period. Light feedings of liquid fertilizer are appreciated when the bulbs start into growth; such feedings should end when the plants are coming into flower.

The bulbs should be planted with the neck just at the soil level or slightly above. The spacing varies with the species, the taller growing should be spaced 10–12 in. apart, with slightly closer plantings for the shorter species.

PROPAGATION Some offsets are produced and these can be removed in the very early spring and grown on in individual containers or lined out in the bed in the greenhouse. Seed should be sown in the spring in a peaty soil mix, the seed being only just covered. Warm temperatures are required, minimum at night should be some 55°F. The seedlings must be kept moist and humidity should be quite high. The seedlings should be transplanted as soon as large enough to handle, planting them in individual pots and grown on till they are of good size and then planted into their final positions and left undisturbed.

PESTS & DISEASES No special problems.

USES Can be used only outdoors in warmer climates where the plants will add a good blue range of colors to the shade garden. Useful plants for borders in the warm greenhouse but must be planted where they can dry out during the dormant season.

SPECIES

G. blumenavia. See: *Hippeastrum blumenavia*.

G. dryades. Native to Brazil; introduced 1868. This species has large bulbs and produces stalked foliage with the distinct lattice-like veining; blades are some 6–10 in. in length. The flower umbel is rather loose and the individual flowers some 3–4 in. in diameter. Flower color is lilac-purple with some white toward the base; flowers lighten in color as they mature. There are some 10 flowers in an umbel, sometimes more, and the flowering stalk will reach up to 18 in. in height in established plants. Early-summer-flowering.

G. hyacinthina. Native to Brazil; the type was introduced 1815. Bulb is of smaller size than *G. dryades*; the foliage is stalked and blades show the lattice-like veining. Leaf shape is ovate to oblong, reaching a length of some 8 in. Up to 10 flowers in an umbel, each some 3 in. in diameter, and reaching a height of some 12 in. The broader upper petals are blue shading to white at the base, the lower are blue. Var. *maxima* has larger flowers, sometimes almost twice the size; the white color at the base of the upper petals is more pronounced and the upper portion of the petals deeper blue. Var. *micrantha* is the opposite of var. *maxima* as the flowers are quite small, barely 1 in. in diameter, and the foliage is also reduced; flower color is closer to the type. Var. *maxima* was introduced some 65 years after the type.

G. intermedia. Native to Brazil. This species has oblong foliage which narrows at the base into the stalk, the length being some 10 in. The flowering stem will reach a little higher, some 12 in., and carries up to 10 flowers. The flowers are not large, with segments up to 2 in. in length which form a small tube at the base. Flower color is lilac tending toward the paler hues; the impression of the plant is that it is a little less robust than the other species. Bulb is ovoid and longer than *G. hyacinthina*.

G. liboniana. Native to Brazil; introduced 1843. This species has a small bulb only about 1 in. in diameter; sessile leaves are some 4–5 in. in length. The flowers are small, with the segments only an inch or a little more in length; the perianth tube is scarcely formed. This species is unusual in that the scape has two distinct edges. Some 6–8 pale lilac flowers in an umbel, reaching a height of 10–12 in.

G. ornata. Native to Brazil; introduced 1876. Has quite distinctive foliage as the edges of the leaves recurve considerably; the flowers are a light blue and turn white as they age. The umbel is large, often over 8 in. in diameter, and rather loose; there can be as many as 20 or more flowers in a single head with each flower some 2 in. in diameter. The flowering stem reaches a height of some 14–18 in.; like *G. libonia*, it has two edges but these are not quite so sharply defined. Bulb is over 4 in. in diameter.

G. parviflora. Native to Brazil; introduced 1815. Species has an ovoid bulb some 3 in., sometimes more, in diameter. The leaf stalk equals the length of the oblong blade, some 6 in., and has the distinct lattice-like veining; the lower portion of the leaf blade seems to extend down the stalk. Some 3–4 leaves are produced but not uncommonly only 2. The flowers are carried to a height of some 12 in. and as many as 15 flowers are held in a fairly compact umbel. The flowers are blue with a hint of violet, with some white showing down the central vein. The ovaries become quite large as the individual flowers fade.

COMMENTS These rather lovely flowers deserve to be more widely grown and their potential as interior plants in heated buildings explored, as for example in established beds together with evergreen bulbous plants.

GYMNOSPERMIUM — BERBERIDACEAE

Genus name derived from Greek *gymnos* ("naked") and *sperma* ("seeds"). This genus is monotypic and represented by the species *G. alberti*. The genus is sometimes classed in the Podophyllaceae family.

This plant is also commonly placed in the genus *Leontice* (which see, page 241) but is indeed different from species of *Leontice*. The leaves are not basal and are found just below the flowers; the rootstock is a corm, not a tuberous rhizome.

G. alberti (syn. *Leontice alberti*). See page 241.

GYNANDRIRIS — IRIDACEAE

There are only 7 species in this genus. One, *G. sisyrinchium*, native to the Mediterranean, was known for many years as *Iris sisyrinchium*. It has little merit, and, if grown at all, it probably is because of its almost orphan-like position among bulbous plants. The other species are South African and quite attractive. The plants are never offered by nurseries and the only way to obtain stock would be from a collector of rare plants.

Gynandriris grow well in poor, rocky ground and like warm, moist winters and hot, dry summers. For this reason, they can be grown outdoors only in those areas with similar climatic conditions.

Unlike the *Iris*, the flowers of this genus have no tube. That part that looks like a tube is in fact a continuation of the ovary. This remains on the plant after the petals fall and becomes quite prominent, sticking up above the papery spathes. The flowers do not last long, often being produced in the afternoon or early evening and finished by the following morning.

CULTURE Plant corm 1 in. deep in well-drained, gritty soil, in full sun. Give water in spring but keep barely moist when in growth. Never expose to cold. Flowers late spring/early summer.

PROPAGATION Easily raised from seed. Sow in sandy soil in early spring. Keep in high light with nighttime temperature of 50°F. When of size, transplant to a pot, allow to rest dry, and plant out in second year. Also, corm will produce offsets which can be grown on.

PESTS & DISEASES No special problems.

USES About only use would be to increase a collection to have a wider range of bulbous plants.

SPECIES

G. setifolia. Native to Cape Peninsula of South Africa. Found in poor soil. 8–10 in. in height. 3 or 4 leaves up to 12–14 in. long. Bluish-white flowers with yellow blotch often with white border on falls. March/April-flowering.

G. sisyrinchium (syn. *Helixyra sisyrinchium*, *Iris maricoides*, *I. sisyrinchium*, *Moraea sisyrinchium*). Native to Mediterranean region; introduced 1854. Corm is subglobose and covered with coarsely netted tunic. 3–4 leaves, length often as great or greater than flower spike, arch up and away from stem. 2 or more flowers produced from papery spathe and carried on stems 18–20 in. in height; blue or violet, often with yellow or white patch in falls. April/May-flowering.

GYNANDRIRIS

G. pritzeliana Native to South Africa. Lavender blue. 12–18 in. in height. Spring flowering.

G. simulans Native to Cape Province. Orchid pink with white/orange centers. 10–14 in. in height. Spring flowering.

HABRANTHUS — AMARYLLIDACEAE

Genus name derived from the Greek *habros* ("delicate") and *anthos* ("flower"). There are several species of *Habranthus*, all of them native to South America. The size of the flowers puts them between *Zephyranthes* and *Hippeastrum*, the *Hippeastrum* being much larger and *Zephyranthes* just a little smaller. There are differences other than size, but these plants are close to each genus.

Habranthus is unusual in that the flowers have stamens of unequal length, generally four different lengths, and differs from *Hippeastrum* by usually having one flower per stalk and the presence of a bract that covers the stalk of the flowers. In *Hippeastrum* the spathe is divided into two equal parts.

These are graceful plants, not hardy but well worth growing in warmer climates or in cool greenhouses where they can receive a certain amount of protection.

CULTURE Must be given winter protection, either of a greenhouse or conservatory with temperature in 45°F range at night, in all but the warmest climates. Plant in a free-draining soil with good fertility. Likes to be on the dry side during the winter months. Plant so that the top of the bulb is some 2–3 in. below the surface. Plant in spring in cooler climates or when being grown in a cool greenhouse. Plant anytime bulbs are available in warmer climates. Moisture must be given while growth is active and until the flower has past, but can be less copious as the season progresses. As some species flower late in the year this requires their being planted in a warm climate or makes protection essential. Can be lifted and stored, but this is not so satisfactory as their being grown in containers and left in the soil undisturbed overwinter.

PROPAGATION Offsets are produced by the bulbs, but these are few in number. Seed can be sown in the spring in a sandy soil mix in gentle heat of 50°F at night. Place the containers outside when the night temperatures are at or above this. Grow on outdoors, then transplant the seedlings to individual pots when large enough to handle. The plants should stay in the pots until they are large enough to plant out in the garden or be placed into 6–8 in. pots for flowering. They will flower in about 4 seasons from seed.

PESTS AND DISEASES No special problems.

USES A fine plant for the cool greenhouse and should be considered for a window sill where high light and warm temperatures, both during the day and at night, are available.

In the garden should be grown only where the climate is frost-free or very nearly so, or at least a cold frame is available to protect the plants from frost.

SPECIES

H. andersonii (syn. *Zephyranthes andersonii*). Native to Argentina and Uruguay; introduced 1829. Bulb is globose with a dark tunic. The single flower seems to pop out of the ground in late spring or early summer, reaching a height of only 6 in.; often comes up quickly after a good rain, especially if the winter has been dry. Flower is a copper color on the outside, just over an inch in length with about the same diameter when open. The leaves may be apparent when it flowers, but only just, and basically develop after flowering, reaching a length of 5–6 in., almost lying on the surface. The flower is upright or nearly so. The plant may on occasion produce another flower in the fall. Several varieties of *andersonii* have been named and include var. *aureus*, golden flowers; var. *cupreus*, copper colored flowers; var. *roseus*, pink flowers. The yellow-flowered var. *texanus* (syn. *Hippeastrum texanum*) may be native to Texas but also may have been an escape.

H. brachyandrus (syn. *Hippeastrum brachyandrum*). Native to Argentina, Brazil, and Paraguay; introduced 1890. Bulb ovoid; plant grows to 12 in. high. Flower color pale rose-pink, veined and deepening to claret-red toward the base. Flower some 3 in. in length and as wide when it opens. Leaves produced before the flower which appears in late summer or early fall.

H. cardinalis (syn. *Zephyranthes bifolia, Z. cardinalis*). Native to South America, but its actual habitat is not known; introduced 1913. As the name implies flowers are bright red, 3 in. in length, and open to about the same diameter. Anthers are lilac. Leaves 6 in. in length, shiny green. Six in. in height. Flowering in June.

H. gracilifolius (syn. *Zephyranthes gracilifolia*). Native to Uruguay; introduced 1821. Reportedly quite hardy. Several flowers to a stalk, pale pink with deeper veining, about an inch long, and open to give a bell-shaped flower of the same diameter. The flowers appear before the leaves, and as it blooms in very late summer or early fall would question its hardiness. Leaves 18 in. long, height 8–9 in. *H.g.* var. *boothianus* has good pink flowers and glaucous leaves.

H. robustus (syn. *H. tubispathus, Zephranthes robusta*). Native to Argentina; introduced 1828. Globose bulb with dark brown tunic. Large flowers, 3 in. in length and as wide in diameter when open, are carried one to a stalk to a height of almost 12 in. Flowers a good pink with darker veining and a green throat; appear in late summer. Leaves are narrow and 12 in. in length, arching gracefully away from the stem. An excellent pot plant that will grow easily indoors or in a cool greenhouse. As an outdoor plant it should be considered only for frost-free areas. It is best cultivated with the neck of the bulb set just above ground level.

H. versicolor. Native to Brazil; introduced 1821. Bulb oblong with a dark tunic. While the stem is only about 6 in. in length, the flower stalk adds another inch or 2, and the flower another 2, giving this plant an overall height of some 10 in. It has an unusual feature in that the color of the flowers changes with maturity. Buds are a deep rose, then turn to white, with the rose color eventually confined to the tips of the petals and the base. The petals also have a thin green stripe on them. Leaves are 10–12 in. in length. Flowers in late fall after having enjoyed a rest and drier conditions during the latter part of the summer. Should be grown only in warm climates or where it is possible to bring it inside after the summer and give it high light and temperatures in the 55°–60°F range at night.

HABRANTHUS

H. advenus see *Hippeastrum advenum*.
H. bagnoldii syn. of *Hippeastrum bagnoldii*.
H. hesperius see *Hippeastrum advenum*.
H. juncifolius Native to Argentina. Flower white flushed pink, tube reddish green.
H. longipes (syn. *Zephyranthes longipes*) Native to Uruguay. Introduced 1898. Pale red.
H. miniatus see *Hippeastrum advenum*.
H. phycelloides syn. of *Hippeastrum phycelloides*. This species is sometimes placed in the genus *Phycella*.
H. pratensis see *Hippeastrum pratense*.
H. roseus syn. of *Hippeastrum roseum* & *Hippeastrum bifidum*.
H. texanus variety of *H. andersonii*.
H. tubispathus see *H. robustus*.

HAEMANTHUS — AMARYLLIDACEAE

From the Greek *haema* ("blood") and *anthos* ("flower"), a reference to the color of the flowers of certain species. *Haemanthus* contains at least 50 species, all from southern Africa, but, perhaps unfortunately, only a few are in cultivation. Their common names are many, nearly all of them referring to the rather distinct flowerheads. Catherine Wheel, Paint Brush, Torch Lily, March Flower, and Snake Plant are the most common, but Blood Flower and April Fool are also names these enjoyable plants are given. This genus has been separated into two genera, *Haemanthus* and *Scadoxus*. This division is discussed under *Scadoxus* and took place circa 1970.

The flowers are produced in a dense umbel. The color is provided not by petals, which are absent, but by the many cylindrical perianth segments being joined in a tube from which the colored stamens protrude. There are 6 stamens per flower, and many flowers in an umbel, giving a host of stamens that gives the appearance of a paint brush or a shaving brush. The place of petals is taken by colored bracts that hold the stamens in a tight cluster.

The leaves in the majority of species are wide, over 2 in., and are a bright glossy green. They are sometimes upright or in some species stay close to the ground. [Those with leaves on the stem are now placed in *Scadoxus*, which see.] The bulbs are very large and in the wild several of the species have the necks aboveground. I have seen this plant growing in the Transvaal and it likes to hide a little from the sun, growing in the shade of a bush, in ground that becomes baked in the summer months. This gives a clue as to their cultural requirements. The bulbs are not able to withstand much frost, a few degrees perhaps but certainly not so much as would make the ground freeze over. It is, therefore, a plant that should be considered as tender. Ideal for milder climates and an excellent pot plant that deserves more attention as it remains in flower for a long time and the flowers are followed by attractive berries.

The flowering stems are thick and fleshy and, when you break them, are found to be full of sap; this is surprising when one sees the parched ground in which they grow in the wild. The stems are on the brittle side and care should be taken when handling. They like to remain in containers for a number of years, the fleshy roots seeming to appreciate crowding.

It is interesting to note the great variations in the bulbs of this genus. Those of *H. albiflos*, *H. magnificus* (syn. *Scadoxus puniceus*), and *H. natalensis* (syn. *Scadoxus puniceus*) are quite large, as much as 4 in. in diameter and consist of many fleshy tunics. *H. albiflos* will produce many young bulbs at the base and these in time form large clusters of plants. The bulbs of *H. katherinae* are smaller, being only 2–3 in. in diameter, while the bulb of *H. sanguineus* is 2–3 in. in diameter one way but flattened to half this diameter the other way. In addition the bulbs of *H. incarnatus* are made up of an equal number of swollen leaf bases, usually 10, and arranged in 5 pairs opposite one another. While the bulbs are stated to be poisonous, they are used medicinally by natives for headaches.

CULTURE   The plants are found in the veldt in South Africa, an area where the tall grass provides a little shade. They also enjoy the light shade of shrubs, with any excess moisture being used by the shrubs. Particularly in warm areas with hot summers and much sunshine *Haemanthus* should be planted in light shade. In areas where they are borderline, i.e., where some but not much frost is experienced, they should be planted in full sun in a protected area.

The soil need not be rich but it must be well-drained. Locate plants where others in the vicinity do not require moisture at the time when the bulbs are resting.

The best species to try in the colder areas would be the bulbs that have their dormant period during the winter. Bulbs should be planted with their tops at or only just below soil level. In heavier soils it is best to have the tops protruding a little from the soil.

In containers they should be given ample room; 1 bulb to a 8–10 in. pot would be correct, but it would be better to have a 3–5 to a 24 in. pot so the bulbs have a good soil depth into which they can send their fleshy roots.

The bulbs send out very long roots. When transplanting, and if left in the ground, they will produce offsets, so a very large hole should be dug for them to place their roots. They must enjoy a period of dryness after the foliage dies down. The time the foliage is produced will vary according to the flowering time of the various species.

PROPAGATION   Offsets are the best way to reproduce the bulbs. These, while not freely produced, are quite numerous if the plants are allowed to stay in one location for several years. This separating of the young bulbs should be done when the plants are just coming into growth.

Seed is freely produced and flowering-sized bulbs are obtained in about 3 years. Sow seed in the spring, cover lightly with a sandy soil mix, and give a little heat while germinating. If this heat is not available, delay sowing until the sun is warm and sow in containers, standing them in a protected, sunny area where it is warm. Keep the seedlings growing. When they go into their dormant state, announced by the waning of the foliage, they should be allowed to become dry. Place in individual pots when plants are completely dormant. They then are started into growth in their normal growing season, at which time some moisture is given to encourage growth. Keep moist throughout their growing season.

PESTS & DISEASES   No special problems.

USES   These plants are well worth growing as pot plants. Grow in the cool greenhouse, not so much because of the warmth, even though they need to be kept frost-free, but due to the necessity of keeping the plants dry during their dormant season. Good in warmer climates for the sunny, well-drained border but should be grown close to footpaths, so the flowers can be looked down upon.

SPECIES

H. albiflos (syn. *H. virescens*). Native to the eastern Cape Province of South Africa and up the coast toward the Transvaal; introduced 1791. Found growing along shady river banks and in the sandy soil of the coastal region. Leaves are quite leathery and spread along the ground, their edges tinged with small white hairs. Under cultivation in a container the leaves are likely to remain evergreen. The flowers are carried on stems to a height of 18 in. and are produced in late summer. The flowers are made up of white to greenish bracts that clothe the stamens that emerge. Stamens are brightly colored orange. After the flowers fade the red berries that are carried are in themselves pretty. This species now contains *H. albo-maculatus* and *H. mackenii*.

H. amarylloides. Native to the Transvaal. Flower stem emerges first and reaches a height of 12 in. The pink bracts are quite narrow; flowers in early summer or late spring. As soon as the flowers fade the leaves emerge and stay green until the fall when the bulb enters its resting period. One of the hardiest.

H. coccineus. Native to Zimbabwe; introduced 1731. In late summer or early fall this plant sends up strong fleshy flower stalks that reach about 9–10 in. in height. They are a brilliant orange-red with scarlet bracts, quite fleshy and larger than the flower umbel. Stamens are lighter in color and topped with orange pollen. After flowering, 2 leaves are produced and continue to grow through the winter and shrivel in the early summer when the bulbs go into their resting period. The leaves reach up to 36 in. in length and are as much as 6 in. wide at their base. Var. *carinatus* has leaves 12 in. long and narrow; var. *coarctatus* has bracts that are shorter than type. *H. coccineus* now contains *H. concolor, H. moschatus, H. splendens,* and *H. tigrinus.*

H. katherinae. See *Scadoxus multiflorus* ssp. *katherinae.*

H. multiflorus. See *Scadoxus multiflorus* ssp. *multiflorus.*

H. puniceus. See *Scadoxus puniceus.*

H. sanguineus (syn. *H. incarnatus*). Native to the eastern shoreline of South Africa. Flowers in the summer; crimson flower stem topped by rose colored flowers with scarlet bracts enclosing them. Leaves come after the flowers have faded; up to 12 in. in length and unusual in having a purple margin and purple flecks on the underside. While not so large a flower as many of the other species, it is an attractive plant for the sunny border. This species is close to *H. tigrinus* but has broader leaves. *H. sanguineus* now contains *H. incarnatus* and *H. rotundifolius.*

HAEMANTHUS

H. albo-maculatus see *H. albiflos.*
H. allisonii Native to Transvaal. Introduced 1894. Pure white. 6–9 in. in height. Early Spring flowering; September in the wild.
H. baurii syn. of *H. deformis.*
H. brevifolius Native to E. Cape Province. Pale pink. 5–8 in. in height. Late Summer flowering; Feb. in the wild.
H. canaliculatus Native to Cape Province. Scarlet, rarely pink. 8 in. in height. Early Summer flowering.
H. carneus Native to South Africa. Introduced 1819. Pink. 6–12 in. in height. Summer flowering.
H. cinnabarinus Native to W. Africa. Introduced 1855. Red. 12 in. in height. Spring flowering.
H. concolor see *H. coccineus.*
H. deformis (syn. *H. baurii*). Native to Natal. Introduced 1869. White, anther pale yellow. 3 in. in height. Spring flowering.
H. incarnatus see *H. sanguineus.*
H. kalbreyeri see *Scadoxus multiflorus* ssp. *multiflorus.*
H. lindenii Native to Congo. Introduced 1890. Scarlet. Spring flowering.
H. lynesii Native to Sudan. Introduced 1921. Yellow with red base. 5 in. in height. June flowering.
H. mackenii see *H. albiflos.*
H. magnificus see *Scadoxus puniceus.*
H. moschatus see *H. coccineus.*
H. natalensis see *Scadoxus puniceus.*
H. pole-evansii see *Scadoxus pole-evansii.*
H. rotundifolius see *H. sanguineus.*
H. rouperi see *Scadoxus puniceus.*
H. sacculus see *Scadoxus multiflorus* ssp. *multiflorus.*
H. splendens see *H. coccineus.*
H. tigrinus see *H. coccineus.*

HELICODICEROS — ARACEAE

From the Greek *helix* ("spiral"), *dis* ("twice"), and *keras* ("horn"), a name that is fully descriptive of the foliage of this monotypic genus. The basal divisions of the leaves twist and turn and then end up erect, looking like horns. The only species is *Helicodiceros muscivorus,* formerly called *Arum crinitum,* and native to the Balearic Islands, Corsica, and Sardinia; it was introduced into cultivation in 1777. The roots are tuberous. The flowers look like an *Arum,* the spathe being dark brownish-purple or grayish-brown, spotted purple, and hairy on the inside. Flower stalk reaches 18 in. in height. The male and female flowers are separated on the spadix by rudimentary flowers, which are produced in April. The flowers are unpleasantly scented to attract flies and beetles, which assist in pollination.

CULTURE The plants can be grown in well-cultivated ground, planting the tubers a couple of inches deep. They like plenty of water during the spring and summer, but should be allowed to dry out toward the end of summer. The plants are not fully hardy and can be harmed by severe cold; the drier they are in the fall the better. Light shade is preferred.

PROPAGATION Division of the tuberous roots in late fall or spring provides the best way of propagation. Seeds are formed, and those saved in summer should be stored overwinter and sown in the spring in a rich soil mix and kept moist with a temperature of around 50°F. After germination they can be placed outside in half shade. Grown on for another season they can be planted out the following spring, i.e., two seasons after the seed has been sown.

PESTS & DISEASES No special problems.

USES An unusual plant for the border or in the shade but not an especially attractive plant. Should be protected against severe cold weather.

SPECIES

H. muscivorus. Described above.

HEMEROCALLIS — LILIACEAE

The name is derived from Greek *hemera* ("day") and *kallos* ("beauty"), an apt name for these great garden plants, as the flowers last only a day, or rarely two. The common name, Daylily, is apt, other common names given usually refer to some type of lily such as Orange Lily, Yellow Lily, etc. A genus of some 15 species, the majority of which have fibrous roots, which are to some extent inclined to be tuberous, and others with distinctly tuberous roots.

They come from eastern Asia and central Europe, but over the years have received much attention from breeders and few of the species are found in gardens today. The foliage is basal, grasslike in appearance, and, in certain species, the leaves can be 48 in. or more in length. The flower stems are sturdy and carry the flowers well above the foliage. While the majority are some 36 in. or so in height, certain of the species are taller, reaching as much as 72 in.

The genus can be divided into two groups: those with an open flowering and branched inflorescence, and others where the flowers are held close together with a distinct, broad bract below the almost sessile flowers. The flowers are trumpet-shaped, the perianth segments form a narrow tube at the base and then flare with the inner segments often wider than the outer. The colors are most commonly in the yellow, orange, or reddish-purple shades. While the flowers are individually not long-lasting, they are produced in such quantity that the plants, during their flowering season, are rarely without flowers.

In certain parts of the world the flowers are gathered, pickled in salt, and then used in soups. Reportedly, in China, the young leaves of *H. minor* are eaten and these appear to have an intoxicating and stimulating effect on persons. The flowers also are prepared as a relish and eaten with meat.

CULTURE Few plants are as easy of culture. They are tolerant of a wide range of soils, almost any garden soil being suitable, and will grow well in some shade but prefer full sun. When established they will withstand dry conditions, yet plants grow and perform better if summer moisture is provided. They can be planted at almost any time of the year, but fall or spring are the better times. The plants should be set so that the base of the foliage is at soil level, spacing them some 18–24 in. apart, the exact distance being determined by the ultimate height of the plants. Some of the newer hybrids are rather lower in height and should be planted at the lesser distance; the taller-growing hybrids and species being more widely spaced. The clumps should be left undisturbed until the production of flowers diminishes, then the clumps should be lifted and divided. The plants are quite hardy and require little or no fertilizer. In poorer soils a spring feeding of a general fertilizer can be given, followed by another in early summer.

PROPAGATION Seed is not freely produced and lifting and dividing the rootstocks provides the best way to increase the stock of these plants. This can be done in either the fall or the spring; in colder climates spring is the better time. The plants pull apart quite readily, and, while fall and spring are recommended, the plants are of such easy culture that such lifting and dividing can be done at almost any time. For the fastest production, the plants can be lifted and divided annually or every two seasons.

PESTS & DISEASES The plants are remarkably free from these problems.

USES These plants can be used in borders, in combination with other perennials, as bold beds in lawns, or to line driveways as they are of such easy culture and seeming robustness under all sorts of conditions. About their only flaw is that individual flowers last such a short time.

SPECIES

H. altissima. Native to China. Strap-shaped leaves 48–60 in. in length and a little over an inch in width. Flowers are carried above the foliage on stout stems which can reach as much as 72 in. The flowers are fragrant, pale yellow, and open at night; the perianth tube is a little over an inch in length with the flowers some 4 in. in diameter. It flowers in late summer and into early fall.

H. aurantiaca. Native to China, but some authorities regard it as being from Japan; introduced prior to 1890. This plant is self-sterile and some feel it is a hybrid. The rootstock is inclined to be rhizomatous. The leaves are up to 36 in. in length, 1 in. in width, with a distinct keel. Up to 15 or more flowers are carried on a strong stalk which reaches some 36 in. in height, with the flowers rather clustered, each up to 4 in. in diameter. Flowers are an intense orange, sometimes with a hint of a purple flush. The perianth tube is about an inch in length or a little less. The flowering time is summer. In warmer areas the foliage remains and thus this plant behaves as an evergreen. This species was much used to create new hybrids. *H. a.* var. *major* has flowers that are a little larger and most commonly with a purple flush. It is more vigorous than the species but is not quite so hardy. The flowers, like the species, have a somewhat starry appearance.

H. baroni. This name is sometimes linked to *H. citrina* and often regarded as a garden name for *H. citrina* or a hybrid closely resembling this. Other authorities regard it as a cross with *H. ochroleuca*.

H. citrina. Native to China; introduced 1902. Unusual in that flowers open in the evening and last throughout the night. Foliage often over 36 in. in length and good dark green. The fragrant flowers are light lemon yellow and quite long, but they remain in a narrow form with the segments rather stiff and narrow; the perianth tube is over an inch in length. It is a vigorous species and will carry as many as 30–40 flowers in the flowerhead, so that while each flower is short-lived the plant is of interest for a considerable time. Flowers in midsummer. In the literature the name Baroni frequently will be found associated with this species as he was the author of the name *H. citrina*.

H. dumortieri (syn. *H. graminea, H. rutilans, H. sieboldii*). Native to Japan; introduced 1832. This is a tufted plant with leaves up to 18 in. in length that form a dense clump. Above this, on an unbranched flower stem, are up to 6–8 tightly clustered flowers with broad, pointed bracts. The flowers are deep yellow-orange and the unopened flower buds are dark brown; flowers rather flat, little over 3 in. in diameter, with a very short tube about ¼ in.

in length. Flowers in late spring. The height of the flowering stem is some 18 in. so that it is nestled in the foliage a little. A rather attractive plant especially valuable for its early flowering.

H. exaltata. Native to Japan; introduced 1934. The leaves are up to 2 in. in width and some 30 in. in length forming a compact plant, above which flowers are carried on a branched inflorescence that can reach 48 in. or more in height. Flowers are slightly fragrant, light orange, about 4 in. in diameter, as many as 12 per spike. Flowers in June–July.

H. flava. Many references will be found in the literature to this plant, which actually is *H. lilioasphodelus*. It also will appear as *H. l.* var. *flava*. The confusion no doubt has arisen due to this plant having been in cultivation since the 16th century.

H. forrestii. Found by George Forrest in the mountains of northwestern Yunnan and Xikang Provinces, China introduced 1910. It is thought by some that this may well be just a variation of *H. plicata*. The foliage is 18 in. in length, narrow, only ½ in. in width; the flowering stem sometimes is unbranched and carries 8–10 orange or reddish-orange flowers on short pedicels. The plant is rather untidy looking and the flowers are mixed in with the foliage.

H. fulva. The origin of this plant is almost certainly Asia, most likely Japan, but its origin has not been established with certainty and some authorities will list it as a European species. It has been long in cultivation (since 1576) and has become naturalized in the United States. Vigorous and spreading by rhizomes, this plant will carry flowers on a 48–72 in. flowering stem that towers over the leaves, which are 36 in. in length and a little over an inch in width. The flowers are 4–5 in. in diameter, 5 in. in length, with the perianth tube some 1 in. long. The color is best described as dull orange-red or orange-buff with a brown tint in the throat and a central apricot band in each perianth segment. This plant is rarely if ever seen with seed. The form most commonly found is known as 'Europa'. A number of selections have been made, including: *flore pleno* (syn. 'Kwanso Flore Pleno,' *H. disticha flore pleno*) with double flowers which are often as much as 6 in. in diameter. Both are old introductions and there are two forms, one with green leaves, the other with leaves that have white stripes in them. It was the form with the white-striped leaves that first had the name 'Kwanso'. *H. f. maculata* has soft coppery colored flowers with a darker zone of color found in the center; *rosea* is a natural variant from China introduced around 1930 and the first of the pink cultivars. The flowers of *rosea* are narrower and recurve a little directly from the perianth tube.

H. lilioasphodelus (Yellow Daylily, Lemon Lily; syn. *H. flava*). Native to China, but some authorities question this. A spring-flowering species with lily-shaped flowers that are a clear yellow with a great fragrance. Plant spreads by rhizomes. Foliage is some 24 in. in length and almost an inch in width. The flowers are carried to a height of 36 in. and as many as 8–12 in a flower head; each flower is some 4 in. in diameter and a little more when fully opened. Flowers in May. The form known as var. *major* has foliage that is a little darker than the type and the petals are more reflexed and inclined to be wavy along the edges. It is unfortunate that over the years many plants have been introduced into the market and distributed under the name of *H. flava* so that there is confusion regarding this plant, which has possibly the best fragrance of any of the genus.

H. × luteola. Hybrid between *H. aurantiaca major* and *H. thunbergii*. There are several forms of this plant; all of them develop dense clumps of foliage up to 30 in. in length with flower spikes reaching 36 in. Flowers are up to 5 in. in diameter and color varies between golden yellow to apricot-yellow.

H. middendorffiana (*H. middendorffi*). Native to Siberia, Japan, China; introduced 1866. Resembles *H. dumortieri* but is distinguished by having flower spikes that are above the foliage, is a little later in flowering in midsummer, and the flowers are orange-yellow. The flowers are carried on an unbranched spike reaching 18–24 in. in height with broad bracts which are just below the sessile (or nearly so) flowers. The flowers are fragrant. *H. m.* var. *major* sometimes listed as cultivar 'Major' has more numerous and somewhat larger flowers and stronger flower spikes which are more erect than in the type. Leaves 18–20 in. in length, 1 in. wide, forming a dense clump.

H. minor. Native to Siberia and Japan; introduced 1759. Forms a compact clump with grasslike leaves up to 20 in. in length. Flowering spike rises above the foliage and carries clear yellow flowers with a hint of brown on the exterior. Flowers are fragrant and only a little over 2 in. in diameter. It is not a free-flowering plant, seldom carrying more than 5 flowers. Flowering time May. This plant has been used to produce several of the low-growing or dwarf hybrids that are presently on the market.

H. multiflora. Native to China; introduced 1934. There is a goodly number of flowers per flowering stalk, as many as 30 not being unusual, so it is well named. Flowers are brownish-red on the exterior and a good yellow on the interior. The flowers are quite small, seldom more than 3 in. in diameter with the inner perianth segments quite narrow. The flowering head is heavily branched, making this an attractive plant when in flower. It is the last of the daylilies to flower in late summer-early fall. Leaves dark green, 30 in. in height, about an inch in width.

H. nana. Native to the Yunnan Province, China; introduced 1914. A dwarf plant, forming a clump with foliage 15 in. in length and seldom over ½ in. in width. The flowering spike will carry up to 3 flowers and is 12–18 in. in height. The flowers are funnel-shaped and of a good orange on the inside with an exterior of reddish-brown. They are fragrant, about 3 in. in diameter, and are produced in early summer/June.

H. × ochroleuca. Hybrid between *H. citrina* and *H. thunbergii*. The flowers are fragrant and a good lemon-yellow, large, 3–4 in. in diameter and starry; flowering time July–August. There are a number of hybrids that have resulted from this cross, sometimes found in the literature as species, but more correctly they should be listed as cultivars. *H. × muelleri* has flowers that are a little larger and some 6 in. more in height, reaching 48 in. with the flowering stem being more branched; a vigorous plant. *H. × o.* 'Baroni' or *H. baroni*, about which there seems some confusion in the literature, probably is most correctly placed in this species (see *H. baroni*).

H. plicata. Native to northwestern Yunnan and Xikang Provinces, China. Very similar to *H. forrestii*, some authorities regarding the latter to be a form of this species. However, the foliage is narrower ¼ in. wide 20" long than *H. forrestii* and is folded; flowers golden yellow and generally solitary. Summer flowering.

H. thunbergii. Native to Japan; introduced 1890. A vigorous plant with flowering spikes that reach 36 in. in height or even a little more. Foliage is 30 in. in length and just under an inch in width. The flowering spike has the habit of branching close to the top; the fragrant, yellow flowers have a hint of apricot to them. Flowers are some 3 in. in diameter and the petals recurve with the flowers being more erect than in many daylilies. Flowering mid- to late summer.

CULTIVARS There has been much hybridizing and nurseries today offer many fine hybrids. The main groups are classed as dwarf, medium, or tall, depending on the height to which the flowers are carried. There are also "miniatures", which are plants with small flowers under 3 in. in diameter. The color range and form of flowers are varied. Some hybrids have narrow petals and thus "starry" flowers, while others have quite broad petals. Certain hybrids also have crinkled petals to a lesser or greater degree. Some hybrids are two-colored, apricot edged with rose, pink, and tangerine blends, pale pink with yellow-green throat, and other attractive combinations of colors, while the single colors vary from plum-red to pastel orchid. The plants are vigorous, free-flowering, and, while a few cultivars are listed, the reader is well advised to seek out catalogs of nurseries that have extensive collections.

'Beautiful Morning'. Creamy yellow and blush-peach; petals crimped.
'Buttercurls'. Diploid; light lemon-yellow; heavily ruffled.
'Dawn Ballet'. Yellow-orange and melon; large-flowered; broad petals.
'Evening Gown'. Broad petals of pink with flush of peach.
'Heavenly Grace'. Ivory pink, ruffled edges, with some tangerine; fragrant.
'Hosanna'. A combination of golden orange and melon; wide petals; fragrant.
'King Alfred'. Large yellow flowers, double; superb plant.
'Morning Dawn'. Scarlet, yellow throat; wide, overlapping petals.
'Satin Clouds'. Closest cultivar to white, being a creamy white with a hint of pink.
'Sombrero Way'. Good deep orange; great award-winning plant.

The following is a series of tetraploid *Hemerocallis* raised by Dr. Robert Griesbach. The plants are robust, free-flowering, with great substance to the flowers; they are outstanding introductions.

'Bald Eagle'. A good deep red with a central eye of intense yellow.
'Big Bird'. Flowers can reach a diameter of 7 in., bright yellow with petals recurving nicely; an attractive, free-flowering plant.
'Cedar Waxwing'. Good orchid-pink with yellow throat.
'Falcon'. One of the darkest reds available; great substance.
'Heron'. A good rose-pink; wide petals; a vigorous plant.
'Lemon Pectin'. Bright lemon-yellow; good form; flowers 6 in. in diameter.
'Meadow Mist'. Large flowers of good pink.
'Nuthatch'. Pastel pink with midrib of ivory, gold and green throat.
'Ruby Throat'. Large flowers; broad petals; one of the best reds available.
'Screech Owl'. Scarlet-carmine petals edged with gold; strong plant.

These represent a sample of the many hundreds of cultivars on the market today. All are worthy of a place in the garden.

COMMENTS Daylilies are great garden plants. Their main attributes are being of easy care and maintenance. It should be noted that in certain climates with a cool summer, such as parts of the coastal region of California, plants do not easily produce many flowers; they prefer greater summer heat. It is surprising that landscape architects do not use more of these great plants, which need so little care and produce flowers over such a long period.

HERBERTIA — IRIDACEAE

This genus is named in honor of Dr. William Herbert (1778–1847), Dean of Manchester University in England, a distinguished botanist, especially famous for his knowledge of bulbous plants. *Herbertia* is a small genus, but the some 14 species have quite a wide distribution, being found in both South and North America.

Although the flowers are made up of 6 segments, the 3 outer segments are the showy ones, being quite large and displayed flat, almost at right angles to the stem or bending down slightly. The flowers are produced on stems that are strong but not very thick, usually about 12 in. in height. The flowers are blue or violet, sometimes marked with another color at the base of the wide-spreading segments. The style is quite prominent and is divided at the top into 3 lobes. Flowers are not long-lasting, usually produced 1 per stalk (sometimes more), opening after each other.

The plants are not fully hardy, needing protection in the winter. This does not prevent them from being suitable for all regions as the corms can be lifted in the fall and stored overwinter in a frost-free location and planted again in the spring.

CULTURE The corms should be planted in the spring after the ground has warmed. Any good garden soil is suitable and it need not be rich. The plants should have full sun. Plant 2–3 in. deep, in a sunny location, and given adequate moisture during the spring and early summer until the bulbs have finished flowering and the foliage has died down. They can be grown in containers in a sandy soil mix. In colder climates, they should be overwintered in sand.

PROPAGATION The bulbs produce a number of offsets, especially if plants have been undisturbed for a number of years. The corms can be divided in the fall when the foliage has died down. Seed is produced and this is best sown in the spring in a sandy soil mix in a temperature of 50°–55°F; germination is quite rapid. Seedlings can be placed into individual containers and grown on that summer. They can be taken out of their pots that fall, stored overwinter, and planted or repotted in the spring. After another season of growth they can be planted out the following spring into flowering locations; they will flower in 2–3 years from seed.

PESTS & DISEASES No special problems.

USES Flowers are pretty and an attractive color, but are not large and more for the keen gardener than general culture. Good for warm, sunny borders. Bulbs should be lifted in the fall and stored overwinter to be planted again in the spring. In warm climates with ample spring moisture they can be left outside in the ground and allowed to increase in their plantings. Interesting container plants but not very spectacular.

SPECIES

H. amatorum. Native to South America; introduced 1903. Found growing in grassy fields of Uruguay. About 12 in. in height. Outer segments of the flower are dark blue with a yellow blotch at the base, the inner segments are short and a darker purple-blue; the total diameter is only about 2–3 in. Sometimes the flower spike is branched. Leaves are only about 8 in. long and narrow, less than ½ in. wide. Early-summer-flowering.

H. drummondii (syn. *H. caerulea, Alophia drummondii, Cypella drummondii*). Native to Texas; introduced 1839. Very similar to *H. amatorum* except the flowers are a paler blue with a white blotch at the base of the petals, with a darker color just between the blotch and the paler blue. The inner segments are small and dark purple. Up to 12 in. in height. Summer-flowering.

H. pulchella (syn. *Alophia pulchella*). Native to Brazil; introduced 1827. The *Herbertia* most commonly found in cultivation. The outer flower segments are held flat on 12 in. high stems; color of the outer segments is pale blue with a white blotch at the base and the petals are flecked with purple. Inner segments small, dark purple in color. Leaves are 9–12 in. long, narrow. Early-summer-flowering.

HERBERTIA

H. amoena (syn. *Alophia amoena*). Native to Argentina, Uruguay. Violet. 12 in. in height. Early Summer flowering.
H. caerulea see *H. drummondii*.
H. platensis Native to La Plata, Argentina. Light china blue. 48 in. in height. May–Sept. flowering.

HERMODACTYLUS — IRIDACEAE

From the Greek *Hermes* ("Mer'cury') and *dactylos* ("finger"), a reference to the rather unique way in which this plant produces its roots. There is only one species, *Hermodactylus tuberosus*, which is native to the Mediterranean region and has spread to many different parts of Europe, including Britain. Has been in cultivation since before 1600 and is a hardy plant.

Commonly known as the Snakes Head Iris, it is distinguished from *Iris* by having the ovary with only 1 cell instead of 3 as in *Iris*. Bulb shape is irregularly and narrow, somewhat tuberous. The root produces a shoot from its tip that comes up to flower. In the meantime another root is produced which will produce the flowering shoot the following year. In this way the plant can creep away from the place where it was originally planted as the older root dies.

CULTURE About the only demands this plant make is to have well-drained soil and full sun. The roots should be set about 3 in. deep. Moisture in the spring is needed, but little during the summer. No other special conditions needed.

PROPAGATION Seed is the best way, sown in the spring in a sandy mix. The seedling should be transplanted to another container after the dormant season. After being grown for 2 seasons they can be planted out in the third.

PESTS & DISEASES No special problems.

USES A plant that is not unattractive in a sunny border for early flowering and, as in its native habitat, can be naturalized in grass such as would be found on a slope where little other vegetation can be grown.

H. tuberosus (syn. *Iris tuberosa*). Native to the Mediterranean region; introduced circa 1600. Flowering in early May; flowers solitary, greenish with dark brownish-purple markings at the tip of the falls. The claw and standards are light yellowish-green. The spathes are very large, light green, and often above the flower. The flower stem has several leaves which sheath it as it grows, while the other foliage reaches a length of some 18 in. or more, continuing to grow after the flowers have passed. The seed pods are pendant when fully ripe.

HESPERANTHA — IRIDACEAE

Names derives from the Greek *hesperos* ("evening") and *anthos* ("flower") as the flowers usually open in the evening. *Hesperantha* is a fairly large genus with perhaps some 30 species, all native to South Africa. Not many of these are in cultivation and only one is sometimes listed in catalogs, *H. vaginata*.

All plants in this genus produce corms. Foliage is grasslike, up to 12 in. in length, depending on the species. Flowers are not large, but several to many are produced in a loose spike on strong stems that are usually some 10 in. in height. The short perianth tube is generally curved a little, and the flowers then open wide with the petals either flat or held in a shallow bowl outward or upward facing. The stamens have anthers that are much longer than the filaments, which are thrust out and distinctive. Flowers generally start to open in the late afternoon or evening, and are quite sweet-scented.

CULTURE A warm, sunny border or containers in a cool greenhouse in colder climates; plants can be harmed by cold weather but are able to stand light frost. Plant corms 2 in. deep in the early fall. At no time do the plants need very high temperatures, but a dry period is necessary after flowering and the foliage has died down. They do require winter rainfall, and when planted in containers for the cool greenhouse should be kept moist throughout the winter with temperatures held above 40°F. The soil mix must be free-draining; under no conditions should the bulbs remain cold and wet.

PROPAGATION Cormels are freely produced and can be separated when the corms are lifted in late summer. Some species occasionally produce bulbils in the axils of the leaves. Seed also is freely produced and seedlings often will flower in their second season. Sow in the spring, grow on until the foliage dies down, then in early fall repot into larger containers or singly in pots. Grown for another season, the plants might flower in these pots, and if they do not they will have reached a size to be planted that fall to flower the following year.

PESTS & DISEASES No special problems.

USES Good container plants for the cool greenhouse and must be protected from frost while in leaf. Good plant for the sunny, well-drained border and should be grown in a location where their fragrance can be appreciated. Not very hardy and protection must be given in areas with frost, but in a sheltered location can withstand temperatures down to 28°F.

SPECIES

H. bachmanii (syn. *H. angusta*). Native to Namaqualand, western South Africa. Found in quite damp clay soils. 12 in. in height. Flowers pure white, on long, wiry, zigzagging stems, well-spaced, outward-facing. Spring-flowering. No bulbils are produced in the leaf axils. Leaves are formed at the base of the flowering stem and are rarely over 10 in. in length.

H. baurii (syn. *H. lactea, H. rupestris, H. schlechteri, H. subexerta*). Native to Transvaal and Natal; introduced 1936. Flowers of this species are pink, open flat, and face outward. The leaves are narrow and erect; generally only 4–5 are produced. Height of flower stem about 10 in., often less; 4–6 flowers per stem. Very

Griffinia parviflora

Plate 145

Gloriosa rothschildiana. No matter which opinion you hold regarding the correct name of this plant, it is a beauty. These plants are now favorites and are grown in quantity in greenhouses for the florist trade. WARD

Gloriosa superba. Even in the wild there is variation in the flower color. If one prefers to maintain that there are different species, (see text), then this is *G. rothschildiana.* AUTHOR

Gloriosa superba. This plant I photographed in Kruger Park in South Africa. Obviously, the many animals around know this plant is poisonous as they left it alone. AUTHOR

Gynandriris pritzeliana. An extremely rare species from Cape Province where it flowers in August-September. Little known in its native land, it is doubtful if this species has been seen outside South Africa. K.N.B.G.

Gynandriris pritzeliana. While the color and form of this plant is good, it, like other species in this genus, is rare in cultivation. BOUSSARD

Gynandriris setifolia. This South African species is quite an attractive plant growing in poor ground and endures considerable heat in the summer. B.R.I.

Gynandriris setifolia. Photographed in the wild where these plants put on quite a show. It is doubtful if these plants are in cultivation. ORNDUFF

Gynandriris simulans. This lovely species of such delicate color belongs in collections of all who love bulbous plants. Unfortunately, it is extremely rare. BOUSSARD

Gynandriris sisyrinchium. This Mediterranean representative of the genus is not commonly grown in gardens, but in the wild, as in this photograph, it is quite attractive. Its appearance makes it easy to understand why it was for many years, considered an *Iris*. WARD

Habranthus andersonii. A rather attractive plant, of easy culture. These bulbs add interest to areas where Daffodils are naturalized, as they provide color, after they had finished flowering. LOVELL

Haemanthus sanguineus. It is a pity this plant is not hardy. The flowers are eye catching, and last quite a long time. It would be worth trying in moderate winter zones, giving it a deep mulch for winter protection. AUTHOR

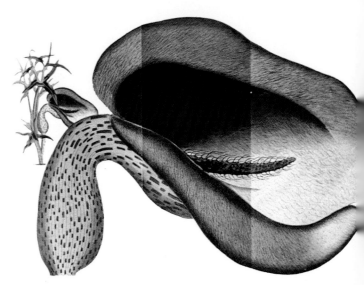

Helicodiceros muscivorus. This illustration is from the 1824 Botanical Register. Perhaps it can be said that age has not increased its beauty, but one must admit, it is different and interesting. J.H.

Hemerocallis. The individual flowers do not last long, the common name 'Day Lily', expresses this well, but each flowering stalk carries many flowers, making this plant good for summer color. COURTRIGHT

'Eenie Weenie'. Dwarf Day lilies, such as this cultivar, make an attractive border plant. KLEHM NURSERY

'Stella de Oro'. A new introduction. A magnificent plant that is very free flowering, of compact habit, and vigorous. KLEHM NURSERY

Hemerocallis. The Tetraploid Day Lilies are strong growing, and have strong colors. KLEHM NURSERY

The vigor of Tetraploid cultivars can easily be seen; these plants are only one year old. KLEHM NURSERY

'Mary Todd'. A clean color, and well-formed flowers, are the mark of a good cultivar. KLEHM NURSERY

Hemerocallis **cultivars.** The modern cultivars of the Day Lily are available in a wide range of colors. Each year new cultivars are introduced onto the market, often in a series, and the names are chosen to catch the attention of the public. Often the plants themselves will have little or nothing to do with the name chosen, for example, 'Bald Eagle' a rich crimson, and 'Falcon' a deeper red, do not show the colors of the birds after which they are named. However, the name given to a new introduction is important, and a great deal of time is spent selecting names for new hybrids.

Hemerocallis **'Bald Eagle'**. Good substance to the petals, good form and clean color, make this cultivar an attractive border plant. KLEHM NURSERY

Hemerocallis **'Beautiful Morning.** This cultivar has not only a clean color and good form, but produces a great number of flowers per stalk. KLEHM NURSERY

Hemerocallis **'Chicago Petticoats'**. The ruffled edges of the petals add another attraction to this cultivar, which is already of a fine color. KLEHM NURSERY

Hemerocallis **'Falcon'**. Attention should be given to the location in which such plants are grown; in partial shade, such deep colors will not stand out. KLEHM NURSERY

Hemerocallis **'Satin Clouds'**. Outstanding, even on a cloudy day, this cultivar with the gentle blending of subtle colors is much admired. KLEHM NURSERY

Hemerocallis **'Sombrero Way'**. Very large flowers are sometimes obtained by sacrificing bud count, but this cultivar has both size and count, and vigor as well. KLEHM NURSERY

Plate 151

Hermodactylus tuberosus. A hardy plant that will wander away from the point where originally planted. Worth considering for grassy sunny slopes where a plant can be ignored and still thrive. WARD

Hesperantha latifolia. With such a diverse flora, it is not surprising that many South African plants are comparatively unknown in cultivation. The clean unusual color of this species makes this a desirable plant for gardens in warmer climates. BOUSSARD

Hesperantha marlothii. This species is at home in the Cape Province of South Africa. BOUSSARD

Hesperantha pauciflora. The warm color of this species commands attention and despite, the small size, it has a dignity of its own. BOUSSARD

Plate 152

Hexaglottis flexuosa

Plate 153

Hesperantha vaginata. An unusual plant for the warm sunny border, and for growing in containers, which is necessary in cooler climates, because these plants will need protection from frosts. ORNDUFF

Hexaglottis virgata. This native of the South Western Cape Province of South Africa, is very similar to *H. flexuosa.* The most visible difference being the absence of the thin green vein in the center of the petals. BOUSSARD

***Hippeastrum* cultivars.** By far the most popular plants in this genus are the modern cultivars, which are always a hit at various flower shows. They are also great indoor plants, commonly sold in nurseries, and offered by mail order houses. The 4 photographs show the wide range of colors now available in these striking flowers, that are of easy culture, and well worth growing. AUTHOR

Hippeastrum rutilum

Hippeastrum 'Masai'. An introduction from South Africa, and named after the famous tribe 'Masai'. BARNHOORN

Hippeastrum hybrid. This shows the color combinations that always attract attention. Many bicolors are listed in catalogs. WARD

Hippeastrum papilio. This plant is offered by a well known nursery in the United States and is also found growing outdoors at the Huntington Botanic Garden. Unfortunately, where it came from I have not been able to determine. AUTHOR

Plate 156

Homeria comptonii. This shows the typical form of the flowers, petals separated but joining at their base to form a cup. A very showy plant with most unusual colors. BOUSSARD

Homeria cookii. With the varied colors in this genus, the hybridizer could have a field day; unfortunately, little work has been done, but the results could be exciting. BOUSSARD

Homeria flaccida. There is always some variation in species. While the majority of this species are light pink, this plant is whiter than normal. The distinct cup formed at the base of the perianth segments, is clearly defined. AUTHOR

Homeria ochroleuca. A rare species that will only be found in collections at botanical gardens, and those of bulb fanciers. AUTHOR

Homeria collina

Plate 158

Homeria pallida. A delightful species that has a delicate fragrance to the flowers and, while rare, does deserve attention. B.R.I.

Homeria pallida. This shows the arrangement of the foliage, plants are of easy culture, but unfortunately are not easy to obtain. B.A.G.S.

Homoglossum huttonii. This lovely species was for many years known as *H. hollandii,* but whatever the name, its beauty does not change. This picture was taken in the wild near Mossel Bay in the Cape Province of South Africa. BOUSSARD

Homoglossum merianellum. Excellent color makes this species worth considering for the sunny border. It is also a superb cut flower. ORNDUFF

Plate 159

Homoglossum watsonium. A species that seems to prefer heavier soils than most. It should be more widely grown, a comment that applies to the entire genus. ORNDUFF

Hyacinthella lineata. The color variation occurs, not only in the flowers, but also in the foliage; as shown here, the plant does not show the grayish green foliage, that is so typical. WARD

Hyacinthoides hispanica. For many, this plant will always be *Scilla campanulata,* the name by which it has been known for so many years. No one, having seen these plants in a natural setting, is likely to forget their beauty. DE HERTOGH

pretty flowers with a very strong fragrance. Summer-flowering.

H. bulbifera. Native to South Africa in Namaqualand. Often produces bulblets in the leaf axils. Leaves, usually 4, are grasslike, 12 in. in length. Flowers are pure white, up to 24 in. in height; usually grouped 4–5 per stem at the top of the spike. Flowers only about 2 in. in diameter, the anthers twice as long as the filaments, bearing light yellow pollen. Early-summer-flowering.

H. radiata. Common to different parts of South Africa; introduced 1794. Flowers white but with a brownish hue on the outside; bowl-shaped but reflexed when fully open. Summer-flowering, often quite late in the season. Foliage linear, 5–6 leaves of dark green.

H. vaginata. Native to South Africa; introduced 1936. Color from bright to dark yellow with various shadings and height somewhat variable. Generally about 12 in. in height but some selections are taller, others shorter. Leaves grasslike, basal, and sickle-shaped. Flower spike wiry with flowers carried in late spring. Flowers almost erect. The best known form is *H. v. inflexa stanfordiae*, often listed in catalogs, which has erect, canary yellow, fragrant flowers that are open from noon till sunset. The spring flowers are quite large, often 2 in., sometimes more, in diameter.

HESPERANTHA

H. angusta see *H. bachmanii*.
H. bicolor Native to E. Cape Province, South Africa. White, shaded pink; small corm is flattened & covered with fiber. 3 in. in height. Early Spring flowering.
H. bracteolata syn. of *H. pilosa*.
H. cinnamomea Native to South Africa. Introduced 1787. Whitish within, reddish-brown without. 6 in. in height. Spring flowering.
H. falcata (syn. *H. linearis, H. lutea, H. pallida, H. pentheri,* & *H. trifolia*). Native to Cape of Good Hope. White within, brown without. 6–12 in. in height. Spring flowering.
H. graminifolia Native to South Africa. Introduced 1808. Greenish-white. 6 in. in height. Autumn flowering.
H. lactea see *H. baurii*.
H. latifolia (syn. *Syringodea latifolia*) Native to South Africa. Cerise. 10–12 in. in height. Spring flowering.
H. linearis syn. of *H. falcata*.
H. longituba Native to South Africa. Introduced 1877. Inner segments white, outer segments tinged reddish brown. 12 in. in height.
H. lutea syn. of *H. falcata*.
H. marlothii Native to South Africa. White. 10–12 in. in height. Spring flowering.
H. pallida syn. of *H. falcata*.
H. pauciflora Native to Cape Province. Rose-purple. 6–8 in. in height. Summer flowering.
H. pentheri syn. of *H. falcata*.
H. pilosa (syn. *H. bracteolata,* & *H. puberula*). Native to South Africa. Introduced 1811. White, speckled red without. 6 in. in height. Spring flowering.
H. puberula syn. of *H. pilosa*.
H. pulchra Native to E. Cape Province, South Africa. Bright pink to lavender; corm small, white, flattened. About 24 in. in height. Summer flowering.
H. rupestris see *H. baurii*.
H. schlechteri see *H. baurii*.
H. stanfordiae see *H. vaginata inflexa stanfordiae*.
H. subexerta see *H. baurii*.
H. trifolia syn. of *H. falcata*.
H. tysonii Native to Amatala Mts., E. Cape Province, South Africa. Whitish. 12–20 in. in height. Summer flowering.

HESPEROCALLIS — LILIACEAE

The name is derived from Greek *hesperos* ("evening") and *kallos* ("beauty"). Some say that *hesperos* indicates "western," however, "evening" would seem to have validity but as the sun sets in the west either meaning could be inferred. *Hesperocallis* is a monotypic genus, its one species, *H. undulata*, was introduced into cultivation in 1882. It is found in the western United States in the desert regions of Colorado and Arizona, as well as the Mojave Desert of California.

The linear leaves are up to 20 in. in length and ½ in. in width, blue-green with white margins. The leaves are basal, wavy along their entire length, and produced in early spring. The fragrant flowers are carried in a raceme and are white with a silvery gray band on the exterior; perianth tube is 1 in. in length, above which the segments flare and produce a funnel-shaped flower, not unlike an Easter lily in appearance. 8 or more flowers per stem, along with various bracts which are paperlike and greenish-white. The height of the stem varies from 12 in. to over 48 in. Flowers March–May.

The bulbs, up to 3 in. in length, are edible and once provided a staple diet for the Indians who inhabitated the U.S. Southwest. The tunicated bulbs are found deep in the sandy soil, often over 18 in. This is quite possibly an adaptation against the bulbs being exposed by wind sweeping the sandy soil away. Its common names are Desert Lily and Ajo Lily.

CULTURE Must have well-draining soil. Bulbs should be set 6–8 in. deep and spaced some 10–18 in. apart. Plant in fall; little or no water required until growth is active, then appreciates a little moisture but care must be taken not to overwater. Able to withstand temperatures down to 30°F or so if deeply planted, but will not withstand frozen soil or much winter moisture. Requires bright sunlight in summer and resting period after growth has stopped and foliage died down.

PROPAGATION Bulbs produce offsets which can be removed during the dormant season. Seed can be sown in the spring, using a sandy soil mix with night temperatures in the 55°F range. Transplant as soon as large enough to handle or sow thinly and transplant after foliage has died down. Plants should be grown on in individual containers or in sunny beds until of a size when they can be transplanted to permanent quarters. For container planting, use deep containers and a sandy soil mix.

PESTS & DISEASES No special problems.

USES Good plants for desert conditions and for warm, dry, sunny borders in gardens in warmer areas of the country. Good container plant for colder climates but must be protected against frost, being brought indoors during the winter months.

SPECIES

H. undulata. Described above.

HEXAGLOTTIS — IRIDACEAE

The name is derived from Greek *hex* ("six") and *glotta* ("tongue") referring to the 6 spreading lobes of the style. In times past this genus has been included in both *Homeria* and *Moraea* but is deservedly separate. It is a small genus which some authorities regard as containing as many as 4 species, others only 2. After reading various books on the genus, I am inclined, to agree that there are only 2 species—*H. flexuosa* and *H. virgata*.

As the name indicates, the style branches into 6 parts. The seeds produced are very small; in *H. virgata* the seed pods are concealed. This and because the flowers are pressed against the stem are the differences between *H. virgata* and *H. flexuosa*. In the literature the name *H. longifolia* is given as a synonym of *Homeria flexuosa*. This inclines me toward the belief that there are only the two species, *H. virgata* and *H. flexuosa*.

Natives of South Africa, *Hexaglottis* are found in the Cape Peninsula, growing in dry areas in well-drained soil. The rootstock is a tunicated corm. The flowers are yellow, sessile, in October–November in the wild, May–June in the Northern Hemisphere. The perianth segments are free to the base, the short filaments of the stamens are flattened and joined at the base. The flowers are clustered at the top of thin stems; the height of the flowering stems seldom exceeds 24 in. and usually is less. Foliage is narrow, rolled, and seldom more than 2 leaves are produced, sometimes leaf is solitary. Plants are not very showy and it is doubtful if many are to be found in cultivation.

CULTURE The genus is not hardy, the plants not not able to withstand soil temperature that falls below some 35°F. They should be grown in areas where temperatures are maintained as in a cool greenhouse. The corms should be planted 2–3 in. deep and spaced 6–10 in. apart. Plants require sun, and moisture should be given in the winter and throughout the growing period, being reduced as the flowers start to appear. In cold areas the corms should be lifted at the end of summer, stored overwinter in a frost-free area, and replanted in the spring. When grown in containers, good drainage must be ensured.

PROPAGATION The best way is by lifting the corms in late summer, dividing them, and replanting then or in the spring. Seed is very fine and if sown should not be covered but sprinkled onto a thin covering of sand over a well-draining soil mix. Small plants should be transplanted only when large enough to handle. The time from seed to flowering size is some 3 seasons.

PESTS & DISEASES No special problems.

USES For the collector only.

SPECIES

H. flexuosa (syn. *H. lewisiae*). Native to the Cape Peninsula. Commonly found growing on the slopes of Table Mountain. Plant reaches 24 in. in height, often less. Flowers in June/July. Flowers are yellow with a thin green vein down the center of the petals, opening to a little over an inch in diameter; segments are narrow, not reaching ¼ in. in width, and the tips of the tepals are inclined to be rolled a little inward. The flowers are carried loosely against the stem. One, or rarely 2, leaf is produced; foliage is thin, somewhat rolled, and 18–24 in. in length.

H. lewisiae. See *H. flexuosa*.

H. longifolia. Described by some authorities as being a separate species but the description is "having a perianth tube" which would seem to be incorrect. Regarded as a synonym of *Moraea flexuosa* and *Homeria flexuosa* but in this author's opinion further study is needed as it is more likely to be a synonym of *Hexaglottis flexuosa*.

H. virgata. Native to the Cape Peninsula. The difference between this species and *H. flexuosa* is that in this species the fruits are hidden and the flowers are pressed against the stem; other characteristics are the same, except that *H. virgata* produces mostly 3 leaves which are cylindrical, 18–20 in. in length.

COMMENTS Rare in cultivation and likely to remain so.

HIPPEASTRUM — AMARYLLIDACEAE

From the Greek *hippeus* ("a knight") and *astron* ("a star"), perhaps due to the flower stem being lateral to the leaves and at the top of which there are very distinguished flowers that could be conjectured to be the colors at the top of a knight's lance. This genus is comprised of some 70 species, all native to South America. These plants are produced from a large true bulb, are very popular, and are commonly known as *Amaryllis*. Breeders have developed many lovely hybrids which are popular indoor plants, being brought into flower at Christmastime. The parentage of the modern hybrids is not known for a certainty.

The flowers are funnel-shaped and the perianth segments are not of equal length, but this in no way detracts from the beauty of the flowers. The stamens also are unequal in length and have the peculiar habit of bending down and then curving upward. The stigma remains below the stamens and usually is much longer, often being equal to the length of the perianth segments.

The flowers are carried on a strong stem which has no leaves but frequently is covered with a "bloom" of fine wax. The number of flowers per stem varies but is generally more than one, and frequently in large bulbs more than one flowering stem is produced. The flowers are often large, over 4 in. and as much as 8–10 in. in diameter in the very large hybrids. Leaves generally are produced after the flowers or starting at the same time, strap-shaped and fleshy.

CULTURE The modern hybrids that are grown for Christmas flowering have to be cultivated indoors in order to get them to flower on time. The bulbs, which are very large, should be grown in a soil mix consisting of equal parts of good topsoil, peat moss or leaf mold, and sharp sand. The bulbs should not be planted with more than a two-thirds of the bulb in the soil. The soil must be well firmed around the bulbs. It takes about 55 days from the time of setting the bulbs in the pots to flowering time. It is an advantage to give the pots a little bottom heat while the roots are being produced, a period of about 21 days. During this time a temperature of some 55°F at night should be maintained.

The bulbs should be kept moist, not wet; more liberal watering can begin when the flower spike emerges. When the flower spike is several inches in height the plants should be given full sun. After the flowers are in bud delay in opening can be obtained by lowering the temperatures.

After flowering the flowers should be removed. The foliage now makes its growth and during this time a weekly feeding of organic liquid fertilizer can be given. This can continue for 6–8 weeks. The plants should show signs of the foliage maturing in late July, and the bulbs should then receive less water. By the beginning of August little or no water should be given and the bulbs allowed to ripen and go into a dormant state for 6–8 weeks. They can then be top-dressed or repotted and started into growth again at the required date. It should be noted that leaves often will stay green if water is given constantly; the bulbs do not go into a dormant state and then the chance of flowering is greatly diminished.

There are no completely hardy species. However, some will take a few degrees of frost if grown in a sheltered and sunny location. They are great plants for warmer climates and for the cool greenhouse.

If being grown outdoors suitable only in Mediterranean climates or subtropical climates, they should be planted in the fall, given as dry a winter as possible, allowing only the rain to provide the moisture; if this is lacking an occasional watering should be given. Once in position the bulbs should not be disturbed but allowed to develop until they become overcrowded, which will take a number of years.

PROPAGATION Seed can be sown as soon as ripe in a light soil mix. Seeds germinate quickly if given a temperature of 60°–65°F at night. When the seedlings have leaves about 4–6 in. in length, they can be planted individually into 4 in. pots. They can be kept watered and growing; only if the foliage shows signs of wanting to become dormant should water be withheld. The bulbs can be brought into flower in these pots, however, it most likely will be necessary to repot them in their second season and allowed to have a period of dormancy before being brought into flower. Offsets often are produced; these should be separated after the foliage dies and repotted into individual pots and grown on. When such separating is done, care should be taken not to damage the roots. Also examine the bottom of the bulbs to make sure that they are sound.

PESTS & DISEASES Thrips can be a problem if plants are grown in a greenhouse, and these insects must be controlled. Red spider mites and mealy bugs also can be a problem, indoors or out, and control must be put into effect as soon as any signs of attack are seen. Also examine bulbs for mealy bugs as they often are harbored in the scales.

Spotted wilt, a fungus disease that causes yellow or whitish spots on the leaves, can be a problem and any foliage with such markings should be removed and the bulbs destroyed if disease is severe.

Many problems in the greenhouse are caused by too dry an atmosphere; spider mites and thrips are not a problem if hygiene and humidity are maintained. Controls used as soon as any damage is seen generally can eradicate the problem. Care should be taken to spot the least sign of any problem as soon as possible, as flower buds can be easily damaged by products used to control pests and diseases.

USES The hybrids, the most commonly grown plants, are ideal for Christmas flowering. In warm climates the species can be grown outside, and indeed the hybrids also, provided the night temperatures do not drop below 40°–45°F.

SPECIES & CULTIVARS

H. advenum (syn. *Amaryllis advena, Habranthus advenus, H. hesperius, H. miniatus, Rhodophiala advena*). Native to Chile. Bulb is ovoid to globular with dark tunic. Produces bright red flowers with a yellow-green center stripe in each of the petals. Height about 15 in., strong stem, 2–6 flowers per stem, held horizontal or a little upward. The flowers are not very large, about 2 in. long with narrow petals. Flowers in August. Leaves strap-shaped, 12 in. in length, and produced after flowering. One of the hardier species; will withstand a little frost but the foliage should be protected and is best in a cool greenhouse in any area where frost can be severe.

H. aulicum (syn. *H. robustum, Amaryllis aulica*). Native to Brazil. 18 in. in height. Scarlet flowers with green center, 5–6 in. in length, 2–4 flowers per spike. Leaves strap-shaped; produced after the flowers, which appear in the winter.

H. blossfeldiae. From Brazil. Name given to a plant introduced by Mr. Harry Blossfeld and offered by Van Tubergen of Holland. Color of flowers soft orange with a greenish-yellow star at the throat. 12–15 in. in height. Late-summer-flowering.

H. blumenavia (syn. *Griffinia blumenavia*). Native to Chile; introduced 1866. Dwarf species, 9–10 in. in height. Leaves short, 6 in. in length, 2 in. in width. Flowers 3 in. in length, 4–5 per flowerhead, white with pale rose lines in the petals. Stem of the flower spike rose colored at base, green above; flowers pendant. Late-summer-flowering.

H. equestre (Barbados Lily; syn. *H. puniceum*). From tropical America; introduced 1698. Scarlet-red flowers; leaves up to 20 in. long 2 in. wide, mostly produced after the flowers. Flowering in the early part of the year or in the spring. Other varieties of this are bright orange with white, orange with green, etc., and double forms are known. Height 12 in.

H. × johnsoni. Introduced in 1799. An early hybrid between *H. reginae* and *H. vittatum*. The large bulb may reach 3 in. across. Deep red flowers with white stripe; quite large, 4 in. in length and a little more in diameter when open. Spring-flowering. Leaves produced mostly after the flowers, 24 in. long, strap-shaped. An easy plant to grow in the cool greenhouse.

H. pratense (syn. *Habranthus pratensis, Rhodophiala pratensis*). Native to Chile; introduced 1840. 12 in. in height. Scarlet flowers with yellow at base, 3–5 flowers in the spike, comes into flower at the same time as the leaves in early summer.

The number of hybrids offered is considerable. The following is a list of some of them.

'African Sunset'. Bright orange-red.
'Appleblossom'. White with a pink flush.
'Athos'. Dark Red.
'Bambara'. Deep red with light green/white midrib.
'Basuto'. Very dark red, possibly the darkest.
'Blushing Bride'. Rose; very large blooms.
'Candy Cane'. Red with white stripes.
'Cicero'. Orange-red.
'Cinderella'. Orange with white stripes.
'Dazzler'. Pure white.
'Desert Dawn'. Light salmon-orange; foliage appears with flowers.
'Fair Lady'. Salmon.
'Helsinki'. White.
'Kalahari'. Rose iridescent.
'Masai'. White flushed with pink stripes.
'Orange Sun'. Orange.
'Queen of Sheba'. Pink.
'Safari'. Red-bearded petals.
'Springtime'. Light rose, white center, white tips to petals.
'Zanzibar'. Dark red with green midrib.
'Zenith'. Pink with a white stripe.

HIPPEASTRUM

H. × *acramanii* (*H. aulicum* × *H. psittacinum*) Garden origin. Introduced before 1870. Rich red. 24 in. in height. Summer flowering.
H. andreanum Native to Colombia. Introduced 1876. Pale red with dark red stripes. 12–18 in. in height. Summer flowering.
H. argentinum syn. of *H. candidum*.
H. bagnoldi (syn. *Amaryllis bagnoldii, Habranthus bagnoldii, Rhodophiala bagnoldii*). Native to Argentina, Chile. Yellow, tinged red. 12 in. in height. Summer flowering.

H. bicolor Native to Chile. Bright red, yellowish-green towards base. 12–18 in. in height. Oct. flowering.

H. bifidum (syn. *Amaryllis bifida, Habranthus roseus, Rhodophiala bifida*). Native to Argentina, Uruguay. Introduced 1825. Bright red. 12 in. in height. March flowering.

H. brachyandrum see *Habranthus brachyandrus*.

H. breviflorum Native to Argentina. Introduced 1836. White, tinged yellowish-green without, & a central red streak. Within the streak has a white line down the middle. April flowering.

H. calyptratum Native to Brazil. Introduced 1816. Pale yellow, netted green. 24 in. in height. April–May flowering.

H. candidum (syn. *H. argentinum*). Similar to *H. solandriflorum*. Native to Argentina. Introduced 1929. White, tinged yellow throat. April–May flowering.

H. correiense syn. of *H. organense*.

H. cybister Native to Bolivia. Introduced 1840. Red below, greenish above, pale striped within. 24 + in. in height. April–May flowering.

H. elegans syn. of *H. solandriflorum*.

H. elwesii Native to Southern Argentina, Chile. Introduced 1903. Pale yellow, claret at inner base. 12 in. in height. July–Aug. flowering.

H. forgetii Native to Peru. Introduced 1909. Dull crimson, green base. 24 in. in height. Winter flowering.

H. leopoldii (syn. *Amaryllis leopoldii*). Native to Peru. Introduced 1869. Bright red, greenish-white in throat. Early Summer flowering.

H. miniatum (syn. *Amaryllis miniata*). Native to Peru. Introduced 1832. Red. 12 in. in height. July flowering.

H. organense (syn. *H. correiense*). Native to Brazil. Introduced 1830. Red, with yellow rays. Early Summer flowering.

H. pardinum Native to Peru. Introduced 1866. Cream, spotted crimson. 18 in. in height. Late Spring flowering.

H. phycelloides (syn. *Habranthus phycelloides, Phycella phycelloides*). Native to Chile. Introduced 1830. Bright red, base yellowish. 6–12 in. in height. May–June flowering.

H. procerum (syn. *Worsleya rayneri*). Native to Brazil. Introduced 1863. Lilac. 12–18 in. in height. Winter flowering.

H. psittacinum (syn. *Amaryllis psittacina*). Native to Brazil. Green and scarlet striped. 24–36 in. in height. May flowering.

H. puniceum see *H. equestre*.

H. reginae (syn. *Amaryllis reginae*). Native to Brazil, Peru, W. Indies. Introduced 1725. Bright scarlet-red with greenish-white star in base. 12 in. in height. Summer flowering.

H. reticulatum (syn. *Amaryllis reticulata*). Native to Brazil. Introduced 1677. Soft pink & white. 12 in. in height. May flowering.

H. robustum see *H. aulicum*.

H. roseum (syn. *Amaryllis barlowii, Habranthus roseus, Rhodophiala rosea, Zephyranthes pumila*). Native to Chile. Introduced 1831. Rose. 18–24 in. in height. September flowering.

H. rutilum (syn. *H. striatum, Amaryllis rutila, A. striata*). Native to Venezuela, Brazil. Bright scarlet-crimson. 18–24 in. in height. Spring–Early Summer flowering.

H. solandriflorum (syn. *H. elegans, Amaryllis solandriflora, A. elegans*). Native to South America. Introduced 1839. Sulphur or cream, greenish or sometimes purple in middle. 24–36 in. in height.

H. striatum syn. of *H. rutilum*.

H. stylosum Native to Guiana, Brazil. Introduced 1821. Pale brownish-pink with deeper coloring. 18–24 in. in height. Summer flowering.

H. texanum see *Habranthus andersonii* var. *texanus*.

H. vittatum (syn. *Amaryllis vittata*). Native to Andes of Peru. White with scarlet stripes. 36 in. in height. Summer flowering.

HOMERIA — IRIDACEAE

Named in honor of Homer, the Greek epic poet. A small genus of about 32 species of South African plants that are produced from corms. While the flowers are not long-lasting, the plants produce so many flowers that they are attractive for a long period of time. Not all of the species have been introduced into culture; most are known in the wild only.

The basal leaves are usually solitary, but the flower stem carries a number of leaves. The 6 segments that make up the flowers are not joined but come together at the base to form a cup. They then spread and form a flattened flower.

The corms can be stored for many months, which is convenient as they can be planted in the spring to flower in the summer in cold climates, or remain in the ground in warmer climates where little or no frost is expected.

CULTURE The corms should be set 1 in. deep and, as the plants are small, spaced only 3–4 in. apart in groups. Plant in well-drained soil with adequate moisture during their growing season, and allow to dry after flowering. They like sun or light shade. In warm areas plant in the fall. In colder climates they should be planted in the spring and lifted late in the summer or early fall before the first frosts arrive. They can be stored overwinter, dry and warm, at about 40°F, and then planted back again in the spring. The success in flowering the second year is not ensured but many will perform. They also can be grown in a cool greenhouse as container plants. In this case plant several corms to a small container and keep moist until well after flowering, yet allow the stems and leaves to ripen before storage.

PESTS & DISEASES No special problems.

USES Good plants for the sunny border where the unusual form and colors can be appreciated. Light and airy, unusual shapes, good container plants.

SPECIES

H. collina (syn. *H. breyniana, Moraea grandiflora*). Native to Cape Peninsula; introduced 1793. Corm slightly with a dark, coarse, fibrous tunic. Reaches a height of 18 in. Flowers are about 2 in. in diameter; color varies from light salmon to a good deep yellow; produced freely so the plant remains attractive for a long period. Flowers in spring from fall planting; later in the summer from spring-planted corms. Leaves are thin and arching. *H. c. aurantiaca* was introduced in 1810 and has soft orange flowers, slightly taller than type, very free-flowering. *H. c. ochroleuca* has soft yellow flowers.

H. comptonii Native to western Cape Province. Leaves about 12 in. in length. Segments widely separated; lemon-yellow in the center, salmon-pink along the edges of the petals. Height 8–10 in. Early-summer-flowering.

H. elegans. Native to western Cape Province; introduced 1797. Bright yellow flowers with splashes of green scattered on the petals, or 3 of the petals can be of another color—mostly a pale orange. Early-spring-flowering. Height 8–10 in.

H. lilacina (syn. *Moraea polyanthos*, which see). Native to Cape Province. Only 12–14 in. in height. Many lilac flowers are produced in early summer.

HOMERIA

H. albida syn. of *H. miniata*.

H. bicolor syn. of *H. flaccida*.

H. breyniana syn. of *H. collina*, & *H. flaccida*.

H. bulbillifera Native to Little Karoo, South Africa. Creamy yellow, sometimes with dark lines at base of tepals or dark yellow splashes. 12 in. in height. Spring flowering.

H. cookii Native to South Africa. Yellow. 8–12 in. in height. Late Summer flowering.

H. flaccida (syn. *H. bicolor*, *H. breyniana*). Native to Cape Peninsula. Peach-pink. 12–24 in. in height. Early Spring flowering.

H. flexuosa see *Hexaglottis longifolia*.

H. glauca syn. of *H. pallida*.

H. humilis syn. of *H. pallida*.

H. lineata syn. of *H. miniata*.

H. lucasii syn. of *H. ochroleuca*.

H. mossii syn. of *H. pallida*.

H. miniata (syn. *H. albida*, *H. lineata*). Native to South Africa. Introduced 1825. Yellow to salmon-red. 24 in. in height. Spring–Early Summer flowering.

H. ochroleuca (syn. *H. lucasii* & *Moraea collina* var. *ochroleuca*). Native to South Africa. Salmon-pink & yellow or golden yellow. To 30 in. in height. Spring flowering.

H. pallida (syn. *H. glauca*, *H. humilis*, *H. mossii*, *H. pura*, *H. townsendiae*). Native to S.W. Africa. Light yellow. Flowers sweetly scented, but leaves & fruit poisonous. To 24 in. in height. Flowering Sept.–Dec in wild; April–July Northern Hemisphere.

H. schlechteri Native to Namaqualand. Yellow. 8 in. in height. Spring–Early Summer flowering.

H. pura syn. of *H. pallida*.

H. townsendiae syn. of *H. pallida*.

HOMOGLOSSUM — IRIDACEAE

The name is derived from the Greek *homos* ("equal" or "alike") and *glossa* ("tongue"), which refers to the nearly equal perianth segments. This genus is found in Africa, both in the tropics and in the Cape Peninsula of South Africa. Some authorities feel the name is derived from the Greek *omoios* = "similar" and *glossa* = "tongue." There are some 11 species. The rootstock is a corm. Few leaves are produced and these are linear. The flowers are somewhat like a gladiolus in shape and these plants are sometimes included in the genus *Antholyza*, but are distinct because the lobes of the perianth segments are similar in form. The flowering stem is unbranched and the flowers carried in a loose terminal spike which will have few or up to 7 flowers. The plants flower in the spring or early summer.

The flowers are tubular at the base, the tepals then form either a narrowly opening flower or the tepals flare outward, then the upper 3 are larger than the lower 3 but the uppermost is not hooded. The base of the tube emerges from leaflike bracts. The stamens are carried close to the uppermost lobe and the stigma is longer and then splits into 3 parts. The flowers are in the red tones with the interior of the tepals sometimes in the orange-yellow hues. The common names are Flames and Red Afrikaner.

CULTURE The corms should be planted in the spring, setting them 2–3 in. deep in well-drained soil and some 4–6 in. apart. They like the sun but will perform well in partial shade and this is actually preferable in the warmest areas. In areas where the winters are cold and the ground freezes they should be lifted and stored in a frost-free area for replanting in the spring. They need moisture during the early part of the season but prefer to be dry once the foliage has started to die down. Some light feedings of organic fertilizer can be given when growth commences but should not continue after the flower spike emerges.

PROPAGATION Seed is produced in quantity and should be sown in the spring in a sandy soil mix with night temperatures in the 45°–50°F range. The seedlings should be kept moist while growth is active and allowed to become on the dry side once the foliage starts to die down. The small corms that have formed in the first season can be lifted and planted again the following spring or left in the flats overwinter and then planted in other containers, spacing them 2 in. apart. They also can be planted out in a warm border. In two seasons, or three at the most, the corms will be of flowering size. Cormels that are produced can be separated from the parent corms and treated as those small corms raised from seed.

PESTS & DISEASES No special problems.

USES Excellent border plants for sun or light shade, these plants are full of color and last quite a while in flower. They are quite good as cut flowers but do not produce so many flowers per stem as Gladiolus and so have a hard time competing. The dwarf species are good rock garden plants.

SPECIES

H. abbreviatum. Native to Cape Province. This is an unusually shaped flower as the perianth segments remain in a tube for their entire length; fire-red in color, including the bracts from which the flowers emerge. It reaches a height of from 12–15 in. and 5–8 flowers are produced. The leaves are narrow and up to 10–12 in. in length. Flowering in Aug.–Sept. in its native habitat; early spring in the Northern Hemisphere, later if it was spring planted. In the wild it is becoming rare due to agricultural development but can still be seen along roadsides near the town of Caledon in Cape Province.

H. aureum. Native to the Cape Peninsula but now lost in the wild. This species perhaps represents a transition stage between this genus and *Gladiolus*, some authorities placing it as *Gladiolus aureus*. It has lovely pure yellow flowers and reaches some 12 in. in height. The tube is not so long as in other *Homoglossum*, the tepals flare back closer to the base of the segments, and the curving of the tube is less pronounced. For these reasons, and also because no other of the species is yellow, it perhaps should be included in *Gladiolus*. In addition to the differences noted, the tepals are much more rounded at the tip than in the other species. It is doubtful if this species still exists in the wild; it grows on gravelly soil and commercial exploitation of the area for gravel has most likely been the demise of this plant. Foliage 8–10 in. in length, sword-shaped. Flowers in early summer.

H. gawleri. Native to South Africa. It is much like *H. watsonium* except that the lower part of the perianth tube is orange-yellow. It also is confused with *H. huttonii* which has yellow streaking. Name of doubtful validity; most likely a form of *H. watsonium*.

H. hollandii (Flames; syn. *H. huttonii*). Native to the area around George in the southern part of South Africa. A lovely species named originally after Frederik Huntley Holland (1874–1955), a well-regarded amateur botanist from Port Elizabeth in South Africa but now correctly named *H. huttoni*. A slender species reaching a height of some 18–20 in., found growing in open grassland. The common name well describes the color of the flowers as they are red streaked with yellow. The tepals are loosely held and the segments separate from each other sharply as they emerge from the tubular portion of the flower. The upper 3 segments are much larger than the lower and the smaller lower segments have streaks of yellow on the interior. The bracts surrounding the lower part of the tube clasp it tightly. Flowering in late winter to early spring and is a species best suited for warmer climates. Var. *zitzikamense* has more yellow streaking than the type but now is included in the species.

G. huttonii. See: *G. hollandii*.

H. merianellum (Flames; syn. *Antholyza merianella, Gladiolus hirsutus roseus*). Native to the Cape of Good Hope; introduced 1795. This plant has a corm of medium size, flowers in winter or early spring, and reaches 12 to 24 in. in height. Flowers are red with a golden interior and flare into a cup shape with the tube some 1½ in. in length; there are some 4–6 flowers per spike, which is one-sided. Unlike most species, which produce only 1 or 2 leaves, this one produces 4–6, which are linear, rigid, and quite short, 8–10 in. in length. In the wild this plant flowers in June–July; early winter in the Northern Hemisphere, or June–July when spring-planted. A gold-colored form is known as *H. m.* var. *aureum*; even in this form the base of the tube is red and there is a flush of red on the interior of the tepals. There is little distinction between the upper and lower tepals in this species and its varieties and thus have a definite cup-like mouth. While the majority of this species have the one-sided flower spike, this is a variable character. There are from 2–5 flowers per spike, a little under an inch in length, in both the species and var. *aureum*.

H. muirii. Native to the area around Caledon near Cape Town in Cape Province of South Africa. It has light carmine-red flowers and produces a narrow tube, the tepals being of equal length and pointed. It reaches 10–16 in. in height and the flowers are 1–1½ in. in length and the number is 1–4. Flowering in the wild July–Sept.; winter–early spring in the Northern Hemisphere. Rare.

H. priorii. Native to the Cape Peninsula. Found growing among bushes on mountain slopes. The tube curves gracefully and the tepals open to be almost flat; upper segments much larger than the lower. Flower color is light red, with white markings found on the throat, these being more prominent on the lower tepals. It is quite tall-growing, up to 20 in. in height. The leaves and stems have a glaucous look. Flowering late fall (May) in its native habitat.

H. quadrangularis. Native to the Ceres area of Cape Province, northwest of Cape Town. Found growing in the foothills of the mountains in damp soils. It is tall, growing often as much as 30 in. in height, and flowering in Sept.–Oct. in the wild, this being spring. The tepals are very pointed, the 3 outer tepals flaring almost to flatness but slightly recurved, the inner hardly being reflexed at all. The flowers are just over an inch in length of which two-third is tube; from 5–9 flowers are carried in the one-sided flowering spike. Flower color is red with hint of purple; a pinker form also is known. Foliage is stiff and erect.

H. revolutum. See: *H. watsonium*

H. vandermerwei. Native to South Africa. The tube is an inch in length and only the very tips of the tepals are not forming the tube; it is a bright shiny red and the uppermost tepals flare slightly while the lowest are held at right angles to the tube. From 1–6 flowers are produced. The flowering time in the wild is September; early spring in the Northern Hemisphere. The height varies between 8–20 in. Foliage stiff, lanceolate, 10–18 in. in length. Rare.

H. watsonioides (syn. *Antholyza watsonioides, Gladiolus watsonioides*). Native to Kenya; introduced 1886. Reaches 24–36 in. in height. The flowers are bright scarlet and from 2–4 are carried in a one-sided flowerhead. The tube is about 1½ in. in length, and the segments which are pointed are some 1 in. in length after the tube, flaring slightly. The stem sometimes has 2 small leaves on it. There are some 4 leaves produced, each one stiff and erect and some 12–18 in. in length.

H. watsonium (syn. *H. revolutum, Antholyza revoluta, Gladiolus praecox, G. watsonius*). Native to the Cape Peninsula; introduced 1880. A fine cut flower and one of the more free-flowering species; produces bright red flowers that are up to 3 in. in length, closely placed on the stem, those above nestling those below. Tepals are pointed but of good width with all being of the same size and shape and overlapping each other. In the wild it flowers from June–Sept.; winter to early spring in the Northern Hemisphere. The leaves are narrow and stiff. It is an easy species to grow and prefers heavier clay soils than most species.

COMMENTS This genus, with its great colors and striking flowers, should be exploited to see what hybrids can be produced. While the plants do not have large numbers of flowers in the spike, breeding might increase this. This is especially important to remember as the *Gladiolus*, now so well regarded as both a cut flower and garden plant, was raised from species many of which were also shy flowering.

HYACINTHELLA — LILIACEAE

The name is a diminutive of *Hyacinthus*. In the past some 10 species were included in genera such as *Bellevalia* and *Hyacinthus* but now are placed in this genus. All species of *Hyacinthella* are native to the Mediterranean region, southeastern Europe, and western Asia. They are found growing in rocky, stony areas among scrub where they receive the best possible drainage and sun, although sometimes they are shaded by the rocks and scrub because of their diminutive size, which is rarely more than 4 in. high.

The most obvious features of these bulbous plants is that the foliage has raised veins, which are fibrous strands, and the bulbs have a powdery coating under the tunic. Only a few basal leaves are produced, seldom more than three. All of the flowers in this genus are in shades of blue; some light, some dark. The anthers are held just inside the mouth of the tubular flowers. None of the species is of great garden merit, although all are easy to grow.

CULTURE Bulbs should be planted in the fall, in well-drained soil, in a sunny location, 1 in. deep and 4–5 in. apart. Water should be given during the winter and early spring, but the bulbs should be left to their own devices for the rest of the year. Once planted, leave them in the same location.

PROPAGATION Seed can be sown in the summer in a sandy loam and small plants raised in pots until the bulb is large enough to plant out. About 2–3 seasons will be needed. There is some natural increase by offsets of the larger bulbs, which can be separated from the parent after foliage has died down.

PESTS & DISEASES No special problems.

USES Plant in rock gardens in the sun. These plants are more for conversation pieces than for their actual beauty.

SPECIES

H. dalmatica (syn. *Hyacinthus dalmaticus*). Native to western Yugoslavia and Dalmatia. Plant is only 3–4 in. high. Linear leaves.

HYACINTHELLA

H. acutiloba Native to Turkey. Introduced 1981. Pale blue. Dwarf. April–June flowering.
H. azurea see Hyacinthus azureus.
H. campanulata Native to Turkey. Introduced 1981. Pale blue. 3–4 in. in height. April–June flowering.
H. glabrescens (syn. *H. hispida* var. *glabrescens*, *H. lineata* var. *glabrescens*). Native to Turkey. Introduced 1981. Deep violet-blue. 3–4 in. in height. April–June flowering.
H. heldreichii Native to Turkey. Introduced 1931. Deep violet-blue, flowers sessile. 3–4 in. in height. April–June flowering.
H. hispida Native to Turkey. Introduced 1931. Deep blue. 3–4 in. in height. April–June flowering.
H. micrantha Native to Turkey. Introduced 1931. Pale blue-white, flowers sessile. 2–3 in. in height. April–June flowering.
H. nervosa Native to Turkey, Israel, Jordan, Syria, Iraq, Lebanon. Introduced 1931. Pale blue, flowers sessile. 4 in. in height. April–June flowering.
H. siirtensis Native to Eastern Turkey. Introduced 1973. Pale blue. 3–4 in. in height. April–June flowering.

Dense spikes of bright blue flowers, each only ¼ in. long. Early-spring-flowering.

H. lineata (syn. *Hyacinthus lineatus*). Native to Turkey; introduced 1887. Found growing in dry areas. The grayish foliage has the distinctive raised-veins characteristic of this genus. The plant is only a few inches high but the dense racemes of flowers are each ¼ in. long and are carried on stalks that are of almost equal length. There is some color variation—the normally light blue flowers may shade into a much darker, almost purplish, blue. Flowers in the spring.

COMMENTS These bulbs are for specialists only. It is not likely that they will ever become popular in gardens, except, perhaps, where collections of bulbous plants are grown.

HYACINTHOIDES — LILIACEAE

The plants in this genus have been transferred from one genus to another over the years. They have been placed in *Scilla*, *Agraphis*, *Endymion*, and now are in *Hyacinthoides*, which is the oldest name and thus has historic precedence. Such changes are wrought by botanists. While rules of nomenclature should be followed, there would appear to be a good case for historic usage among plants that have enjoyed a name for many, many years. However, the common English bluebell now is placed in this genus. May it stay there forever and not be changed!!

This genus only contains a few species, including the bluebell as mentioned. These plants have a curious growth pattern in that the bulb is completely replaced each year. In addition these flowers have two bracts for each flower.

CULTURE The bluebells love the open deciduous woodlands; leafy, humus-rich soil; with light shade in the spring, when they are in flower, followed by deeper shade provided by the trees as they break out into leaf. The bulb should be planted 4 in. deep. They will multiply quite quickly and indeed can become almost a weed in some gardens. They are, however, a delight, especially when planted with rhododendrons and azaleas providing a perfect blue framework for their colors.

PROPAGATION The easiest is to lift and divide the bulbs that have been in place for a few seasons. Seed can be raised, gathered when ripe, sown in a leafy soil mix, placed in the shade, watered, and then (except for keeping the soil mix moist) ignored. No special care until the little bulbs are of size to be placed outside. Should larger bulbs be required, sow seed thinly in drills, about ½ in. deep, and grown for 2 seasons in that location; keep them moist during the growing season and then hold off the moisture in late summer.

PESTS & DISEASES No special problems.

USES Ideal plants for open woodlands, where the great drifts of flowers look splendid in the spring; and for use as a border for spring-flowering shrubs such as rhododendrons and azaleas. They can be planted in the open but should have some shade during the hottest spring days, not only for the health of the plants but also because the sunlight can make the flowers look paler and less attractive. One of the few bulbs that is transplanted, as it were, by gophers and not eaten. Always plant in large quantities, not a few at a time, and allow to stay in same location for a number of years. Transplant after the foliage dies down.

SPECIES & CULTIVARS

H. hispanica (Spanish Bluebell; syn. *Endymion hispanicus*, *Scilla hispanica*). Native to the Iberian Peninsula. Long, bell-shaped flowers, each nearly ¾ in. in length. More than 15 flowers per stem is not unusual, and the height can be as much as 20 in. Leaves are strap-shaped, produced with the flowers, and continue to grow after the flowers fade, up to 24 in. in length, about an inch in width, and quite fleshy. Flowers in spring, usually in April or May, but can be earlier or delayed according to the weather conditions. Very hardy plant and one that, despite its popularity, deserves to be grown even more. There are a number of named cultivars that have been selected and are offered by various nurseries. Among them are the following, often found listed under *Scilla campanulata*, *S. hispanica*, or *Endymion campanulatus*.

'Alba'. White flowers.
'Arnold Prinsen'. A fine pink-flowering form.
'Blue Bird'. Early-flowering; dark blue.
'Blue Queen'. Lighter blue and later flowering than some.
'City of Haarlem'. Soft violet flowers.
'Excelsior'. Violet-blue with marine-blue edge.

'Rose Queen'. Pure pink.
'Sky Blue'. Late flowering; dark blue.
'White Triumphator'. Robust spikes of pure white.
   In all of the above the anthers are blue; in the English bluebell they are cream colored.

H. italica (syn. *Scilla italica*, which see). Native to southwestern Europe. Height 4–12 in. Flowers starry rather than bell-shaped, deep lilac-blue or violet-blue, carried in a conical flowerhead. Spring-flowering, March–May. Leaves 4–10, linear, deep green, less than ½ in. wide.

H. non-scriptus (English Bluebell; syn. *Endymion non-scripta, E. nutans, Scilla non-scripta*). Native of Europe, including Britain; long in cultivation. Reaches a height of some 18 in. with the top of the flower spike bending over. The flowers tend to be carried on one side of the spike; color can vary from shades of blue to pink to white. An excellent plant for the woodlands and, if planted with the Spanish bluebells, will often hybridize naturally, resulting in many interesting shades of color.

HYACINTHUS — LILIACEAE

The name is derived from an ancient Greek name used by Homer, among others; a flower is said to have sprung from the blood of the dead Hyakinthos, who, according to legend, was killed accidentally by Apollo.

This is another genus much subjected to revision in which botanists have moved plants from one genus to another. While at one time there were some 30 species in this genus, there now remain but 3—*H. litwinowii, H. orientalis* and *H. tabrizianus. H. tabrizianus* is only about 2 in. high and it is not certain that it is a *Hyacinthus*; the second is a recent introduction by plant-hunter Paul Furze. By far the most important is *H. orientalis* (described below), from which are derived the many cultivars we enjoy in our gardens and see in use by florists and commercial growers.

The bulbs of *Hyacinthus* are tunicated and inclined to be flattish or ovate. All are spring-flowering plants. While a powdery substance, such as in *Hyacinthella*, is not noticed when examining the bulbs of cultivars and selections of *H. orientalis*, persons should be aware that prolonged contact and handling of the bulbs can cause skin irritation. This is generally noticed around the fingernails. Persons with sensitive skin should wear gloves when handling.

Hyacinthus has tubular, bell-shaped flowers; the stamens are attached to the tube, except near the tip. The flower spike is a raceme, and the stigma and style are short. The foliage is linear. It should be mentioned that the popular hyacinth of the florist and gardens is the most likely member of this genus that gardeners will encounter.

The majority of the other species formerly in this genus are now in either *Bellevalia, Brimeura, Hyacinthella,* or *Pseudomuscari*. A brief description of the differences among each of these perhaps is worth noting here. (See also separate entries for these genera.)

Bellevalia has taller flower spikes, often up to 8 in., but the flowers are not of great interest, being rather dull in color. It differs from *Hyacinthus* in that the anthers are carried in the mouth next to the petals; in *Hyacinthus* the anthers are held well within the tube.

Brimeura has much smaller flowers than *Hyacinthus* and, in addition, the bracts that subtend the flower stalks are almost unnoticeable in *Hyacinthus* but in *Brimeura* are significant, being long and tapering.

Hyacinthella is very dwarf, reaching only about 4 in. in height. Flowers are more erect and the foliage has noticeable raised veins, a feature not found in the other genera mentioned here or in *Hyacinthus*. In addition the bulbs have a powdery substance under their papery tunic, which is white in color, also very distinctive. Few if any of the species are of any garden value.

Pseudomuscari has dense flower spikes, compared with *Hyacinthus*, which has a rose spike and the mouth of the tube formed by the perianth segments is not restricted, or not nearly so as in *Hyacinthus*.

CULTURE Bulbs for outdoor growing can be planted in the fall—in early October in colder climates where frost is experienced, early November in warmer areas where little frost is experienced—and in very warm climates plant in early December. The bulbs should be set with 3–5 in. of soil over them (8 in. in warmest areas) and spaced some 6–9 in. apart. A well-drained, sunny location in average garden soil is fine but some well-rotted compost or manure could be added to the soil prior to planting. While the bulbs are hardy, late spring frosts, especially if severe, can harm the flower spikes and protection should then be given. The bulbs can be left in the ground after flowering but their performance the second year will not be quite so good as in their first season; the following year only some of them will produce flowers. Thus it is better to lift after the foliage has started to die back. Dry the bulbs and store in a well-ventilated area, on racks or in paper sacks with holes. Placing the bulbs into temperatures of 40°F for a period of 6–8 weeks prior to replanting is necessary. However, once flowered, the bulbs are not to be counted upon to perform as well as in the first season. For outdoor planting, smaller-sized bulbs can be used, as compared with those to be forced for indoors. It should be remembered that hyacinth bulbs do not enjoy being heavily fertilized.

If cultivated in containers to be brought into flower early, "prepared" bulbs should be used. These are sold as such in nurseries. For the preparation of the bulbs for early forcing, see chapter on "Preparation of Bulbs for Forcing, etc." Prepared bulbs are planted in September and require up to 10–12 weeks for good root development. This means a low temperature, as supplied by growing rooms or by being plunged into soil or sand/ashes outdoors in a cool location. After root development is completed, the roots will fill the containers, the top of the bulb being level with the soil at planting time. This will allow you to see the flower spike in the neck of the bulb; it must be visible before being brought into warmer temperatures. In growing the bulbs in containers, they can be close together but should not touch each other or the sides of the container.

As soon as the roots are developed and the flower spike showing in the neck of the bulb, the bulbs can be brought indoors to subdued light. When the flower spike begins to lengthen and has grown about 2 in., temperature should be increased to around 65°F and increased light given. During the entire time the bulbs should be kept moist but good drainage always must be provided. Bulbs that are not prepared will take several weeks longer to develop their roots. For such forcing the largest-sized bulbs obtainable should be used. The same culture is required, except for the length of time in root production, as for prepared bulbs. By obtaining a good selection of prepared bulbs—both earlier- and later-flowering cultivars—hyacinths can be had in flower for a long period of time.

After being "forced," the bulbs either should be discarded or planted outside in the border, where they will begin to produce some flowers in a season or two. They should not be "forced" again.

Hyacinths also can be grown in water, placing them in the neck of a vase designed for this purpose. Some pieces of charcoal in the water is an advantage but not essential. The water should be just below the bottom of the bulb but never touching. Place the bulbs in a dark area with a low temperature and keep them there

Hyacinthoides non-scripta. When seen in woodland settings, this plant is superb. That there is a slight color variation in the species, is evident in these photographs. But if of deep blue or a little more violet, these plants make a grand display in the spring. As a boy I used to gather bunches of these flowers and take them home, but they never look as pretty as in the wild, and do not last long as a cut flower. AUTHOR

Hyacinth 'Gypsy Queen'. A good cultivar for both bedding out and container growing. The more delicate colors are best given some shade when in flower. AUTHOR

Hyacinthus 'L'Innocence'. One of the purest whites, and a long time favorite for both forcing and the garden. I.B.

Hyacinthus. While expensive, a bed of hyacinths makes a grand sight, and the fragrance from such mass planting is almost overwhelming. AUTHOR

Hyacinthus 'Marconi'. A lovely deep pink, more at home in the garden than being forced. The stems are strong and hold the flower spikes erect, even when very heavy after rain. I.B.

Hyacinthus. Red, white and blue, not an original planting scheme, but the cultivars having been correctly selected, as they are in flower at the same time and are all the same height. DE HERTOGH

Hymenocallis amancaes. In all other species the flowers are white, but as shown here, the yellow color is only found in the center of the flower. BOUSSARD

Hymenocallis littoralis. While this plant needs warm temperatures that, for the most part, can only be given in greenhouses, there are few more spectactular flowers. Despite flowering in mid-summer, they last a long time. TRAGER

Hypoxis latifolia. Unfortunately, this species from South Africa is not fully hardy, and is rare in cultivation. Despite a wide geographic distribution, *Hypoxis* have never been greatly cultivated. They deserve more attention. B.R.I.

Hyacinthoides hispanica

Hypoxis obtusa

Plate 164

until the vase is filled with roots and the flower spikes have emerged. Then give subdued light for a week or so, then full light and temperatures in the 65°F range. Water will need to be added from time to time to the vase. Such bulbs are exhausted after they have flowered and should be discarded.

The commercial forcing of hyacinths is discussed in another chapter. The home gardener is advised to read that chapter as a supplement to the general cultural directions given above.

PROPAGATION The bulbs will produce some offsets naturally at their bases. To increase the number produced two methods of "scaring" the bulbs are used. The easiest is to slice into the bottom of the bulb, either making a cross or three large slits, and then placing the bulbs on wire trays with moderate temperatures and controlled humidity. Bulblets will form, often as many as 70–100 on the areas of the cuts. The other method is to cut out a cone-shaped portion from the bottom of the bulb; again, bulblets will be formed at the base of the individual leaves that make up the bulb.

These bulbs with the bulbils that are now formed, although small, are planted late in the fall. By spring they are growing well and producing much, almost grasslike, foliage but few flowers. In early summer, June or July, these bulbs are lifted and the bulbils removed. They are then planted back in the fall, lifted again, replanted and lifted until they reach commercial size. This process will take 3 or 4 years before commercial sizes are reached.

New methods using tissue culture now are being employed and these are somewhat quicker but also quite expensive. Thus the cost of hyacinth bulbs is always somewhat higher than other spring-flowering bulbs due to the amount of handwork involved.

Seed can be sown, and germinates quite easily in a sandy soil mix. However, it takes some 6 years to see the flowers and as much as another 12–15 years to have enough stock for commercial sales.

PESTS & DISEASES
Yellows. *Xanthomonas hyacinthi* can be a problem. The bulbs rot before or soon after they are planted. Upon cutting through a bulb, yellow spots are seen on the closely packed scales; if cut lengthwise, stripes are noticed. Oozing slime from these areas causes the entire bulb to rot. Such bulbs must be discarded.

Grey Bulb Rot. Caused by the fungus *Sclerotium tuliparum*, a disease common in tulips which also may attack hyacinths. For treatment see section on tulip diseases, Vol. 2, p. 00.

Soft Rot. Another bacterial disease, *Erwinia carotovora*, has an even quicker action on the bulbs. Growth stops and all growing points shrivel and die. In a few days the bulbs are just an evil-smelling mess. All attacked bulbs must be discarded. Over-watering and the use of high-nitrogen fertilizers and manures contribute to the problem. Bulbs also are susceptible to soft rot if they are brought into too high a temperature when being forced. Thus, while the average home gardener is unlikely to encounter the problem, it points out the need to follow the exact temperature requirements when forcing the bulbs commercially. Good hygiene and careful examination of stocks throughout the growing season are essential in commercial production.

A blue mold, caused by *Penicillium*, sometimes is seen on bulbs in storage. This is not serious unless the basal plate is attacked. Drying the bulbs in well-ventilated areas, dusting prior to planting, or spraying in the storage areas and providing good ventilation during storage will take care of this mold. Bulbs that have been attacked and turn soft or where the basal plate is damaged must be discarded.

Aphids can be a problem but are controlled by using a pesticide, both in the storage places and if noticed in the field.

USES Hyacinths are much used in the florist trade as they are easy to force and to grow and have a good shelf life. They can be grown by home gardeners and be in flower for the holiday season and early spring. Outdoors they are good bedding plants, lasting a long time. However, as the bulbs are not inexpensive, groups planted where the fragrance can be enjoyed, such as by a door or window, is suggested.

SPECIES & CULTIVARS
H. orientalis. Native to the eastern Mediterranean and the southern part of Turkey; introduced 1753. It is quite unlike the cultivars and selections that are on the market today. As mentioned above, this is the species from which the common hyacinth of the gardens is derived. Flower color varies from white to mauve to pinkish-mauve. The flower spike is up to 12 in. in height and carries up to 15 flowers. Flowers are extremely fragrant, almost an inch in length, and widely scattered on the stem; tepals open wide to give an almost starlike flower. There are 4–6 leaves almost an inch wide, bright green, with a distinctive hook at the tips. Flowers in March/April. *H. o.* var. *albulus* is the Roman hyacinth, native to the south of France, smaller-flowered and more like the species, however, is now doubtfully in cultivation.

CULTIVARS
There are many of these. As mentioned, they will flower well outdoors but, as the years go by, the number of flowers per spike becomes reduced but the plants themselves will persist for a number of years. As the number of cultivars listed is often quite large, only a selection is presented below.

Cultivars often prepared for early forcing

'Ann Mary'. Soft pink.
'Carnegie'. Pure white.
'City of Haarlem'. Yellow.
'Delft Blue'. Soft blue.
'Jan Bos'. Deep red.
'L' Innocence'. White.
'Ostara'. Bright but deep blue.
'Pink Pearl'. Deep pink.
'Tubergen's Scarlet'. Bright scarlet.
'White Pearl'. Pure white.

Double Hyacinths

'Chestnut Flower'. Soft pink; good for bedding.
'General Kohler'. Clear blue; good for bedding.
'Hollyhock'. Dark carmine-red; can be forced.
'Madame Haubensak'. A large number of flowers per stem, over 80 being recorded.
'Madame Sophie'. Sport of 'L' Innocence' but with double flowers; pure white.

Single Hyacinths
Blue-
'Bismark'. Large-flowered; clear blue.
'Blue Jacket'. Dark blue with small purple stripe.
'Delft Blue'. A favorite; blue flushed with mauve on the exterior.
'Marie'. Very large spikes; dark blue.
'Ostara'. Bright blue with band of deeper blue in center of each tepal.
'Perle Brillante'. Outside deeper blue than sky-blue interior.
White-
'Carnegie'. Pure white; large flowers.
'L' Innocence'. Pure white; an older but still good cultivar.
'White Pearl'. Good white; strong grower.
Yellow-
'City of Haarlem'. Primrose-yellow; compact flowerhead.
'Yellow Hammer'. Softer yellow.
Orange-
'Gipsy Queen'. A good yellow, flushed with apricot-tangerine.

'Nankeen'. Very large spikes of orange-yellow.
'Oranje Boven'. A salmon color; very lovely; sometimes known as 'Salmonetta'.

Red and Pink-
'Amsterdam'. Deep salmon pink with a darker midrib; old but still good cultivar.
'Anna Marie'. Very light pink; sometimes known as 'Christmas Joy.'
'Eros'. Very deep pink.
'Jan Bos'. Good carmine-red with white center.
'La Victoire'. One of the best red hyacinths.
'Marconi'. Deep pink.
'Pink Pearl'. An older cultivar and one of the best deep pinks.
'Queen of the Pinks'. A fine pink; sometimes called 'Rebecca.'

Purple and Violet-
'Amethyst'. Good violet with shades of lilac-mauve.
'Attila'. Violet with an edge of amethyst to the tepals.
'Lord Balfour'. Rose-purple; flowers a little more loosely arranged than some.
'Queen of the Violets'. A good, true violet.

Multiflora Hyacinths

Certain firms will offer these by color. They produce a number of smaller spikes but often many are produced per bulb. These are suitable for indoor culture as well as for use in the garden.

Many catalogs still list the genera below in *Hyacinthus*, although they have been placed in other genera now.

H. amethystinus (now *Brimeura amethystina*, which see). Native to the Pyrenees. Height 10 in. A good blue without any violet tones. The bulb is small, about an inch thick or a little less. Loose spike of funnel-shaped flowers, some 5–15 to a head, about ½ in. in length. Narrow leaves. Flowers in April/May. A good little plant for the shrub border where it will frequently naturalize. There is also a white form known.

H. azureus (now *Pseudomuscari azureum*; syn. *Hyacinthella azurea, Muscari azureum*). Native to Asia Minor. Appearance very much like a *Muscari* but the flowers are not restricted at the mouth; now placed in *Pseudomuscari*. Stem 4–8 in. in height; flowers bright blue with darker stripe in center, about ¼ in. long; the spike being about an inch or just over in length. Easy to grow and good rock garden plant. Fully hardy.

H. candicans. See: *Galtonia candicans*.

H. dalmaticus (now *Hyacinthella dalmatica*). Native to Dalmatia. Short, only 3–4 in. in height; dense spike of bright blue flowers, each only ¼ in. long; leaves linear. Early flowering.

H. fastigiatus (now *Brimeura fastigiata*; syn. *Hyacinthus pouzolzii*). Native to Corsica and Sardinia. Up to 4 in. in height, with flowers that are almost erect. Up to 8–10 flowers in a dense raceme, pale blue, almost ⅓ in. in length. Leaves number 4–6, up to 6 in. in length, ½ in. in width, erect then bending over. Outer flower segments broader than inner and sometimes twisted a little and slightly recurving, but more or less flat when they break from the tube. Flowers April–May.

H. romanus (now *Bellevalia romana*; erroneously called the Roman hyacinth, which, however, is the var. *albulus* of *H. orientalis*). Native to the south of France and eastward to the Balkans in countries around the Mediterranean. Up to 12 in. in height, with a dense raceme of greenish-white or brownish, dirty white flowers. The petals flare out. Not an attractive species.

COMMENTS It is certain that the cost of these bulbs will limit their application in mass planting in the garden, yet they are ideal for such a purpose. No garden that has other spring-flowering bulbs should be without at least a unit of 7–9 of these bulbs planted in small groupings.

HYACINTHUS

H. corymbosus syn. of *Polyxena corymbosa*.
H. lineatus see *Hyacinthella lineata*.
H. litwinowii Native to N. Iran, & Kopet Dag in W. Central Asia. Introduced early 1970s. Pale blue with darker stripe on each segment (distinguished from *H. orientalis* by having fewer, but broader leaves & narrower petals). 6 in. in height. March–April flowering.
H. orientalis ssp. *chionophilus* Native to Turkey. As *H. orientalis* with broader leaves, perianth lobes same length as tube. April–May flowering.
H. pouzolzii see *H. fastigiatus*.
H. spicatus see *Strangweia spicata*.
H. tabrizianus Native to Iran. Whitish, blue at apex. 2 in. in height. March flowering.
H. transcaspicus M. Rix describes as having a longer tube than the free part of the petals. The validity of this being a separate species, is questionable & is most likely a form of *H. litwinowii*.

HYMENOCALLIS — AMARYLLIDACEAE

The name is derived from the Greek *hymen* ("membrane") and *kallos* ("beauty"), in reference to the membrane that unites the stamens and forms the staminal cup. There are some 30 species in the genus, all native to the warmer regions of the Americas, extending from the southern United States to the Andes in South America. For a number of years the genus *Ismene* was considered a separate genus but now is placed in *Hymenocallis*. The flowers closely resemble those of *Pancratium*, which, however, is native to the Mediterranean region. The basic difference, though, between these two genera is that in *Pancratium* the flowers are striped with green and the outer perianth segments are shorter and wider, not twisting and curling as in *Hymenocallis*.

This is another genus where, over the years, there has been a great deal of discussion with regard to the nomenclature. In the genus can be found plants known in times past as *Pancratium* and *Ismene*, as noted. An example of this process is *Hymenocallis narcissiflora*, which was known as *Ismene calathina*; when incorporated into *Hymenocallis*, it became *H. calathina*. The earliest name for this species, however, was found to be *Pancratium narcissiflorum*; today the correct name, based on historic precedence, is *Hymenocallis narcissiflora*. There are a number of common names—Spider Flower, Basket Flower, and Summer Daffodil are most often used.

In the wild, the globose and rather fleshy bulbs are found in grassy areas and among rocks. The evergreen species are from those parts that have moisture throughout the year; the deciduous species are from areas with a drier climate throughout the winter months. The leaves vary in number from a few to many. All are strap-shaped and sheathe one another at the base. In some species this sheathing gives the impression of a false stem and on occasion reaches a height of 12 in.

The flowers are carried in umbels on a strong, frequently two-edged, spike. The 6 perianth segments of the commonly sweet-scented flowers are fused at the base to form a long, narrow tube, which is most often green, sometimes a little whiter. The segments themselves flare outward and often twist and curl. They surround the "cup" found in the center of the flowers. The cup is made up of a membrane that unites the lower parts of the stamens, which are usually white. The stamens are erect and spreading where not joined by the membrane. The margin of the cup formed by the membrane is variously entire, lobed, or cut, depending on the species. The style is long and projects above the other flower parts.

CULTURE   *H. amancaes, H. narcissiflora,* and *H. pedunculata,* all native to Peru, are the hardiest species of this genus, which contains both evergreen and deciduous species. These three can be grown in areas that are practically frost-free but they should be given the added warmth of a south-facing wall. Except for the evergreen species (see descriptions), the other species can be planted in the spring and will flower that summer. All of the evergreen species need a temperature of at least 45°F at night throughout the year; warmer temperatures will not bother them. They do need moisture at all times, especially during the warmer months. The deciduous types need a dry resting period after the foliage has died down. No species should be grown where it will be exposed to more than very minor frosts.

The bulbs should be planted some 2–4 in. deep and spaced according to species, the taller species being spaced farther apart than the dwarf. They all like well-drained soils, rich in organic matter, and at least 3–4 hours of sun per day during the growing season. Those that are planted in the spring for summer flowering should be lifted in the fall and stored in a sandy soil mix in a well-protected area, with nighttime temperatures in the 40°F range. Little water should be given during this resting period. They should be planted out when the days become warm with a minimum of 40°–45°F at night.

PROPAGATION   The bulbs will form small bulblets alongside the parents. These can be removed when of fair size and planted. This natural method of increase is a slow process, taking about 3 years. It is due to the small number of offsets produced that the price of these lovely bulbs remains high.

Seed offers the best method of increasing stock, except in the case of the hybrids offered by nurseries. The seed should be gathered when ripe and stored until the following spring. It then can be sown in sandy soil mix and barely covered, with warm temperatures, in the 55°F range at night. As soon as large enough to handle, the seedlings should be potted up individually and grown on.

The evergreen species should be kept growing in warm temperatures, minimum at night in the 65°F range. The deciduous species are allowed to go into their resting period at the end of summer/early fall. Plants can be set outside during the summer but should not be exposed to cool winds. After 2 growing seasons the new plants can be set out in their permanent positions.

If the soil is rich in organic matter, little or no feeding is required. When being grown in containers, however, weak feedings of liquid organic fertilizer can be given every 2 weeks once active growth is taking place. The feeding should cease after flowering but evergreen plants being grown in temperatures in the 70°F range can receive feedings through to the fall.

PESTS & DISEASES   Few problems will be experienced; however, rotting will occur if too much moisture is given to the deciduous kinds during the winter months. Evergreens should not be grown in waterlogged soil to avoid any problems with rot.

USES   The flowers are striking and make for an unusual and special summer display. While not a "must" for every garden, they are worthy of consideration by those who like to grow something different and where facilities are available. They perform well in containers and, if given the proper winter protection, the containers can be moved in and out as needed. Where the possibility exists of planting in a greenhouse border, the larger flowers that are more commonly scented than not and those species that are from the high altitudes of Peru should be considered, even if temperatures drop into the lower 30s at night during the winter. They make excellent plants for a well-lighted area in foyers or atriums, provided they are moved to more protected quarters if any chance exists of their being exposed to cold winds. In frost-free areas they are good for indoor culture in office buildings and hotels but should be planted with other types of plants and not relied upon by themselves for decoration, even though the evergreens are still attractive when not in flower. While they do not have a great deal of interest when not in flower, do remember that, when in flower, they are lovely plants and the fragrance of many of the species is outstanding. As cut flowers they also are spectacular but cannot be relied upon to last for very long.

SPECIES

H. amancaes. Native to Peru; introduced 1808. Found growing mainly in stony and rocky areas. The deciduous, strap-shaped leaves, varying in length from 10–12 in. and 1–2 in. wide, are sheathed at the base, forming a false stem often up to 9 in. high. The flower stalk can reach a height of 20 in. but is more commonly 12–14 in. The very fragrant flowers are distinctive in that only in this species are they yellow; in all other species in this genus the flowers are white. The number of flowers per stalk varies, from solitary to as many as 8. The perianth tube is almost 3 in. long, and the flowers above the tube are 2 in. long and of the same diameter. This species has been crossed with *H. narcissiflora* and the resulting hybrid, *H.* ×*spofforthiae,* is named for the home of W. Herbert, an eminent botanist and hybridizer of bulbous plants in the early 1800s, who lived in Spofforth, England. Summer-flowering, June/July.

H. caribaea. Native to the West Indies, hence its name; one of the earliest species introduced, prior to 1701. For a number of years it was known as *Pancratium caribaeum.* This evergreen produces many strap-shaped leaves. They are 12–20 in. long, 3 in. wide, and narrowed at the base. The flower spike reaches 18–24 in. in height and carries 8 or more of the white flowers, which are sessile. The perianth tube is short, 1–2 in., with 3–4 in. long segments; margins of the staminal cup are erect and funnel-shaped. The stamens are longer than the cup by ½ in. or more. June-flowering.

H. harrisiana. Native to Mexico; introduced before 1840. Found growing on grassy slopes, often at an altitude of more than 6,000 ft., which makes it worthy of being considered in gardens with little or no frost. This deciduous plant has 3–5 leaves, 8–12 in. long and often shorter at flowering time. The white flowers, 3–4 per stalk, are scented and spidery in appearance due to the narrow segments which curl. The perianth tube is longer than the flower segments, 4–5 in. as compared to 2–3 in. June-flowering.

H. littoralis (syn. *H. adnata, Pancratium littorale*). Native to Mexico and Guatemala; introduced prior to 1752. It has been introduced to other tropical areas and has escaped into the wild. Widely cultivated and one of the few species of this genus that is listed in commercial catalogs. The leaves of this evergreen plant are distinctive for their bright green color. They are 2 in. wide, strap-shaped, and narrowed at the base. The flower spike is more than 20 in. high, carrying 6–12 white, sessile, and fragrant flowers. The perianth tube is usually 5 in. long, sometimes longer. The segments are close to the staminal cup at their base and then spread up to 5 in. in length; staminal cup is 1½ in. high and the stamens some 2½ in. Although this species flowers in midsummer, it needs considerable heat throughout the year for good growth and flower production, with minimum nighttime temperatures in the upper 50s during the winter. Var. *dryandri* has a short tube; var. *acutifolia* has shorter leaves and flowers.

H. narcissiflora (syn. *H. calathina, Pancratium calathinum, P. narcissiflorum,*). Native to Peru; introduced 1796. Often found above 8,000 ft. growing in fields and among rocks. Because of its native habitat, this species should be able to be grown outdoors where winter temperatures do not drop below 35°F at night. The leaves are few, deciduous, 12–20 in. long and up to 2 in. wide; they sheathe at the base, forming a false stem that can be as much as 12 in. tall but more commonly 6–9 in. The flower stalk rises some 18–20 in. above the false stalk, bearing 2–5 white, fragrant flowers. The perianth tube is as long, or nearly as long, as the free floral segments, which are 3 in.; perianth segments, with the same greenish hue of the tube at their base, are narrow and only slightly longer than the 2-in. staminal cup, which is of the same diameter and has narrow green bands. Summer-flowering.

H. pedunculata (syn. *H. macleana, H. virescens, Ismene pedunculuta*). Native to Peru; introduced before 1837. The deciduous leaves sheathe the lower part of the flowering stem to produce a false stem which reaches some 6 in. above the ground. The leaves are strap-shaped or sword-shaped after they stop sheathing, and can vary in length from 9–18 in. Width also variable, from an inch to over 2 in. Flower stalk can reach over 24 in. but more usually 18–20. The fragrant flowers have short pedicels and the number of flowers per stem varies from 2–6. The perianth tube is 2 in. long, deep green; segments up to 2 in., white with flush of green on reverse. The staminal cup is funnel-shaped with spreading lobes up to 2 in. long, greenish-white with white lobes. Lobes are sometimes green with white margins. Flowering in midsummer.

H. speciosa (syn. *Pancratium fragrans, P. speciosum*). Native to the West Indies; introduced prior to 1782. This evergreen species should be considered for culture only in tropical climates or in a greenhouse where tropical temperatures (55°F minimum at night in winter) can be maintained. The leaves are up to 24 in. long and 6 in. wide. Each stalk carries 5–7 white, fragrant flowers about 12–16 in. above the leaves. The perianth tube is 3 in. long, and the segments are up to twice as long. The staminal cup is funnel-shaped and 1–2 in. long, the same length as the stamen filaments. This species is listed occasionally in commercial catalogs. Summer-flowering.

HYMNENOCALLIS

H. acutifolia variety of *H. littoralis.*
H. adnata see *H. littoralis.*
H. amoena syn. of *H. ovata.*
H. andreana see *Leptochiton quitoensis.*
H. arenicola (syn. *H. crassifolia*). Native to Bahamas. Introduced before 1872. White. 16–22 in. in height. Summer flowering.
H. borskiana syn. of *H. tubiflora.*
H. calathina see *H. narcissiflora.*
H. caymanensis syn. of *H. latifolia.*
H. choretis (syn. *H. glauca*). Native to Mexico. Introduced before 1837. White. 14 in. in height. August flowering.
H. concinna syn. of *H. dillenii.*
H. cordifolia Native to Mexico, Venezuela. Introduced 1899. White. 17 in. in height. Summer flowering.
H. crassifolia syn. of *H. arenicola.*
H. deleuillii syn. of *H. expansa.*
H. dillenii (syn. *H. concinna*). Native to Mexico. Introduced before 1732. White. 8–12 in. in height. Summer flowering.
H. eucharidifolia Native to Tropical America. Introduced before 1884. White. 12 in. in height. May flowering.
H. expansa (syn. *H. deleuillii, Pancratium expansum*). Native to West Indies. Introduced before 1941. Pale green. 16–32 in. in height. November flowering.
H. fragrans Native to West Indies. Introduced 1730. White. 18 in. in height. Summer flowering.
H. glauca syn. of *H. choretis.*
H. guianensis syn. of *H. tubiflora.*
H. horsmannii Native to Mexico. Introduced 1883. White. 4–6 in. in height. July flowering.
H. lacera syn. of *H. rotata.*
H. latifolia (syn. *H. caymanensis, Pancratium latifolium*). Native to Florida, W. Indies. Introduced before 1768. White. 24–36 in. in height. Summer flowering.
H. macleana see *H. pedunculata.*
H. × macrostephana (syn. *Pancratium macrostephanum*) (*H. narcissiflora × H. speciosa*). Of garden origin. Introduced before 1879. White. 12–18 in. in height.
H. moritziana syn. of *H. tubiflora.*
H. ovata (syn. *H. amoena, Pancratium fragrans*). Native to West Indies. Introduced before 1752. White. 12–18 in. in height. October flowering.
H. pedalis Native to South America. Introduced before 1821. White. 24 in. in height. Summer flowering.
H. quitoensis see *Leptochiton quitoensis.*
H. rotata (syn. *H. lacera, Pancratium rotatum*). Native to Southern U.S.A. Introduced 1790. White. 18–24 in. in height. Summer flowering.
H. schizostephana Native to Brazil. Introduced circa 1894. White. 12 in. in height. Summer flowering.
H. × spofforthiae (*H. narcissiflora × H. amancaes* by Dean Wm. Herbert). Garden origin. Introduced 1830. Flows green, segments sulfur-yellow. 12 in. in height.
H. tenuifolia see *Leptochiton quitoensis.*
H. tubiflora (syn. *H. borskiana, H. guianensis, H. moritziana, H. undulata, Pancratium guianense*). Native to N. South America. Introduced 1803. White. 24 in. in height. Summer flowering.
H. undulata syn. of *H. tubiflora.*
H. virescens see *H. pedunculata.*

HYPOXIS — HYPOXIDACEAE

The name is derived from the Greek *hypo* ("below") and *oxy* ("pointed"), alluding to the sharp points of the inferior petals. It is a genus that contains many species, which are found worldwide—in the United States from Maine, south to Florida, west to Texas, and north to North Dakota; in the tropics of the Americas, Asia, and Africa, as well as Australia, Madagascar, and South Africa. *Hypoxis* are either corms or rhizomes. The flowers are yellow and star-shaped: perianth segments are spreading and almost equal, and the foliage is grasslike. Frequently the stems are flattened and the foliage hairy. The majority are summer-flowering, often late into the season. Common names vary but usually refer to the grasslike foliage and flower color, such as Yellow Star or Yellow Star Grass. For the most part they are found in grasslands where, unless the yellow flower is seen, they will often go undetected. Not many of them are of horticultural importance. A number of authorities write that it is difficult to find criteria to separate out the various species and of the need for a revision of the genus. It is doubtful that this has high priority due to the insignificance of the plants as garden subjects.

CULTURE Hardiness depends on the habitat of the various species. *H. hirsuta*, the North American species, is widely distributed along the eastern seaboard and is obviously quite hardy. Those from warmer climes are less so, but, except for the species from the tropics, most would be hardy in areas where the ground does not freeze in winter. Corms or rhizomes should be planted 2–4 in. deep, spaced 6–8 in. apart for the dwarf species, a little wider for the taller growing. They like bright light, such as they would experience in grasslands, with only short periods of exposure to full sun. They require moisture in the winter and spring and into the early summer.

PROPAGATION Seed is freely produced. This should be sown in the spring in a sandy, porous mix, with temperatures in the 45°–50°F range at night. Broadcasting seed in the area where the plants are to grow also is successful, but, as the foliage is much like grass, weeding can be a problem.

PESTS & DISEASES No special problems.

USES More for the fancier of bulbs than for the average gardener. Rock garden areas and dry, sunny borders are most suitable locations.

SPECIES

H. argentea (Small Yellow Star, Star of Bethlehem). Native to the eastern part of Cape Province, South Africa. Common in coastal grasslands, and in the wild it flowers June–Nov.; in the Northern Hemisphere Feb.–June. It has a tunicated corm, numerous fibrous leaves, and a flowering stem that reaches 8–10 in. in height. The many flowers are produced over a long period, opening only in the sun. The Xhosa tribe uses the oil from the corms for chafes on horses and in periods of poor harvest the corms are dried and pounded into meal.

H. hirsuta (Yellow Star Grass; syn. *H. erecta*). Native to the United States and is found from Maine, south to Florida, west to Texas, and north to North Dakota; introduced 1752. Leaves are basal, grasslike, somewhat hairy, up to 12 in. in length and some 4 in. wide. Leaves number from 3–4 to many. The flowers are almost an inch in width with several produced in a rather loose flowerhead; height of the flowers varies from 3–8 in. Flowers appear in late spring and last through summer, March–Sept., depending on habitat. Those from warmer climates flower earlier; those from colder climes, later. The majority of flowers can be expected in May–July.

H. multiceps. Native to the eastern part of Cape Province, South Africa, north to Natal. It grows in grasslands. The corm is covered with fibrous leaf bases. The foliage is basal, linear, hairy on both surfaces, arranged in a triangular fashion, and curls backward; length is up to 12 in. The flowering stem is flattened, hairy, and usually carries about 5 flowers, each bright yellow with pale green and hairy on the reverse. 6–10 in. in height. This is only one of many species found in the grassy areas but is the most common. It is not unlikely that others are but varieties of the species or geographic forms.

H. obtusa. Native to the Transvaal and northward to the Victoria Falls area of Zimbabwe. Found growing in open grassland and at the lower mountain elevations in rocky places. The rootstock is rhizomatous with yellow flesh. Up to 12 in. in height, the flowering stems carry 3–7 flowers, which are a shiny yellow with a green reverse. Strap-shaped foliage is basal up to 15 in. long. In the wild they flower from Aug.–Dec., frequently before the onslaught of the summer rains in Zimbabwe. Flowers late spring to midsummer in the Northern Hemisphere.

H. rooperi. Native to the Transvaal. Found growing in rocky areas and regarded as the showiest of the South African species. Plants reach 12–18 in. in height. Arching, keeled leaves are neatly arranged in ranks of 3. The yellow flowers are up to 2 in. in diameter, with as many as 10 or more per stem. Flowers are produced in early to midsummer; June–July in the Northern Hemisphere. The fibrous leaves often are used as binding material by the natives. This species does deserve more attention and has potential as a garden plant.

H. setosa. Native to the eastern part of Cape Province. A low-growing plant, only 3–6 in. in height, found springing into flower after grassland burning when the grass is still short. The flowers are bright yellow, only 2–4 per stem. The foliage is quite wide, often more than an inch, and 5–8 in. in length, held close to the ground. It flowers in the wild from Oct.–April; March–Sept. in the Northern Hemisphere.

H. villosa (Golden Winter Star). Native to eastern Cape Province. Found growing on level land near the coast. Height 12–15 in. The yellow flowers remain closed during cloudy weather and, as the reverse of the petals is green, they are inconspicuous, except during sunny weather. Flowers are produced in winter and early spring. 3–5 flowers per inflorescence, generally only 1 opening up at a time. Numerous hairy leaves are produced, which clasp the lower part of the flowering stem and persist for almost a year before the corm goes into a brief resting period in late summer/early fall.

COMMENTS This genus was for many years in the Amaryllidaceae family. It was separated from this family, for one reason, because the flower spikes are not umbels.

HYPOXIS

H. "colchicifolia" Not a valid name. Native to South Africa. Introduced 1884. Bright yellow within, greenish-yellow without. 12 in. in height. Fall flowering.
H. elata syn. of *H. hemerocallidea*.
H. erecta syn. of *H. hirsuta*.
H. glabella Native to Western Australia. Yellow. 6–8 in. in height. August flowering.
H. hemerocallidea (syn. *H. elata*). Native to Natal. Introduced 1862. Golden yellow. June flowering.

H. hygrometrica Native to Eastern Australia & Tasmania. Bright yellow, greenish yellow without. 2–3 in. in height. Summer flowering.
H. latifolia (syn. *H. oligotricha*). Native to Natal. Introduced 1854. Bright yellow, green exterior. Summer flowering.
H. leptantha Native to Western Australia. Yellow. 6–8 in. in height. September flowering.
H. longifolia Native to Algoa Bay, Cape Province. Introduced 1871. Golden yellow, greenish outside. 18 in. in height. August flowering.
H. membranacea Native to Cape Province. Creamy & yellow. Dwarf. May–Oct. flowering.
H. occidentalis Native to Western Australia. Yellow. 6–8 in. in height.
H. oligotricha syn. of *H. latifolia*.
H. stellata (see also *Spiloxene capensis*). Native to S.W. Cape of South Africa. Introduced 1752. White, cyclamen pink, or yellow with blue or green splashes. 6–18 in. in height. Late Winter, Early Spring flowering.